**Magnetic Resonance Microscopy**

*Edited by*
*Sarah L. Codd and Joseph D. Seymour*

## Further Reading

Weil, J. A., Bolton, J. R.

**Electron Paramagnetic Resonance**

**Elementary Theory and Practical Applications**

2007
ISBN: 978-3-471-75496-1


Stapf, S., Han, S.-I. (Eds.)

**NMR Imaging in Chemical Engineering**

2006
ISBN: 978-3-527-31234-4


Stapf, S. (Ed.)

**Introduction to NMR Imaging**

**Fundamentals and Examples**

2010
ISBN: 978-3-527-31951-0

# Magnetic Resonance Microscopy

*Spatially Resolved NMR Techniques and Applications*

*Edited by*
*Sarah L. Codd and Joseph D. Seymour*

**WILEY-VCH**

WILEY-VCH Verlag GmbH & Co. KGaA

**The Editors**

**Prof. Dr. Sarah L. Codd**
Montana State University
Mechanical and Industrial Engineering
220 Roberts Hall
Bozeman, MT 59717-3800
USA

**Prof. Dr. Joseph D. Seymour**
Montana State University
Chemical and Biological Engineering
306 Cobleigh Hall
Bozeman, MT 59717-3920
USA

**Library of Congress Card No.:** applied for

**British Library Cataloguing-in-Publication Data**
A catalogue record for this book is available from the British Library.

**Bibliographic information published by the Deutsche Nationalbibliothek**
The Deutsche Nationalbibliothek lists this publication in the Deutsche Nationalbibliografie; detailed bibliographic data are available on the Internet at http://dnb.d-nb.de.

**Composition**  SNP Best-set Typesetter Ltd., Hong Kong
**Printing**  betz-druck GmbH, Darmstadt
**Bookbinding**  Litges & Dopf GmbH, Heppenheim

Printed in the Federal Republic of Germany
Printed on acid-free paper

**ISBN:**  978-3-527-32008-0

# Foreword

When the first Magnetic Resonance Microscopy meeting was held in Heidelberg in 1991, those of us then present hoped that the meeting would be the first of many. Not only was that hope realized, but the series has gone on to develop in two remarkable ways. First, it has resulted in a series of edited volumes, in which the latest developments in the field have been made available, by the foremost practitioners, to an international audience. Second, spatially resolved magnetic resonance has moved outside the laboratory, to include the field of mobile NMR. This volume, which emerges from the 2007 Aachen meeting in the Heidelberg series, attests to that emergence. Indeed, the scope of the subject matter now is quite extraordinary, encompassing hyperpolarization, scanning probe microscopy, earth-field NMR, MRI under acoustic waves, chemical mapping, sensor technology and compact electronics. Search the pages of this book and you will see the subjects of physics, chemistry, biomedicine and engineering brought together. While other branches of magnetic resonance press deeper into the specializations of particular disciplines, the specialties reflected in this volume are inherently interdisciplinary.

Of course, the discipline breadth of which we are rightly proud, is in the best spirit of the NMR pioneers. And one cannot write in general terms about magnetic resonance without reflecting on past heritage. One year ago, we lost one of the pioneers of our conference series, Nobel Laureate and codiscoverer of MRI, Paul Lauterbur. Paul wrote the foreword to the 1998 volume in this series, and in doing so challenged us to show that NMR spectroscopy and imaging were more than answers looking for questions. For those of us who dream up new NMR measurement tricks the challenge is apposite. Can we show that what we do will make a difference to the world around us? If there is anything that our 60-odd years of NMR experience tells us, it is that there is a high probability of relevance, of significant application. But nonetheless, a duty to actively seek application is one we cannot shirk.

And what of future prospects? Look at the new topics herein, unforeseen in 1991, and ask what has driven them. In addition to noting the obvious, the boundless ingenuity of the human mind, we find the following list: new magnet technologies; new cell phone electronics; new computing power; new optical pumping methods; new micro- and nanotechnologies. Magnetic resonance has the capacity

*Magnetic Resonance Microscopy.* Edited by Sarah L. Codd and Joseph D. Seymour
Copyright © 2009 WILEY-VCH Verlag GmbH & Co. KGaA, Weinheim
ISBN: 978-3-527-32008-0

to regenerate and to expand in capacity with every new wave of influential technology. It is for that reason, and because of the depth of measurement insight that the spin Hamiltonian provides, that the future is bright for our field of research. The contents of this volume make that point with abundant clarity.

*Paul T. Callaghan*
*Alan MacDiarmid Professor of Physical Sciences*
*School of Chemical and Physical Sciences*
*Victoria University of Wellington*
*Wellington*
*New Zealand*

# Contents

*Magnetic Resonance Microscopy.* Edited by Sarah L. Codd and Joseph D. Seymour
Copyright © 2009 WILEY-VCH Verlag GmbH & Co. KGaA, Weinheim
ISBN: 978-3-527-32008-0

# Preface

*C'est par la logique qu'on démontre, c'est par l'intuition qu'on invente.*
(It is by logic that we prove, but by intuition that we discover.)
*Henri Poincaré, Science et méthode (1908)*

The works collected in this volume represent the third edited volume based on the International Conference on Magnetic Resonance Microscopy (ICMRM). This conference, which is organized biannually by the Division of Spatially Resolved Magnetic Resonance of the AMPERE (Atomes et Molécules Par Études Radio-Électriques) Society, explores the frontiers of technique development and application of methods primarily in the field of magnetic resonance microscopy (MRM). However, as the range of subjects covered in this volume demonstrates, the meeting includes some macroscale MR applications and developments of mobile nuclear magnetic resonance (NMR) methods and technology. Typically, some 20 countries are represented, and the meeting–which originated in Heidelberg, Germany–now rotates between Europe, North America and Asia, a situation which is indicative of the highly international nature of the field.

Readers will find that the chapters of this book provide an up-to-date perspective on the current state of the art. In the intervening ten years since publication of the last volume in this series, we have seen some new directions emerge by practitioners in this field, a wealth of creative applications of old techniques to important systems, remarkable hardware advances, and a blossoming of technique development and applications in the newer subfield of mobile NMR. The developments in this latter area deserve special mention, with their commensurate ability to move spatially resolved magnetic resonance into processing and field environments.

The interdisciplinary nature of the field is evident from a quick perusal of the contributions in this volume from several commercial companies, a number of industrial laboratories and a range of academic laboratories in physics, chemistry and engineering. Another indicator of the strength of the field is the scientific approach of scientists who practice spatially resolved MR, as evidenced by the quality of the present chapters and the care with which the review of these texts was conducted. The editors thank the authors not only for their participation, but

*Magnetic Resonance Microscopy.* Edited by Sarah L. Codd and Joseph D. Seymour
Copyright © 2009 WILEY-VCH Verlag GmbH & Co. KGaA, Weinheim
ISBN: 978-3-527-32008-0

also for the motivation provided by the lucid reviews and descriptions of the cutting-edge research contained within their texts.

The results reviewed within these pages are a tribute primarily to the talents and expertise of the authors, and also – importantly – to the collegiality and extensive collaborations that exist between the authors. Whether you are just starting out in the field of MRM, are an expert NMR scientist wishing to implement a new technique, or a specialist in another field who recognizes the potential of these techniques to impact your field of study, you will find in these chapters the 'tricks of the trade' behind all these elegant experimental results. One hope of the editors is that this volume will lead to a continued expansion in the number of researchers who realize that the richness of the technique is not the high spatial resolution that optical or electron microscopy techniques pride themselves on, but instead the richness provided by the noninvasive molecular contrast mechanisms and the inherent three-dimensional nature of the methods.

Rather than making a claim as to the general infancy or maturity of the current state of the field, this volume demonstrates the diversity of subfields – some mature and making inroads into industrial applications, and others just emerging from the laboratories of their founders – which compose this unique experimental method.

*September 2008*
*Sarah L. Codd*

*Bozeman, Montana USA*
*Joseph D. Seymour*

# Editor's Biographies

Just as hospitals use magnetic resonance imaging (MRI) scans to examine patients' bodies with millimeter resolution, scientists wish to examine materials nondestructively and noninvasively, on the microscopic scale. Today, spatial resolution down to the micrometer range has been achieved, and major efforts are being made to develop magnetic resonance technology further. Characterization of the structure and transport function of materials is important in applications, ranging all the way from biomedicine and food science to geophysics and alternative energy.

This Handbook and ready reference covers materials science applications as well as microfluidic, biomedical and dental applications and the monitoring of physicochemical processes. It includes the latest in hardware, methodology and applications of spatially resolved magnetic resonance, such as portable imaging and single-sided spectroscopy, for materials scientists, spectroscopists, chemists, physicists and medicinal chemists alike.

*Sarah L. Codd* is codirector of the Magnetic Resonance Microscopy (MRM) laboratory and an associate Professor in the Department of Mechanical and Industrial Engineering at Montana State University. Her research focuses on technique development, spatially resolved studies of gas in ceramics, flow and diffusion studies in porous media, and the investigation of fluid dynamics in hydrogels, biofilms, cellular suspensions and polymer electrolyte membranes. Sarah's doctoral studies were carried out at the University of Kent at Canterbury, UK. She then held postdoctoral positions at Massey University, New Zealand, Ulm University, Germany and New Mexico Resonance, USA, before moving to Montana in 2002.

*Joseph Seymour* is codirector of the Magnetic Resonance Microscopy (MRM) laboratory and an Associate Professor in the Department of Chemical and Biological Engineering at Montana State University. His primary area of research interest is in transport imaging using MRM. Prior and future research includes laboratory and field studies of transport phenomena using MRM's ability to measure both coherent motion, or velocity, and random motion, or diffusion. Joseph's doctoral studies were carried out at the University of California at Davis, USA. He then held post-doctoral positions at Massey University, New Zealand, Universität Ulm, Germany and was a staff scientist at New Mexico Resonance, USA, before moving to Montana in 2001.

*Magnetic Resonance Microscopy*. Edited by Sarah L. Codd and Joseph D. Seymour
Copyright © 2009 WILEY-VCH Verlag GmbH & Co. KGaA, Weinheim
ISBN: 978-3-527-32008-0

# List of Contributors

*Stephen A. Altobelli*
New Mexico Resonance
2301 yale SE
Albuquerque, NM 97106
USA

*Andrea Amar*
RWTH Aachen
Institute of Technical Chemistry
and Macromolecular Chemistry
Worringerweg 1
52074 Aachen
Germany

*Brandon D. Armstrong*
University of California
Santa Barbara
Department of Physics
Santa Barbara, CA 93106
USA

*Mladen Barbic*
California State University
Long Beach
Department of Physics and
Astronomy
1250 Bellflower Boulevard
Long Beach, CA 90840
USA

*Michael J. Barlow*
University of Nottingham
Sir Peter Mansfield Magnetic
Resonance Centre
School of Physics and Astronomy
University Park
Nottingham, NG7 2RD
UK

*Haskell W. Beckham*
Georgia Institute of Technology
School of Polymer, Textile and
Fiber Engineering
Atlanta, GA 30332-0295
USA

*Bernhard Blümich*
RWTH Aachen
Institute of Technical Chemistry and
Macromolecular Chemistry
Worringerweg 1
52074 Aachen
Germany

*Louis-Serge Bouchard*
Department of Chemistry and
Biochemistry
University of California
Los Angeles, CA 90095
USA

*Magnetic Resonance Microscopy*. Edited by Sarah L. Codd and Joseph D. Seymour
Copyright © 2009 WILEY-VCH Verlag GmbH & Co. KGaA, Weinheim
ISBN: 978-3-527-32008-0

**Melanie Britton**
University of Birmingham
School of Chemistry
Edgbaston
Birmingham, B15 2TT
UK

**Jennifer R. Brown**
Montana State University
Chemical and Biological
Engineering
306 Cobleigh Hall
Bozeman, MT 59717
USA

**Bogdan Buhai**
Universität Ulm
Sektion
Kernresonanzspektroskopie
89069 Ulm
Germany

**Lisandro Buljubasich**
RWTH Aachen
Institute of Technical Chemistry
and Macromolecular Chemistry
Worringerweg 1
52074 Aachen
Germany

**Paul T. Callaghan**
Victoria University of Wellington
School of Chemical and
Physical Sciences
MacDiarmid Institute
Wellington
New Zealand

**Federico Casanova**
RWTH Aachen University
Institut für Technische Chemie und
Makromolekulare Chemie
Worringerweg 1
52074 Aachen
Germany

**Hsing-Wei Chang**
National Taiwan University
Department of Chemistry
Taipei
Taiwan

**Hyung Joon Cho**
Memorial Sloan-Kettering Cancer
Center
1275 York Avenue
New York, NY 10021
USA

**Luisa Ciobanu**
Commissariat a l'Energy Atomique
NeuroSpin
Bat 145, Point Courrier 156
91191 Gif sur Yvette
France

**Sarah L. Codd**
Montana State University
Mechanical and Industrial
Engineering
220 Roberts Hall
Bozeman, MT 59717-3800
USA

**Andrew Coy**
Magritek Limited
32 Salamanca Road
Wellington
New Zealand

**Ernesto Danieli**
RWTH Aachen
Institut für Technische Chemie
und Makromolekulare Chemie
Worringerweg 1
52074 Aachen
Germany

**John P.M. van Duynhoven**
Unilever Food and Health
Research Institute
Olivier van Noortlaan 120
3130 AC Vlaardingen
The Netherlands

**Robin Dykstra**
Massey University
Institute of Fundamental
Sciences
Palmerston North
New Zealand

**Craig D. Eccles**
Magritek Limited
32 Salamanca Road
Wellington
New Zealand

**Einar O. Fridjonsson**
Montana State University
Chemical and Biological
Engineering
306 Cobleigh Hall
Bozeman, MT 59717
USA

**Eiichi Fukushima**
ABQMR
2301 Yale Boulevard
SE (Suite C2)
Albuquerque, NM 87106
USA

**Jon K. Furuyama**
University of California
Department of Chemistry and
Biochemistry
Los Angeles, CA 90095
USA

**Lynn F. Gladden**
University of Cambridge
Department of Chemical Engineering
Pembroke Street
Cambridge, CB2 3RA
UK

**Gert-Jan W. Goudappel**
Unilever Food and Health Research
Institute
Olivier van Noortlaan 120
3130 AC Vlaardingen
The Netherlands

**Josef Granwehr**
University of Nottingham
Sir Peter Mansfield Magnetic
Resonance Center
School of Physics and Astronomy
University Park
Nottingham, NG7 2RD
UK

**Farida Grinberg**
University of Leipzig
Department of Physics
Linnéstraße 5
04103 Leipzig
Germany

**Tomoyuki Haishi**
MRTechnology
Tsukuba Research Center B-5
2-1-6 Sengen
Tsukuba 305-0047
Japan

**Meghan E. Halse**
Victoria University of Wellington
MacDiarmid Institute
Wellington
New Zealand

**Songi Han**
University of California
Santa Barbara
Department of Chemistry and
Biochemistry
Santa Barbara, CA 93106
USA

**Shinya Handa**
University of Tsukuba
Institute of Applied Physics
1-1-1 Tennoudai
Tsukuba 305-8573
Japan

**Jason P. Hindmarsh**
Massey University
Riddet Center
Private Bag 11
222 Palmerston North
New Zealand

**Shuichiro Hirai**
Tokyo Institute of Technology
2-12-1 O-okayama
Meguro-ku
Tokyo
Japan

**Natalia Homan**
Wageningen University
Wageningen NMR Center and
Laboratory of Biophysics
Dreijenlaan 3
6703 HA Wageningen
The Netherlands

**Susie Y. Huang**
University of California
Department of Chemistry and
Biochemistry
Los Angeles, CA 90095
USA

**Henk Huinink**
Eindhoven University of Technology
Department of Applied Physics
Den Dolech 2
5600 MB Eindhoven
The Netherlands

**Mark W. Hunter**
Victoria University of Wellington
MacDiarmid Institute
Wellington
New Zealand

**Dennis W. Hwang**
University of California
Department of Chemistry and
Biochemistry
Los Angeles, CA 90095
USA

**Lian-Pin Hwang**
National Taiwan University
Department of Chemistry
Taipei
Taiwan
Harvard Medical School
Boston, MA 02115
USA

**Michael L. Johns**
University of Cambridge
Department of Chemical Engineering
Pembroke Street
Cambridge, CB2 3RA
UK

**Rainer Kimmich**
Universität Ulm
Fakultät für Naturwissenschaften
Sektion
Kernresonanzspektroskopie
Albert-Einstein-Allee 11
89069 Ulm
Germany

**Walter Köckenberger**
University of Nottingham
Sir Peter Mansfield Magnetic
Resonance Center
School of Physics & Astronomy
University Park
Nottingham, NG7 2RD
UK

**Igor V. Koptyug**
International Tomography Center
SB RAS
3A Institutskaya Street
Novosibirsk 630090
Russia

**Katsumi Kose**
University of Tsukuba
Institute of Applied Physics
Tsukuba 305-8573
Japan

**Kirill V. Kovtunov**
International Tomography Center
SB RAS
3A Institutskaya Street
Novosibirsk 630090
Russia

**James Leggett**
University of Nottingham
Sir Peter Mansfield Magnetic
Resonance Center
School of Physics & Astronomy
University Park
Nottingham, NG7 2RD
UK

**Johannes Leisen**
Georgia Institute of Technology
School of Polymer, Textile and Fiber
Engineering
Atlanta, GA 30332-0295
USA

**Chih-Hao Li**
Harvard-Smithsonian Center for
Astrophysics
60 Garden St
Cambridge, MA 02138
USA

Harvard University
Department of Physics
60 Garden St
Cambridge, MA 02138
USA

**Yujie Li**
Universität Ulm
Sektion Kernresonanzspektroskopie
89069 Ulm
Germany

**Guangzhi Liao**
China University of Petroleum
School of Resource & Information
NMR Lab.
Changping District
Beijing 102249
China

**Yung-Ya Lin**
University of California
Department of Chemistry and
Biochemistry
Los Angeles, CA 90095
USA

**Angel J. Perez Linde**
University of Nottingham
Sir Peter Mansfield Magnetic
Resonance Center
School of Physics & Astronomy
University Park
Nottingham, NG7 2RD
UK

**Mark D. Lingwood**
University of California Santa
Barbara
Department of Chemistry and
Biochemistry
Santa Barbara, CA 93106
USA

**Evan R. McCarney**
University of California Santa
Barbara
Department of Chemistry and
Biochemistry
Santa Barbara, CA 93106
USA

**Michael J. McCarthy**
University of California
Department of Food Science and
Technology and Department of
Biological and Agricultural
Engineering
One Shields Ave
Davis, CA 95616
USA

**Jeffrey S. McLean**
J. Craig Venter Institute
La Jolla, CA 92121
USA

**Ross W. Mair**
Harvard-Smithsonian Center for
Astrophysics
60 Garden St
Cambridge, MA 02138
USA

**Paul D. Majors**
Pacific Northwest National Laboratory
Biological Sciences Division
Richland, WA 99352
USA

**Igor V. Mastikhin**
University of New Brunswick
UNB MRI Center
Department of Physics
8, Bailey Drive
Fredericton, NB E3B 5A3
Canada

**Thomas Meersmann**
Colorado State University
Department of Chemistry
Fort Collins, CO 80523
USA

**Rebecca R. Milczarek**
University of California
Department of Food Science and
Technology and Department of
Biological and Agricultural
Engineering
One Shields Ave
Davis, CA 95616
USA

**Ales Mohoric**
University of Ljubljana
Jadranska 19
1000 Ljubljana
Slovenia

**Benedict Newling**
University of New Brunswick
UNB MRI Center
Department of Physics
8, Bailey Drive
Fredericton, NB E3B 5A3
Canada

**Rafal Panek**
University of Nottingham
Sir Peter Mansfield Magnetic
Resonance Center
School of Physics & Astronomy
University Park
Nottingham, NG7 2RD
UK

**Samuel Patz**
Brigham and Women's Hospital
Department of Radiology
221 Longwood Avenue
Boston, MA 02115
USA

Harvard Medical School
25 Shattuck Street
Boston, MA 02115
USA

**Galina E. Pavlovskaya**
Colorado State University
Department of Chemistry
Fort Collins, CO 80523
USA

**Leo Pel**
Eindhoven University of
Technology
Department of Applied Physics
Den Dolech 2
5600 MB Eindhoven
The Netherlands

**Juan Perlo**
RWTH Aachen
Institut für Technische Chemie und
Makromolekulare Chemie
Worringerweg 1
52074 Aachen
Germany

**L. Guy Raguin**
Michigan State University
Department of Mechanical
Engineering
2555 Engineering Building
East Lansing, MI 48824-1226
USA

Michigan State University
Department of Radiology
East Lansing, MI 48824
USA

**Pedro Ramos-Cabrer**
Universidade de Santiago de
Compostela
Hospital Clinico Universitario
Laboratorio de Investigación en
Neurociencias Clinicas
Trv. Choupana s/n
15796 Santiago de Compostela
Spain

**Matthew S. Rosen**
Harvard-Smithsonian Center for
Astrophysics
60 Garden St
Cambridge, MA 02138
USA

Harvard University
Department of Physics
60 Garden St
Cambridge, MA 02138
USA

**Nikolas Salisbury-Andersen**
University of Nottingham
Sir Peter Mansfield Magnetic
Resonance Center
School of Physics & Astronomy
University Park
Nottingham, NG7 2RD
UK

**Mark H. Sankey**
University of Cambridge
Department of Chemical
Engineering
Pembroke Street
Cambridge, CB2 3RA
UK

**Rachel N. Scheidegger**
Harvard-Smithsonian Center for
Astrophysics
60 Garden St
Cambridge, MA 02138
USA

Harvard-MIT Division of Health
Sciences and Technology
77 Massachusettes Avenue
Cambridge, MA 02139
USA

Montana State University
Chemical and Biological
Engineering
306 Cobleigh Hall
Bozeman, MT 59717
USA

**Eric E. Sigmund**
New York University
Department of Radiology
660 First Avenue
New York, NY 10016
USA

**Yi-Qiao Song**
Schlumberger-Doll Research
1 Hampshire Street
Cambridge, MA 02139
USA

**Siegfried Stapf**
TU Ilmenau
Department of Technical Physics
Unterpörlitzer Str. 38
98693 Ilmenau
Germany

**Christiane Timmel**
University of Oxford
Department of Chemistry
Inorganic Chemistry Laboratory
South Parks Rd.
Oxford OX1 3QR
UK

**Leo L. Tsai**
Harvard-Smithsonian Center for
Astrophysics
60 Garden St
Cambridge, MA 02138
USA

Harvard Medical School
Brigham and Women's Hospital
Department of Surgery
75 Francis Street
Boston, MA 02115
USA

**Shohji Tsushima**
Tokyo Institute of Technology
2-12-1 O-okayama
Meguro-ku
Tokyo
Japan

**Henk Van As**
Wageningen University
Wageningen NMR Center and
Laboratory of Biophysics
Dreijenlaan 3
6703 HA Wageningen
The Netherlands

**Frank J. Vergeldt**
Wageningen University
Wageningen NMR Center and
Laboratory of Biophysics
Dreijenlaan 3
6703 HA Wageningen
The Netherlands

**Jamie D. Walls**
University of California
Department of Chemistry and
Biochemistry
Los Angeles, CA 90095
USA

**Ronald L. Walsworth**
Harvard-Smithsonian Center for
Astrophysics
60 Garden St
Cambridge, MA 02138
USA

Harvard University
Department of Physics
60 Garden St
Cambridge, MA 02138
USA

**Zhongdong Wang**
China University of Petroleum
School of Resource &
Information NMR Lab
Changping District
Beijing 102249
China

**Wladyslaw P. Weglarz**
Polish Academy of Sciences
Institute of Nuclear Physics
Radzikowskiego 15231-342
31-342 Kraków
Poland

**David I. Wilson**
University of Cambridge
Department of Chemical Engineering
Pembroke Street
Cambridge, CB2 3RA
UK

**Carel W. Windt**
Wageningen University and Research
Centre
Laboratory of Biophysics
Dreijenlaan 3
6703 HA Wageningen
The Netherlands

**Yang Xia**
Oakland University
Department of Physics
Rochester, MI 48309
USA

**Lizhi Xiao**
China University of Petroleum
School of Resource & Information
NMR Lab
Changping District
Beijing 102249
China

**Ranhong Xie**
China University of Petroleum
School of Resource & Information
NMR Lab
Changping District
Beijing 102249
China

**Zhi Yang**
University of New Brunswick
UNB MRI Centre
Department of Physics
8, Bailey Drive
Fredericton NB E3B 5A3
Canada

**ShaoKuan Zheng**
Oakland University
Department of Physics
Rochester, MI 48309
USA

# 1
# Musings on Hardware Advances and New Directions

*Eiichi Fukushima*

## 1.1
## Scope and Introduction

This chapter is an edited version of a rambling tutorial talk given at the 9th ICMRM in 2007, followed by a longer discussion of modern permanent magnets being used for NMR. All of this is supposed to be couched in 'nuts and bolts' terms, but that is a phrase which is often applied to my presentations regardless of the subject – it should not be taken too literally.

NMR/MRI is an extremely broad field, both in terms of the physics that is the basis for the field as well as its applications. In these senses, it is truly unique among all techniques. It was the aim of the tutorial lectures to provide sufficient background for a better understanding of the varied presentations given at the conference.

One dominant discipline that uses NMR is NMR spectroscopy for analytical chemistry and biochemistry. This was made possible when chemical shift was discovered shortly after the inception of the field. MRI, which can be thought of as spatially resolved NMR as opposed to chemically resolved NMR of NMR spectroscopy, is a relative latecomer to the field; it has been around for roughly 50% of the approximately 60-year history of NMR. Then, there is the original 'physically resolved' NMR, where the contrast mechanism consists of any number of relaxation parameters that are affected by molecular structure and motion. Thus, the parameter space for NMR/MRI can be thought of as shown in Figure 1.1. The three regions are the three overall contrast mechanisms (chemical, physical, spatial), and the intersections of the regions contain some of the most interesting applications. For this audience, they include velocity and diffusion imaging (as described by some of the chapters in the Transport section of this book), NMR well-logging (as described by Lizhi Xiao in Chapter 31), and NMR elastography.

*Magnetic Resonance Microscopy.* Edited by Sarah L. Codd and Joseph D. Seymour
Copyright © 2009 WILEY-VCH Verlag GmbH & Co. KGaA, Weinheim
ISBN: 978-3-527-32008-0

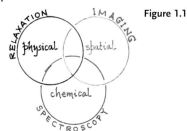

**Figure 1.1**

## 1.2
## NMR Building Blocks

NMR experiments have broad analogy to radio/television broadcast and reception. As shown in Figure 1.2, the objective of radio/TV broadcast and reception is to receive what was broadcast with as few changes as possible. However, if there is any change in the quality of reception, it is possible to characterize the medium that changed the signal if the input and output (i.e. what was transmitted and received) are known. In the same way, NMR starts with a known transmitted signal – for example, a rectangular pulse or a sinc pulse – and analyzes the sample (the intervening medium) by deconvoluting the received signal by the transmitted signal.

Thus, the NMR/MRI apparatus consists broadly of source of radiofrequency (RF) excitation, an NMR probe that interacts with the sample, a signal receiver, a

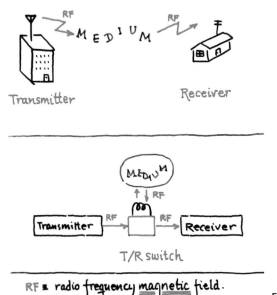

**Figure 1.2**

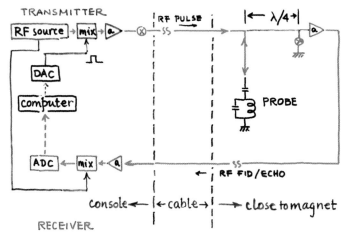

Figure 1.3

source of the magnetic field, and a controller that keeps order. Figure 1.3 is a block diagram of a typical NMR apparatus. Some common requirements are dictated by the typical signal size of few to tens of microvolts that needs to be amplified 100 000 to 1 000 000-fold in order to be detectable.

Many of the requirements vary over many orders, depending on the application. The magnetic field strength can vary from the Earth's field (e.g. Halse, *et al.* in Chapter 2) to few tens of Tesla, a range of five orders of magnitude. (Some of the strongest field magnetic resonance experiments have been carried out, not in magnets but in the vicinity of strongly polarized atoms.)

One important parameter is the recovery time of the NMR apparatus after an excitation, as this determines how soon the signal can be seen. The difficulty of obtaining short recovery times is the main reason for the relative slowness in the progress of imaging objects with short $T_2$. The common MRI sequences that use spin echoes rely on the samples having a long spin–spin relaxation time $T_2$, so they do not rely on short-recovery-time hardware. Thus, they are most suitable for low-viscosity liquids – which happen to be abundant in many systems, especially in biology. The question of how to study short(er) $T_2$ samples was the subject of several presentations at this conference.

## 1.3
## NMR of Short-$T_2$ Samples

Liquids tend to have $T_2 \sim T_1$ while solids tend to have $T_2 \ll T_1$. Therefore, it is much easier to obtain big signals from liquids than solids (because the short $T_2$ may make the signal decay in time short compared to the recovery time of the NMR apparatus). The general rule is that the more something is like a solid, the shorter $T_2$ becomes. This means normal tissues might have a quite long $T_2$, whereas cartilages, bones and lungs have very short $T_2$ and are not commonly imaged in clinical MRI units. Today, hardware and pulse sequence improvements are available that address this issue, and some of the presentations at the conference have dealt with this problem. Thus, there will be an increasing number of NMR (and imaging) of shorter $T_2$ samples such as soft bones, lungs, plastics, and so on, performed in the near future as we improve our ability to measure short-$T_2$ samples easily, and especially for larger samples.

## 1.4
### Field Dependence of Signal Strength

The signal-to-noise ratio (SNR) of NMR signals depends strongly on the strength of the static magnetic field. Conventional wisdom states that the SNR goes up as frequency to the 7/4th power. This fact is the driving force behind the push to perform experiments at ever stronger magnetic fields. (It might be interesting to know that it was believed early in MRI that whole-body imaging in fields higher than 1 T would not be possible because of electromagnetic absorption and other problems, although I have now heard of a 7 T whole-body scanner which should have a SNR that is enhanced from the original estimate by around 30-fold.) Today, the instrument manufacturers now pushing to raise the field above 23 T for high-resolution NMR which means a proton resonance frequency of 1 GHz.

So, how is it possible for some of us, especially at this conference, to consider performing NMR (and MRI) at significantly weaker fields? One helping factor is that it is usually possible to use solenoid coils at the weaker fields because the magnet geometry is different at the lower fields than at the axial fields of normal, superconducting magnets. Solenoids are better coils than the normal type used in common cylindrical superconducting magnets with an axial field (e.g. birdcage, saddle and Alderman–Grant coils) because the effective turns density is higher. Thus, the penalty of moving to weaker fields is partly offset by the availability of solenoid coils.

Another approach is simply to make the sample larger to compensate for the loss of signal at the weaker field, and this is done in many geophysical NMR applications, including borehole logging.

## 1.5
### Sample Size Dependence

NMR spans the range from the subsurface detection of water with 100 m surface coils to microcoil studies with samples that are fractions of a nanoliter. In units of milliliters, the sample for the surface coil is 10 or 11 orders of magnitude larger, while the microcoil samples are six orders or more smaller. This total range of 16 to 17 orders is difficult and unusual for any measuring method.

## 1.6
## Transmitter and Receiver Coils

Coils are the all-important link between the sample and the measuring apparatus – as is well-known, a bad coil can completely ruin an experiment. So, there is
a choice of either using separate coils for transmitting and receiving, or the same
coil can be multiplexed to perform both tasks. Separate coils are used often in
clinical imagers where the transmitter coil is built into the magnet housing while
the receiver coil is tailored to the sample. This is because a bad transmitter coil
only increases the need for more input power, whereas a bad receiver coil will lead
to loss in SNR that is not recoverable. Therefore, we can usually tolerate inefficien-
cies in transmission, but not in reception.

One criterion for a good (receiver) coil is that it generates an intense magnetic
field at the sample per unit current in the coil. David Hoult proved this using
the reciprocity principle [1]. This seemingly simple principle leads to some interesting conclusions that are not always appreciated. For a given overall coil geometry, the turns density is important because the field generated is directly
proportional to the number of turns. For identical windings, the quality factor
$Q = \omega L/r$ can be maximized to limit the bandwidth, although the bandwidth does
not have to be set by the NMR coil. Andrew McDowell has a scheme to tune
microcoils with terrible $Q$-values due to thin wires that have substantial resistivity
[2] at frequencies that are too low to deal with convenient values of tuning capacitors. He obtains excellent SNRs by connecting the microcoils in series with large,
high-$Q$ dummy coils. The microcoil acts as a resistor to the tank circuit, while
the dummy coil sets the resonance condition. Additional descriptions of these
coils will be provided later, in conjunction with the accompanying ultracompact
magnet.

## 1.7
## Shrinking Magnets

Here again, I will discuss a 'fringe' area of NMR and MRI. The most common
magnet in NMR/MRI is the persistent-mode superconducting magnet that generate fields along the axis, having replaced the iron core electromagnet as the dominant magnet in the 1970s, for good reasons. These magnets create very strong and
uniform fields that are stable, but they are expensive to buy and neither cheap nor
easy to maintain. This latter fact may be the single most important factor in these
magnets not becoming the basis for common uses of NMR/MRI outside laboratories, for example in factories, stores and outdoors. However, this situation is
slowly changing with the modern permanent magnet, which performs better than
ever before, are easy to maintain, require small installation spaces, and are relatively inexpensive to purchase and to maintain. The re-emergence of these magnets
is due, in part, to the discovery of new rare-earth-based magnetic materials such
as SmCo and NdFeB. However, another important reason is the realization that

**Table 1.1** The advantages and disadvantages of the various magnet types.

| Type of magnet | Strengths | Weaknesses |
| --- | --- | --- |
| Permanent | Small size; low maintenance; cheapest | Weaker field |
| Electromagnet | Variable field strength | Weaker field |
| Superconducting | Strong field; most stable field strength | High maintenance; most expensive |

**Figure 1.4**

NMR can achieve much at the relatively low field, and this was not realized in the past.

In Table 1.1 the advantages and disadvantages are listed of a variety of magnets, including those electromagnets not mentioned previously. Although electromagnets are convenient for changing field strengths, they otherwise are bulky, expensive (due to the need for fairly elaborate power supplies), and they suffer from short-term field fluctuations.

The effect of the coil that offsets some of the loss in SNR in going from a high-field superconducting magnet to a lower-field permanent magnet was mentioned previously. Therefore, these permanent magnets – which are relatively compact, inexpensive and maintenance free – will become increasingly popular. Figure 1.4 shows several small permanent 1 T magnets (plus a coffee cup for scale comparison) that are used in our laboratory. The two larger magnets (manufactured by NEOMAX, a division of Hitachi Metals Co. in Japan) are NdFeB magnets with gaps of 40 and 25 mm, and have homogeneity of 10 ppm over half of the gap, without shim coils. The smallest unit is a SmCo magnet (made by Aster) which

**Figure 1.5**

has a gap of 5 mm, weighs 700 g, and has an unshimmed homogeneity of ¼ ppm with a 300 μm microcoil sample. The NdFeB material is inexpensive and strongly magnetized, but has the disadvantage for NMR of a large temperature coefficient of the field ($\sim$−1000 ppm K$^{-1}$). Therefore, temperature regulation and a field-frequency lock are advisable. In order to take advantage of ¼ ppm homogeneity, the temperature would have to be regulated to within 250 microdegrees if there were no field-frequency lock present. Unfortunately, this is a daunting task and will limit sample access because of the significant insulation that would be needed. Thus, it makes sense to divide the task of field stability between temperature regulation and field-frequency lock.

For larger gaps, the magnet in Figure 1.5 should be considered. This 0.27 T/150 mm magnet, which is produced by Aster, is big and heavy and has vertical and horizontal access, which is a good feature to have for flow NMR. There is no fringe field due to the substantial return paths provided by the four cylindrical posts, and the gradient coils take up almost 25 mm, so that the space available for the probe is approximately 125 mm.

The NEOMAX magnets such as those shown in Figure 1.4 are also available in larger sizes. For example, the 1 T/60 mm magnet which we have used in our laboratory for a number of years weighs approximately 220 kg and can easily be moved around on a wheeled stand. Although it has been used for mouse imaging, the small gap (the 60 mm measurement does not include the gradient coils) limits its usefulness when compared to more modern NEOMAX magnets that leave a 60 mm gap, even with the gradient coils installed. The newer magnets also have crossed access slots at right-angles so that it is possible to have both vertical and horizontal access at the same time.

The largest magnet in this family is a modern 1 T/100 mm magnet described by Kose in Chapter 23 (see Figure 23.6). This magnet has a footprint of approximately 1 m², and the entire NMR/MRI system can be fitted into a space twice that size. The magnet's homogeneity can be shimmed to 16.4 ppm over a 60 mm diameter spherical volume, while the magnet temperature can be regulated at around 30 °C to minimize temperature drift of the Larmor frequency.

The NEOMAX magnets are what I refer to as 'modified' Halbach magnets. The standard Halbach magnets are annular arrangements of magnetic materials which generate transverse fields that are uniform, provided that the annulus is long compared to the diameter [3]. It is difficult to obtain superb uniformity with a standard Halbach magnet because the field homogeneity depends on the uniformity of the magnetic material, which is difficult to achieve at present. The NEOMAX magnets alleviate these difficulties by 'squaring' the geometry, thus enabling the use of steel pole caps to define the gap [4]. Now, the burden shifts to the accuracy of the pole caps' geometry plus balancing the flux on the two sides – which are much more manageable tasks. In addition, the magnetized blocks are arranged to shorten the magnet, resulting in an almost cubic aspect ratio.

## 1.8
## Shrinking NMR

In several presentations at the conference the details of small instruments that, by necessity, used small magnets and miniaturized electronics, were discussed. In fact, a consortium has now been set up which is headed by Peter Blümler and called the Virtual Institute for Portable NMR (http://www.portable-nmr.eu/index.php?index=2). Several commercial efforts have also been undertaken to create a compact NMR, mainly because a significant proportion of the cost (and weight) of any NMR or MRI is due to the magnet, so the instrument would be both cheaper and lighter if the magnet were to be made very small. The fact that there may be a 100- to 1000-fold reduction in the cost of the magnet harbors great promise for the future of ultra-compact (even hand-held?) NMR/MRI.

Andrew McDowell has pioneered the use of microcoils (coils smaller than 500 μm) with permanent magnets [2, 5]. As already mentioned, a major challenge here was in determining how to tune a coil with such a small inductance at low frequency, but this was overcome nicely with an auxiliary (normal size) coil that set the resonance condition with a normal tuning capacitor [2].

Figure 1.6 shows the first proton spectrum acquired using the smallest magnet in Figure 1.4. The drawn curve is a Lorentzian fit with width 9.9 Hz. The magnet used weighs 685 g, has a volume of 125 cm³, requires no adjustments, is nearly impervious to external magnetic fields, and responds gracefully to temperature variations – that is, the field profile does not change as the center value of the field changes. Most impressively, a microcoil probe built for this magnet achieves 0.25 ppm resolution without the use of current shims from 21 nl of water to provide a SNR of 61 in a single data acquisition.

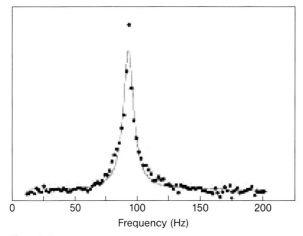

Frequency (Hz)

**Figure 1.6**

The eight-turn coil was hand-wound on a glass capillary with a 400 µm outer diameter and 300 µm inner diameter. Enameled magnet wire (50 gauge, total diameter 37 mm) was used for the winding. To date, the diameters of the capillaries used have ranged from 100 to 550 µm. The 400 µm size was chosen as a compromise between enclosing enough sample volume to provide adequate SNR performance and the need to probe the smallest practical region of the magnetic field in order to achieve a narrow line width.

When combined with existing single-board NMR spectrometer technology, such as a SpinCore board [6], this magnet and probe would be a hand-held, frequency-domain NMR spectrometer. The beauty of such a small NMR is that no power amplifier is needed, although the RF output of the transmitter may have to be attenuated to make this apparatus work.

Another area in which small and portable magnets will be used is unilateral NMR. This is a field that has been developed to a high degree of sophistication by the Aachen group in the form of NMR MOUSE [7]. A major application has been to the cultural heritage [8], and we at New Mexico Resonance and ABQMR have also been dabbling in this field. Our idea is to trade the high sensitivity of the MOUSE for a greater distance of detection with a barrel magnet [9]. Without going into details, this magnet is better at farther distances because it creates a 'sweet spot' where the field is reasonably uniform, thus enlarging the sensitive volume. Because of this uniform field spot, it should also perform well when measuring $T_2$ and the parameters derived from it, such as the diffusion constant. McDonald [10] and Marble [11] have presented different versions of such a unilateral magnet, while Callaghan's group has also developed a device that is closely related to ours [12].

Finally, the petroleum industry uses bore-hole well loggers which detect NMR signals from the material outside boreholes from tools within bore-holes, as described by Xiao [13]. However, even though these incredible devices are obviously mobile, they are neither small nor cheap.

## 1.9
## Future Prospects

Who knows what the future holds? Nobody predicted many of the recent advances in this field, such as even chemical shift, NMR imaging or the Earth's field J-coupling spectroscopy experiments [14]. Who would have thought NMR would be performed in the grasslands of Siberia [15] or on the Ross Ice Shelf in Antarctica [16]?

Having said that, there are certain points for the future which must surely come true. First, there will be a future! Besides the predictable directions of using higher fields and developing more sophisticated ways in which to manipulate spins, there will be diversification in new directions, as I have implied here. There will also be a spread of NMR/MRI techniques to areas that have not yet been exposed to such technology, or have been to only very limited degrees. This whole advance will be aided by the miniaturization of the magnets and electronics – it may not be too long before there is a cell-phone version of NMR!

## References

**1** Hoult, D.I. and Richards, R.E. (1976) The signal-to-noise ratio of the nuclear magnetic resonance experiment. *Journal of Magnetic Resonance*, **24**, 71–85.

**2** Sillerud, L.O., McDowell, A.F., Adolphi, N.L., Serda, R.E., Adams, D.P., Vasile, M.J. and Alam, T.M. (2006) 1H NMR detection of superparamagnetic nanoparticles at 1 T using a microcoil and novel tuning circuit. *Journal of Magnetic Resonance*, **181**, 181–90.

**3** Halbach, K. (1980) Design of permanent multipole magnets with oriented rare earth cobalt material. *Nuclear Instruments and Methods*, **169**, 1–10.

**4** Aoki, M. and Tsuzaki, T. (2006) Magnetic field generating device and MRI equipment using the device, U. S. Patent No. 7,084,633 (August 1).

**5** McDowell, A.F. and Adolphi, N.L. (2007) Operating nanoliter scale NMR microcoils in a 1 Tesla field. *Journal of Magnetic Resonance*, **188**, 74–82.

**6** SpinCore Technologies, Inc. (http://www.spincore.com).

**7** Eidmann, G., Savelsberg, R., Blümler, P. and Blümich, B. (1996) The NMR MOUSE, a mobile universal surface explorer. *Journal of Magnetic Resonance*, A**122**, 104–9.

**8** Proietti, N., Capitani, D., Rossi, E., Cozzolino, S. and Segre, A.L. (2007) Unilateral NMR study of a XVI century wall painted. *Journal of Magnetic Resonance*, **186**, 311–18.

**9** Fukushima, E. and Jackson, J.A. (2004) Unilateral magnet having a remote uniform field region for nuclear magnetic resonance, U.S. Patent No. 6,828,892 (December 7).

**10** McDonald, P.J., Aptaker, P.S., Mitchell, J. and Mulheron, M. (2007) A unilateral NMR magnet for sub-structure analysis in the built environment: the surface GARField. *Journal of Magnetic Resonance*, **185**, 1–11.

**11** Marble, A.E., Mastikhin, I.V., Colpitts, B.G. and Balcom, B.J. (2007) A compact permanent magnet array with a remote homogeneous field. *Journal of Magnetic Resonance*, **186**, 100–4.

12 Manz, B., Coy, A., Dykstra, R., Eccles, C.D., Hunter, M.W., Parkingso, B.J. and Callaghan, P.T. (2006) A mobile one-sided NMR sensor with a homogeneous magnetic field: The NMR-MOLE. *Journal of Magnetic Resonance*, **183**, 25–31.

13 Coates, G.C., Xiao, L. and Prammer, M.G. (1999) *NMR Logging Principles and Applications*, Halliburton Energy Services, Houston.

14 Appelt, S., Kühn, H., Häsing, F.W. and Blucmich, B. (2006) Chemical analysis by ultrahigh-resolution nuclear magnetic resonance in the Earth's magnetic field. *Nature Physics*, **2**, 105–9.

15 Shushakov, O.A. (1996) Groundwater NMR in conductive water, *Geophysics*, **61**, 998–1006.

16 Callaghan, P.T., Eccles, C.D. and Seymour, J.D. (1997) An Earth's field NMR apparatus suitable for pulsed gradient spin echo measurements of self-diffusion under Antarctic conditions. *Review of Scientific Instruments*, **68**, 4263–70.

# Part One    Novel Techniques

# 2
# Multidimensional Earth's-Field NMR

*Meghan E. Halse, Andrew Coy, Robin Dykstra, Craig D. Eccles, Mark W. Hunter and Paul T. Callaghan*

## 2.1
## Introduction

In 1954, Packard and Varian [1] made the suggestion that the Earth's magnetic field could be used to detect nuclear precession, thus making Earth's-field NMR (EFNMR) almost as old as NMR itself [2, 3]. The most extensive use of Earth's-field NMR to date is in the measurement of geomagnetism. Earth's-field NMR magnetometry owes its precision to the fact that the local magnetic field intensity is measured via the nuclear spin precession frequency. As every good physics student knows, when precision and accuracy are required, frequency measurements are the preferred mode. Due to the low cost of the instrumentation and the safe nature of the magnetic fields involved, EFNMR is a great teaching tool for people learning the principles of NMR [4].

However, the potential for Earth's-field NMR to be used for a great deal more than magnetometry, and indeed more than teaching elementary NMR principles to undergraduates, has only begun to be explored in recent years. For example, large-scale NMR transmit and receive coils have been used to detect ground water at depths of hundreds of meters [5]. It has also been shown that single-shot free induction decays (FIDs) obtained using Earth's-field NMR can yield spectral resolution which is significantly higher than that available with high-field laboratory magnets [6, 7], thus opening up prospects for new applications in NMR spectroscopy. Furthermore, it has been shown possible to obtain MR images in the Earth's magnetic field [8–10]. The extension of Earth's-field NMR to imaging and spectroscopy thus begs the question as to whether the method might be able to exhibit some of the versatility and power of its high-field laboratory equivalents.

Achieving this versatility and power hinges on the ability to program multipulse sequences combining radiofrequency (RF), or in the Earth's-field case, ultra-low-frequency (ULF) pulses of controlled phase, along with magnetic field gradient pulses, and to multiplex the acquisitions and evolution processes to yield multidi-

*Magnetic Resonance Microscopy.* Edited by Sarah L. Codd and Joseph D. Seymour
Copyright © 2009 WILEY-VCH Verlag GmbH & Co. KGaA, Weinheim
ISBN: 978-3-527-32008-0

mensional data sets. Equally important is the need to optimize the homogeneity and stability of the polarizing (and precession) magnetic field, $B_0$, and to ensure that optimal signal-to-noise ratios (SNRs) are achieved. The conventional perspective in NMR is that sensitivity follows a $B_0^2$ dependence of SNR, with both the equilibrium polarization and the Faraday detection sensitivity being proportional to $B_0$. This suggests a very poor performance for EFNMR where $B_0$ is about $50\,\mu T$. The $B_0^2$ problem can be partly obviated by using a large prepolarizing magnetic field, in which the spins are allowed to come to a prior thermal equilibrium at much higher levels of magnetization than would prevail in the Earth's field alone. Provided that this prepolarizing field (which need not be particularly homogeneous) is switched off adiabatically over a time shorter than $T_1$, the large magnetization persists for use in the subsequent EFNMR observation. The $B_0$ dependence of the detection stage can be circumvented by the use of non-Faraday detection methods, for example, by means of SQUID flux detectors [11]. Finally, the high homogeneity of the Earth's magnetic field can be exploited by using very large sample volumes relative to those used in high-field NMR, thereby achieving a higher signal level.

In this chapter, we focus on improvements to Earth's-field NMR that lie within the realm of Faraday detection combined with prepolarization. These improvements permit a wide class of imaging and spectroscopy experiments to be performed and, importantly, permit the use of the method inside ferrocement buildings, where the magnetic field used for detection may exhibit inhomogeneities normally absent in outdoor environments, and where significant ULF interference may arise. These advances are possible because of four principal technical enhancements:

- The first enhancement concerns the use of Faraday screening, which is a relatively obvious tool for reducing ULF noise pickup.

- The second concerns the use of three-axis gradient coils as shims, so as to restore magnetic field homogeneity so that sub-hertz resolution is possible.

- The third involves the use of a field-frequency lock, to compensate the slow drifts in Earth's-field intensity associated with the geophysical diurnal cycle and other more localized field shifts.

- The fourth, and most significant feature, is the use of a purpose-designed ultra-low-field spectrometer which provides fully digital operation with multidimensional loop control, ULF phase control and flexible ULF and magnetic field gradient pulse sequence timing.

It is this feature, along with clever magnetic field gradient coil designs, that permits a wide range of experiments: from gradient and spin-echo MRI, to pulsed gradient spin-echo (PGSE) diffusivity measurements, to 2-D NMR spectroscopy. This chapter will review these advances and demonstrate the applications that have become possible.

## 2.2
## Apparatus Developments

### 2.2.1
### Shimming and Screening

Historically, one of the most significant practical limitations of NMR experiments carried out using the Earth's magnetic field was the need to set up the instrument in a remote, nonurban environment. The reason for this was two-fold. First, in order to benefit from the high degree of homogeneity of the Earth's magnetic field the apparatus must be far removed from any ferrous or magnetic materials. Generally speaking, it is very difficult to achieve this within ferrocement buildings. Second, the Larmor frequency of protons in the Earth's magnetic field is approximately 2 kHz, a frequency at which external noise pick-up in urban environments exceeds the Johnson noise of the receive coil by several orders of magnitude.

The issue of the external noise pick-up of the receive coil can be overcome through the use of a grounded Faraday cage. Skin depths in common conductors such as copper and aluminum are quite significant at 2 kHz (ca. 1.5 mm for copper and 2 mm for aluminum). Therefore, the wall thickness of the Faraday cage needs to be very large. A copper box with a wall thickness of 10 mm and a mass of approximately 50 kg was found to reduce the amplitude of the external pick-up noise in the time domain by a factor of about 25, for example from 400 to 17 μV. An additional reduction of the noise amplitude through Faraday screening to 2.5 μV was achieved by shorting the prepolarizing coil during signal acquisition, as described previously [12]. For comparison, the expected Johnson noise from this coil is about 0.6 μV. It should be noted that, when designing such a Faraday cage, care must be taken that the box is sufficiently large that any eddy currents induced by the switching of the polarizing or gradient fields do no disrupt the homogeneity of the field during signal detection.

The effectiveness of the Faraday cage was found to be only slightly reduced when the end plates of the copper box were removed such that the resultant open-ended box was oriented collinear with the sinusoidal receiver coil (Figure 2.1). The substantial weight reduction of the Faraday cage achieved by removing the end caps is a very significant advantage, and therefore the open-ended design is a good compromise between weight and effectiveness. It was also found that an open-ended box of 12 mm-thick aluminum achieved similar screening results as the 10 mm-thick copper box. The aluminum box is a much more attractive solution to the screening problem because, despite the increased wall thickness, the overall weight and cost of the aluminum Faraday cage will be greatly reduced when compared to the equivalent in copper.

The issue of degradation in the natural homogeneity of the Earth's magnetic field due to the proximity of ferrous or magnetic materials in an indoor environment can be overcome by the combination of a judicious placement of the apparatus and the use of first-order shimming. In order to ensure a high degree of

**Figure 2.1** A photograph of a 10 mm-thick, open-ended copper box acting as a Faraday screen around the Terranova EFNMR probe, which is oriented such that the $B_1$ transmit/receive coil is coaxial with the open-ended copper box.

field homogeneity, it is important to place the apparatus on a table or stand that is free of ferrous materials and to remove to a distance of 1–2 m any objects likely to disturb the field homogeneity. It is often advisable to place the probe roughly in the center of a room, at a significant distance from all walls, including the ceiling and the floor. Once this isolation of the apparatus is achieved, it has been found that the dominant inhomogeneities in the field are linear and so can be effectively countered by first-order shimming. The three orthogonal gradients used for imaging, the design of which is detailed later in this chapter, were found to be well suited for the purpose of first-order shimming.

### 2.2.2
### Field Stabilization

Though highly homogeneous in a spatial sense, the Earth's magnetic field undergoes significant temporal variations, on the order of tens of nT, on a timescale of hours and minutes. In imaging and relaxometry applications, it may be possible to choose gradients strengths and total experiment times such that no degradation in the experiment quality is observed. However, in lengthy multidimensional spectroscopy experiments, where sub-hertz resolutions are required, these temporal fluctuations cannot be ignored.

High-resolution NMR is naturally well suited to detecting small changes in frequency. Therefore, it is possible to track any changes in the Earth's magnetic field with a high degree of precision through the use of a reference scan interleaved into any multistep pulse sequence. Any detected change in the

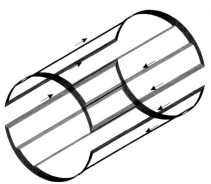

**Figure 2.2** A schematic diagram of the multiloop saddle coil used as a $B_0$ lock coil for frequency stabilization.

Earth's field can then be offset by the field from a specially designed $B_0$ lock coil.

The design requirements for a $B_0$ lock coil for the Earth's field are three-fold: (i) it must be tunable by means of a current value under software control; (ii) it must be able to generate a $B_0$ field offset equal to the largest observed fluctuation in the Earth's field (100 nT); and (iii) the homogeneity of the $B_0$ offset field must be such that linewidths of better than 0.1 Hz (2 nT) can be achieved. These conditions require only a 2% field homogeneity, which is easily realized using modern coil design techniques. The arrangement of one such coil, a multiturn saddle coil, is illustrated in Figure 2.2. This coil is designed to generate a field that is homogeneous to better than 2 nT (0.1 Hz) for fields of ±117 nT (±5 Hz) over a 100 mm sample diameter. The linearity of the relationship between the current and the $B_0$ lock field produced over the range of offset fields required is such that no iteration is necessary in order to achieve the appropriate current value for the $B_0$ lock coil. The field shift is simply quantified using a single reference scan, and the current through the $B_0$ lock coil is adjusted according to a calibration constant.

In Figure 2.3 the black dots show the temporal variations in the proton Larmor frequency due to drift in the magnitude of the Earth's field over a period of 24 h. The gray dots show the observed proton frequency of a concurrent EFNMR measurement carried out using a reference scan and a $B_0$ lock field to counter the observed drift in the proton Larmor frequency. From this plot it can be seen that the Larmor frequency is stabilized to within about ±0.3 Hz. The linewidth of the EFNMR spectra, as show in the inset in Figure 2.3, was approximately 0.3 Hz.

### 2.2.3
### Ultra-Low-Field Spectrometer

In some sense, the core of the Earth's-field NMR apparatus is the ultra-low-field spectrometer which executes and controls the pulse sequence, from turning on and off the polarizing coil to recording and processing the FID signal. The

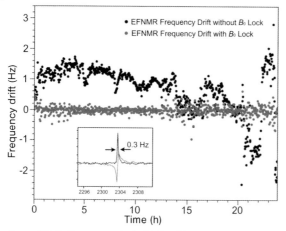

**Figure 2.3** Drift over a period of 24 h of the EFNMR proton frequency from spectra acquired with (gray dots) and without (black dots) the $B_0$ lock field. The spectral resolution, as shown in the inset, was 0.3 Hz.

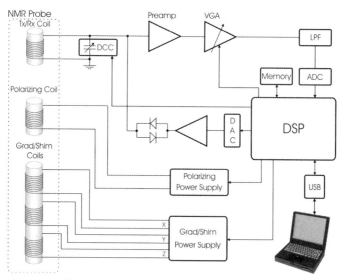

**Figure 2.4** A block diagram of the Earth's-field ultra-low-field spectrometer.

ultra-low-field spectrometer described herein is very similar to a conventional high-field laboratory system except that it operates in the kHz range. Therefore, relatively simple transceiver circuitry can be used and the signal can be sampled directly, eliminating the need for intermediate frequency (IF) stages.

The block diagram in Figure 2.4 provides an overview of the spectrometer system. The control and signal processing part of the Earth's-field apparatus consists of a digital signal processor (DSP)-based pulse programmer and a data acquisition unit that uses a universal serial bus (USB) interface to communicate with a host computer. By using a DSP, many of the functional blocks can be implemented in software instead of hardware, and therefore a greater flexibility can be achieved.

The transmit (Tx) signal is generated using a digital oscillator algorithm running on the DSP. Multiple oscillators are implemented for multiple phases. As the power required for the $B_1$ excitation is minimal, a low-noise preamplifier can be connected permanently to the probe, while ordinary operational amplifiers can be used to drive the probe. A series of capacitors, under software control, are switched in for probe tuning. Probe impedance matching is not required due to the very long wavelength; the probe operates with a resonant impedance of several tens of kilo-ohms.

The spectrometer allows for two types of signal acquisition. In the first method, the incoming signal is sampled at a rate several times higher than the Larmor frequency, and thus the entire carrier is digitized and stored. While simple, this approach requires a large amount of memory to store the entire waveform. A more efficient sampling methodology is to implement a quadrature receiver in the manner of a conventional high-frequency NMR system. Here, the incoming signal is digitized at a constant rate of 100 kHz. The digital data are then mixed with two digital oscillators, 90° out of phase, to produce two quadrature channels at baseband. This high sample rate data is then low-pass filtered and decimated down to a lower sample rate. Conceptually, this is exactly the same technique as used in digital receivers, but in this implementation all steps are carried out in the software on the DSP. Since oversampling is used, only a first-order analogue filter is required to prevent aliasing.

In addition to controlling the transmitting and receiving portions of the NMR experiment, the ultra-low-field spectrometer also contains five current-controlled amplifiers which are used to drive the prepolarizing coil, the three imaging/shimming gradient coils, and either the PGSE gradient coil or the $B_0$ lock coil, depending on the requirements of the given experiment.

The entire ultra-low-field spectrometer is housed within a box with physical dimensions of $340 \times 240 \times 160$ mm and a mass of less than 5 kg. As this spectrometer can be run off a 24 V, 10 A power supply, or from two series connected 12 V car batteries, it is useful for both outdoor operation and laboratory use.

## 2.2.4
### Gradient Coil Design

In any NMR experiment, the richness of the information content comes about through the manipulation of various terms in the interaction Hamiltonian via a combination of RF and gradient pulses. Therefore, the design of high-quality gradient coils is essential for a versatile and powerful Earth's-field NMR apparatus. In many ways the design of gradient coils for ultra-low field NMR applications is analogous to the high-field NMR approach, but with one very important difference – that the Earth's field is relatively weak (on the order of 50 µT) – and so care must be taken to ensure that concomitant gradient effects are taken into account where necessary.

For imaging applications, a quick calculation of $B_0/G$ [13] shows that any artificial curvature due to concomitant gradients at the Earth's field will be around 1 m, and so can be neglected. However, there remains the subtle issue of the relative orientation of the imaging gradient set with respect to the Earth's field ($B_E$) itself. It is very important to ensure that the 'z'-axis of the gradient coil set is aligned with the Earth's field in order to maintain the orthogonality of the set of three imaging gradients. More detailed information on how the relative orientation of the Earth's field and the 'z'-axis of a three-axis gradient coil set affects the gradient fields generated can be found in Ref. [10].

Figure 2.5 presents a line drawing of a gradient coil set, comprised of a saddle coil, which generates a gradient field, $G_x$, along the axis of the probe and two quadrupolar coils 45° apart, which generate two orthogonal gradient fields, $G_z$ and $G_y$, in the plane orthogonal to the axis of the probe, where 'z' is aligned with the Earth's magnetic field. These gradient coils adhere to the traditional electromagnet geometry employed at high fields but, as discussed above, are designed to generate very weak gradient fields (0.27 nT mm$^{-1}$ mA$^{-1}$).

For PGSE applications, the gradient field strengths required are much larger than those needed for imaging. Therefore, the effects of concomitant gradients cannot be dismissed but rather must be included in the gradient coil design process. In the case of PGSE gradient design in ultra-low fields, the relevant gradient is that in the magnitude of **B** rather than just the gradient in the component of the field parallel to a strong, dominant $B_0$ field [14]. A simple Maxwell pair, which is commonly used to generate $\mathbf{G}_z$ where the prevailing strong field is $\mathbf{B} = B_0 \hat{z}$, will not produce a pure longitudinal gradient in $|\mathbf{B}|$ in the ultra-low $B_0$ field case because of the divergence-free nature of **B** [15]. A solution to this problem has been proposed [16] whereby a combined Maxwell–Helmholtz configuration is employed to produce a gradient in $|\mathbf{B}|$ along the axis of the coil, $x$. The superposition of the Maxwell pairs (opposed currents) and Helmholtz pairs (parallel currents) produces a gradient field in which the null point of the field is shifted out of the sample volume. This results in a near-uniform $d|\mathbf{B}|/dx$ gradient which varies by less than 10% over a 100 mm diameter sample. This net field

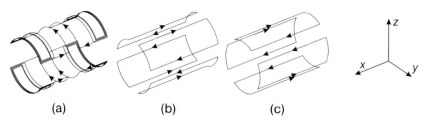

(a)   (b)   (c)

**Figure 2.5** Drawings of the three Earth's-field MRI gradient coils used to generate (a) $\mathbf{G}_x$, (b) $\mathbf{G}_y$ and (c) $\mathbf{G}_z$.

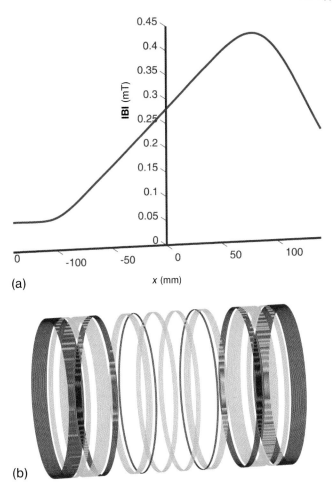

(a)

(b)

**Figure 2.6** (a) The field along $x$ which results from the superposition of the Earth's field and the field generated by the PGSE gradient coil; (b) A line drawing of the PGSE gradient coil; the black turns are the Maxwell pairs and the gray turns the Helmholtz pairs.

is oriented along $x$, not along the direction of the Earth's field, and so the gradient pulse must be switched off adiabatically–that is, slowly on the time scale of the precessing magnetization. Figure 2.6b presents a line drawing of the PGSE gradient coil design, showing the combination of Maxwell pairs (black turns) and Helmholtz pairs (gray turns). The total field along $x$, which results from the superposition of the Earth's field and the field generated by the PGSE gradient coil is plotted in Figure 2.6a. The gradient strength produced by this coil is $2.35 \, \mathrm{mT \, m^{-1} \, A^{-1}}$.

## 2.3
## Applications

### 2.3.1
### Pulsed-Gradient Spin-Echo (PGSE) NMR

During the 1970s it was shown that, by appropriately tailoring the magnetic field, NMR could be used for the precise measurement of molecular diffusion. The idea that diffusion could be measured in the Earth's magnetic field using PGSE NMR led to a project in which an EFNMR apparatus was developed for use in Antarctic sea ice studies. The Antarctic EFNMR project resulted, in 1994, in the acquisition of the first proton NMR signals obtained from unfrozen brine in extracted sea-ice cores [17]. This was followed by successive measurements, including PGSE measurements of brine diffusivity, made on the sea ice of three further spring seasons [16, 18, 19]. Figure 2.7 presents a sample brine diffusivity measurement acquired of an *in situ* sea-ice core at a depth of 143 cm using Earth's-field PGSE NMR in Antarctica in 2002.

### 2.3.2
### Magnetic Resonance Imaging

Over the past few decades, magnetic resonance imaging (MRI) has become one of the most widely used and successful applications of NMR. Two-dimensional

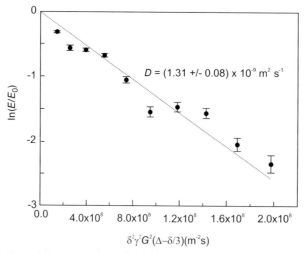

**Figure 2.7** An example Earth's-field PGSE NMR measurement of the diffusivity of sea ice at a depth of 143 cm, acquired in Antarctica in 2002. The dotted line shows a weighted fit to the data giving a diffusion coefficient of $(1.31 \pm 0.08) \times 10^{-9} \, m^2 s^{-1}$. (After Ref. [15].)

MRI was first performed in the Earth's magnetic field by Stepisnik *et al.* in 1990 [8], and diffusion and flow imaging [9] and 3-D MRI [10] were later also demonstrated to be possible. Most of the high-field imaging pulse sequences, including gradient-echo and spin-echo imaging, can be implemented in the Earth's magnetic field. The most significant limitation of Earth's-field MRI is the long minimum echo time (~100 ms) which is a consequence of the long $B_1$ coil dead times (~20 ms). A method for reducing the dead time of a $B_1$ coil which resonates at ultra-low frequencies has been suggested by Ward *et al.* [20]. A variant on this technique has been implemented to reduce the effective dead-time to about 1 ms allowing echo times as short as 10 ms. This method may greatly increase the range of samples which can successfully be imaged using Earth's-field NMR.

As mentioned in Section 2.2.4, although concomitant gradients are not a concern for imaging applications in the Earth's magnetic field, drift in the magnitude of the Earth's field occurring between phase-encode steps can cause blurring or distortions in the image. As detailed in Ref. [10], it is possible to choose a pixel size (in frequency units) which is greater than the frequency drift anticipated within the total imaging time. In this way artifacts can be avoided. Alternatively, the $B_0$ lock coil described in the apparatus section and a reference scan interleaved between phase-encode steps can be used to prevent any artifacts in the image which might otherwise occur due to field drift.

Figure 2.8 presents a 2-D spin-echo image acquired of a six-tube water phantom in the Earth's magnetic field using a Terranova EFNMR system (Magritek Ltd, Wellington, New Zealand) with the probe inside a Faraday cage made from 12 mm-thick aluminum plate. Each tube within the phantom is 15 mm in diameter and 50 mm long. The pixel size in the image is just over 1 mm and the resolution is about 3 mm. The image was acquired in 17 min with four signal averages.

### 2.3.3
### Multi-Dimensional Spectroscopy

Until recently it was generally accepted that, because the chemical shifts for most nuclei are vanishingly small at the Earth's field, the only chemical information

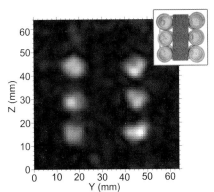

**Figure 2.8** A 2-D spin-echo image of a six-tube water phantom acquired in the Earth's field using the Terranova EFNMR apparatus.

available at ultra-low field strengths was heteronuclear J coupling. However, Appelt *et al.* have recently shown [6] that both heteronuclear and homonuclear J coupling constants are observable when the heteronuclear coupling breaks the magnetic equivalence of two groups of homonuclear spins within a molecule. In addition, Appelt *et al.* have demonstrated that many of the heteronuclear J coupling constants fall into the strong coupling regime at the Earth's field, thus providing a further mechanism for differentiating groups of homonuclear spins which would otherwise appear identical in purely J-coupled NMR spectra acquired in the weak coupling regime.

These surprising 1-D spectroscopy results beg the question of whether or not the wide range of 2-D NMR spectroscopy techniques, which have proven to be so useful at high field, are possible in the Earth's magnetic field. In 2006, Robinson *et al.* [21] presented heteronuclear $^1$H–$^{19}$F COSY (correlation spectroscopy) spectra acquired in the Earth's field, thus demonstrating the viability of multidimensional EFNMR spectroscopy.

Figure 2.9a presents a heteronuclear $^1$H–$^{19}$F COSY spectrum of 1,4-difluoroben-zene acquired using a Terranova EFNMR system along with the screening, shimming and frequency stabilization techniques described in Section 2.2. A calculated spectrum is presented in Figure 2.9b for comparison. All of the J coupling constants used for the calculation were obtained from Paterson and Wells [22]. The three-bond heteronuclear coupling constant, $^3J_{HF} = 7.6$ Hz, and the four-bond heteronuclear J coupling constant, $^4J_{HF} = 4.6$ Hz, are significantly different. Therefore, although the protons – and, similarly, the fluorine nuclei – are chemically equivalent, they are not magnetically equivalent. Thus the homonuclear proton–proton and fluorine–fluorine couplings must be included in the calculation. The homonuclear coupling constants used were as follows: $^5J_{FF} = 12$ Hz; $^3J_{HH} = 8$ Hz; $^4J_{HH} = 2$ Hz; and $^5J_{HH} = 0$ Hz. Good agreement is observed between the calculated and observed spectra.

## 2.4
## Conclusions and Future Outlook

The recent developments, both technical and methodological, in the area of ultra-low field NMR have revealed that the possible applications of EFNMR, from the measurement of molecular dynamics to NMR imaging and spectroscopy, are much broader than were originally anticipated. Indeed, the information content of an Earth's-field NMR measurement is, in a sense, as rich as that encountered in high-field NMR, due largely to the exceptionally high spectral resolutions achievable because of the highly homogeneous nature of the Earth's field itself and also the very long $T_2$ values of many organic molecules at such a low field.

Despite the richness of its information content, it is highly unlikely that Earth's-field NMR will achieve the power and versatility of a high-field NMR system unless the problem of low sensitivity can be overcome. Therefore, it is critical that new methodologies be employed to overcome the $B_0^2$ sensitivity dependence of tradi-

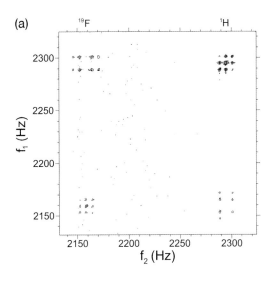

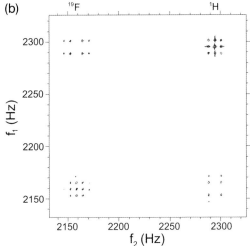

**Figure 2.9** (a) A $^1$H–$^{19}$F heteronuclear COSY spectrum of 1,4-difluorobenzene acquired with a Terranova EFNMR system; (b) A simulated $^1$H–$^{19}$F heteronuclear COSY spectrum of 1,4-difluorobenzene calculated using the homonuclear and heteronuclear J coupling constants [20].

tional NMR methods. There are several exciting possibilities for increasing the sensitivity of ultra-low field NMR, both in terms of enhancing polarization and also in terms of using a more efficient signal detection scheme. In terms of detection, both superconducting quantum interference devices (SQUIDs) [11] and atomic magnetometers [23] are emerging as promising technologies for enhancing the detection efficiency of NMR at ultra-low fields. In terms of enhancing polarization, the spin-polarized induced nuclear Overhauser effect (SPINOE) has been demonstrated by Appelt *et al.* [24] to greatly enhance signals at fields as low as 1 mT, and shows promise for use in Earth's-field NMR. Dynamic nuclear polarization (DNP) has also been successfully implemented to greatly increase the sensitivity of EFNMR magnetometers [25], and in the future may provide an effective mechanism for significant sensitivity enhancement in other EFNMR applications.

Another possible adaptation of the Earth's-field approach to NMR, which could potentially extend its power and versatility, would be to continue performing the NMR detection in an ultra-low field but, instead of using the Earth's magnetic field, the experiment could be performed in a magnetically shielded room where the detection field is generated by a coil arrangement, such as a Helmholtz pair. There are several advantages to this scheme. First, there would no longer be a need to compensate for the diurnal variations in the Earth's field. Frequency stability would be a function of the stability of the power supply used to generate the detection field. Second, the detection field could be oriented in a more convenient direction than the naturally occurring Earth's magnetic field, and could be switched in polarity in a spin-echo sequence to refocus dephasing due to detection-field inhomogeneity. In addition, the strength of the detection field would be under user control. However, to move away from the use of the Earth's magnetic field for detection would be to give up the extremely high degree of homogeneity of that field, which allows for the use of large samples and produces astonishingly good spectral resolution. The generation of even a moderately homogeneous detection field would also require relatively large Helmholtz coils, which increases the complexity and overall size and weight of the otherwise small and compact ultra-low field NMR system.

## Acknowledgments

The authors are grateful to the New Zealand Foundation of Research Science and Technology for financial support.

## References

1 Packard, M. and Varian, R. (1954) *Physical Review*, **93**, 941.
2 Bene, G.J., (1971) 1st International Society of Magnetic Resonance Conference, Rehovot, Israel.
3 Bene, G.J. (1980) *Physics Reports – Review Section of Physics Letters*, **58**, 213.
4 Callaghan, P.T. and Legros, M. (1982) *American Journal of Physics*, **50**, 709–13.
5 Shushakov, O.V. (1996) *Geophysics*, **61**, 998–1006.

6 Appelt, S., Kuhn, H., Wolfgang Hasing, F. and Blumich, B. (2006) *Nature Physics*, **2**, 105–9.
7 Appelt, S., Wolfgang Hasing, F., Kuhn, H., Perlo, J. and Blumich, B. (2005) *Physical Review Letters*, **94**, 197602.
8 Stepisnik, J., Erzen, V. and Kox, M. (1990) *Magnetic Resonance in Medicine*, **15**, 386–91.
9 Mohoric, A., Stepisnik, J., Kos, M. and Planinsic, G. (1999) *Journal of Magnetic Resonance*, **136**, 22–6.

**10** Halse, M.E., Coy, A., Dykstra, R., Eccles, C., Hunter, M., Ward, R. and Callaghan, P.T. (2006) *Journal of Magnetic Resonance*, **182**, 75–83.

**11** McDermott, R., Trabesinger, A.H., Muck, M., Hahn, E.L., Pines, A. and Clarke, J. (2002) *Science*, **295**, 2247–9.

**12** Callaghan, P.T. and Eccles, C.D. (1996) *Bulletin of Magnetic Resonance*, **18**, 62–4.

**13** Yablonskiy, D.A., Sukstanskii, A.L. and Ackerman, J.J.H. (2005) *Journal of Magnetic Resonance*, **174**, 279–86.

**14** Callaghan, P.T. and Stepisnik, J. (1996) *Advances in Magnetic and Optical Resonance*, Academic Press, San Diego, pp. 326–89.

**15** Stepisnik, J. (1995) *Zeitschrift für Physikalische Chemie*, **190**, 51–62.

**16** Mercier, O.R., Hunter, M.W. and Callaghan, P.T. (2005) *Cold Regions Science and Technology*, **42**, 96–105.

**17** Callaghan, P.T., Eccles, C.D. and Seymour, J.D. (1997) *Review of Scientific Instruments*, **68**, 4263–70.

**18** Callaghan, P.T., Eccles, C.D., Haskell, T.G., Langhorne, P.J. and Seymour, J.D. (1998) *Journal of Magnetic Resonance*, **133**, 148–54.

**19** Callaghan, P.T., Dykstra, R., Eccles, C.D., Haskell, T.G., Seymour, J.D. and Regions, C. (1999) *Science and Technology*, **29**, 153–71.

**20** Ward, R.L. and Dykstra, R. (2006) *Proceedings of the 13th Electronics New Zealand Conference*, Christchurch, New Zealand.

**21** Robinson, J.N., Coy, A., Dykstra, R., Eccles, C.D., Hunter, M.W. and Callaghan, P.T. (2006) *Journal of Magnetic Resonance*, **182**, 343–7.

**22** Paterson, W.G. and Wells, E.J. (1964) *Journal of Molecular Spectroscopy*, **14**, 101–11.

**23** Savukov, I.M. (2005) *Physical Review Letters*, **94**, 123001.

**24** Appelt, S., Haesing, F.W., Baer-Lang, S., Shah, N.J. and Blumich, B. (2001) *Chemical Physics Letters*, **348**, 263–9.

**25** Kernevez, N. and Glenat, H. (1991) *IEEE Transactions on Magnetics*, **27**, 5401–4.

# 3
# Multiple-Echo Magnetic Resonance

*Yi-Qiao Song, Eric E. Sigmund and Hyung Joon Cho*

## 3.1
## Introduction

Many nuclear magnetic resonance (NMR) experiments are executed by preparing some initial state of the spin system and then measuring the resulting magnetization. Through variation of the initial state or its evolution over repeated acquisitions, a series of data is accumulated as a function of the dynamics of the spin system. Such experiments require multiple scans, and often also a significant waiting time between them. For many applications, an acceleration of the acquisition is critical, either due to practical time constraints (e.g. in medical imaging) or because of the rapid variation in the underlying physical phenomena.

One of the best examples of accelerated acquisition is echo-planar imaging (EPI) [1], which uses multiple echoes for phase encoding. Multiple echoes of the Carr–Purcell–Meiboom–Gill (CPMG) [2, 3] sequence are used to measure relaxation, diffusion, flow or imaging [4]. The difftrain technique [5] employs a train of low-angle radiofrequency (RF) pulses to measure diffusion. Another class of techniques achieves different encoding at different spatial locations in a uniform sample to accelerate the measurements of spin relaxation [6], diffusion [7] or multidimensional NMR spectra [8].

The application of $n$ RF pulses can create $(3^{n-1} - 1)/2$ spin echoes in a spin-1/2 system. A recent series of reports [9–13] has described methods to create and to selectively encode all of these echoes for diffusion, flow or imaging applications. The basic sequence, termed multiple modulation–multiple echo (MMME), consists of a set of high-angle RF pulses with unequal time spacings that generate a maximal number of spin echoes. In particular, these recent reports have shown that such encoding can be achieved for the simultaneous measurement of diffusion or flow along different directions. Previously, similar RF sequences have been used for rapid imaging [14, 15]. In this chapter we will discuss the spin dynamics of the MMME sequences and their applications.

*Magnetic Resonance Microscopy.* Edited by Sarah L. Codd and Joseph D. Seymour
Copyright © 2009 WILEY-VCH Verlag GmbH & Co. KGaA, Weinheim
ISBN: 978-3-527-32008-0

## 3.2
## MMME Technique

### 3.2.1
### Multiple Modulation Multiple Echoes

Consider a static magnetic field gradient ($g$) and a train of three pulses with tipping angles $\alpha_1$, $\alpha_2$, and $\alpha_3$, and time spacings between them to be $\tau_1$, $\tau_2$,

$$\alpha_1 - \tau_1 - \alpha_2 - \tau_2 - \alpha_3 - \text{acquisition}\,(\tau_3) \tag{3.1}$$

where the time of the echo after the last pulse is $\tau_3$. The nutation angles of the pulses do not have to be multiples of 90°. Neglecting recovery due to $T_1$ relaxation for the moment, the above sequence will allow a total of five coherence pathways ($Q$) [16] to be observed, and will thus create five signals:

| Signal number | $q_0$ | $q_1$ | $q_2$ | $q_3$ | $b_Q$ | $c_{1Q}$ | $c_{2Q}$ |
|---|---|---|---|---|---|---|---|
| 1 | 0 | 0 | 0 | −1 | 0 | 4 | 0 |
| 2 | 0 | 1 | 0 | −1 | 11/3 | 3 | 2 |
| 3 | 0 | −1 | 1 | −1 | 6 | 0 | 6 |
| 4 | 0 | 0 | 1 | −1 | 18 | 1 | 6 |
| 5 | 0 | 1 | 1 | −1 | 128/3 | 0 | 8 |

where $q_3 = -1$ for detection, as is the convention. The first coherence pathway $(0,0,0,-1)$ gives rise to a free induction decay (FID) signal, and all others produce echoes. The second signal is in fact a stimulated echo, and the third is a spin-echo that has been refocused twice. When the number of RF pulses increases, other types of echo appear [17]. The symbols $b_Q$, $c_{1Q}$ and $c_{2Q}$ are factors due to diffusion and relaxation, and will be explained later.

During the periods of $\tau_1$ and $\tau_2$, transverse magnetization ($q = \pm 1$) will acquire a phase of $q_i g \tau_i$ ($i = 1$ and 2). The echo appears when the total phase is zero ($\sum_{i=1}^{3} q_i g \tau_i = 0$),

$$q_1 \tau_1 + q_2 \tau_2 - \tau_3 = 0 \tag{3.2}$$

since $q_3 = -1$. When the ratio of $\tau_1$ and $\tau_2$ is set to $1:3$, all echoes are separated by $\tau_1$, as first suggested by Hennig [18]. For sequences with more RF pulses, the time periods with the proportion of $1:3:9:27 \ldots$ will maintain equal echo spacing of $\tau_1$.

3.2.2
**Echo Shape and Amplitude**

The overall amplitude variation of the echoes can be understood by considering the on-resonant signal alone [9]. For the first RF pulse, the flip angle dependence is either $\sin\alpha_1/\sqrt{2}$ (when rotating $q_0 = 0$ to $q_1 = \pm 1$) or $\cos\alpha_1$ (for $q_0 = 0$ to $q_1 = 0$). If $\alpha_1 = 90°$, the second term is zero. On the other hand, these two terms are equal when $\sin\alpha_1/\sqrt{2} = \cos\alpha_1 = 1/\sqrt{3}$ ($\alpha_1 = 54.74°$). For the last pulse, the flip angle dependence is either $\sin\alpha_4/\sqrt{2}$ or $(1 - \cos\alpha_4)/2$. Similarly, they are equal when $\alpha_4 = \arccos(-1/3) = 109.47°$.

All other RF pulses involve four types of the flip-angle dependence, that is $\sin\alpha/\sqrt{2}$, $\cos\alpha$, $(1 + \cos\alpha)/2$ and $(1 - \cos\alpha)/2$. The minimum spread of the signals is achieved at $\alpha = \arccos(1/3) = 70.53°$ when

$$\frac{1 + \cos\alpha}{1 - \cos\alpha} = \frac{\sin\alpha}{\sqrt{2}\cos\alpha}, \tag{3.3}$$

and also for $\alpha = \arccos(-1/3) = 109.47°$ when

$$\frac{1 - \cos\alpha}{1 + \cos\alpha} = \frac{\sin\alpha}{\sqrt{2}\cos\alpha}. \tag{3.4}$$

For both cases, $\sin\alpha/\sqrt{2} = 2/3$, $|\cos\alpha| = 1/3$ and $|1 \pm \cos\alpha|/2$ is either $1/3$ or $2/3$. From this analysis, the optimal tipping angles for minimal echo variation are found: $\alpha_1 = 54.74°$, $\alpha_2 = \alpha_3 = 70.53°$, and $\alpha_4 = 109.47°$.

Several interesting features are of note. First, the sum of the magnitude of the 13 echoes is $74\sqrt{3}/81$, which is 2.24 times the magnitude of a Hahn echo (generated by the 90°–180° sequence) as $\sin(\pi/2)(1 - \cos\pi)/2\sqrt{2} = 1/\sqrt{2}$. The magnitude of each of the six largest echoes is about one-quarter of that of the Hahn echo. Except for the seventh echo, the magnitude difference among all echoes is only a factor of 2.

For experiments in a constant field gradient and extended sample size along the gradient direction, the off-resonance signals cannot be neglected. In this case, each pulse is slice-selective and its frequency dependence needs to be included explicitly in order to understand the spin dynamics and evaluate the echo shapes. The detailed formalism for evaluating off-resonance effects and calculating the echo shapes in general and specific to MMME has been discussed previously [13, 17]. It is interesting to note that, although the above analysis of the tipping angles does not directly optimize echo shape, it nevertheless reduces the echo-shape variation significantly compared to the tipping angles [90°-90°-90°-180°] used in early MMME experiments. The echo shapes for the optimal tipping angles are illustrated in Figure 3.1.

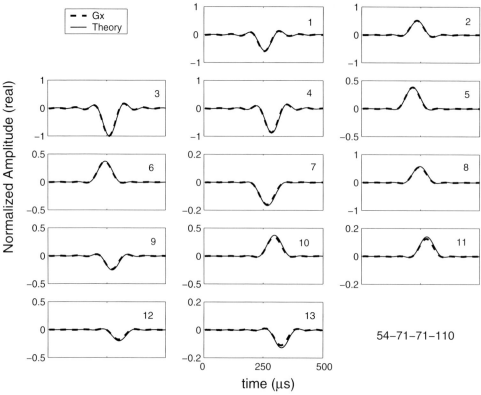

**Figure 3.1** Theoretical (line) and experimental (dashed line) echo shapes of MMME4 with the optimal flip angles: [54°-71°-71°-110°] [13]. The applied gradient is $10\,\text{G}\,\text{cm}^{-1}$ and $\tau_1 = 1\,\text{ms}$.

## 3.2.3
## Echo Phases

The benefit of the complete separation of coherence pathways in the time domain is that no phase cycling is needed. However, different RF pulse–phase combinations still affect the individual echo phases.

For a given coherence pathway, the relative phase of the resulting signal for different phases of the RF pulses can be obtained by:

$$\Phi = \sum_{i=1}^{N}(q_{i-1} - q_i)\phi_i + \frac{\pi}{2}, \tag{3.5}$$

where $\phi_i$ is the phase of the $i$-th pulse. Since $q_0 = 0$ and $q_N = -1$ for all coherence pathways, there is always an odd number of $|q_{i-1} - q_i| = 1$. For the commonly used MMME sequence with all pulses at equal phase, for example zero

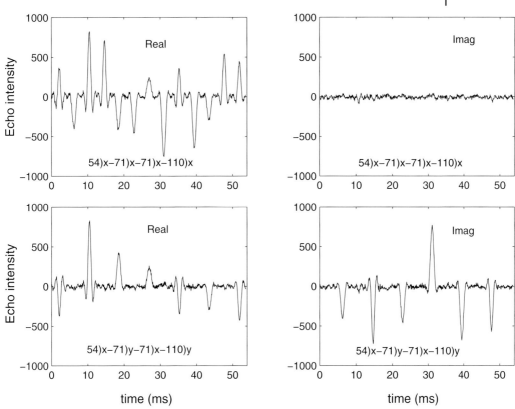

**Figure 3.2** Experimental echoes of MMME4 with [54°-71°-71°-110°] flip angles with constant and alternating phase programs [19]. Note the alternating phase of the echoes for the phase-alternating sequence.

(or $x$), the echoes will appear at an equal phase 90° (or $y$) away from the RF pulse phase. A change of $\phi_i$ by 180° may only change the echo phases by 0 or 180°.

For sequences with a 90° phase shift between adjacent RF pulses, such as xyxy, the subsequent echo will also alternate between in-phase (0 or 180°) and out-of-phase (90 or 270°), as shown in Figure 3.2. The key benefit of the phase-alternating echoes is that the signals of the adjacent echoes are 90° out of phase, effectively doubling the time separation of the echoes. This allows the use of shorter echo spacing for ultrafast measurements.

### 3.2.4
### Echo Sensitivity to Diffusion, Relaxation and Flow

The complex magnetization at the point of echo formation for each coherence pathway can be written as a product of four factors:

$$M_Q = A_Q \cdot B_Q \cdot C_Q \cdot F_Q \tag{3.6}$$

Here, $A_Q$ is the frequency spectrum of the resulting signal and depends only on the RF pulses, $B_Q$ reflects attenuation due to molecular diffusion, $C_Q$ contains the relaxation attenuation factors, and $F_Q$ reflects phase shift and attenuations due to flow. The three factors, $B$, $C$ and $F$ are properties of the specific coherence pathway and are independent of offset frequency.

If a single gradient direction is used to detect displacements, the attenuation induced by unrestricted, isotropic diffusion for a given coherence pathway can then be evaluated [20–22]:

$$B_Q = \exp\left\{-D\int_0^T k^2(t)\,dt\right\}, \tag{3.7}$$

where $D$ is the bulk diffusion constant, time $t = 0$ is defined as the beginning of the sequence and $T$ is the echo time, and

$$\vec{k}(t) = \int_0^t \gamma q(t')g(t')\,dt'. \tag{3.8}$$

Here, $q(t)$ and $g(t)$ are the explicit time-dependent forms of $q$ and $g$. For a constant gradient $g$ and $1:3:9$ time spacings, $B_Q$ is only a function of $Dg^2\tau_1^3$:

$$B_Q = \exp\left(-b_Q D\gamma^2 g^2 \tau_1^3\right). \tag{3.9}$$

Each echo will be associated with a unique $b_Q$ (see the table in Section 3.2.1) and thus a different diffusion weighting factor. However, the variation in echo amplitudes due to the RF pulses needs to be taken in consideration in order to quantify the diffusion decay. The $B_Q$ factor due to anisotropic diffusion will be discussed later, in the respective sections.

The relaxation factor $C_Q$ reflects signal decays with a time constant $T_1$ during the periods when $q_i = 0$ and with $T_2$ when $q_i = \pm 1$:

$$C_Q = \exp\left[-\sum_{k=1}^N \left(q_k^2/T_2 + \frac{1-q_k^2}{T_1}\right)t_k\right] \tag{3.10}$$

$$= \exp\left[-c_{1Q}\frac{\tau_1}{T_1} - c_{2Q}\frac{\tau_1}{T_2}\right]. \tag{3.11}$$

For molecules moving at a constant velocity $\mathbf{v}$, the phase accumulation can be rewritten in an integral form [23]:

$$\phi(\mathbf{x}_0) = \int_0^T q(t)\gamma\mathbf{g}\cdot(\mathbf{v}t + \mathbf{x}_0)\,dt \tag{3.12}$$

$$= \phi_1 + \phi_0. \tag{3.13}$$

Here, $T$ is the echo time and the position of the spin is linear in time, $\mathbf{x} = \mathbf{v} \cdot t + \mathbf{x}_0$, where $\mathbf{x}_0$ is the initial position at time zero and is distributed within the entire sensitive region of the sample.

The first term is a function of $\mathbf{v}$:

$$\phi_1 = \int_0^T q(t)\gamma\mathbf{g}\cdot\mathbf{v}t\,dt \tag{3.14}$$

$$= \gamma v \int_0^T q(t)g_v(t)t\,dt \tag{3.15}$$

$$= v \cdot f_Q \tag{3.16}$$

Here, $v$ is the magnitude of the velocity and $g_v$ is the gradient along the velocity. Note that the time dependence of the gradient is explicit in Equation 3.15. We introduce the coefficient $f_Q$ to denote the sensitivity of the coherence pathway ($Q$) to the velocity,

$$f_Q \equiv \gamma \int_0^T q(t)g_v(t)t\,dt. \tag{3.17}$$

For flow at velocity $v$, $F_Q = \exp(-iv \cdot f_Q)$. Note also that this term can be conveniently related to the coherence pathway wavevector $k_v$ along the velocity through integration of Equation 3.17 by parts.

The second term in Equation 3.12, $\phi_0 = \gamma\mathbf{x}_0 \cdot \int_0^T \mathbf{g}q(t)dt$, does not depend on $\mathbf{v}$, but it is spatially dependent. This phase factor often causes the FID signal to decay rapidly and is the basis for magnetic resonance imaging (MRI). An echo will form when $\phi_0$ becomes zero for all positions and the residual phase of the echo ($\phi_1$) will be proportional to the velocity.

For the pulse sequence with $\tau_1 : \tau_2 = 1 : 3$ and constant gradient g, the values of $b_Q$, $c_{1Q}$ and $c_{2Q}$ are listed in the table in Section 3.2.1. Note that these expressions assume no fresh magnetization regrowth during the course of the sequence.

The remainder of this chapter will describe the applications of MMME sequence variants for measurements of diffusion, flow and spin density images in multiple dimensions. The extension of $B_Q$ and $F_Q$ to three dimensions will be discussed in the respective sections.

## 3.3
**Diffusion Measurement**

### 3.3.1
**One-Dimensional (1-D) Diffusion**

One way in which to use MMME sequences to determine diffusion is by repeating the sequence, for instance, MMME4, with two sets of $\tau_l$ with $\tau_1 = \tau$ and $\tau'$. The shapes of the corresponding echoes will then be identical and their amplitude ratio can be found to be

$$\frac{S_Q(\tau)}{S_Q(\tau')} = \exp[-b_Q D\gamma^2 g^2(\tau^3 - \tau'^3)].\tag{3.18}$$

Each pair of the echoes will provide one datum point for the diffusion decay, with a total of 13 data points being be obtained by the two acquisitions, as shown in Figure 3.3 for water at room temperature.

Many variations of these experiments can be performed for specific needs. For example, if measurement speed is of concern, the echo shapes can be obtained beforehand using a standard sample of small or known diffusion constant, and a small $\tau$ to further minimize the diffusion effects. The desired

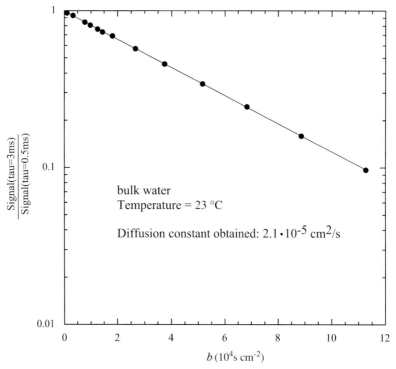

**Figure 3.3** MMME4 result for measuring the diffusion constant in bulk water [9]. MMME4 echoes were obtained with $\tau = 0.5$ and $\tau' = 3$ ms. The ratio of the amplitudes of corresponding echo signals is shown for all 13 echoes as a function of the $b = \gamma^2 g^2 b_Q (\tau'^3 - \tau^3)$. The data shows an exponential decay that determines the diffusion constant to be $2.1 \times 10^{-5}$ cm$^2$ s$^{-1}$.

sample can then be scanned with a selected $\tau$ that is sufficient for obtaining data of multiple $b$.

When $\tau_1$ is close to $T_1$ or $T_2$, the ratio of two scans should include relaxation decay together with the diffusion decay. If the trends of $c_{1Q}$, $c_{2Q}$ and $b_Q$ with echo number are sufficiently different, one can conceivably obtain $D$, $T_1$ and $T_2$ from the MMME data. Studies to obtain the relaxation rate and diffusion constant in a single scan are currently ongoing at the authors' laboratory.

### 3.3.2
### Two-Dimensional (2-D) Diffusion

With a single direction of the field gradient, the spatial modulation wavevector is parallel to **g**; thus diffusion along that direction is measured. The conventional method for determining an anisotropic diffusion tensor is to repeat this experiment with **g** in several independent directions. In contrast, the MMME method can produce simultaneous spatial modulations along multiple directions and thus sensitize the signals to multiple diffusion tensor elements simultaneously in one scan.

To illustrate the method, let us first consider the three-pulse sequence with a constant $z$-gradient during the entire sequence and only one $x$-gradient pulse $G_{x2}$ during the $\tau_2$ period. As only the latter three echoes exhibit the transverse magnetization during the $\tau_2$ period, they alone are modulated by $G_{x2}$. In order to form the latter three echoes, another identical $G_{x2}$ gradient pulse must be applied after the second signal to unwind the modulation. In this case, diffusion along $x$ induces an identical amount of decay only for the latter three echoes. On the other hand, all echoes experience different amount of decay due to diffusion along $z$. As a result, one scan produces four echoes with a different range of decay along $x$ and $z$, and thus allows measurement of the 2-D diffusion tensor. This type of sequence is referred to as MMME2D.

Another example of the MMME2D sequence is shown in Figure 3.4, where two $x$-gradient pulses ($-G_x$ and $+ G_x$) were employed. Correspondingly, four $x$-gradient pulses are needed after the last RF pulse for refocusing. An experiment using such a pulse sequence was reported to obtain the 2-D diffusion tensor [11].

For multiple gradient directions used to probe an anisotropic diffusion tensor $D_{ij}$, Equation 3.7 generalizes to the matrix form [24]:

$$B_Q = \exp\left(-\sum_{ij} b_{ij} D_{ij}\right) \qquad (3.19)$$

where

$$b_{ij} = \int_0^T k_i(t) k_j(t) dt \qquad (3.20)$$

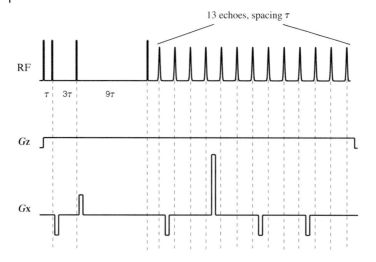

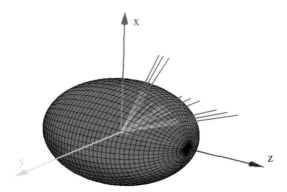

**Figure 3.4** An MMME2D sequence for the measurement of a 2-D diffusion tensor, and a vector diagram showing the effective range of diffusion sensitization directions for this sequence [11].

and

$$\vec{k}(t) = \int_0^t \gamma q(t')\vec{g}(t')dt'. \tag{3.21}$$

Note that the conventional definition of $b_{ij}$ is used here, which is different from the unitless $b_Q$.

Figure 3.5a shows the variation of $b_{ij}$ for the MMME2D (Figure 3.4) sequence. The two-$\tau$ method was used here to scale out the echo-shape variation by measuring the ratio of two experiments (S1/S2). Here, we used a pattern of diffusion

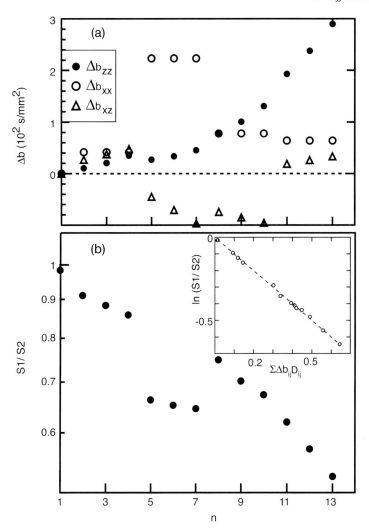

**Figure 3.5** The MMME2D results on the asparagus sample [11]. (a) Plot of the calculated three $\Delta b$ elements versus the echo number $n$; (b) Plot of the measured amplitude ratios versus $n$. The gradual signal decay is due mostly to longitudinal diffusion ($D_{zz}$), whereas the abrupt drop of echoes 5–7 reflects transverse diffusion ($D_{xx}$). Inset: plot of the amplitude ratios versus the full diffusion weighting argument, demonstrating the quality of the fit.

weighting ($\Delta b$, the difference of $b_{ij}$ for the two experiments with different $\tau$) containing different trends versus the echo number $n$ for different weighting factors, which is desirable for reducing correlated errors between the corresponding diffusion tensor elements. The sample used to test this sequence was an asparagus stalk. Asparagus is known to exhibit anisotropic diffusion due to the elongated cells which occur along its stalk [25]. The cell membrane reduces water permeation such that diffusion is limited particularly transverse to the stalk axis. The principal axes of the diffusion tensor are expected to be parallel to the laboratory frame axes in these experiments as the stalk was oriented along the applied field [11].

The echo amplitude ratios shown in Figure 3.5b show a gradual reduction due to diffusion along $z$ and two steps between echoes 4 and 5, 7 and 8, due to diffusion along $x$, qualitatively correlated to the $\Delta b$ patterns in Figure 3.5a. A least-squares fit of the echo amplitude ratios by Equation 3.19 obtains: $D_{zz}$ / $D_{xx}$ / $D_{zx} = 1.5/1.2/0.01 \times 10^{-5}\,\mathrm{cm^2\,s^{-1}}$, which is in good agreement with the values obtained by conventional PFG measurements [11].

The multiple directions of the diffusion sensitivity of MMME2D can be visualized through a 3-D display of the $b$ vector, $\Delta \mathbf{b} = \left( \sqrt{\Delta b_{xx}}, \sqrt{\Delta b_{zz}} \right)$ for each echo (see Figure 3.4). The length of the vectors is proportional to magnitude of $\Delta b$ and the angle is determined $\tan\left( \sqrt{(\Delta b_{xx}/\Delta b_{zz})} \right)$. Figure 3.4 shows that all 13 echoes exhibit noncolinear $\Delta \mathbf{b}$ vectors within the $x$–$z$ plane. This noncolinearity is important for determining the diffusion tensor.

### 3.3.3
### Three-Dimensional (3-D) Diffusion

Compared to 1-D MMME, the 2-D tensor experiment extracts more information from the same number of echoes, and thus the uncertainty of the result may be higher. Furthermore, the uncertainty of individual tensor elements can be substantially different. For example, if two diffusion weighting factors cause similar echo decay patterns, the error on one diffusion element will cause a corresponding error on the related element. Quantitative metrics such as the condition number [26] have also been employed to show that it is difficult to use MMME2D to obtain the 3-D diffusion tensor [12]. The condition number of the diffusion weighting matrix $A$ (which is defined to be $||A|| \cdot ||A^{-1}||$, where $|| \cdot ||$ is the matrix norm and $A^{-1}$ is matrix inversion or pseudoinversion for nonsquare matrices) relates the relative error of the input data (attenuations) to that in the output parameters (diffusion tensor elements) and is also equal to the ratio of the maximum and minimum singular values. Clearly, a lower condition number is favorable.

Recently, Cho et al. have shown that the inclusion of an extra gradient pulse during the $\tau_1$ period can reduce the condition number for 3-D diffusion tensor measurement significantly [27]; the sequence is shown in Figure 3.6. A six-dimensional search was conducted to find the best configuration of gradient pulses along the two axes during periods '1', '2' and '3' which has the lowest condition number. The minimum condition number was found to be 4.9. In comparison, the minimum condition number of a similar MMME sequence without the gradient pulse during period '1' was 318. The application of this sequence on both isotropic and anisotropic samples produced good agreement with conventional PFG measurements [27].

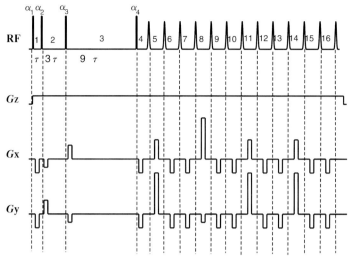

**Figure 3.6** MMME pulse sequence for rapid 3-D diffusion tensor measurement [27]. Note that it is important to include a gradient pulse during period '1'.

## 3.4
## Application: Flow

### 3.4.1
### One-Dimensional (1-D) Flow

For macroscopic flow, the velocity is often distributed around the average velocity. The flow-based magnetization factor then can be written as

$$F_Q = \int P(v)\exp(-iv \cdot f_Q)\,dv \tag{3.22}$$

where $P(v)$ is the normalized velocity distribution. In Poiseuille pipe flow, the distribution of velocities is rectangular and symmetric about the mean velocity $\bar{v}$, and has a width of $2\bar{v}$. One can integrate Equation 3.22 over this distribution and find that a distribution of velocities results in a nonzero $\phi_1$ and a reduction in the echo amplitude. When the phase shift is small, $\phi_1 \ll 1$, $\phi_1$ will always measure the mean velocity: $F_Q \approx 1 - i\bar{v}f_Q$ and $\phi_1 \approx \bar{v}f_Q$.

In Figure 3.7, the phase shift is shown as a function of $f_Q$ for three flow velocities in a MMME1D experiment performed on water flowing through a pipe of inner diameter 4.2 mm [10]; $f_Q$ is then evaluated using Equation 3.17. For each flow velocity, the phase shift is shown to be proportional to $f_Q$, consistent with Equation 3.16. The proportionality coefficient obtained from the plot for each flow rate agrees (within a few percent) with the mean flow velocity calculated from the volume flow rate. It should be noted that the 13 data points for each flow rate were

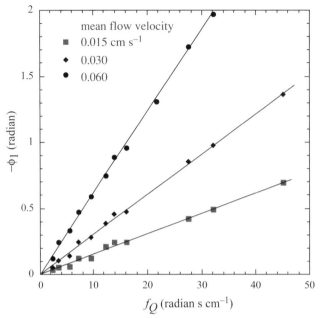

**Figure 3.7** Plots of flow induced phase shift at three mean flow velocities [10]. The MMME4 experiments were performed with the $\tau_1/\tau_2/\tau_3 = 1/3/9$ ms. The field gradient was along the flow direction, $g = 10\,G\,cm^{-1} \cdot f_Q$ was calculated using Equation 3.17. The data points for each flow rate fell on the straight line, and the slope of the lines agreed with the mean flow velocities, respectively. The lines indicate linear fits to the data.

obtained in one scan. This method has been used in recent studies [19] to measure flow velocities of up to $50\,cm\,s^{-1}$.

It is also worth noting that, when the distribution of velocities and displacements is not symmetric about the mean value [28, 29], the phase of the echoes may no longer be exactly linear in $f_Q$. Higher-order cumulants of the displacement distribution can then, in principle, be inferred from the phase of the signal. If a sufficient number of suitable $f_Q$-values can be probed, the velocity distribution can be obtained by this single-shot method.

### 3.4.2
### Three-Dimensional (3-D) Flow

Similar to the principle of MMME2D for diffusion tensor measurement, gradient pulses along one or two orthogonal directions to that of the static gradient will encode flow-induced phases for multiple components of the 3-D velocity vector. The MMME2D-type sequence in Figure 3.4 can also be used for flow velocity measurement.

In such a sequence where the gradient changes orientation within the sequence, $f_Q$ of Equation 3.17 becomes a vector,

$$\mathbf{f}_Q \equiv \gamma \int_0^T q(t)\mathbf{g}(t)\,t\,dt. \tag{3.23}$$

In a sense, the scaler $f_Q$ (Equation 3.17) is a component of the general flow sensitivity of the sequence. By adjusting the gradient pattern, one can modify the vector $f_Q$ to expand the 3-D space of flow velocity.

For numerical analysis, $\mathbf{f}_Q$ is written in the phase shift coefficient matrix, the measured flow induced phase shift is a vector $\Phi_1$, and the velocity a vector, $V$:

$$\Phi_1 = F \times V, \tag{3.24}$$

where $F^{ij} = \left[\gamma \int_0^T g_j t\,dt\right]_i$, $V^j = v_j$ and $\Phi_1^i = \phi_1^i$ ($j = x, y, z$, and $i =$ echo number).

Clearly, it is favorable to have a lower condition number of matrix $F$ if possible as it improves the accuracy of $V$ with a given measurement error of $\Phi_1$. On the other hand, the aim also is to reduce the number gradient switchings, so it is preferable to use lower gradient values to minimize the effect of eddy currents. These can be important issues as the MMME sequence may require many refocusing gradient pulses to acquire the full train of echoes. Two schemes of pulse sequences capable of obtaining the flow vector [19] are shown in Figure 3.8.

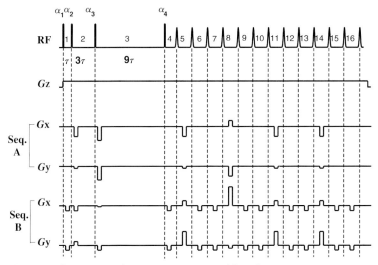

**Figure 3.8** The MMME pulse sequences optimized for a flow velocity measurement along an arbitrary direction [19]. Seq. A has an advantage of less gradient switchings, while Seq. B provides a lower condition number. Each number in the RF section labels the corresponding gradient pulse during each time interval.

In Figure 3.8, Seq. A has an advantage of fewer gradient switchings by having no pulsed gradient field during the period '1', and provides the condition number of 10. Omitting the pulsed-field gradient during the period '1' is useful from a practical point of view because refocusing gradients are only needed before every three echoes. Also in the figure, Seq. B can reach closer to the condition number of 1, but requires more intensive gradient switchings before every echo. These sequences are representative for both pulsed-gradient schemes with reasonable condition numbers but not necessarily the best gradient configuration for every case, because the sequence optimization was limited in the strength ($\leq 14\,\mathrm{G\,cm^{-1}}$) and the duration (1 ms) of gradient pulses for both cases. Cho et al. showed that both sequences worked well to obtain flow velocity vectors [19].

## 3.5
## Summary

It is well known that multiple echoes can be created by the simple application of a few RF pulses. This chapter has outlined the usefulness of such multiple echoes for the individual encoding of information in order to achieve rapid measurements of diffusion, flow and imaging. The ability of this sequence to determine a diffusion tensor or flow vector quickly rather than one scalar component is quite unique. In addition, each echo can be phase-encoded individually to accelerate the imaging experiment; in fact, it was reported that this scheme obtains a 2-D image in 5 ms [13].

One important aspect of these sequences is the presence of a constant field gradient. Such a gradient makes it necessary to correctly handle, both theoretically and experimentally, the off-resonance effects due to the slice-selection of all RF pulses. The ability to correctly handle the constant field gradient can be very important in applications using inside-out or *ex situ* NMR, such as the NMR-MOUSE [30] and NMR well-logging tools [31, 32]. These instruments are becoming increasingly useful in the field of materials characterization.

## References

1 Mansfield, P. and Grannell, P.K. (1975) "Diffraction" and microscopy in solids and liquids by NMR. *Physical Review B*, 12, 3618–34.

2 Carr, H.Y. and Purcell, E.M. (1954) Effects of diffusion on free precession in NMR experiments. *Physical Review*, 94, 630–8.

3 Meiboom, S. and Gill, D. (1958) Modified spin-echo method for measuring nuclear relaxation times.

*Review of Scientific Instruments*, 29, 688–91.

4 Hennig, J. (1986) Rare imaging: a fast imaging method for clinical MR. *Magnetic Resonance in Medicine*, 3, 823.

5 Stamps, J.P., Ottink, B., Visser, J.M., van Duynhoven, J.P. and Hulst, R. (2001) Difftrain: a novel approach to a true spectroscopic single-scan diffusion measurement. *Journal of Magnetic Resonance*, 151, 28–31.

6 Loening, N.M., Thrippleton, M.J., Keeler, J. and Griffin, R.G. (2003) Single-scan longitudinal relaxation measurements in high-resolution NMR spectroscopy. *Magnetic Resonance Imaging*, **164**(*2*), 321–8.

7 Sigmund, E.E. and Halperin, W.P. (2003) Hole-burning diffusion measurements in high magnetic field gradients. *Journal of Magnetic Resonance*, **163**, 99–104.

8 Frydman, L. (2006) Single-scan multidimensional NMR. *Comptes Rendus Physique*, **9**(*3–4*), 336–45.

9 Song, Y.-Q. and Tang, X. (2004) A one-shot method for measurement of diffusion. *Journal of Magnetic Resonance*, **170**, 136–48.

10 Song, Y.-Q. and Scheven, U.M. (2005) An NMR technique for rapid measurement of flow. *Journal of Magnetic Resonance*, **172**(*1*), 31–5.

11 Tang, X.-P., Sigmund, E.E. and Song, Y.-Q. (2004) Simultaneous measurement of diffusion along multiple directions. *Journal of the American Chemical Society*, **126**(*50*), 16336–7.

12 Sigmund, E.E. and Song, Y.-Q. (2006) Multiple echo diffusion tensor acquisition technique. *Magnetic Resonance Imaging*, **24**, 7–18.

13 Cho, H., Chavez, L., Sigmund, E.E., Madio, D.P. and Song, Y.-Q. (2006) Fast imaging with the MMME sequence. *Journal of Magnetic Resonance*, **180**, 18–28.

14 Heid, O., Deimling, M. and Huk, W. (1993) Quest–a quick echo split NMR imaging technique. *Magnetic Resonance in Medicine*, **29**, 280–3.

15 Counsell, C.J. (1993) Preview: a new ultrafast imaging sequence requiring minimal gradient switching. *Magnetic Resonance Imaging*, **11**, 603–16.

16 Shriver, J. (1992) Product operators, coherence transfer in multi-pulse NMR experiments. *Concepts in Magnetic Resonance*, **4**, 1–33.

17 Song, Y.-Q. (2002) Categories of coherence pathways in the CPMG sequence. *Journal of Magnetic Resonance*, **157**, 82–91.

18 Hennig, J. (1991) Echoes-how to generate, recognize, use or avoid them in MR-imaging sequences. *Concepts in Magnetic Resonance*, **3**, 179–92.

19 Cho, H., Ren, X.H., Sigmund, E.E. and Song, Y.-Q. (2007) A single-scan method for measuring flow along an arbitrary direction. *Journal of Magnetic Resonance*, **186**, 11–16.

20 Stejskal, E.O. and Tanner, J.E. (1965) Spin diffusion measurements: spin echoes in the presence of a time-dependent field gradient. *Journal of Chemical Physics*, **42**, 288–92.

21 Callaghan, P.T. (1993) *Principles of Nuclear Magnetic Resonance Microscopy*, Oxford University Press, New York.

22 Sodickson, A. and Cory, D.G. (1998) A generalized k-space formalism for treating the spatial aspects of a variety of NMR experiments. *Progress in NMR Spectroscopy*, **33**, 77–108.

23 Haacke, E.M., Brown, R.W., Thompson, M.R. and Venkatesan, R. (1999) *Magnetic Resonance Imaging: Physical Principles and Sequence Design*, Springer-Verlag, New York.

24 Basser, P.J. (1995) Inferring microstructural features and the physiological state of tissues from diffusion-weighted images. *NMR in Biomedicine*, **8**(*7–8*), 333–44.

25 Heyes, J.A. and Clark, C.J. (2003) Magnetic resonance imaging of water movement through asparagus. *Functional Plant Biology*, **30**, 1089–95.

26 Skare, S., Hedehus, M., Moseley, M.E. and Li, T.Q. (2000) Condition number as a measure of noise performance of diffusion tensor data acquisition schemes with MRI. *Journal of Magnetic Resonance*, **147**(*2*), 340–52.

27 Cho, H., Ren, X.-H., Sigmund, E.E. and Song, Y.-Q. (2007) Rapid measurement of three-dimensional diffusion tensor. *Journal of Chemical Physics*, **126**, 154–501.

28 Kubo, R., Toda, M. and Hashitsume, N. (1991) *Statistical Physics, II*, Springer-Verlag, New York.

29 Scheven, U.M. and Sen, P.N. (2002) Spatial and temporal coarse graining for dispersion in randomly packed spheres. *Physical Review Letters*, **89**(*25*), 254–501.

30 Eidmann, G., Savelsberg, R., Blumler, P. and Blumich, B. (1996) The NMR mouse,

a mobile universal surface explorer. *Journal of Magnetic Resonance. A*, **122**, 104–9.

**31** Locatelli, M., Mathieu, H., Bobroff, S., Guillot, G. and Zinszner, B. (1998) Comparative measurements between a new logging tool and a reference instrument. *Magnetic Resonance Imaging*, **16**(5–6), 593–6.

**32** Kleinberg, R.L., Sezginer, A., Griffin, D.D. and Fukuhara, M. (1992) Novel NMR apparatus for investigating an external sample. *Journal of Magnetic Resonance*, **97**, 466–85.

# 4
# Magnetic Resonance Force Microscopy

*Mladen Barbic*

## 4.1
## Introduction

Magnetic resonance force microscopy (MRFM) [1–9] is a scanning probe microscopy technique that can be viewed as the culmination of a long history of instrumentation development towards more sensitive and higher spatial resolution magnetic moment detection and imaging. It is based on a fundamental physics principle that a force will be exerted on a magnetic moment in a spatially varying magnetic field (magnetic field gradient), and that the magnetic moment can therefore be detected mechanically. The well-known experiment performed by Stern and Gerlach [10] utilized this principle to split a beam of magnetic atoms into a discrete number of divergent beams in an inhomogeneous magnetic field, while in another classic experiment Einstein and de Haas [11] demonstrated the relationship between a magnetic moment and its angular momentum through mechanical detection. The extension of these concepts into the field of magnetic resonance also has a long history. Rabi and coworkers were the first to utilize gradient-based force detection of nuclear magnetic resonance (NMR) in the molecular beam deflection method [12], while Evans [13] and Alzetta and coworkers [14] were the first to detect magnetic resonance by using a mechanical detector. It is interesting that the two critical technologies used in MRFM, namely magnetic resonance imaging (MRI) [15, 16] and scanning probe microscopy (SPM) [17], were developed independently and in parallel at around the same time. Both, of these imaging technologies have advanced tremendously over the past three decades, such that today MRFM represents a unique microscopy technology that aims to unify the 3-D, nondestructive and chemically specific imaging capability of MRI with the atomic spatial resolution surface imaging capability of SPM as a 3-D atomic resolution force-detection magnetic resonance microscope with single proton spin sensitivity. Although, surface atomic resolution electron spin and resonance microscopy have been demonstrated with scanning tunneling microscopy (STM) [18–20], the prospect of MRFM as an atomic resolution subsurface magnetic resonance microscopy (MRM) technique

*Magnetic Resonance Microscopy.* Edited by Sarah L. Codd and Joseph D. Seymour
Copyright © 2009 WILEY-VCH Verlag GmbH & Co. KGaA, Weinheim
ISBN: 978-3-527-32008-0

gives it a particular appeal that is attracting an increasing number of research groups.

## 4.2
## MRFM Instrumentation

The basic hardware components of a typical magnetic resonance force microscope are shown in Figure 4.1a [21]. A microscopic ferromagnetic particle/tip is placed on an end of a long and flexible micromechanical cantilever, and then brought into close proximity with a sample that is placed on a high-resolution mechanical positioning stage. The sample is also exposed to an alternating current (ac) field

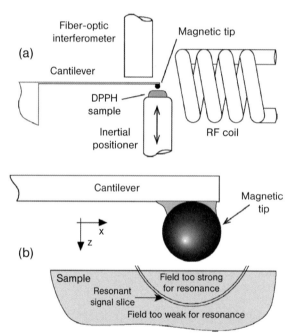

**Figure 4.1** (a) The basic hardware components of a typical magnetic resonance force microscope in a tip-on-cantilever mode. A microscopic ferromagnetic particle is placed on an end of a long and flexible micromechanical cantilever. The tip is brought into close proximity with a sample placed on a high-resolution mechanical positioning stage. The sample is also exposed to a radiofrequency (RF) field from a coil placed near the assembly for application of time-varying magnetic field sequences common in magnetic resonance; (b) The ferromagnetic tip generates a large magnetic field gradient at the sample such that only those spins within a narrow slice of the sample are in resonance. The resonant spins of the sample experience a mechanical force due to their presence in a magnetic field gradient. The time-varying magnetic force between resonant spins and a magnetic tip causes mechanical vibration of the cantilever which is detected by a fiber-optic interferometer. Reprinted with permission from Ref. [21]. Copyright 1998 by the American Institute of Physics.

from a coil placed near the assembly for the application of time-varying magnetic field sequences common in magnetic resonance practice. The purpose of the ferromagnetic tip is to generate a large magnetic field gradient at the sample, which in turn provides two critical features in MRFM: (i) only those spins within a narrow slice of the sample are in resonance (see Figure 4.1b [21]); and (ii) resonant spins of the sample experience a mechanical force due to their presence in a magnetic field gradient. The time-varying magnetic force between resonant spins and a magnetic tip causes mechanical vibration of the cantilever that is typically detected using a fiber-optic interferometer [22].

In this brief description, several features of the MRFM instrument should be emphasized as contrasting with the conventional, inductive-coil-based MRM [23, 24]. First, the field gradients from a ferromagnetic particle in MRFM can reach values on the order of $1 \times 10^7 \, T \, m^{-1}$ for a 100 nm-diameter iron sphere [intensity of magnetization $(\mu_0 M) = 2.26 \, T$]. This is six orders of magnitude larger than the strongest field gradients used in the MRM technique of stray-field magnetic resonance imaging (STRAFI) [25]. In addition, gradients in MRFM are constant and cannot be rapidly changed in time, as is common in conventional MRM, although some time dependence can be introduced by mechanical motion of the sample relative to the ferromagnetic tip. Such large gradients also place MRFM into the category of solid-state MRI, since common diffusion parameters of liquid samples (the diffusion coefficient, $D$, is typically in the range of $0.1–10 \, \mu m^2 \, ms^{-1}$) render MRFM ineffective. A second feature unique to MRFM is that the characteristic natural frequency of the cantilever detector (1–100 kHz range) is typically several orders of magnitude lower than the characteristic frequency of the spin magnetic resonance processes (MHz for protons and GHz for electrons). Therefore, additional means must be provided to couple the resonant spins of the sample to the mechanical motion of the MRFM cantilever detector – something which typically is not required in conventional, inductive coil-based MRM detection. These unique conditions ensured that novel magnetic resonance sequences, detection methods and imaging protocols had to be developed to successfully realize the potential of MRFM.

## 4.3
## Spin Manipulation in MRFM

The specific operating conditions mentioned above prompted the development of spin manipulation protocols unique to MRFM. The most common MRFM measurement method to date has been detection of the z-component of resonant magnetization, $M_Z$. This ability renders MRFM uniquely different from conventional MRM where, typically, the transverse component of resonant spin precession is detected with an inductive coil. The initial proof-of-concept demonstration of MRFM was performed with an electron spin resonant system of paramagnetic diphenylpicrylhydrazil (DPPH) [26]. For DPPH, $M_Z$ is a linear function of $B_Z$, except in the vicinity of ESR, where longitudinal magnetization is sharply suppressed. Rugar and coworkers used an external magnetic field modulation along the z-axis of the form $B_Z(t) = B_0 + B_M\sin(\omega_M t)$, where the modulation frequency $\omega_M$ is half of the cantilever resonant frequency $\omega_C$ ($\omega_M = \omega_C/2$). The nonlinearity of $M_Z$ near ESR caused a time-varying spin magnetization $M_Z(t)$ that was frequency doubled to the cantilever resonant frequency, $\omega_C$, that resonantly excited cantilever vibration [26]. In this initial experiment, a DPPH volume sensitivity of $3\,\mu m^3$ was demonstrated.

Following the improvements in MRFM instrumentation through the use of more sensitive cantilevers and higher magnetic fields – and therefore a higher spin polarization – Rugar and coworkers demonstrated the mechanical detection of NMR of ammonium nitrate [27]. This was a more challenging experiment due to the much smaller (by approximately three orders of magnitude) magnetic moment of the proton as compared to the electron. The MRFM was still able to detect the longitudinal component of magnetization, $M_Z$, but the spin modulation technique used was based on cyclic adiabatic inversion [28]. Under the correct conditions the nuclear magnetic moment can be made to follow the direction of the effective field $B_{EFF}$ in the rotating reference frame [29]. By frequency modulating the applied radiofrequency (RF) field at the cantilever resonant frequency under adiabatic conditions, an oscillating magnetic moment $M_Z$ was generated that in turn created a cyclic force $F_Z$ on the resonant cantilever [27]. By using this approach, a single-shot sensitivity on the order of $10^{13}$ proton spins was achieved [27].

These two pioneering MRFM proof-of-concept experiments on electron and proton spin systems validated the concept of cantilever-detected magnetic resonance in a strong magnetic field gradient, and established the path for MRFM improvements through lower operating temperatures, more sensitive cantilevers, and smaller magnetic tips for further advances in force sensitivity and imaging resolution. However, as the miniaturization of magnetic tips (and thus increased magnetic field gradients) in MRFM continued, another fundamental concept in spin physics – that of statistical spin excess (spin self-polarization) [30] – became prominent. For a system of $N$ magnetic moments $\mu$, statistical arguments show [30] that there will occur a spin excess or spin self-polarization with the rms value on the order of $\sqrt{N} \cdot \mu$. In conventional magnetic resonance experiments with large spin ensembles this fraction of spins is negligible compared to the thermal (Boltzmann) polarization in common laboratory fields and temperatures. However, for small spin ensembles, this self-polarization exceeds the Boltzmann polarization at the crossover number on the order of $10^7$ spins [31]. With the gradients of 5 (Gauss nm$^{-1}$) using ferromagnetic tips on mass-loaded cantilevers [32], as shown in Figure 4.2 [33], MRFM has reached the operating range where statistical spin self-polarization dominates the Boltzmann polarization.

Rugar and coworkers have developed a technique which they termed interrupted oscillating cantilever-driven adiabatic reversal (iOSCAR) that very effectively manipulates and detects resonant properties of the small self-polarizing spin ensembles [33]. The study protocol is described in Figure 4.3. A cantilever with a magnetic particle as its tip is set to oscillate at its resonant frequency. The effective field $B_{EFF}$ at the sample location is modulated due to this cantilever oscillation and

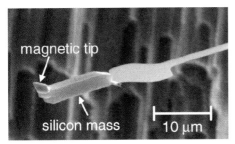

**Figure 4.2** A scanning electron micrograph of the single-crystal silicon cantilever with the micron-sized magnetic tip. The cantilever has a mass-loaded geometry in order to suppress motion of the tip for higher-order flexural modes. Reprinted with permission from Ref. [33]. Copyright 2003 by the American Physical Society.

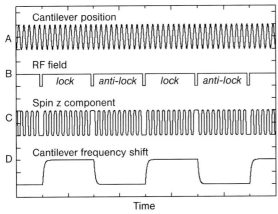

**Figure 4.3** Timing diagram for the interrupted OSCAR spin manipulation protocol in MRFM. The cantilever is oscillated continuously. The RF magnetic field (trace B) is normally on, but is periodically interrupted for one-half cantilever cycle. The z-component of the magnetization (trace C) oscillates in response to the cantilever motion due to adiabatic rapid passage when the RF field is on, but is left static when it is off. The oscillating magnetization reverses phase with respect to the cantilever for each RF field interruption, giving a cantilever frequency shift (trace D) that oscillates at one-half the RF field interrupt frequency. Reprinted with permission from Ref. [33]. Copyright 2003 by the American Physical Society.

field gradient from the tip. If $B_{EFF}$ changes sufficiently slowly (adiabatic condition), the statistical spin polarization will be locked to the effective field $B_{EFF}$, and the z-component of the magnetization in the rotating frame $m_{EFF}$ will oscillate synchronously with the cantilever position. Due to the force between the magnetic tip and the spins, the oscillating magnetization causes the resonant frequency shift of the cantilever that is proportional to $m_{EFF}$ [34]. If the microwave magnetic field $B_1$, which is normally on, is interrupted periodically for one-half of the cantilever cycle

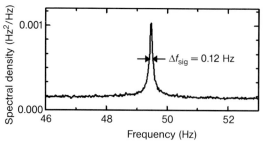

**Figure 4.4** Power spectral density of the frequency demodulated signal (trace D of Figure 4.3). The peak at 49.5 Hz is the statistical spin signal. The integrated signal in the peak is equivalent to that from about six spins, and the baseline noise is 1.8 spins in the 0.12 Hz natural bandwidth. Reprinted with permission from Ref. [33]. Copyright 2003 by the American Physical Society.

(trace B in Figure 4.3), the $B_{EFF}$ will have reversed orientation and the statistical spin polarization will have changed from a locked to an anti-locked orientation (trace C in Figure 4.3); this causes the cantilever resonant frequency to shift periodically at the interrupt frequency (trace D in Figure 4.3). The final measurement is performed by taking the power spectrum of the frequency-demodulated signal (trace D in Figure 4.3), resulting in the data shown in Figure 4.4 at a temperature of 200 mK for about six electron spins with the baseline noise floor of 1.8 electron spins in a 0.12 Hz bandwidth [33].

The same spin manipulation protocol was subsequently applied to nuclear spins [35] with 2000 spins sensitivity at 7 K. It was found that the statistical polarization has a rotating-frame relaxation time on the order of seconds for electron spins [33], and on the order of 100 ms for nuclear spins [35]. The attractive features of the iOSCAR statistical spin polarization manipulation protocol for MRFM are its apparently robust immunity to external noise sources, its successful coupling of spin resonance at high frequencies to low resonance frequencies of mechanical cantilevers, and the elimination of the need to wait for the spin-lattice relaxation time $T_1$ between measurements in order for the sample to re-polarize. By utilizing all of these advantages, the IBM Almaden research group demonstrated single electron spin detection in 2004, using the iOSCAR protocol on a sample of well-isolated unpaired electrons [8]; however, the detection of single nuclear spin resonance remains the ultimate challenge for MRFM.

**4.4**
**Imaging with MRFM**

Soon after the experimental proof-of-concept demonstrations of mechanical detection of electron and nuclear spin resonance, MRFM was implemented in the

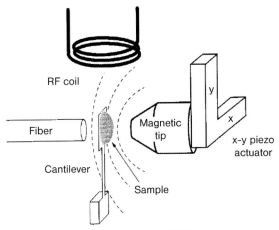

**Figure 4.5** Schematic diagram of the MRFM microscope in a sample-on-cantilever mode. The dashed lines represent the contours of a constant magnetic field. Reprinted with permission from Ref. [37]. Copyright 1996 by the American Institute of Physics.

electron and nuclear spin MRI mode [36, 37]. Unlike the magnet-on-cantilever configuration of Figure 4.1 that has been used to demonstrate spin sensitivity, the most commonly used MRFM imaging configuration has been of the sample-on-cantilever type [37] (see Figure 4.5). The same spin manipulation protocols as described earlier were used for electron and nuclear spin resonance imaging. The only additional procedures required for imaging are the mechanical positioning in two or three dimensions, and the image reconstruction process to obtain the spin density maps.

Figure 4.6 is an example of demonstrated ESR microscopy [36], where Figure 4.6a is an optical image of two DPPH particles placed on a cantilever and separated by approximately 35 μm, while Figure 4.6b shows the reconstructed MRFM image [36]. As the MRFM signal at a particular position is typically obtained from a parabolic resonant imaging shell, simple inverse filtering was used to reconstruct the image. The outer ring structure in the reconstructed image is an artifact of the deconvolution procedure [36].

In the initial MRFM imaging reports, the magnetic tip (as shown schematically in Figure 4.5) was millimeter-scale in size and the imaging resolution achieved was in the micrometer range. Recently, multiple reports have been made on submicron imaging resolution by several groups [9, 38–40]. The highest nuclear spin MRFM resolution was obtained by utilizing a micromachined array of silicon tips [9] (see Figure 4.7), coated with a 100 nm-thick layer of ferromagnetic material that provided imaging shells with magnetic field gradients on the order of $10^6 \, \mathrm{T \, m^{-1}}$. The spin manipulation protocol was almost identical to the iOSCAR method shown in Figure 4.3, except that the RF pulse sequence in line B of Figure 4.3 is

(a)  (b)

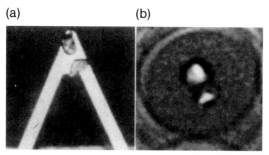

**Figure 4.6** Imaging by MRFM. (a) Optical micrograph showing two DPPH particles (20 μm wide) attached to a silicon nitride cantilever; (b) The reconstructed image, with the two DPPH particles appearing as two bright features. The large rings slightly visible in the image are artifacts of the deconvolution procedure. Reprinted with permission from Ref. [36]. Copyright 1993 by the American Institute of Physics.

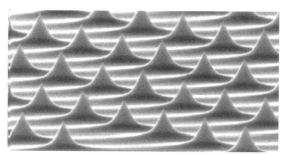

**Figure 4.7** Scanning electron micrograph of an array of etched silicon tips used to form magnetic tips. The silicon tips were coated with a thin film of CoFe alloy to generate strong magnetic field gradients near the tip apexes. Reprinted with permission from Ref. [9]. Copyright 2007 by the Nature Publishing Group.

inverted. The RF signal was mostly off, and pulsed on in short bursts to induce spin inversion; this resulted in the frequency shifts required for sensitive detection.

In addition to the demonstrated single electron spin MRFM detection capability in systems of spatially well-isolated spins [8], a nanomagnetic planar design (Figure 4.8) was recently proposed [41] that creates a localized Angstrom-scale point in 3-D space above the nanostructure with a nonzero minimum of the magnetic field magnitude. This consists of a 120 nm-diameter thin circular disk of magnetic material in the x–y plane with a perpendicular anisotropy axis so that it is permanently magnetized along the z-direction (out of the page). In addition, two quarter-circle cuts are made in the disk, diagonally opposed, and with a smaller radius.

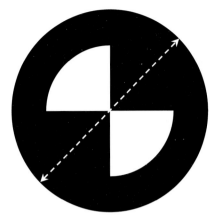

**Figure 4.8** The magnetic resonance microscopy 'lens' design consists of a 10 nm-thick perpendicular anisotropy magnetic material disk (magnetized out of the page) with two inside quarter-circle diagonally opposed cuts. The outside diameter, indicated by the white dashed markers, is 120 nm, and the inner radius of the cuts is 40 nm. A bias field opposite to the magnetization direction (into the page) is also required for obtaining a localized magnetic field magnitude minimum above the structure. Reprinted with permission from Ref. [41]. Copyright 2005 by the American Chemical Society.

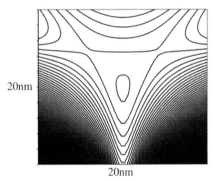

20nm

20nm

**Figure 4.9** Contours of constant magnitude of the magnetic field above the 'lens' structure along the +45° symmetry plane, as indicated by the white dashed line in Figure 4.8. The 'focus' or the localized magnetic field magnitude minimum is located at $z = 23.8$ nm above the plane, and has a value of 99.5 Gauss. A 20 nm × 20 nm area centered around the minimum point is shown. The contours are drawn at 6 Gauss intervals, with the central contour at 100.5 Gauss. Only the spins within the central contour satisfy the magnetic resonance condition, and would potentially be detected by a narrow bandwidth resonant cantilever detector. Reprinted with permission from Ref. [41]. Copyright 2005 by the American Chemical Society.

An example of the calculated contours of constant magnitude of magnetic field **B** centered at 24 nm above the structure along one of the symmetry planes is shown in Figure 4.9. The contours are 6 Gauss apart, with the center minimum contour at 100.5 Gauss. The localized point minimum of the magnetic field magnitude only occurs if a bias magnetic field is applied in this case, although novel designs have recently been described that are self-biased [42]. A spin resonance linewidth of ~1 Gauss, typical of nuclear spins in a solid-state environment, would mean that this 'lens' would be able to frequency-separate different spins located approximately 1 nm apart, as only the spin located within the central contour in Figure 4.9 would be resonant and detected in the presence of other spins in its natural dense spin environment with spins only few Angstroms apart [41]. Finally, the case was presented that nanometer-scale recording in perpendicular anisotropy thin magnetic films, using presently available data storage technology, could potentially provide the ultimate miniaturization of these 'lens' structures [42].

Although it remains to be seen whether MRFM can provide atomic resolution single nuclear spin imaging, it has also been shown [43, 44] that, for the case of crystals, certain MRFM configurations of Figure 4.1 significantly relax the challenging technical requirements for single spin MRFM by allowing many spins to coherently contribute to the magnetic resonance signal, while still providing atomic resolution imaging of the crystal lattice planes. The proposed approach assumes a 100 nm-diameter spherical ferromagnetic tip, as shown in Figure 4.1, with a large external bias field $B_0$ on the order of 10 T applied parallel to the cantilever long axis and the crystal surface. This field saturates the sphere magnetization and polarizes the sample spins. As the external direct current (dc) polarizing magnetic field $B_0$ is much larger than the field from the ferromagnetic sphere, only the z-component of the magnetic field from the ferromagnetic sphere is included when considering the resonant spins of the atomic lattice [23, 24]. The imaging contours of the constant $B_z$ field from the sphere are shown in Figure 4.10, and have the azimuthally symmetric form around the z-axis.

Numerical summation was computed to construct a histogram of the number of resonant spin sites in the sample within a 1 Gauss-wide imaging shell of con-

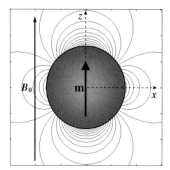

**Figure 4.10** Azimuthally symmetric contours of constant value of the z-component of the magnetic field $B_z$ from a ferromagnetic sphere in a large external magnetic field $B_0$. The spins of the sample in proximity with the sphere laying on the same contour have the same magnetic resonance frequency. For a submicron diameter ferromagnetic sphere, the magnetic field gradients are sufficiently large that the discrete nature of the spins of the sample becomes important. Reprinted with permission from Ref. [44]. Copyright 2002 by the American Institute of Physics.

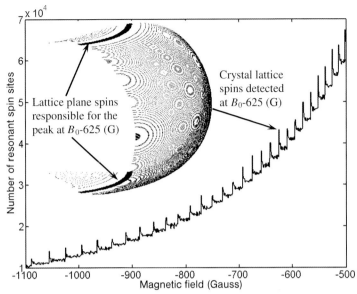

**Figure 4.11** Model of a magnetic resonance spectrum for a crystalline sample next to a 100 nm-diameter cobalt ferromagnetic sphere. The discrete nature of the spins in the crystal lattice of this model gives rise to sharp magnetic resonance spectral peaks. The inset shows a 3-D plot of the spins in the crystal lattice that are resonant at the magnetic field value of $B_0$-625 G, the location of one of the sharp spectral peaks. A black dot is placed for every lattice site for which the field at that site is within 1 G of the resonant magnetic field. Dark bands of spins from the crystal lattice planes intersected by the top and bottom sections of the resonant contour (indicated by the arrows) are responsible for the observed sharp peak above the background in the spectrum. Reprinted with permission from Ref. [44]. Copyright 2002 by the American Institute of Physics.

stant $B_Z$. Distinct spectral peaks were discovered in the number of resonant spin sites with respect to the applied magnetic field; the spectrum between the field range of $B_0$-1100 Gauss and $B_0$-500 Gauss is shown in Figure 4.11. The appearance of the magnetic resonance spectral peaks was a direct signature of the discrete atomic lattice spin sites, since the magnetic resonance of a continuous medium would result in a monotonic spectrum.

The appearance of spectral peaks was explained using the 3-D plots of the resonant spins of the crystal lattice [43, 44]. The inset in Figure 4.11 shows one such 3-D plot at the magnetic field value of $B_0$-625 Gauss, the location of one of the sharp resonant peaks in the spectrum of Figure 4.11. The shell of constant $B_Z = -625$ Gauss intersects the crystal lattice such that a large number of spin sites from the two lattice planes at the top and bottom sections of the resonant shell satisfy the resonance condition. The two bands of the resonant atoms from the lattice planes are clearly visible in the inset of Figure 4.11, and resonant ring bands such as these at the distinct magnetic field values are responsible for the sharp peaks in the magnetic resonance spectrum of Figure 4.11. The number of spins

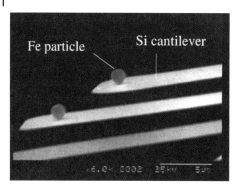

**Figure 4.12** Example of a spherical iron microparticle mounted on a silicon cantilever. Reprinted with permission from Ref. [46]. Copyright 2003 by the American Institute of Physics.

that would have to be detected ranges between $10^4$ and $10^5$, and the number of spins in a spectral peak above the background level is on the order of $10^3$ (see Figure 4.11). With the demonstrated nuclear spin detection capability on the order of $10^3$ spins [35], as well as the availability of the ferromagnetic spherical ultra-high-field gradient sources mounted on sensitive cantilevers [45, 46] (as shown in the example of Figure 4.12 [46]), the experimental groundwork has been laid out for an opportunity to detect NMR from individual crystal lattice planes with atomic resolution. (It is an interesting historical note that the atomic resolution imaging of crystal lattice planes was pondered upon in the first pioneering studies on MRI by Mansfield and Grannell [16], for the case of linear magnetic field gradients.)

## 4.5
## Conclusions

Further progress in MRFM places challenging demands on the technical requirements. The mechanical detection of a single electron magnetic resonance was performed on a sample with spatially well-isolated spins with a significant averaging time of 13 h per point [8]. The signal-to-noise considerations [4, 26, 47] show that, in order to improve the sensitivity and therefore reduce the averaging time, further miniaturization of the mechanical detector and the magnetic field gradient source is required, whilst simultaneously allowing for the efficient optical detection of cantilever vibration. Towards this end, a nanowire-based methodology for the fabrication of ultra-high-sensitivity and resolution probes for atomic resolution MRFM was recently proposed [48]. This fabrication technique combines the electrochemical deposition of multifunctional metals into nanoporous polycarbonate membranes and the chemically selective deposition of the optical nanoreflector onto the nanowire, so that all of the crucial components of the MRFM sensor are miniaturized simultaneously. The fabricated composite nanowire structure (see Figure 4.13) contains: (i) a magnetic nickel nanowire segment which provides atomic resolution magnetic field imaging gradients, as well as large gradients for high force sensitivity; (ii) a silver nanoparticle-enhanced nanowire segment provid-

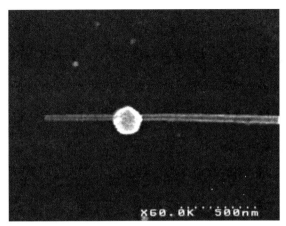

**Figure 4.13** Scanning electron micrograph of the fabricated nanowire-based composite magnetic resonance force microscopy sensor. The magnetic nickel section is on the left, followed by the subwavelength silver nanoreflector, and completed by the platinum nanowire-based cantilever structure extending to the right of the figure. Reprinted with permission from Ref. [48]. Copyright 2005 by the American Chemical Society.

ing an efficient plasmon resonant scattering cross-section from a subwavelength source [49] for the optical readout of nanowire vibration; and (iii) a platinum nanowire segment providing the cantilever structure for the mechanical detection of magnetic resonance.

It is evident that MRFM will advance in all other relevant aspects, including new spin, cantilever and field manipulation protocols [50–52], image reconstruction procedures [53] and spin quantum measurement theory [7] in order to reach the ultimate MRFM goal of single nuclear spin sensitivity and 3-D atomic imaging resolution. It remains to be seen whether the technique will become sufficiently easy to use such that it achieves wider acceptance in the scientific community, especially in view of the progress that is concurrently being made in conventional MRM technology [54, 55] and alternative NMR sensing methods [56–63]. Whichever pathway this might be, single nuclear spin detection and imaging will continue to motivate and challenge scientists for quite some time.

## Acknowledgments

The chapter content is based upon the results of studies supported by the National Science Foundation under the NSF-CAREER Award Grant No. 0349319 and NSF-ECS Award Grant No. 0622228, as well as the California State University Long Beach Scholarly and Creative Activities Award (SCAC). I thank Professor Axel Scherer and Dr Joyce Wong for many stimulating discussions over the years on the topic of MRFM.

## References

1 Sidles, J.A. (1991) *Applied Physics Letters*, **58**, 2854–6.

2 Sidles, J.A. (1992) *Physical Review Letters*, **68**, 1124–7.

3 Sidles, J.A., Garbini, J.L. and Drobny, G.P. (1992) *Review of Scientific Instruments*, **63**, 3881–99.

4 Sidles, J.A., Garbini, J.L., Bruland, K.J., Rugar, D., Zuger, O., Hoen, S. and Yannoni, C.S. (1995) *Reviews of Modern Physics*, **67**, 249–65.

5 Nestle, N., Schaff, A. and Veeman, W.S. (2001) *Progress in Nuclear Magnetic Resonance Spectroscopy*, **38**, 1–35.

6 Suter, A. (2004) *Progress in Nuclear Magnetic Resonance Spectroscopy*, **45**, 239–74.

7 Berman, G.P., Borgonovi, F., Gorshkov, V.N. and Tsifrinovich, V.I. (2006) *Magnetic Resonance Force Microscopy and Single-Spin Measurement*, World Scientific Publishing Co., Singapore.

8 Rugar, D., Budakian, R., Mamin, H.J. and Chui, B.W. (2004) *Nature (London)*, **430**, 329–32.

9 Mamin, H.J., Poggio, M., Degen, C.L. and Rugar, D. (2007) *Nature Nanotechnology*, **2**, 301–6.

10 Gerlach, W. and Stern, O. (1922) *Zeitschrift fur Physik*, **9**, 349–52.

11 Einstein, A. and de Haas, W.J. (1916) Proceedings of the Section of Sciences, Koninklijke Akademie van Wetenschappen te Amsterdam, vol. **18**, pp. 696–711.

12 Rabi, I.I., Zacharias, J.R., Millman, S. and Kusch, P. (1938) *Physical Review*, **53**, 318.

13 Evans, D.F. (1956) *Philosophical Magazine*, **1**, 370–3.

14 Alzetta, G., Arimondo, E., Ascoli, C. and Gozzini, A. (1967) *Il Nuovo Cimento*, **52**B, 392–402.

15 Lauterbur, P.C. (1973) *Nature (London)*, **242**, 190–1.

16 Mansfield, P. and Grannell, P.K. (1973) *Journal of Physics C: Solid State Physics*, **6**, L422–6.

17 Young, R., Ward, J. and Scire, F. (1972) *Review of Scientific Instruments*, **43**, 999–1011.

18 Wiesendanger, R., Güntherodt, H.-J., Güntherodt, G. and Gambino, R.J. and Ruf, R. (1990) *Physical Review Letters*, **65**, 247–50.

19 Manassen, Y., Hamers, R.J., Demuth, J.E. and Castellano, A.J., Jr (1989) *Physical Review Letters*, **62**, 2531–4.

20 Durkan, C. and Welland, M.E. (2002) *Applied Physics Letters*, **80**, 458–60.

21 Bruland, K.J., Dougherty, W.M., Garbini, J.L., Sidles, J.A. and Chao, S.H. (1998) *Applied Physics Letters*, **73**, 3159–61.

22 Rugar, D., Mamin, H.J. and Guethner, P. (1989) *Applied Physics Letters*, **55**, 2588–90.

23 Callaghan, P.T. (1991) *Principles of Nuclear Magnetic Resonance Microscopy*, Oxford University Press, New York.

24 Blumich, B. (2000) *NMR Imaging of Materials*, Oxford University Press, New York.

25 McDonald, P.J. and Newling, B. (1998) *Reports on Progress in Physics*, **61**, 1441–93.

26 Rugar, D., Yannoni, C.S. and Sidles, J.A. (1992) *Nature (London)*, **360**, 563–6.

27 Rugar, D., Zuger, O., Hoen, S., Yannoni, C.S., Vieth, H.-M. and Kendrick, R.D. (1994) *Science*, **264**, 1560–3.

28 Abragam, A. (1983) *Principles of Nuclear Magnetism*, Oxford University Press, New York.

29 Rabi, I.I., Ramsey, N.F. and Schwinger, J. (1954) *Reviews of Modern Physics*, **26**, 167–71.

30 Kittel, C. and Kroemer, H. (1980) *Thermal Physics*, 2nd edn, W.H. Freeman Company, New York.

31 Muller, N. and Jerschow, A. (2006) *Proceedings of the National Academy of Sciences of the United States of America*, **103**, 6790–2.

32 Mozyrsky, D., Martin, I., Pelekhov, D. and Hammel, P.C. (2002) *Applied Physics Letters*, **82**, 1278–80.

33 Mamin, H.J., Budakian, R., Chui, B.W. and Rugar, D. (2003) *Physical Review Letters*, **91**, 207604.

34 Bergman, G.P., Kamenev, D.I. and Tsifrinovich, V.I. (2002) *Physical Review Part A*, **66**, 023405.

**35** Mamin, H.J., Budakian, R., Chui, B.W. and Rugar, D. (2005) *Physical Review Part B*, **72**, 024413.

**36** Zuger, O. and Rugar, D. (1993) *Applied Physics Letters*, **63**, 2496–8.

**37** Zuger, O., Hoen, S.T., Yannoni, C.S. and Rugar, D. (1996) *Journal of Applied Physics*, **79**, 1881–4.

**38** Thurber, K.R., Harrell, L.E. and Smith, D.D. (2003) *Journal of Magnetic Resonance*, **162**, 336–40.

**39** Chao, S.-H., Dougherty, W.M., Garbini, J.L. and Sidles, J.A. (2004) *Review of Scientific Instruments*, **75**, 1175–81.

**40** Tsuji, S., Yoshinari, Y., Park, H.S. and Shindo, D. (2006) *Journal of Magnetic Resonance*, **178**, 325–8.

**41** Barbic, M. and Scherer, A. (2005) *Nano Letters*, **5**, 787–92.

**42** Barbic, M., Barrett, C.P., Vltava, L., Emery, T.H., Walker, C. and Scherer, A. (2008) *Concepts in Magnetic Resonance*, **33B**, 21–31.

**43** Barbic, M. (2002) *Journal of Applied Physics*, **91**, 9987–94.

**44** Barbic, M. and Scherer, A. (2002) *Journal of Applied Physics*, **92**, 7345–54.

**45** Lantz, M.A., Jarvis, S.P. and Tokumoto, H. (2001) *Applied Physics Letters*, **78**, 383–5.

**46** Ono, T. and Esashi, M. (2003) *Review of Scientific Instruments*, **74**, 5141–6.

**47** Sidles, J.A. and Rugar, D. (1993) *Physical Review Letters*, **70**, 3506–9.

**48** Barbic, M. and Scherer, A. (2005) *Nano Letters*, **5**, 187–90.

**49** Mock, J.J., Barbic, M., Smith, D.R., Schultz, D.A. and Schultz, S. (2002) *Journal of Chemical Physics*, **116**, 6755–9.

**50** Kempf, J.G. and Marohn, J.A. (2003) *Physical Review Letters*, **90**, 087601.

**51** Degen, C.L., Lin, Q., Hunkeler, A., Meier, U., Tomaselli, M. and Meier, B.H. (2005) *Physical Review Letters*, **94**, 207601.

**52** Barbic, M. and Scherer, A. (2006) *Journal of Magnetic Resonance*, **181**, 223–8.

**53** Hammel, P.C., Pelekhov, D.V., Wigen, P.E., Gosnell, T.R., Midzor, M.M. and Roukes, M.L. (2003) *IEEE Proceedings*, **91**, 789–98.

**54** Glover, P. and Mansfield, P. (2002) *Reports on Progress in Physics*, **65**, 1489–511.

**55** Ciobanu, L., Webb, A.G. and Pennington, C.H. (2003) *Progress in Nuclear Magnetic Resonance Spectroscopy*, **42**, 69–93.

**56** Savukov, I.M. and Romalis, M.V. (2005) *Physical Review Letters*, **94**, 123001.

**57** Webb, A.G. (1997) *Progress in Nuclear Magnetic Resonance Spectroscopy*, **31**, 1–42.

**58** Barbic, M. and Scherer, A. (2005) *Solid-State Nuclear Magnetic Resonance*, **28**, 91–105.

**59** Maguire, Y., Chuang, I.L., Zhang, S. and Gershenfeld, N. (2007) *Proceedings of the National Academy of Sciences of the United States of America*, **104**, 9198–203.

**60** Greenberg, Y.S. (1998) *Reviews of Modern Physics*, **70**, 175–222.

**61** Boero, G., Besse, P.A. and Popovic, R. (2001) *Applied Physics Letters*, **79**, 1498–500.

**62** Jin, J. and Li, X.-Q. (2005) *Applied Physics Letters*, **86**, 143504.

**63** Black, R.D., Early, T.A., Roemer, P.B., Mueller, O.M., Mogro-Campero, A., Turner, L.G. and Johnson, G.A. (1993) *Science*, **259**, 793–5.

# 5
# Dynamic Fixed-Point Generation Using Non-Linear Feedback Fields – with Applications in MR Contrast Enhancement

*Jon K. Furuyama, Dennis W. Hwang, Susie Y. Huang, Jamie D. Walls, Hsing-Wei Chang, Lian-Pin Hwang and Yung-Ya Lin*

## 5.1
## Introduction

An important goal of magnetic resonance (MR) microscopy is to differentiate materials or tissues on the basis of physical and chemical properties. Some properties, such as molecular dynamics or microscopic inhomogeneity, can manifest themselves in the form of $T_1$, $T_2$ and $T_2^*$ [1] relaxation, respectively. Regions that differ in relaxation parameters result in varying signal intensities, and thus provide contrast in the resulting images. While relaxation-based contrast can provide good soft-tissue differentiation, these contrast mechanisms may be limited when the relaxation parameters do not vary significantly. This scenario is characteristic of the early stages of tumor growth and the onset of certain pathologies, in which differences between healthy and cancerous tissues are often difficult to detect.

Aggressive tumors are characterized by rapid cell growth and angiogenesis. This increase in blood vessels can lead to a slight bulk magnetic field variation, shifting the average local resonance offset of the bulk water protons. This shift in the precessional frequency leads to a change in the phase, but not magnitude, of the magnetization [2]. Consequently, bulk susceptibility variations may not be readily detected by relaxation-based sequences, which typically influence the magnitude of the magnetization vector. Because differences in susceptibility are magnetic in nature, contrast must be developed from the dynamic control of the orientation of the magnetization vector. As a result, there has been much interest in using nonlinear feedback fields, which have been shown to be sensitive to small susceptibility variations [3–5]. Two particular feedback fields relevant to magnetic resonance imaging (MRI) and MR spectroscopy are the distant dipolar field, and the radiation damping feedback field. The nonlinearity of these two feedback fields makes the dynamic evolution highly sensitive to the initial conditions which can, in some cases, lead to chaotic dynamics [6]. The use of radiation damping alone has also been shown to produce unique contrasts between tissues, with only slight differences in susceptibility [3].

*Magnetic Resonance Microscopy.* Edited by Sarah L. Codd and Joseph D. Seymour
Copyright © 2009 WILEY-VCH Verlag GmbH & Co. KGaA, Weinheim
ISBN: 978-3-527-32008-0

The radiation damping feedback field was first studied by Bloembergen and Pound while nuclear magnetic resonance (NMR) was still in its infancy [7]. By Lenz's law, the signal-induced current in the receiver coil is opposed by a reaction field, which is fed back into the sample, resulting in the magnetization being rotated back towards the equilibrium position (+z axis) [8]. The corresponding effect results in a diminished free induction decay (FID) which has the appearance of an additionally damped oscillator; hence the term, radiation damping. Consequently, all of the spins in a given sample are indirectly coupled together through the radiation damping field. The net effect of radiation damping is to lower the total Zeeman energy of the system by bringing the total magnetization back to the equilibrium state, limiting the extent of this indirect coupling.

With the advent of pulsed Fourier-transform NMR [9], continuous wave (CW) irradiation has been used primarily for secondary dynamics such as presaturation or decoupling purposes [10]. However, the application of CW irradiation in the presence of radiation damping allows the magnetization to evolve continuously under the nonlinear coupling with the receiver coil. The constant, perturbative nature of CW irradiation mixed with radiation damping forces the magnetization into nonequilibrium fixed points (i.e. not along the +z axis) which, for a multiple component system, is capable of producing imaging contrast that differs fundamentally from conventional, relaxation-based methods.

## 5.2
## Quantum Mechanical Derivation of Fixed Points

From the classical Zeeman energy, $E_{zeemna} = -\vec{\mu} \cdot \vec{B}$, the quantum mechanical Hamiltonian operator for a dipole placed into an external magnetic field can be constructed

$$\hat{H}_{zeeman} = -\gamma\hbar\left\{\hat{I}_x B_x + \hat{I}_y B_y + \hat{I}_z B_z\right\} \tag{5.1}$$

where $\gamma$ is the gyromagnetic ratio, $\vec{\mu} = \gamma\hbar\vec{I}$ and $\vec{I}$ the nuclear spin operator. For simplicity, consider a two-component system with equal spin density in a reference frame rotating at approximately the Larmor frequency, $\omega_0 = \gamma B_0$, where $B_0$ is the strength of the externally applied Zeeman field along the +z axis. For a system having a difference in precession frequency between the components of $\Delta\omega_{ij} = \gamma\Delta B_{ij} = 10\,\text{Hz}$, the rotating frame can be set such that the resonance offset for both components within this frame is $\delta\omega_{i/j} = \omega_{i/j} - \omega_0 = \pm 5\,\text{Hz}$. Consequently, a resonance offset from the Larmor frequency manifests itself as a magnetic field in the Zeeman direction such that $\gamma B_{z,i/j} = \delta\omega_{i/j}$. Within the classical framework, a continuous wave manifests itself as an oscillating magnetic field in the transverse plane

$$\gamma B_{cw,+}(t) = 2\omega_1 e^{i\omega_0 t} \tag{5.2}$$

where the complex notation is used for $B_+ = B_x + iB_y$, $\omega_1$ is the effective strength of the continuous wave, often referred to as the Rabi frequency [11], and $\omega_0$ is the carrier frequency of the applied CW. In the case that the carrier frequency matches the Larmor frequency, the CW becomes a static magnetic field vector in the transverse plane, offset by a particular phase, $\vartheta$. With these two static fields, the Hamiltonian can be rewritten as

$$\hat{H}_{tot} = \hbar \left[ \delta\omega_i \hat{I}_{z,i} + \delta\omega_j \hat{I}_{z,j} \right] + \hbar\omega_1 \left[ \cos(\vartheta)\left(\hat{I}_{x,i} + \hat{I}_{x,j}\right) + \sin(\vartheta)\left(\hat{I}_{y,i} + \hat{I}_{y,j}\right) \right] = \hat{H}_i + \hat{H}_j \tag{5.3}$$

where $\hat{I}_{x,y,z}$ are products of the familiar Pauli matrices. The constants of motion for the total Hamiltonian, $\hat{H}_{tot} = \hat{H}_i + \hat{H}_j$, are those that satisfy the equation

$$\frac{\partial \hat{\rho}_{ij}}{\partial t} = -\frac{i}{h}\left[\hat{H}_{tot}, \hat{\rho}_{ij}\right] = 0 \tag{5.4}$$

where this is simply when the Liouville–Von Neumann equation is equal to zero. Since both $\hat{H}_{tot}$ and $\hat{\rho}_{ij}$ commute in Equation 5.4, the constants of motion are those that are either parallel or anti-parallel to the total Hamiltonian. The four physically relevant density matrices that satisfy Equation 5.4 are thus

$$\hat{\rho}_{++} = \frac{1}{4}\hat{1} + \left(\hat{H}_i + \hat{H}_j\right)$$

$$\hat{\rho}_{--} = \frac{1}{4}\hat{1} - \left(\hat{H}_i + \hat{H}_j\right)$$

$$\hat{\rho}_{+-} = \frac{1}{4}\hat{1} + \left(\hat{H}_i - \hat{H}_j\right) \tag{5.5}$$

$$\hat{\rho}_{++} = \frac{1}{4}\hat{1} - \left(\hat{H}_i - \hat{H}_j\right)$$

where the $\pm$ index corresponds to the sign in front of the $\hat{H}_{ij}$ terms, and $\hat{1}$ is the identity matrix. It can be seen that $\hat{\rho}_{++}$ has both components aligned parallel with the total field, $\hat{\rho}_{--}$ has both components anti-parallel to the field, and $\hat{\rho}_{+-/-+}$ has one component parallel and one anti-parallel to the total field. The expectation value of the energy is $\langle E_{ij} \rangle = Tr\left[\hat{H}_{tot}, \hat{\rho}_{ij}\right]$, where $Tr$ indicates the matrix trace. It can be seen that $\langle E_{++} \rangle < \langle E_{+-/-+} \rangle < \langle E_{--} \rangle$.

Since $\delta\omega_{i/j} = \pm 5\,Hz$, there is little contrast between the magnetization vectors in both the $\hat{\rho}_{++}$ and $\hat{\rho}_{--}$ states. The $\hat{\rho}_{+-}$ and $\hat{\rho}_{-+}$ states are characterized by a 180° phase-shift in the transverse plane, and considerable contrast between the magnetization vectors would be generated if these states were readily accessible. Unfortunately, the small value of $\Delta\omega_{ij} = 10\,Hz$ between both components makes it practically challenging to selectively excite one of the components in the $\hat{\rho}_{++}$ state to either of the $\hat{\rho}_{+-}$ or $\hat{\rho}_{-+}$ states in the presence of strong radiation damping. Remarkably, with the addition of radiation damping to the total Hamiltonian, the

$\hat{\rho}_{++}$ and $\hat{\rho}_{--}$ states are no longer constants of motion, while Equation 5.4 is still satisfied by both $\hat{\rho}_{+-}$ and $\hat{\rho}_{-+}$.

Radiation damping behaves like a transverse magnetic field applied orthogonally to the average transverse state of the entire sample. Thus, a new term needs to be added to Equation 5.3 of the form

$$\hat{H}_{rd,ij} = \hbar\omega_r \left( \Re \left[ \langle i\hat{I}_{ij}^+ \rangle \right] \hat{I}_x + \Im \left[ \langle i\hat{I}_{ij}^+ \rangle \right] \hat{I}_y \right) \tag{5.6}$$

where $\Re$ and $\Im$ are the real and imaginary components respectively, $\omega_r$ is a constant dependent on the equilibrium polarization of the sample as well as the coupling efficiency between the receiver coil and the sample, $\langle i\hat{I}_{ij}^+ \rangle = iTr\left[ \hat{I}^+, \hat{\rho}_{ij} \right]$, and $\hat{I}^+ = \hat{I}_x + i\hat{I}_y$ is the raising operator. For $\langle \hat{I}_{ij}^+ \rangle \neq 0$, Equation 5.6 contains pseudo-bilinear terms $\left( \text{such as } \langle \hat{I}_{ij}^+ \rangle \hat{I}_{x,j} \right)$. It can be seen that for the $\hat{\rho}_{++}$ and $\hat{\rho}_{--}$ states, $\langle I_{\pm\pm}^+ \rangle = \pm\omega_1 e^{i\vartheta}$, resulting in a new total Hamiltonian, $\hat{H}_{tot}' = \hat{H}_{tot} + \hat{H}_{rd}$. By Equation 5.6, $\lfloor \hat{H}_{tot}, \hat{H}_{rd} \rfloor \neq 0$, and as a result, $\lfloor \hat{H}_{tot}', \hat{\rho}_{++/--} \rfloor \neq 0$. Consequently, $\hat{\rho}_{++}$ and $\hat{\rho}_{--}$ are no longer constants of motion for Equation 5.4. Surprisingly, for the $\hat{\rho}_{+-}$ and $\hat{\rho}_{-+}$ states, $\langle \hat{I}_{+-}^+ \rangle = \langle \hat{I}_{-+}^+ \rangle = 0$, and thus $\hat{H}_{rd,+-} = \hat{H}_{rd,-+} = 0$. As a result, in the presence of radiation damping, the total Hamiltonian for the $\hat{\rho}_{+-}$ and $\hat{\rho}_{-+}$ states still satisfies Equation 5.4, thus preserving these states as constants of motion. Consequently, the presence of strong radiation damping produces constants of motion, or fixed points, where both components of the system are oriented anti-parallel to each other. For this two component system, the anti-parallel orientation manifests itself as a phase difference in the transverse plane. Prior to CW irradiation, both components of the magnetization are initially at the equilibrium $+z$ position. The continuous excitation from the CW irradiation allows the magnetization for both components to evolve dynamically to the fixed points. Insight into the dynamic generation of the fixed points is best understood within the classical picture.

## 5.3
## Classical Derivation of Fixed Points

The evolution of a normalized magnetization vector (normalized with respect to the equilibrium magnetization, $m = M / M_0$) under CW irradiation and radiation damping is governed by the classical Bloch equations [12]:

$$\frac{\partial \vec{m}(r,t)}{\partial t} = \gamma \vec{m}(r,t) \times \left[ \frac{\delta\omega(r)}{\gamma}\hat{z} + \vec{B}(r,t) \right] - \frac{m_x\hat{x} + m_y\hat{y}}{T_2} - \frac{m_z - 1}{T_1}\hat{z} + D\nabla^2 \vec{m}(r,t) \tag{5.7}$$

where $\vec{B}(r,t) = \vec{B}_{cw} + \vec{B}_{rd,+}$ is comprised of both CW irradiation and radiation damping, $T_1$ and $T_2$ are the respective longitudinal and transverse relaxation time constants, and $D$ is the self-diffusion constant. The CW irradiation is represented by $\vec{B}_{cw} = \omega_1\{\cos(\vartheta), \sin(\vartheta), 0\}$, and the radiation damping can be described compactly as [13]:

$$\gamma \vec{B}_{rd,+} = \frac{ie^{-i\varphi}}{V\tau_r} \int_V m_+ d^3r \qquad (5.8)$$

where $\varphi$ is a tuning-dependent phase of the radiation damping field, and $\tau_r = \omega_r^{-1}$ is the radiation damping time constant. For a perfectly tuned probe, $\varphi = 0$, which corresponds to the radiation damping vector lagging 90° behind the total transverse magnetization. For dynamical considerations, relaxation and diffusion are ignored in the following discussion.

The explicit dependence of $\vec{B}_{rd,+}$ (Equation 5.8) on the total magnetization renders Equation 5.7 nonlinear, and an analytical solution to Equation 5.7 for multiple components is, in general, not possible. For the two-component case with equal density, each component has an equal contribution to $\vec{B}_{rd,+}$. The radiation damping generates a mean-field coupling between all components in the sample. As a result, Equation 5.7 must be solved simultaneously for each component, making an analytical solution nearly impossible. Despite the lack of an analytical solution for the total evolution, the fixed points under CW irradiation and radiation damping can be determined by setting Equation 5.7 equal to zero for both components. It can be shown that the fixed points for a simple two-component case are

$$m_i = -\aleph^{-1}\{\omega_1 \cos(\vartheta), \omega_1 \sin(\vartheta), \delta\omega_i\}$$
$$m_j = \aleph^{-1}\{\omega_1 \cos(\vartheta), \omega_1 \sin(\vartheta), \delta\omega_j\} \qquad (5.9)$$

and

$$m_i = \aleph^{-1}\{\omega_1 \cos(\vartheta), \omega_1 \sin(\vartheta), \delta\omega_i\}$$
$$m_j = -\aleph^{-1}\{\omega_1 \cos(\vartheta), \omega_1 \sin(\vartheta), \delta\omega_j\} \qquad (5.10)$$

in agreement with the quantum calculations given in Equation 5.5, where $\aleph_{i/j} = \sqrt{\omega_1^2 + \delta\omega_{i/j}^2}$, is just a normalizing factor. By quick inspection, it can be seen that Equation 5.9 corresponds to the $\hat{\rho}_{-+}$ state, while Equation 5.10 corresponds to the $\hat{\rho}_{+-}$ state. A remarkable feature of Equations 5.9 and 5.10 is the absence of the radiation damping strength. The reason for this is analogous to the quantum picture: at the fixed points, the integral in Equation 5.8 is zero when both components align anti-parallel to each other.

## 5.4
## Evolution of the Fixed Points

The evolution of magnetization in the presence of a CW and radiation damping can be determined by the numerical integration of Equation 5.7. Despite not having a role in the location of the fixed points, the dynamic magnitude and orientation of the radiation damping field compared to the static CW indicates that the strength of the radiation damping, $1/\tau_r$, plays a role in the system evolving towards the fixed points. Figure 5.1 demonstrates the dynamic evolution of a

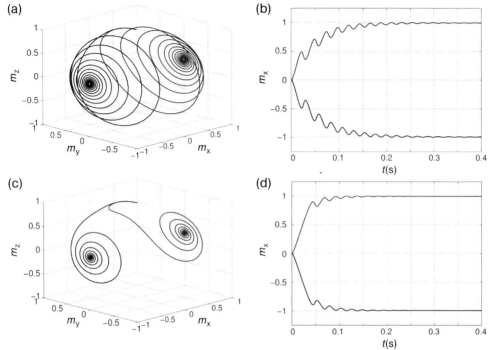

**Figure 5.1** (a) Three-dimensional evolution of a two-component system with $\delta\omega_{ij} = \pm 5$ Hz, $\omega_1 = 45$ Hz, and $\tau_r = 5$ ms, and allowed to evolve for 400 ms; (b) Evolution of the $m_x$ component for the same system in panel (a) showing the effect of the imbalanced Rabi cycle; (c) 3-D evolution of the same two- component system in panel (a) with a stronger radiation damping strength, with $\tau_r = 3$ ms; (d) Evolution of the $m_x$ component for the same system in panel (c) showing the difference in the fixed point with an even more imbalanced Rabi cycle.

simple two-component case in the presence of a CW and strong radiation damping. A comparison between two different strengths of the radiation damping field for a constant CW strength can be seen between Figure 5.1a–b and c–d. The phase of the CW is set to $\vartheta = 0$ to correspond to the magnetic field vector being applied along the +x axis. The 3-D evolution is clearly affected from simply changing the radiation damping time constant, from $\tau_r = 5$ ms (Figure 5.1a) to $\tau_r = 3$ ms (Figure 5.1c). The number of rotations required before reaching the fixed point appears to be directly proportional to $\tau_r$. From Figure 5.1b and d, it can be seen that the time required to reach the fixed points is greatly reduced by increasing the strength of the radiation damping. When comparing Figure 5.1b and 5.1d, it can be seen that simply changing the time constant from $\tau_r = 5$ ms to $\tau_r = 3$ ms can shorten the time to the fixed points by ~150 ms.

The development of the fixed points for a two-component system is fairly straightforward. In terms of dynamics, Equation 5.8 can be restated for a single component as [8, 14]

$$\frac{\partial \theta}{\partial t} = -\frac{1}{\tau_r}\sin(\theta) \tag{5.11}$$

where $\vartheta$ is the angle of the magnetization vector with respect to the +z-axis with $0 \leq \vartheta \leq \pi$. Due to radiation damping, the magnetization vector will experience a torque back towards the +z-axis at any time when $\vartheta \neq 0, \pi$. For magnetization starting about the +z-axis, the application of weak CW irradiation along the +x-axis will cause the magnetization to precess about the CW field. However, radiation damping will oppose the CW field and act to tilt the magnetization back to the +z-axis as the magnetization is rotated towards the −z-axis. However, once the magnetization passes through the −z-axis, radiation damping will now act in concert with the CW irradiation and accelerate the magnetization's return to the +z-axis. In the absence of radiation damping, the magnetization simply precesses about the CW field, and such a precession is often referred to as the Rabi cycle. The time for the first half of the Rabi cycle, $\tau_1$ (the time to rotate from the +z-axis to the −z-axis), is equal to the time for the second half of the Rabi cycle, $\tau_2$ (the time to rotate from the −z-axis to the +z-axis). However, it can be seen pictorially that the effect of Equation 5.11 is to create an imbalanced Rabi cycle, where the first half takes place on a time scale where, $\tau_1 > (2\omega_1)^{-1}$, and the second half where $\tau_2 < (2\omega_1)^{-1}$. The generation of the contrast between two components depends explicitly on this imbalance. As the two components are excited, they will acquire a phase between them which depends on the difference in precession frequency

$$\phi_{ij}(\tau) = \pm\int_{\tau}\Delta\omega_{ij}d\tau' \tag{5.12}$$

where the $\pm$ sign depends on whether or not the overall magnetization is being excited away from or towards the +z-axis. With no radiation damping it can easily be seen that, after one full revolution, the net phase $\phi_{ij}(\omega_1^{-1}) = \phi_1 + \phi_2 = 0$ by Equation 5.12, because $\tau_1 = \tau_2$. In contrast, after one full revolution in the presence of strong radiation damping, $\phi_{ij}(\omega_1^{-1}) > 0$ because $\tau_1 > \tau_2$. This can be seen specifically in Figure 5.1b and 5.1d, where the two components move out away from each other during the excitation away from the +z-axis. The two components then begin to precess back towards each other but, as the rotation time back is shorter, the two vectors cannot fully converge back towards each other. This process is allowed to repeat itself as the CW continuously rotates the magnetization. With each successive rotation, the magnetization is allowed to further separate itself until it eventually reaches the fixed points. From a comparison of Figure 5.1b and 5.1d, it can be seen that increasing the imbalance in the Rabi cycle creates a larger $\phi_{ij}(\omega_1^{-1})$, which is how the increased radiation damping strength decreases the amount of time required to reach the fixed points.

In order for these dynamics to be applied to imaging, it is useful to consider systems where the components have unequal magnetization magnitudes. Within the quantum approach, the same sets of eigenstates can be found in the absence of radiation damping. The general approach for the equal components falls apart, however, with the addition of radiation damping. The differing volume ratios between the two components leads to one component having a stronger contribution to the radiation damping than the other. As a result, it can be seen that $\langle \hat{I}_{ij}^{+} \rangle$ is finite for all possible constants of motion for Equation 5.3, making Equation 5.6 orthogonal to the original Hamiltonian. Consequently, none of the four possible states can be found which commutes with the total Hamiltonian. Thus, as the perfect symmetry is broken, it would be expected that the location of the fixed points should have an explicit dependence on the strength of the radiation damping since the nonzero Hamiltonian in Equation 5.6, scaled by $\omega_r$, tips the total effective Hamiltonian away from the CW-axis. This tip of the effective Hamiltonian thus requires a rotation of the fixed points to align with the effective field in order to minimize the total Zeeman energy of the system. For small values of $\delta\omega_{ij}$ compared with $\sqrt{\omega_1^2 + \omega_r^2}$, by simple geometric argument (Figure 5.2), the tip angle of the effective field away from the CW-axis can be approximated as

$$\theta_t \cong \tan^{-1}\left[\frac{\Delta p\omega_r \cos(\theta_t)}{\omega_1 - \Delta p\omega_r \sin(\theta_t)}\right] = \sin^{-1}\left[\frac{\Delta p\omega_r}{\omega_1}\right] \qquad (5.13)$$

where $\vartheta_t$ is the shifted angle of the fixed points away from the CW-axis, and $\Delta p = |p_i - p_j|$ represents the difference in proportion of both components. It can be seen simply that $\vartheta_t = 0$ for the equal two-component case, consistent with Equations 5.5, 5.9 and 5.10.

The effect of the broken symmetry can be seen in Figure 5.3b–d. By increasing the volume ratio from 1 (Figure 5.3a) to 4 (Figure 5.3b), the effect of the imbalanced radiation damping can be seen to tip the transverse location of the fixed points away from the CW-axis. From Equation 5.13, the tip angle would be expected

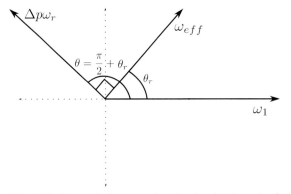

**Figure 5.2** Geometric demonstration showing the tip angle of the effective field away from the CW-axis, where $\vec{\omega}_{eff} = \vec{\omega}_1 - \Delta p\vec{\omega}_r$. An increase in $\Delta p\omega_r$ is balanced out by tipping the fixed points away from the CW-axis, and consequently, tipping the effective field by $\vartheta_r$.

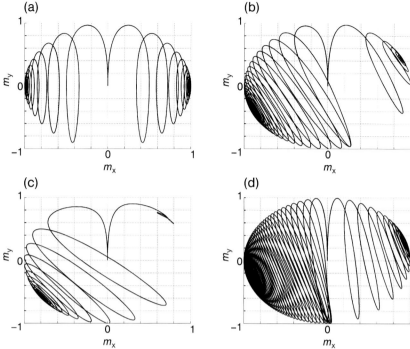

**Figure 5.3** Transverse evolution of a two-component case demonstrating effect of the strength of the radiation damping, CW irradiation (applied along the +x-axis) and volume ratio on the imbalance of the Rabi cycle and location of the fixed points, with $\delta\omega_{ij} = \pm 5$ Hz, where (a) the volume ratio is 1:1, $\tau_r = 6$ ms and $\omega_1 = 40$ Hz; (b) The volume ratio is 4:1, with the same values of $\tau_r$ and $\omega_1$ as in panel (a); (c) The same volume ratio and value of $\omega_1$ as in panel (b), with an increased radiation damping strength, with $\tau_r = 4$ ms; (d) The same volume ratio and value of $\tau_r$ as in panel (b), with an increased CW irradiation strength with $\omega_1 = 60$ Hz.

to increase with a stronger radiation damping strength. Figure 5.3c shows the shift in evolution as well as the fixed points by simply increasing the strength of the radiation damping for the same system. It can be seen that stronger radiation damping has the effect of further tipping the fixed points away from the CW-axis, following Equation 5.13. By increasing the strength of the CW irradiation, the fixed points in Figure 5.3b would thus expect to shift back towards the CW-axis. This shift in fixed-points back towards the CW-axis is seen in Figure 5.3d. The effect of the imbalanced Rabi cycle can be seen to apply even with an unequal volume ratio. By increasing the radiation damping strength in Figure 5.3b, the time to reach the fixed points is reduced with an increased imbalance. Increasing the CW irradiation strength decreases the imbalance and thus increases the time as well as the number of cycles required to reach the fixed points. There is therefore a trade-off between the time required to reach the fixed points and the amount that the fixed points deviate from CW-axis.

The discussion up to this point has ignored the presence of the $T_1$ and $T_2$ relaxation terms present in Equation 5.7. Of course with enough time, constant CW irradiation will saturate the magnetization, effectively destroying the fixed points. The longitudinal and transverse relaxation times *in vivo* are on the order of a second, where for this chapter, the time scales for one Rabi cycle are on the order of ~25 ms. While relaxation cannot be completely ignored, it can be seen that since $T_{1/2} \gg \omega_1^{-1}$, most of the initial dynamics towards the fixed points will still occur. The net effect of relaxation is to eradicate the long-term stability of the fixed points, and relaxation has little effect on the short-term dynamics. Consequently, while the long-term fixed-points are unlikely to be observed, the acquired phase difference resulting from the imbalanced Rabi cycle should still be measurable and useful for imaging contrast between regions with slightly different susceptibilities.

As can be seen in Figures 5.1 and 5.3, the effect of placing the CW directly between the resonance offset of the two components resembles a sort of repelling of the two orientations, regardless of the volume ratio. As for imaging, the developed contrast is thus dependent on the resonance offset of each component instead of the any type of relaxation parameter. The dynamics herein have been described for systems with only two unique components. Based on the underlining principles required for two components, it would be expected that the same dynamics would apply for multiple components, given a strong enough radiation damping field to create the necessary asymmetric Rabi cycle. The presence of more than two components greatly complicates the physical picture. The placement of the CW will inevitably break the symmetry, as well as add more terms to consider in Equations 5.6 and 5.8. As a rough approximation, the system can be modeled in two-component form, with one component having an offset above the CW frequency, and one below. The contrast would thus be expected to develop depending upon the location of the CW frequency with respect to the other components.

Using the pulse sequence featured in Figure 5.4a, the contrast for a multiple component phantom under CW irradiation and radiation damping can be seen in Figure 5.4b. The phantom consists of a standard NMR tube with capillaries of varying acetone concentrations. The presence of the ketone can shift the water resonance by a few parts per billion (ppb), depending on the concentration of the acetone. This yields a multicomponent sample with differing water resonances which can be resolved using a CW with radiation damping. The contrast in Figure 5.4b can be seen to reflect the dynamics presented in the theory, as some of the components can have negative values, while others have positive values. This is uncharacteristic of relaxation-based contrast methods, which strictly measure the magnitude of the magnetization at each coordinate and cannot have negative values. The extent of the contrast can be compared directly with a phase image (Figure 5.4c), which measures the phase difference between two $T_2^*$ images taken with different time echoes. The change in phase, weighted by the difference in echo time, gives an approximate frequency mapping of the sample. When comparing Figure 5.4b and c, it can be seen that the use of CW irradiation in the presence

(a)

$90^\circ_y$ $90^\circ_y$

$\tau$ $GT$

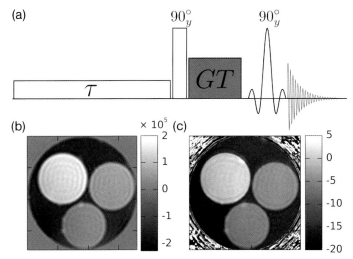

(b) $\times 10^5$ (c)

2
1
0
-1
-2

5
0
-5
-10
-15
-20

**Figure 5.4** (a) General pulse sequence used to obtain the fixed-point image. The crusher gradient sandwiched by the two $90^\circ_y$ pulses are used to measure the $m_x$ component of the magnetization at the end of the evolution; (b) Fixed-point image taken on a Bruker AVANCE 600 with a microimaging probe where $\tau_r \approx 6\,\text{ms}$. For this image, $\omega_1 \approx 50\,\text{Hz}$, and $\tau = 20\,\text{ms}$; (c) Phase image comparing the phase difference between two $T_2^*$ images with $TE_1 = 3\,\text{ms}$ and $TE_2 = 6\,\text{ms}$.

of radiation damping can give an accurate mapping of the frequency. One advantage of this approach is that the frequency mapping can be obtained with a single image acquisition, whereas the phase analysis requires the comparison of two separate images. This is particularly useful for *in vivo* measurements where it is difficult to ensure that the patient remains absolutely still for both images such that the phase at each point can be compared precisely. This method also avoids the need for complicated phase-unwrapping algorithms that are required to process the acquired images.

## 5.5
## Applications

The lack of differences in relaxation parameters between cancerous and healthy tissues complicates *in vivo* tumor detection at its earliest stages. Tumors are often accompanied by angiogenesis, leading to changes in blood oxygenation levels in the tissue and slight shifts in the resonance frequency of water. The use of CW irradiation in conjunction with radiation damping should be sensitive to small differences in resonance offset between cancerous and healthy tissues. Unfortunately, in clinical scanners the radiation damping is fairly weak, making it complicated to adapt for *in vivo* applications. However, with the use of an active feedback circuit, radiation damping can effectively be used at lower fields. During

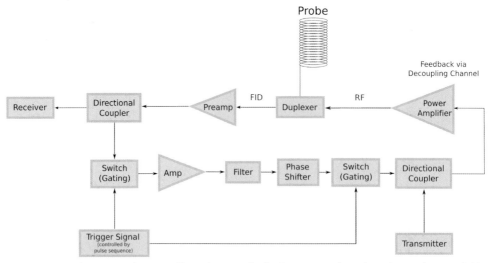

**Figure 5.5** Schematic of how the active feedback circuit emulates the radiation damping field.

the evolution, the FID is passed through the active feedback circuit, where the signal is amplified and retransmitted back to the sample and the noise is filtered through an active bandpass filter, resembling the natural radiation damping feedback field (Figure 5.5). The phase of the radiation damping, $\varphi$, can be arbitrarily adjusted to achieve the constructive or deconstructive interference.

With the use of an active feedback circuit, contrast between healthy and cancerous tissues can be detected on an *in vivo* mouse tumor, such as that shown in Figure 5.6a. The tumor can clearly be seen as the large mass in the sagittal images in Figure 5.6b–f. However, little contrast between the conventional proton density (b), $T_1$-weighted (c), $T_2$-weighted (d) and $T_2^*$ (e) images is seen compared with the

**Figure 5.6** (a) Photograph of a mouse showing the external location of the tumor on the right leg; (b) Sagittal proton-density image; (c) Sagittal $T_1$-weighted image, with inversion time (TI) = 0.1 s; (d) Sagittal $T_2$-weighted image, with TE = 50 ms; (e) Sagittal $T_2^*$ image with TE = 10 ms; (f) Sagittal fast spin-echo fixed-point image with active feedback circuit, and $\tau$ = 45 ms; (g) Transverse $T_1$-weighted image, with inversion time, TI = 0.9 s; (h) Transverse $T_2$-weighted image with TE = 30 ms; (i) Transverse $T_2^*$-weighted image with TE = 7 ms; and (j) Transverse susceptibility-weighted image taken by comparing the phase change between two $T_2^*$ images taken with $TE_1$ = 4 ms and $TE_2$ = 5 ms; (k) Transverse fast spin-echo fixed-point image taken with the use of an active feedback circuit where $\tau$ = 45 ms; (l) Pathology of a transverse slice of the leg, verifying the presence of a tumor in the mouse. The lower region (indicated by the arrow) contains necrosis as well as hemorrhage and correlates well with features in panel (k).

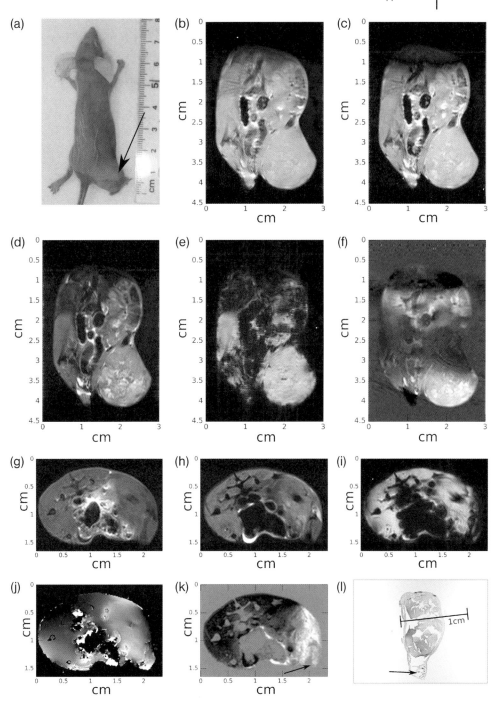

highlighted region taken using CW irradiation and radiation damping (f). The contrast enhancement can be seen further in the transverse images (Figure 5.6g–k), which show the cross-section of the mouse. Once again, the conventional $T_1$-weighted (g), $T_2$-weighted (h) and $T_2^*$ (i) images yield little information about the presence of the tumor. The contrast resulting from the CW with radiation damping (Figure 5.1k) can be compared with a susceptibility-weighted image (Figure 5.6j, which compares the acquired phase difference between two images taken with different echo times). Both images indicate a susceptibility variation on the right, but little detail can be extracted from the susceptibility-weighted image. Histologic examination confirmed the presence of tumor cells in the region of interest, seen in Figure 5.6l. The pathology also revealed a mass at the bottom of the tumor that consisted both of necrosis and hemorrhage, which might be visible in the CW + radiation damping image. The use of a CW with radiation damping is successful at differentiating the regions with cancerous tissue, and may serve as a valuable supplement for tumor detection *in vivo*.

## 5.6
## Conclusions

In the presence of radiation damping, the use of CW irradiation was shown to create unique fixed-points between magnetization components, with slight differences in resonance offset. The equilibrium-restoring nature of the radiation damping interferes with the CW irradiation, creating an imbalanced Rabi cycle and allowing the components to precess away from each other. While relaxation prevents the development of the analytical fixed points, the dynamics *en route* to the fixed points still apply and are useful for contrast development. This method was shown to be useful in tumor detection, where current relaxation-based methods are not sufficiently sensitive to differentiate between healthy and cancerous tissues. The use of an active feedback circuit allows for this method to be applied in clinical scanners, with only minor hardware modification.

## References

1 Haacke, E.M., Brown, R.W., Thompson, M.R. and Venkatesan, R. (1999) *Magnetic Resonance Imaging – Physical Principles and Sequence Design*, John Wiley & Sons, Inc.

2 Duyn, J.H. *et al.* (2007) *Proceedings of the National Academy of Sciences*, **104**, 11796–801.

3 Huang, S.Y., Furuyama, J.K. and Lin, Y.-Y. (2006) *Magnetic Resonance Materials in Physics*, **19**, 333–46.

4 Datta, S., Huang, S.Y. and Lin, Y.-Y. (2006) *The Journal of Physical Chemistry B*, **110**, 22071–8.

5 Huang, S.Y. *et al.* (2006) *Magnetic Resonance in Medicine*, **56**, 776–86.

6 Lin, Y.-Y., Lisitza, N., Ahn, S. and Warren, W.S. (2000) *Science*, **290**, 118–21.

7 Bloembergen, N. and Pound, R.V. (1954) *Physical Review*, **95**, 8–12.

8 Bloom, S. (1957) *Journal of Applied Physics*, **28**, 800–5.

**9** Ernst, R.R. and Anderson, W.A. (1966) *Review of Scientific Instruments*, **37**, 93–102.

**10** Ernst, R.R., Bodenhausen, G. and Wokaun, A. (1987) *Principles of Nuclear Magnetic Resonance in One and Two Dimensions*, Oxford Science, New York.

**11** Rabi, I.I. (1937) *Physical Review*, **51**, 652–4.

**12** Abragam, A. (1961) *Principles of Nuclear Magnetism*, Oxford Science, New York.

**13** Vlassenbroek, A., Jeener, J. and Broekaert, P. (1995) *Journal of Chemical Physics*, **103**, 5886–97.

**14** Warren, W.S., Hammes, S.L. and Bates, J.L. (1989) *Journal of Chemical Physics*, **91**, 5895–904.

# 6
# Shimming Pulses

*Louis-Serge Bouchard*

## 6.1
## Introduction

The energy splittings of greatest interest to liquid-state NMR and MRI are the scalar (J) coupling and Zeeman interactions.[1] At low enough magnetic fields, all chemical shift information is lost, and consequently J-coupling spectroscopy has emerged as a promising new alternative to molecular structure determination. Recent studies [1–4] have demonstrated spectacular homonuclear and heteronuclear fine structures, the observation of which is made possible by the very sharp lines that result from a highly homogeneous Earth's magnetic field. Although, if we increase the magnetic field the homonuclear chemical shifts eventually become distinguishable, their observation may be hampered by inhomogeneities in the applied field. This is often the case with portable sensors (e.g. Ref. [5]). In this context, so-called shim pulse methodologies have been proposed to recover spectra by applying spatially dependent phase corrections that are proportional to the local static field so that inhomogeneously broadened lines can be made narrow [6–8]. These methods were introduced with the aim of relaxing the constraints placed on hardware design. This enables NMR and MRI over larger volumes using low-cost, portable, single-sided systems. An important challenge to MRI at low fields arises when the gradient field becomes comparable to or greater than the static field, and the Zeeman splitting from the static field is no longer the dominant interaction. In this chapter some of the fundamental limits to low-field MRI are reviewed. Of particular interest here are weak and inhomogeneous Zeeman fields. In addition, the problems of Fourier encoding, spectroscopy and volume selection are addressed. These problems may be termed shimming pulses because the methods correct for nonideal static-field or gradient-field conditions. These nonidealities force a reconsideration of the approach to MRI in low fields based on conventional Fourier imaging principles.

---

1) We include chemical shielding as part of the Zeeman interaction.

*Magnetic Resonance Microscopy.* Edited by Sarah L. Codd and Joseph D. Seymour
Copyright © 2009 WILEY-VCH Verlag GmbH & Co. KGaA, Weinheim
ISBN: 978-3-527-32008-0

## 6.2
## The Low Magnetic Field Regime

Liberal use is made of the terminology low magnetic field when referring to the condition where the Zeeman field no longer truncates the perpendicular components of an applied gradient field. Precession in the rotating frame[2] under an applied gradient $\mathbf{G} = (G_x, G_y, G_z)$ is typically expressed as

$$\frac{\partial \mathbf{M}}{\partial t} = \mathbf{M} \times \hat{\mathbf{z}}(\mathbf{G} \cdot \mathbf{r}), \tag{6.1}$$

in high fields, but at low fields, the gradient takes a different form,

$$\frac{\partial \mathbf{M}}{\partial t} = \mathbf{M} \times (\mathbf{G} \cdot \mathbf{r}), \tag{6.2}$$

where $\mathbf{G}$ is a tensor. This is because the applied gradient can no longer be treated as a vector. Indeed, a multivariate Taylor expansion of the magnetic field at a point $\mathbf{r} = (x, y, z)$ near the origin $\mathbf{r} = 0$,

$$\mathbf{B}(\mathbf{r}) = \mathbf{B}(\mathbf{r} = 0) + \mathbf{r} \cdot \left. \frac{\partial \mathbf{B}}{\partial \mathbf{r}} \right|_{\mathbf{r}=0} + \mathcal{O}\left(|\mathbf{r}|^2\right), \tag{6.3}$$

shows[3] that the gradient $\partial \mathbf{B}/\partial \mathbf{r}$ is a tensor field $\mathbf{G}$. The remainder term $\mathcal{O}(|\mathbf{r}|^2) = 0$ if no higher-order gradient fields are applied.

The rotating-frame transformation can also be done using unitary rotations, which are more useful for analyzing and constructing pulse sequences. The Zeeman interaction Hamiltonian $\mathcal{H} = -\mathbf{I} \cdot \mathbf{B}$ for a nuclear spin $\mathbf{I}$ coupled to a magnetic field $\mathbf{B}$ (static component $B_z$ along $\hat{\mathbf{z}}$ plus a first-order gradient component $\partial \mathbf{B}/\partial \mathbf{r}$), is written, using the summation convention[4] on $i$ and $j$:

$$\mathcal{H} = -B_z(\mathbf{r} = 0)I_z - I_i r_j \partial_j B_i(\mathbf{r} = 0). \tag{6.4}$$

Effecting a transformation to the rotating frame with the unitary operator $e^{-i\omega I_z t}$, where $\Omega = B_z(\mathbf{r} = 0)$, we have

$$\mathcal{H}' = e^{i\omega I_z t}[-I_i r_j \partial_j B_i(\mathbf{r} = 0)]e^{-i\omega I_z t}. \tag{6.5}$$

---

**2)** From here on the nuclear gyromagnetic ratio $\gamma$ constant is included as part of the magnetic field and gradients, which are then reported in units of $\text{rad s}^{-1}$ and $\text{rad s}^{-1}\text{cm}^{-1}$, respectively. Occasionally, values are reported in Hz.

**3)** The big-oh notation, $\mathcal{O}(x)$, denotes a quantity for which $\mathcal{O}(x) \leq M|x|$ for some finite positive number $M$.

**4)** For every pair of repeated indices, such as $a_i b_i$, a summation is implied, that is, $\Sigma_i a_i b_i$.

The terms containing $I_z$ are invariant to this rotation transformation, while $I_x$ and $I_y$ become time-dependent. The coefficients of $I_x$ and $I_y$ are called concomitant gradient fields. At high fields, that is, $|\Omega| \gg |r_j \partial_j B_i(\mathbf{r} = 0)|$, rapid oscillations lead to their averaging and we are left with

$$\mathcal{H}' = -I_z r_j \partial_j B_z(\mathbf{r} = 0).$$

This phenomenon, called truncation, is characterized by spin dynamics which are determined solely by the terms in $I_z$. At low fields, the truncation condition is not satisfied. The components in $I_x$ and $I_y$ perturb the motion of the spins significantly, and must be accounted for.

## 6.2.1
### Concomitant Fields and Berry's Phase

The magnetic field $\mathbf{B}$ (or $\mathbf{H}$) in a current-free region can be obtained from a vector potential $\mathbf{A}$, as $\mathbf{B} = \nabla \times \mathbf{A}$, or a scalar potential $\Phi_M$, via $\mathbf{H} = -\nabla \Phi_M$. This field $\mathbf{B}$ automatically satisfies Maxwell's equations. The statement that only the $z$ component, $B_z$, of the gradient of the magnetic field survives the rapid averaging means that the gradient tensor components $\partial B_x / \partial r_j$ and $\partial B_y / \partial r_j$ ($j = 1, 2, 3$) are ignored and only the gradient in $B_z$ is left. This truncation procedure leads to the concept of the gradient vector

$$\vec{G} = (\partial_x B_z, \partial_y B_z, \partial_z B_z).$$

Knowledge of only the gradient vector is insufficient when one deals with low magnetic fields because it does not specify the remaining components. These are important at low fields when truncation is incomplete, and they depend on the type of coil used to generate the gradient fields. It should be noted that pure gradient fields such as $\mathbf{B} = y\hat{z}$ cannot exist in an inertial frame of reference because Maxwell's equations,

$$\nabla \cdot \mathbf{B} = 0 \quad \text{and} \quad \nabla \times \mathbf{B} = 0,$$

would be violated. On the other hand, a field $\mathbf{B} = y\hat{z} + z\hat{y}$ is physical.

Yablonskiy et al. [9] have described the effects of concomitant fields in terms of distortions in magnetic resonance images observed at low magnetic fields. Recently, these distortions have been explained in terms of the geometric (Berry's) phase [10]. The concept of Berry's phase in MRI is particularly important because, as will be shown, it enables a solution to be found to this fundamental problem of Fourier encoding in low magnetic fields. When the Hamiltonian (effective field) traces a closed path $C$ in parameter space, the adiabatic geometric phase is given by the solid angle $\Omega(C)$ subtended by the curve $C$. This phase, for a transverse magnetization component evolving under this effective field (i.e. all gradient components), is imparted in addition to the dynamic phase which would result if the

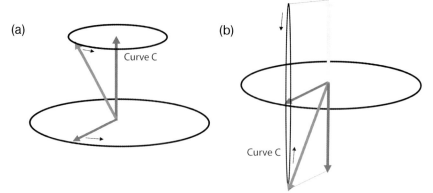

**Figure 6.1** The rotating-frame Hamiltonian (effective field) traces a closed curve (C) in parameter space. (a) A conventional static imaging gradient. This field has a static $\hat{z}$ component and a rotating transverse component due to the concomitant field. The resulting motion is a rotation about the $\hat{z}$ axis which subtends a nonzero solid angle $\Omega(C)$.

In high fields, rotation is so rapid that the transverse component averages to zero; (b) A type I rotating-frame gradient has a stationary transverse component plus a linearly polarized concomitant component oscillating between $+\hat{z}$ and $-\hat{z}$. The solid angle $\Omega(C)$ of this motion is zero. The small arrows indicate the direction of motion for the closed curve C.

same effective field were to be applied along a static axis perpendicular to the magnetization. For MRI in low fields, minimization of the geometric-phase error $\Omega(C)$ is essential.[5]

The transformation in Equation 6.5 leads to a Hamiltonian of the form:

$$\mathcal{H}' = I_x(G_x \cos \omega t - G_y \sin \omega t) + I_y(G_y \cos \omega t + G_x \sin \omega t) + G_z I_z. \tag{6.6}$$

where $G_i = r_j \partial_j B_i$. The first two terms describe concomitant fields and associated oscillating components in the rotating frame, while the last term is the conventional gradient term. A gradient field of the form $a(z\hat{x} + x\hat{z})$, leads to a net magnetic field which traces a conical motion about $z$ if and only if $z \neq 0$, as shown in Figure 6.1a. In the language of geometric phases, the Fourier encoding which is intended in MRI is the $\hat{z}$ component of this field, and the phase accrued due to evolution under this component is the dynamic phase. A consequence of this lack of truncation is that an additional, but unwanted, phase is accumulated. The geometric phase is the difference between dynamic and total phases. In the case of Fourier encoding, this geometric phase contributes a spatially dependent error to the intended phase encoding. In Ref. [10], a method based on rotating-frame gradients which minimizes $\Omega(C)$ is shown to provide a significantly improved performance for Fourier-encoded MRI in low magnetic fields, and this is discussed below.

---

[5] On the other hand, there may be cases where this geometric phase could be used advantageously, to provide spatial phase corrections which are not normally available from truncated fields.

## 6.2.2
## Rotating-Frame Gradients

In this section, we explain why the use of rotating-frame gradients for various encodings in low fields, instead of a conventional static MRI gradient, significantly improves the performance. Consider two gradient coils, each driven by an ac current at the NMR resonance (Larmor) frequency, but with the current in the first coil being 90° out of phase with the current in the second coil. The total gradient field therefore rotates at the Larmor frequency $\Omega$ and can be used to generate time-independent field components in the rotating-frame for use in Fourier encoding. In the following discussion, we describe and discuss type I and II gradient configurations (this terminology for ac gradients originates in Ref. [10]).

Apart from an overall rotation of the coordinate system, a type I gradient is defined as follows. A gradient of the form $a(z\hat{\mathbf{x}} + x\hat{\mathbf{z}})$ is added to another gradient field rotated by 90° about the z-axis, with respect to the first one, $b(z\hat{\mathbf{y}} + y\hat{\mathbf{z}})$, but with the second field driven by a current that is 90° out of phase with respect to the first field; that is,

$$a(t) = g\cos(\omega t + \varphi) \quad \text{and} \quad b(t) = g\sin(\omega t + \varphi)$$

where $g$ is a constant representing the gradient amplitude. Thus, the two gradient coils are geometrically orthogonal to each other, whereas their currents are phase-orthogonal. The reader will recognize these two fields as being the fields generated by two different Golay pairs rotated by 90° with respect to one another. The rotating-frame Hamiltonian is:

$$\mathcal{H}' = zg\cos\varphi I_x + zg\sin\varphi I_y + g[x\cos(\omega t + \varphi) + y\sin(\omega t + \varphi)]I_z. \qquad (6.7)$$

Taking $\varphi = 0°$ gives a time-independent z gradient field in $I_x$, while $\varphi = 90°$ gives a time-independent z gradient field in $I_y$. The time-dependence of the gradient has been relinquished to an oscillating field along $I_z$ (see Figure 6.1b). Because $\Omega(C) = 0$, this type I gradient possesses better averaging properties than one with rotating components (conventional static MRI gradient). In Ref. [10], significant improvements in Fourier encoding were demonstrated. Here, we show that volume selection in low fields can be performed with gradients so strong that conventional MRI gradients would not allow any meaningful selectivity. The improved performance is due to the fact that the motion of this Hamiltonian over one period traces a closed path whose solid angle is zero, whereas the conical motion has nonzero solid angle (for a conventional gradient). In fact, this specific gradient configuration was arrived at by using reasoning based on the concept of geometric phase. A type II gradient is a field,

$$a(y\hat{\mathbf{x}} + x\hat{\mathbf{y}}),$$

superimposed with a second field

$$b(-x\hat{\mathbf{x}} - y\hat{\mathbf{y}} + 2z\hat{\mathbf{z}})$$

that is scaled by an adjustable parameter $\varepsilon$. The corresponding Hamiltonian for the type II ac gradient is discussed in Ref. [10].

### 6.2.3
### Spatial Selectivity in Low Fields

As will be seen in Section 6.2.6, slice selection in low fields is seriously hampered by the presence of concomitant gradient fields. While algorithms could be devised to acquire full 3-D data sets and to reconstruct undistorted slices, such algorithms do not currently exist and 3-D acquisitions are prohibitively long at low fields due to the lower signal-to-noise ratio (SNR). Thus, it is imperative to find solutions to the general problem of slice/volume selection. Here, novel techniques will be described for slice and volume selection in low magnetic fields, which overcome many of the current limits imposed by concomitant fields.

In many of the examples that follow, the Hamiltonian possesses two widely different time scales. We average over the rapid scale to produce effective Hamiltonians that describe the dynamics. Assuming a radiofrequency (RF) field polarized along $I_x$, we examine two different schemes for selective excitation. The first scheme uses a type I ac gradient, the Hamiltonian of which (Equation 6.7) can be made to provide a stationary $Z$ rotating-frame gradient along $I_y$. The second scheme uses a type II gradient to produce a stationary $X$ or $Y$ gradient along $I_y$ in the rotating frame.

Consider the following sequence of events: (i) A nonselective (hard) $90°$ pulse rotates the equilibrium $z$ magnetization towards $I_y$: $I_z \rightarrow I_y$; (ii) a soft $90°$ pulse along $I_x$ in the rotating frame is applied in the presence of a $z$-gradient (a term $g \cdot z\ I_y$). The soft pulse will rotate $I_y$ towards $-I_z$ within its bandwidth, and leave spins unaffected (i.e. pointed along $I_y$) outside its bandwidth. The nuclear spin moments outside the pulse bandwidth will remain oriented along $I_y$ as the rotating frame gradients induce nutations mostly about $I_y$. With the excited spins along $I_z$, spatial encoding (along $x$- and $y$-directions) is then performed with a rotating-frame gradient the field of which is oriented about $I_y$. The MRI signal is contained in the $X$ and $Z$ components of the magnetization. A readout of $M_x$ provides the in-phase NMR signal while permitting further subsequent encoding of this magnetization. To obtain the quadrature component, the $XZ$ plane can be rotated into the $XY$ plane, followed by a readout of the missing component. Alternatively, a second acquisition (phase cycle) with a $\pi/2$ phase shift in the excitation would provide the quadrature component.

There are two regimes of interest when considering such a selective excitation: (i) when the rotating-frame frequency is rapid compared to variations in the RF pulse envelope; and (ii) when the rotating frame is slow (low field limit). In the first case, the RF pulse is perceived as being a constant field (in the rotating frame) with a time interval $\Delta t$, during which the $I_z$ term oscillates back and forth possibly many times. The oscillating $I_z$ term will then experience a significant amount of self-averaging. In the second case, the field from the $I_z$ term appears stationary on the time scale of RF field envelope fluctuations. In this case, multipulse techniques can be used to eliminate the $I_z$ term. In both cases, the result is justified by average Hamiltonian theory [11]. We now take a moment to explain how the averaging principle works.

## 6.2.4
### Coherent Averaging in Composite Selective Pulses

Coherent averaging theory [11] is a useful tool for designing new RF pulses based on intuitive principles of crafting an effective Hamiltonian. Here, we consider composite pulses made of a soft pulse mixed with a rapid train of hard pulses. This introduces two time scales: a slow scale (soft pulse), and a rapid scale (hard pulses). On the slow scale, we may average the rapid fluctuations and consider the effective field which results from this averaging. This enables us to eliminate unwanted terms in the Hamiltonian.

For example, if we want to eliminate the $I_y$ term and use the $I_x$ term for spatial encoding, two possible approaches can be considered. The first method eliminates $I_y$ by a fast train of 180° pulses applied along $I_x$. This transforms $I_z \rightarrow -I_z$ and $I_y \rightarrow -I_y$ and the $I_y$ and $I_z$ terms of the effective Hamiltonian vanish in the limit of short inter-pulse spacings. The second method eliminates only the $I_y$ term and is required in the following situation: RF or dc field along $I_x$, slice-selective gradient field along $I_z$, and concomitant gradient along $I_y$. Consider the repeated sequence of four delta pulses:

$$\{2\tau - (\pi_x) - \tau - (\pi_y) - 2\tau - (\pi_{-y}) - \tau - (\pi_{-x}) - 2\tau\}_n. \tag{6.8}$$

The refocusing pulses are short, hard pulses, in between which the spins evolve under an arbitrary Hamiltonian of the form

$$\mathcal{H}' = a(t)I_x + b(t)I_y + c(t)I_z.$$

Over this period of duration $8\tau$, this pulse sequence produces a zeroth order average Hamiltonian

$$\langle \mathcal{H}' \rangle = \frac{1}{2}(\bar{a}I_x + \bar{c}I_z),$$

where $\bar{a}$ is the time average of $a$. In the Magnus expansion [11], the $I_y$ term is $\mathcal{O}(|\tau|^2)$ and the scaling factor [12] of $1/2$ for $I_x$ and $I_z$ causes a time dilation.

### 6.2.5
### Composite Selective Pulses

As an example, consider a band-selective uniform-response pure-phase (BURP)-selective pulse, such as in Ref. [13]. The envelope of the pulse is shown in Figure 6.2a. In order to obtain a 90° flip angle with $T_p = 6.28\,\mathrm{ms}$ on the proton resonance, this pulse requires a peak $B_1$ amplitude of approximately $2040\,\mathrm{rad\,s^{-1}}$. To produce a selective pulse which averages out the $I_z$ term to zero during its course, the soft pulse is modified by inserting the following coherent train of hard pulses

$$\{2\tau - (\pi_y) - \tau - (\pi_z) - 2\tau - (\pi_{-z}) - \tau - (\pi_{-y}) - 2\tau\}_n. \tag{6.9}$$

This subunit is repeated $n$ times, for the entire duration of the soft pulse, and its duration should be short compared to the time scale of fluctuations in the soft pulse envelope. It is shown later that, for this particular pulse, $\tau$ should be roughly three orders of magnitude less than the pulse duration; the exact figure ultimately depends on the shape of the soft pulse, with smoother shapes generally less demanding of the coherent train. The new soft/hard composite pulse will be slightly longer; its length increases by the total duration of the hard pulses added. The new pulse is sketched in Figure 6.2b. This procedure introduces two widely different time scales to the pulse, the fine structure of which naturally averages out.

(a)

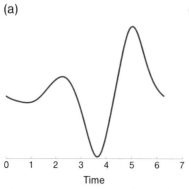

0  1  2  3  4  5  6  7
Time

(b) Basic subunit

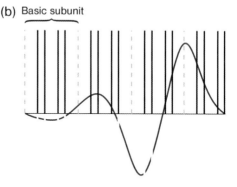

**Figure 6.2** (a) Amplitude modulation for a BURP pulse of duration $T_p = 6.28\,\mathrm{s}$; (b) A composite selective pulse for low-field imaging takes the soft pulse (a) and inserts a coherent train of hard pulses (such as Equation 6.9) for the entire duration of the soft pulse. The duration of the experimentally realized pulse is increased by the total length of all the hard pulses. For clarity, only a few subunits are sketched. In (b) the soft part of the pulse is shown in gray and the hard pulses are shown in black. Interruptions in the soft pulse are shown as cuts in the gray curve.

## 6.2.6
### Slice Selection in Low Fields

In static fields that are smaller than or comparable to the applied gradient field, slice selection appears to be impossible because the gradient approaches a quadrupole field (such as the previous example, $\mathbf{B} = y\hat{z} + z\hat{y}$) and no longer offers the possibility of mapping the frequency linearly in space along a single direction. The relevant dimensionless parameter is the ratio $\Delta B_{max}/B_0$, which should be as small as possible for undistorted slice/volume selection and Fourier encoding. $\Delta B_{max}$ is the maximum gradient field over the field of view (FOV) or sample volume, that is, the quantity

$$\max_{\mathbf{r} \in FOV} \|(\mathbf{r} - 0) \cdot \nabla \mathbf{B}\|$$

with the origin 0 fixed at the center of the FOV. In this section, it is shown how pulses can be constructed to perform slice/volume selection under conditions of large ratios $\Delta B_{max}/B_0$.

The first case to be examined is that of $\Delta B_{max}/B_0 \gtrsim 1.6$, where conventional MRI slice selection schemes are incapable of producing slices without a severe amount of distortion. Whether we look at the excitation profile at ($y = 0$ cm) or away ($|y| = 10$ cm) from the origin for a slice selection along $X$ makes no difference, as seen in Figure 6.3a and d, in the sense that the slice profile is heavily distorted and there is a significant amount of excitation occurring outside the intended slice. Figure 6.3b and c illustrate slice selection on-axis ($y = 0$ cm) along $X$ or $Y$ using a type II rotating-frame gradient with $\epsilon = 1.0$. While the performance is slightly degraded when going off-axis ($|y| = 10$ cm), as seen in Figure 6.3e and f, the degradation is far less important than the conventional case of Figure 6.3a and d, and such distortions are only significant near the edges ($|y| > 8$ cm) of the volume.

The effects of concomitant gradient fields on slice selection were first analyzed by Gao et al. [14]. Yablonskiy et al. [9] reported a slice curvature effect when the surfaces of constant field are excited by the RF pulse. In the case of an applied $X$ gradient of strength $G_x$ from a Golay pair [15], these become cylindrical surfaces

$$(x - x_c)^2 + z^2 = R_x^2$$

of radius

$$R_x = |x + B_0/G_x|$$

and $x_c = -B_0/G_x$ rather than a plane at $x = x_0$, where

$$x_0 = (B - B_0)/G_x,$$

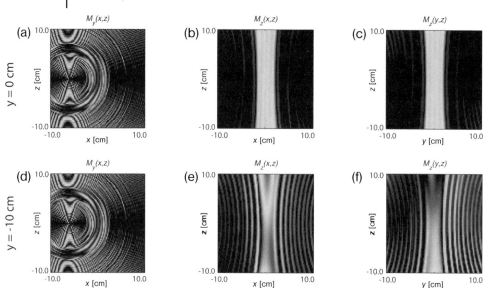

**Figure 6.3** Comparison of static and type II rotating-frame gradients for slice selection in the case $\Delta B_{max}/B_0 \sim 1.6$. The intended slice is centered at the origin $\mathbf{r} = 0$. Plots (a,d) are X slice-selection profiles for a saddle pair gradient along X for XY planes at $y = 0$ cm and $y = -10$ cm, respectively. This is for a conventional MRI gradient and no correction is applied. (b,e) are X slice selections using type II gradients employing a stationary rotating-frame X gradient component for the nutation. (c,f) is the Y slice selection for a type II gradient with stationary Y gradient field. Parameters were: $\tau = 15.75\,\mu s$, length of subunit was 126 $\mu s$, $T_p = 10$ ms, 79 subunits per soft pulse, $g = 1605\,\mathrm{rad\,s^{-1}\,cm^{-1}}$, $B_0 = 10080\,\mathrm{rad\,s^{-1}}$ (one Larmor precession period is 623 $\mu s$; there are 16 such cycles across this pulse's duration). All hard pulses are δ-pulses. (See Ref. [10] for the color map.)

where B is the center frequency of the RF pulse. For applied Z gradients, the surfaces of constant frequency are ellipsoids of revolution [9]. It is generally difficult to predict analytically the effects of nonsecular concomitant components in the rotating frame. These components lead to excitation of spins outside the slice region, as seen in Figures 6.3a,d and 6.4a,b.

The case of conventional slice selection with an applied Z gradient from a Maxwell coil [15] ($\Delta B_{max}/B_0 \sim 1.6$) is shown in Figure 6.4a and b. The performance is slightly better than a Golay pair generating an orthogonal slice on axis (compare Figures 6.3a and 6.4b). Unfortunately, the slice profile suffers from a heavy elliptical curvature and strong contamination originating from outside the intended volume. At any rate, it is clear that conventional MRI gradient encoding performs poor slice selection when $\Delta B_{max}/B_0 \sim 1.6$. In contrast, a type I rotating-frame gradient provides clean slice selection both on- and off-axis, as seen in Figure 6.4c and d. This type I gradient performs equally well in the asymptotic regime $\Delta B_{max}/B_0 \gtrsim 25$ (data not shown).

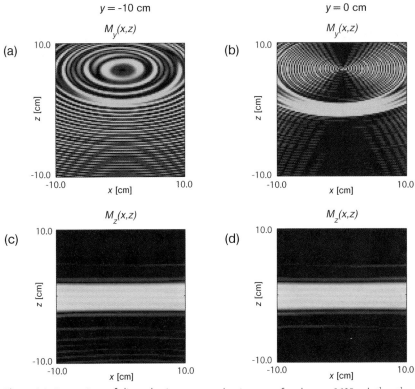

**Figure 6.4** Comparison of slice selection along $Z$ for a conventional Maxwell coil and a type I rotating-frame gradient. Plots (a,b) are for the static Maxwell coil; plots (c,d) are for a type I gradient. $XZ$ planes are aligned according to: (a,c) 10 cm off-center and (b,d) on-axis. Parameters were: $\tau = 15.75\,\mu s$, subunit duration is $126\,\mu s$, $T_p = 10\,ms$, 79 subunits per soft pulse, $g = 1605\,rad\,s^{-1}\,cm^{-1}$, $B_0 = 10\,080\,rad\,s^{-1}$ (one Larmor period lasts $623\,\mu s$; $T_p$ contains 16 such cycles) so that $\Delta B_{max}/B_0 \sim 1.6$ and FOV = 20 cm. Near-identical performance is obtained at $\Delta B_{max}/B_0 \sim 25$. All hard pulses are $\delta$-pulses. (See Ref. [10] for the color map.)

## 6.2.7
### Slice Selection in Zero Fields

Slice selection in zero field[6] is a relatively simple matter as there are no time-dependent gradient components. The role of RF pulses is played by dc pulses [16], while signal detection can be achieved by sample shuttling into high field [16] or by using a magnetometer to detect flux. An applied gradient in zero field is a pure quadrupole field. For volume or slice selection, it can be viewed as the limit $\Delta B_{max}/B_0 \to \infty$ of the previous selection schemes, but with the important difference that only a single gradient field is required for slice selection because excitation

---

6) This really means zero field and not the Earth's field. This requires compensating the residual field.

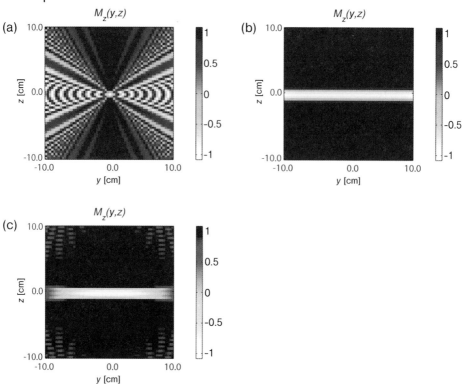

**Figure 6.5** Slice selection in zero field. (a) A soft pulse of duration $T_p = 10$ ms in zero field is unable to excite a slice due to the presence of concomitant gradient fields. (b, c) A soft pulse combined with a train of hard pulses to eliminate the concomitant component for an interpulse spacing of (b) $\tau = 1$ μs and (c) $\tau = 20$ μs. A longer $\tau$ corresponds to composite pulse trains that contain fewer hard pulses, for a fixed soft pulse duration. All hard pulses are δ-pulses. (See Ref. [10] for the color map.)

with a quadrature RF field is not needed.[7] Moreover, a two-component gradient field should be used, rather than three-components, because the third component no longer self-averages.

A field such as $\mathbf{B} = a(z\hat{\mathbf{x}} + x\hat{\mathbf{z}})$ can be used to provide slice selection along $X$ or $Z$ in the following manner. For a $Z$ slice, a selective pulse field is applied along $I_y$ and the $I_z$ component is eliminated using the pulse train of Equation 6.9. For an $X$ slice, a selective pulse is applied along $I_x$ with a coherent train of hard pulses to eliminate the $I_x$ term. Likewise, a field $\mathbf{B} = a(z\hat{\mathbf{y}} + y\hat{\mathbf{z}})$ can provide $Y$ and $Z$ slice selection. The $Y$ slice is obtained, for example, by applying the selective pulse along $I_x$ and eliminating the $I_y$ term using the coherent train of Equation 6.8.

In Figure 6.5a it is clear that conventional soft pulses in zero field with concomitant gradient fields are incapable of slice selection. The magnetization plots in

---

7) 'Quadrature' here refers to two coils driven by sinusoidal currents that are 90° out-of-phase.

Figure 6.5b and c show that good slice selection can be achieved for inter-pulse spacings $\tau$ less than 20 µs. At $\tau = 20$ µs and beyond, the slice profile begins to break down, as excited magnetization outside the slice of interest begins to contaminate the signal.

Thus, a soft pulse can be altered by using a train of hard, composite pulses to eliminate the effects of concomitant fields during the slice-selection process of an MRI experiment in zero field. The calculations suggest that, for good performance, the inter-pulse spacing $\tau$ should be three orders of magnitude shorter than the soft pulse duration, $T_p$. Ultimately, this depends on the pulse shape, with flatter soft pulse shapes requiring less frequent refocusing.

## 6.3
## The Inhomogeneous Field Regime

### 6.3.1
### Slice Selection

We now explore the possibility of executing slice-selective pulses in inhomogeneous fields. The fundamental problem is the presence of an additional 'permanent' static field[8] $\delta B_P$ to the applied slice-select gradient $\mathbf{G}_S$:

$$U_{pulse} = \vec{T} \exp\left\{-i \int_0^{T_p} [B_1(t) I_x + (\mathbf{G}_S \cdot \mathbf{r} + \delta B_P) I_z] dt\right\}. \tag{6.10}$$

where $B_1(t)$ is the RF waveform, $\vec{T}$ indicates time-ordering of the exponential and $T_p$ is the pulse duration. It is the term $\delta B_P(\mathbf{r})$ that leads to distortions in the slice profile. Slice selection could be achieved if the coefficient of $I_z$ were linear in $\mathbf{r}$. We also note that a necessary condition for slice selection is that the field be monotonically[9] increasing or decreasing [17]. This condition is satisfied if the gradient $\mathbf{G}_S \cdot \mathbf{r}$ is large enough to overwhelm the inhomogeneities $\delta B_P$. This is an important problem because, in mobile NMR systems, the inhomogeneities $\delta B_P$ are typically much larger than the available gradients.

A possible – but less than satisfactory – approach would be to acquire a 3-D imaging data set, followed by a reconstruction that includes geometric distortion corrections and the extraction of undistorted slices. True slice selection for 2-D imaging in an otherwise inhomogeneous field, however, has not been achieved. Such pulses would have advantages for fast imaging in situations where the acquisition of a complete 3-D scan would take too long. A solution to this problem lies in the fact that the applied gradient $\mathbf{G}_S$ can be made time-dependent, whereas $\delta B_P$ cannot. We may therefore capitalize on the well-known idea that a spin echo

8) To simplify the notation, the spatial dependence of the field is not written explicitly as $\delta B_P = \delta B_P(\mathbf{r})$.

9) The author is presently unaware of slice selection schemes for use in nonmonotonic fields.

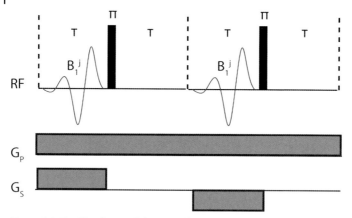

**Figure 6.6** The *j*th subunit of slice- or volume-selective pulse for use in inhomogeneous fields, in the presence of an unknown gradient $G_p$ which would otherwise distort the slice selection process. $G_p$ indicates a permanent gradient, or more generally, this corresponds to an arbitrary local field offset $\delta B_p(\mathbf{r})$. $G_s$ is the slice-select gradient. The flip angle of each soft pulse shown in this *j*th subunit is typically on the order of the total flip angle of the entire pulse divided by the number of segments.

refocuses only the time-symmetric component of a gradient. The gradient asymmetry survives the averaging and provides the necessary slice selection.

Figure 6.6 is an example of such a pulse train, where only the basic (*j*th) subunit is shown. As before, a soft (selective) RF pulse is subdivided into $N$ subunits, each with a segment waveform $B_1^j$. Each subunit applies a small flip angle, and the sum of all subunits leads to a larger cumulative flip angle. This RF pulse achieves undistorted slice selection in the presence of inhomogeneous fields. The rotation operator corresponding to the *j*th subunit is:

$$U_j = e^{-iI_z\delta B_P\tau}e^{-i\pi I_x}e^{-iI_x B_1^j\tau - iI_z(\delta B_P - G_S\cdot\mathbf{r})\tau}e^{-iI_z\delta B_P\tau}\,e^{-i\pi I_x}e^{-iI_x B_1^j\tau - iI_z(\delta B_P + G_S\cdot\mathbf{r})\tau}$$

$$= e^{-iI_z\delta B_P\tau}e^{-iI_x B_1^j\tau + iI_z(\delta B_P - G_S\cdot\mathbf{r})\tau}e^{iI_z\delta B_P\tau}e^{-iI_x B_1^j\tau - iI_z(\delta B_P + G_S\cdot\mathbf{r})\tau}$$

$$(6.11)$$

In the limit of short $\tau$, this rotation operator approximates to:[10]

$$U_j \sim e^{-i2\tau\left(I_x B_1^j + I_z G_S\cdot\mathbf{r}\right)}.$$

$$(6.12)$$

For this approximation to be valid, the rate of the pulse train $1/\tau$ must exceed all three amplitude parameters $\delta B_P$, $B_1^j$ and $G_S\cdot\mathbf{r}$. In practice, this means that if $\tau = 50\,\mu s$, inhomogeneity effects $\delta B_P$ of no more than $20\,kHz$ can be corrected by this pulse train. Synthesis of the entire pulse by time-ordering of all the subunits yields an undistorted slice selection pulse:

---

10) Short here means that the product $\omega\tau$ is less than 1, where $\omega$ is the strength of any nonaveraged magnetic field components in the rotating frame.

$$U_{pulse} = \vec{T} \exp\left\{-i \int_0^{T_p} [B_1(t)I_x + \mathbf{G}_S \cdot \mathbf{r}I_z] dt\right\}. \tag{6.13}$$

We now check that the time scales involved are accessible experimentally. Let us assume a pulse length $T_p$ of 10 ms and peak gradient amplitude of $2.5\,\mathrm{G\,cm^{-1}}$. A slew rate realized experimentally in our laboratory using flat gradient coils designed for a portable MRI sensor is $50\,\mathrm{G\,cm^{-1}\,ms^{-1}}$. For a trapezoid gradient waveform, the risetime to $2.5\,\mathrm{G\,cm^{-1}}$ is 50 μs. In the absence of inhomogeneities in $B_0$, a 1 cm slice thickness to be excited in a $2.5\,\mathrm{G\,cm^{-1}}$ gradient would see a 10 kHz spread in its resonance frequencies across the thickness dimension. This means that the RF pulse should have a bandwidth of 10 kHz. The reader may recognize the above pulse waveform to be similar to a spectral-spatial pulse, the bandwidth of which is bounded from above by the inverse of the short pulse width, and from below by the inverse of the total pulse width. We consider the most challenging scenario where the small pulse segment determines the pulse bandwidth. In this case, the small pulse segment should be less than 100 μs (= 1/10 kHz); this yields a minimum τ value of 200 μs. Thus, in 10 ms the selective pulse may contain at most 12 such subunits, each of which is 800 μs long. In practice, the pulse bandwidth will be less than the small pulse segment's bandwidth (the actual bandwidth may be calculated by simulating the pulse) and more subunits can be applied. The condition that $1/\tau = 5\,\mathrm{kHz}$ exceeds $B_1^j$ is met, for instance, by dividing a 90° excitation pulse into 12 subunits (each of 7.5° flip). Each small RF pulse segment will have a peak RF amplitude on the order of 210 Hz, which is much less than $1/\tau$.

This scheme proves that undistorted slice selection in inhomogeneous fields is possible, at least in theory. It should be noted here that coherent averaging is used to present the basic ideas in their simplest form in order to illustrate general principles. A foreseeable complication of this pulse – as is also the case with spectra-spatial pulses – is the appearance of undesired sidebands outside the slice bandwidth. There could also be distortions outside the slice if the term $\mathbf{G}_S \cdot \mathbf{r}$ is too large compared to $1/\tau$. However, these effects could be mitigated using more sophisticated approaches to pulse design such as optimal control.

## 6.3.2
### Restoring Spectroscopic Resolution

The method of shim pulses was introduced by Meriles and coworkers [6] and Topgaard *et al.* [7]. In its first incarnation, an RF field was used for which the spatial dependence was correlated to that of the static field. In this manner, periodic trains of corrective pulses could be applied to correct the spin evolution between the acquisition of consecutive points in a free induction decay (FID) [6]. Experimental demonstrations of the method achieved modest (2 kHz) linear field inhomogeneity corrections across the sample, most likely owing to the fact that a single RF coil was used to construct composite $z$ rotations. This $z$ rotation requires a homogeneous $B_1$ field to be used in conjunction with the inhomogeneous pulse, which is not possible if only one RF coil is used throughout. Moreover, while a linear gradient field correction was demonstrated, different types of correction would require special RF coil designs tailored to the needs of each experiment. A generalization of this method was recently published [18] in which a transmitter array was used to synthesize a high-fidelity, matched RF field that was rapidly switched between homogeneous and inhomogeneous patterns during the same $z$ rotation. With this approach, an order of magnitude performance improvement was easily obtained.

In the second approach [7], an adiabatic double-passage RF pulse is applied during a time-modulated gradient waveform, the latter of which is optimized in such a way as to impart a desired phase-correction to different regions in space. The corrections thus obtained were shown to correct for nonlinearities in the static and imaging gradient fields. This remains an active area of research with ongoing development [19–21]. These novel ideas and future improvements are likely to play an important role for miniaturized and portable NMR applications. Shim pulses could also be of great interest for *in vivo* MR spectroscopy near the tissue–air susceptibility interfaces that severely limit the maximum achievable spectral resolution. One aspect which remains unexplored is the possibility of imparting phase corrections that are strongly localized in space, which would be relevant in cases where hardware-based shimming is unable to correct for inhomogeneities over small spatial extents. To that effect, it is encouraging to know that RF pulses can be encoded into gradient-modulated k-space trajectories for the purpose of exciting arbitrary spatial patterns (see, for example, Pauly *et al.* [22]).

## 6.4
## Conclusions

In this chapter some of the fundamental limits to NMR and MRI in low and/or inhomogeneous magnetic fields have been outlined. Overcoming these limits will offer new possibilities for NMR and MRI using portable and low-cost hardware. In addition, several new approaches to undistorted Fourier encoding, slice and volume selection in zero and low magnetic fields, and selectivity in inhomogeneous magnetic fields, have been presented. The pulses are derived from intuitive principles, and make use of coherent averaging to provide control over the rotations on a fine scale. This provides motivation for designing new classes of pulses that perform well in less-than-ideal conditions.

## Acknowledgments

These studies were supported by the Director, Office of Science, Office of Basic Energy Sciences, Materials Sciences and Engineering Division, of the US Department of Energy under Contract DE-AC03-76SF00098. I acknowledge several stimulating discussions with and generous support from Alex Pines, as well as contributions from M. Sabieh Anwar (School of Sciences and Engineering, LUMS,

Pakistan) regarding investigations on field arrays and geometric phases. Grateful thanks are also given to my colleagues Sabieh Anwar, Scott Burt, Dominic Graziani, David Michalak and Jeffrey Paulsen, and to my wife Rennie Tang for careful proofreading of the manuscript and useful suggestions. Discussions with John Franck and Vasiliki Demas are also acknowledged.

## References

1 Appelt, S., Kühn, H., Häsing, F.W. and Blümich, B. (2006) Chemical analysis by ultrahigh-resolution nuclear magnetic resonance in the earth's magnetic field. *Nature Physics*, **2**, 105–9.

2 Appelt, S., Häsing, F.W., Kühn, H. and Blümich, B. (2007) Phenomena in J-coupled nuclear magnetic resonance spectroscopy in low magnetic fields. *Physical Review A*, **76**, 023420.

3 Appelt, S., Häsing, F.W., Kühn, H., Sieling, U. and Blümich, B. (2007) Analysis of molecular structures by homo- and hetero-nuclear J-coupled NMR in ultra-low field. *Chemical Physics Letters*, **440**, 308–12.

4 Robinson, J.N., Coy, A., Dykstra, R., Eccles, C.D., Hunter, M.W. and Callaghan, P.T. (2006) Two-dimensional NMR spectroscopy in Earth's magnetic field. *Journal of Magnetic Resonance*, **182**, 343–7.

5 Eidmann, G., Savelsberg, R., Blümler, P. and Blümich, B. (1996) The NMR mouse, a mobile universal surface explorer. *Journal of Magnetic Resonance Series A*, **122**, 104–9.

6 Meriles, C.A., Sakellariou, D., Heise, H., Moulé, A.J. and Pines, A. (2002) Approach to high-resolution ex situ NMR spectroscopy. *Science*, **293**, 82–5.

7 Topgaard, D., Martin, R.W., Sakellariou, D., Meriles, C.A. and Pines, A. (2004) Shim pulses for NMR spectroscopy and imaging. *Proceedings of the National Academy of Sciences of the United States of America*, **101**, 17576–81.

8 Sakellariou, D., Meriles, C.A. and Pines, A. (2004) Advances in ex situ NMR. *Comptes Rendus Physique*, **5**, 337.

9 Yablonskiy, D.A., Sustanskii, A.L. and Ackerman, J.J. (2005) Image artifacts in very low magnetic field MRI: the role of concomitant gradients. *Journal of Magnetic Resonance*, **174**, 279–86.

10 Bouchard, L.-S. (2006) Unidirectional magnetic-field gradients and geometric phase errors during Fourier encoding using orthogonal ac fields. *Physical Review B*, **74**, 054103.

11 Haeberlen, U. and Waugh, J.S. (1968) Coherent averaging effects in magnetic resonance. *Physical Review*, **175**, 453–67.

12 Llor, A., Olejniczak, Z., Sachleben, J. and Pines, A. (1991) Scaling and time reversal of spin couplings in zero-field NMR. *Physical Review Letters*, **67**, 1989–92.

13 Geen, H., Wimperis, S. and Freeman, R. (1989) Band-selective pulses without phase distortion. a simulated annealing approach. *Journal of Magnetic Resonance*, **85**, 620–7.

14 Gao, J.-H., Anderson, A.W. and Gore, J.C. (1992) Effects on selective excitation and phase uniformity of concomitant field gradients components. *Physics in Medicine and Biology*, **37**, 1705–15.

15 Turner, R. (1993) Gradient coil design: a review of methods. *Magnetic Resonance Imaging*, **11**, 903–20.

16 Emsley, L., Laws, D.D. and Pines, A. (1999) Lectures on pulsed NMR. 3rd edn (eds S. I. du Fisica, V. Moastero and B. Maraviglia), Proceedings of the International School of Physics, Enrico Fermi, Course CXXXIX.

17 Epstein, C.L. (2004) Magnetic resonance imaging in inhomogeneous fields. *Inverse Problems*, **20**, 753–80.

18 Bouchard, L.-S. and Anwar, M.S. (2007) Synthesis of matched magnetic fields for controlled spin precession. *Physical Review B*, **76**, 014430.

19 Pryor, B. and Khaneja, N. (2006) Fourier decompositions and pulse sequence design

algorithms for nuclear magnetic
resonance in inhomogeneous fields.
*Journal of Chemical Physics*, **125**, 194111.

**20** Bouchard, L.-S. (2007) RF shimming
pulses for ex-situ NMR spectroscopy and
imaging using B1 inhomogeneities.
arxiv: physics. chem-ph, 0706.3528v1
(accessed 10 September).

**21** Magland, J. and Epstein, C.L. (2006) A
novel technique for imaging with
inhomogeneous fields. *Journal of Magnetic
Resonance*, **183**, 183–92.

**22** Pauly, J., Nishimura, D. and Macovski, A.
(1989) A k-space analysis of small-tip-angle
excitation. *Journal of Magnetic Resonance*,
**81**, 43–6.

**Part Two   Polarization Enhancement**

# 7
# Parahydrogen-Induced Polarization in Heterogeneous Catalytic Hydrogenations

*Kirill V. Kovtunov and Igor V. Koptyug*

## 7.1
### Introduction

Both, NMR and MRI techniques often suffer from low intrinsic sensitivity, and consequently significant efforts are devoted to the development of various nuclear spin polarization schemes. One of the approaches is based on the parahydrogen-induced polarization (PHIP) produced in hydrogenation reactions [1, 2]. The PHIP phenomenon has long been recognized as a powerful tool for studying homogeneous hydrogenation processes in solution [3–8]. Recently, a major step towards a new objective in the use of PHIP has been made [9], wherein a homogeneously catalyzed hydrogenation with parahydrogen was followed by the transfer of polarization to $^{13}$C nuclei in the product molecule, and the solution was then injected into the tail vein of a rat. A $^{13}$C MRI experiment was then performed which yielded a $^{13}$C angiographic image of the blood vessels in an imaging time of less than 1 s. The procedures used were later significantly refined. In particular, in the original study the dissolved catalyst was injected into the bloodstream along with the polarized product. This is clearly unacceptable for potential medical MRI applications, and therefore removal of the catalyst using a cation-exchange filter was implemented at the preinjection stage [10]. With the use of the improved technology, the dynamic imaging of blood circulation and the perfusion of various organs has been addressed [10, 11]. At present, efforts are being directed towards further challenges such as metabolic $^{13}$C imaging [11–13].

Despite the impressive results achieved with the development and application of PHIP technology, there is room for significant further progress. In particular, we believe that much can be gained by combining PHIP with heterogeneous catalytic processes. This combination can be advantageous for developing novel ways of producing polarized gases as well as catalyst-free polarized liquids for MRI applications. In addition, PHIP is a unique and a highly sensitive tool which could be useful in the studies of mechanisms and kinetics of heterogeneous catalytic reactions, and of processes in operating reactors.

*Magnetic Resonance Microscopy.* Edited by Sarah L. Codd and Joseph D. Seymour
Copyright © 2009 WILEY-VCH Verlag GmbH & Co. KGaA, Weinheim
ISBN: 978-3-527-32008-0

## 7.2
## Background

Molecular hydrogen has two nuclear spin isomers: orthohydrogen (o-$H_2$) with the total nuclear spin of the two hydrogen atoms $I = 1$, and parahydrogen (p-$H_2$) with $I = 0$. At room temperature, the composition of the equilibrium mixture ('normal' $H_2$, n-$H_2$) comprises approximately three parts of o-$H_2$ and one part of p-$H_2$; that is, it is characterized by an almost statistical ortho:para ratio. However, mixtures with different compositions can be produced. In particular, parahydrogen enrichment is a straightforward procedure that can be achieved by cooling $H_2$ and storing or passing it over a suitable ortho–para conversion catalyst kept at a low temperature. For this purpose, FeO(OH), activated charcoal or certain other paramagnetic materials are often used. Conversion at 77 K is easy to perform and produces a mixture with an almost 1:1 ortho:para ratio. Conversion at liquid hydrogen temperature allows one to obtain almost pure p-$H_2$, but is more demanding technically. Liquid hydrogen, which is now becoming available commercially, is preconverted into p-$H_2$ upon liquefaction. At the same time, for most experiments the 1:1 o:p mixture is sufficient and provides an NMR signal amplification of several orders of magnitude. Although the use of pure p-$H_2$ adds only another factor of three to signal enhancement, pure spin states can be important in some applications such as quantum computing [14, 15]. Mixtures enriched in o-$H_2$ can be produced using low-temperature hydrogen adsorption on appropriate adsorbents [16]. The use of o-$H_2$ instead of p-$H_2$ should produce similar polarization patterns in the NMR spectra of reaction products, but of the opposite sign [17]. This fact can be employed to confirm that any polarization observed is due to p-$H_2$ and is not caused by other known polarization mechanisms. In the following discussions, 'parahydrogen' in fact implies any mixture with an o:p ratio which differs from 3:1.

Conventionally, the o:p ratio in a hydrogen mixture is evaluated by thermal conductivity measurements at low temperature. We have developed an NMR-based procedure for this purpose. Since p-$H_2$ possesses no spin, only o-$H_2$ gives an observable NMR signal. Under normal conditions, the latter is characterized by very short relaxation times ($T_2 = 0.15$ ms, $T_1 = 0.3$ ms) [18] and is therefore not easy to detect. In order to increase the relaxation times of hydrogen without increasing the gas pressure above 1 bar, it can be admitted into a cell filled with a porous material, such as dehydrated $Al_2O_3$ pellets. After subtraction of the residual background $^1H$ NMR signal, the measured signal intensity is proportional to the amount of o-$H_2$ in the cell. The same measurement, when performed with 1 bar of n-$H_2$, provides the necessary calibration. To measure the residual background signal, the cell can be purged with a hydrogen-free gas such as $N_2$ or Ar (air should not be used for safety reasons).

As ortho–para conversion is very slow in the absence of the conversion catalyst, when p-$H_2$ is produced it can be kept at room temperature (RT) for days or even months, without any substantial change in the o:p ratio. This is in fact an example of the long-lived spin states which are now being actively explored in molecules larger than $H_2$ [19–23]. Interestingly, o–p conversion can be slow at RT even in the presence of a paramagnetic material. We expected that when p-$H_2$ produced in our experiments is passed through the same amount of FeO(OH) kept at RT and at the same flow rate, it will be converted back to n-$H_2$. However, no back-conversion at RT was observed at all. The details of o–p conversion of $H_2$ have been discussed in detail elsewhere [24, 25].

An excess of one of the spin isomers in the mixture provides a nonzero pairwise correlation of nuclear spins in the ensemble of $H_2$ molecules. This correlation is not immediately observable by NMR as long as the nuclei of the $H_2$ molecule remain equivalent. The conventional way to eliminate this equivalence is to use molecular hydrogen in a homogeneous hydrogenation reaction, whereby a suitable substrate (reactant) with a double or a triple bond is hydrogenated in solution in the presence of a transition metal complex as the hydrogenation catalyst (e.g. Wilkinson's catalyst, $ClRh(PPh_3)_3$). In such homogeneous hydrogenations, the catalytic cycle shown in Scheme 7.1 is known to ensure a pairwise addition of both hydrogen atoms of a $H_2$ molecule, first to the catalyst and eventually to the substrate molecule [26]. This means that when parahydrogen is used in a homogeneous hydrogenation reaction, the original correlation of nuclear spins of the p-$H_2$ molecule will be preserved throughout the entire catalytic cycle, all the way to the product molecule. If the two H atoms then become nonequivalent in the product, and if the duration of the catalytic cycle is not too long compared to the nuclear spin relaxation times, the spin correlation is converted into nuclear spin alignment, which is easy to convert into a strong signal enhancement of the NMR lines in the spectrum of the reaction product. Furthermore, equivalence of the two H atoms is often broken early in the catalytic cycle upon the oxidative addition of an $H_2$ molecule to the catalyst to yield the intermediate metal dihydride complex (Scheme 7.1). This means that the signals of the intermediate dihydride, and also of the other intermediate species, will be amplified significantly in the NMR spectrum. As a result, PHIP has become a very sensitive spectroscopic technique that has significantly advanced studies of the mechanisms and kinetics of homogeneous catalytic hydrogenation processes in solution [3–7].

The polarization patterns observed in the $^1H$ NMR spectra depend on how the experiment is performed. If the hydrogenation is carried out in the probe of an NMR spectrometer (i.e. in a high magnetic field), the two strongly enhanced anti-phase multiplets are commonly observed in the $^1H$ NMR spectra of the reaction products. This type of experiment is often referred to as PASADENA (parahydrogen and synthesis allow dramatic enhancement of nuclear alignment) [1, 2, 27].

$$ML \xrightarrow{\text{H}_2} M(H)_2L \xrightarrow{\text{CH}_2=\text{CHR}} M(H)_2(\eta^2\text{-CH}_2=\text{CHR})L' \rightarrow$$
$$\longrightarrow M(H)(CH_2CH_2R)L' \longrightarrow ML + CH_3CH_2R$$

**Scheme 7.1** The catalytic cycle for a homogeneous hydrogenation reaction. M is the metal atom, L and L' are different combinations of ligands. For $ClRh(PPh_3)_3$ catalyst, M = Rh, L = (Cl)(PPh$_3$)$_3$, L' = (Cl)(PPh$_3$)$_2$.

At the same time, if the hydrogenation is carried out outside the NMR magnet (e.g. in the Earth's magnetic field) and the reaction products are then adiabatically transferred to the magnet for detection, the two multiplets show a net signal enhancement of the opposite sign, and the experimental scheme is termed ALTADENA (adiabatic longitudinal transport after dissociation engenders net alignment) [28]. For mechanistic studies of homogeneous hydrogenations, the PASADENA experiment is most widely used, especially if the short-lived reaction intermediates are to be detected. ALTADENA, in principle, can provide twice the signal enhancement of PASADENA, but this advantage can be lost due to spin relaxation of the reaction product while the sample is in transit from the low field to the NMR probe. In PASADENA experiments, the observed signal is largest for a $\pi/4$-pulse and zero for a $\pi/2$-pulse [8], whereas in ALTADENA the largest signal is observed for a $\pi/2$-pulse. The analysis of polarization patterns is often compli-cated by the fact that it is impossible to stop homogeneous hydrogenation instan-taneously. When the bubbling of a solution with p-H$_2$ is terminated, the concentration of dissolved hydrogen does not immediately drop to zero, and the reaction continues for a poorly defined period of time. One consequence of this is that, in ALTADENA-type experiments, a mixture of ALTADENA and PASA-DENA polarization patterns can be observed.

Parahydrogen-induced polarization can be transferred to heteronuclei, which significantly broadens the scope of potential applications. In certain cases, it is enough just to apply a single pulse to a heteronucleus to observe its polarization. In particular, polarization can be transferred spontaneously to a $^{13}$C nucleus if it forms an AA'X spin system with the two p-H$_2$ derived $^1$H nuclei in the reaction product [29–31]. This polarization transfer requires evolution of the coupled spin system, which can take place while the reaction is being carried out and/or while the sample is being transferred to the NMR probe. At the same time, this approach does not work for A$_2$X or A$_2$A'$_2$X spin systems [30]. It has been shown that the hydrogenation of substituted alkynes can lead to the polarization of virtually all carbon atoms in the corresponding alkenes in an ALTADENA experiment, includ-ing those which are remote from the original location of the multiple bond [32]. For an efficient polarization transfer to a $^{13}$C nucleus, $^{13}$C and $^1$H nuclei can be made strongly coupled by reducing the magnetic field nonadiabatically from the Earth's field (0.5 G) to a value of circa $10^{-3}$ G or lower, and subsequently returning the field back adiabatically [9, 10, 33, 34]. In recent studies, polarization transfer to $^{13}$C has been achieved in a low field (17.6 G, proton frequency 75 kHz) by apply-ing an appropriate polarization transfer pulse sequence to an AA'X spin system [34, 35]. Polarization transfer pulse sequences can be also used in the high field of the spectrometer for weakly coupled $^1$H nuclei to achieve the polarization of a heteronucleus in a PASADENA experiment [36–38]. Polarization transfer to nuclei other than $^{13}$C (e.g. $^2$H, $^{19}$F, $^{29}$Si, $^{31}$P) in product molecules and in dihydride metal complexes has been reported [36, 39–41]. In certain cases, cross-relaxation effects are apparently responsible for polarization transfer.

Until recently, PHIP was considered only in the context of homogeneous cata-lytic hydrogenation and some related processes (e.g. hydroformylation [42, 43]). At the same time, industrial catalytic processes including catalytic hydrogenations are usually heterogeneous [44]–that is, the catalyst represents a phase which is separate from the reactants and products. Quite often, heterogeneous catalysts are solid, while the reactants are supplied as liquids and/or gases. Heterogeneous hydrogenations usually use supported metal catalysts, for example, $Pt/Al_2O_3$ or $Pd/Al_2O_3$. The reaction mechanism is very different from that of the homogeneous hydrogenations discussed above, with the reaction proceeding on the surface of metal particles and involving the dissociative chemisorption of hydrogen. Hence, heterogeneous hydrogenation apparently cannot match the key requirement for the PHIP effects to be observed: the pairwise addition of the two H atoms of the same p-$H_2$ molecule to the same substrate molecule to yield the product. Thus, heterogeneous hydrogenations were not expected to produce any PHIP effects [45].

Despite that, the aim of these studies was to combine PHIP with heterogeneous catalysis, as the potential advantages are too significant to be ignored. The first advantage is that it is much easier to remove heterogeneous catalysts from the reaction products than the dissolved catalysts. In industrial processes, for instance, this fact by far outweighs the higher activity and selectivity of an homogeneous catalyst. This could lead to a better way of producing catalyst-free hyperpolarized liquids; moreover, it could be used to produce hyperpolarized gases – something which is hardly possible with homogeneous hydrogenations. A second advantage is that the observation of PHIP effects in heterogeneous catalytic reactions could be used as a highly sensitive tool for studying these reactions, in analogy to the PHIP studies of homogeneous hydrogenations. Clearly, if successful, this combination would benefit both catalytic research and NMR/MRI applications.

## 7.3
## PHIP Using Immobilized Transition Metal Complexes

The advantages and disadvantages of homogeneous and heterogeneous catalysts are well known in modern catalysis. One of the existing trends is to bridge the gap between homogeneous and heterogeneous catalysis by combining the advantages of both [46]. One of the possibilities presently being pursued is the heterogenization or immobilization of homogeneous catalysts on a suitable support [46–48]. This approach is also widely applied to produce novel hydrogenation catalysts, and in particular is based upon the immobilization of transition metal complexes (homogeneous hydrogenation catalysts) on various supports. This was the first possibility to be explored in these investigations.

In order for a catalyst to be suitable for PHIP experiments, the catalytic cycle should ensure a pairwise addition of the two H atoms, and the nuclear spin relaxation should not be too fast. Metal complexes are expected to retain their reaction mechanism upon immobilization (cf. Scheme 7.1). Therefore, the key issue with immobilized catalysts is not the reaction mechanism but rather the relaxation of nuclear spins. Although parahydrogen is largely immune to relaxation processes, this immunity is lost when it reaches the reaction center and enters the catalytic cycle. 'Immobilization' does not necessarily imply that a heterogenized complex becomes absolutely immobile, as it can be attached to a support in a variety of ways, for instance with a long flexible tether. Nevertheless, the mobility of the reaction intermediates involved (dihydride complex, etc.) can be significantly reduced if they are attached to a solid support, which should significantly affect their relaxation times. Furthermore, the lifetime of reaction intermediates could be altered by modifications of the complex necessary to attach it to a surface, which can further enhance polarization losses if the lifetime of the intermediates becomes longer. Another potential complication is the enhanced relaxation of the product due to its interaction with the solid support. At the same time, PHIP has been successfully observed in hydrogen chemisorption on a surface of solid ZnO [49], indicating that $T_1$ in solids can be long enough to allow the observation of PHIP effects. Therefore, given the variety of the possible immobilization strategies [46] and the multitude of supports with different properties, this avenue appeared to be worth exploring.

The feasibility of this approach was demonstrated recently [50, 51], when rhodium complexes often used as homogeneous hydrogenation catalysts were used after immobilization on a polymer or silica gel. Suspensions of these hetero-genized catalysts in benzene solutions of styrene yielded the polarized hydrogenation product ethylbenzene upon bubbling with parahydrogen; this represented the first example of PHIP in a heterogeneous hydrogenation process (Figure 7.1). Furthermore, unlike their homogeneous analogues, these catalysts can also be used to produce polarized gases, as shown by the heterogeneous hydrogenation of propylene into propane (Figure 7.2).

One important issue with immobilized catalysts is the leaching of a catalyst from its support into solution. Indeed, at present this is one of the key problems with using immobilized catalysts in various industrial applications. To demonstrate that PHIP was produced in a heterogeneous reaction, hydrogenation experiments were repeated with the same solutions after removing the solid fraction [50]. In the case of the polymer-immobilized catalyst, the solutions exhibited some weak residual activity, but for the two catalysts immobilized on silica gel no product or polarization were detected. These experiments prove that the polarization observed (Figure 7.1) is produced exclusively in a heterogeneous reaction. Nevertheless, the issue of catalyst leaching remains important, particularly if potential biomedical applications are to be addressed, and this problem will have to be addressed in more detail in the future. At the same time, clear-cut proof that PHIP can be observed in heterogeneous catalytic processes is provided by the experiments on the gas-phase hydrogenation of propylene (Figure 7.2), as leaching of the catalyst is impossible in the absence of a liquid phase.

Apart from being the first example of PHIP in heterogeneous hydrogenation reactions, the results obtained can serve as a verification of the pairwise mechanism for hydrogenations catalyzed by immobilized metal complexes. While this is in agreement with expectations, the observation of PHIP effects is the direct proof of the reaction mechanism. At the same time, these results have raised many

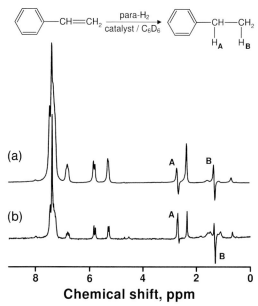

**Figure 7.1** PASADENA (a) and ALTADENA (b) polarization patterns observed in the hydrogenation of styrene in $C_6D_6$ using rhodium complex immobilized on silica gel, $RhCl(PPh_3)_2PPh_2(CH_2)_2$-$SiO_2$, as the catalyst. The two polarized hydrogen atoms in the product which originate from the p-$H_2$ molecule are labeled A and B. All experiments used a 1:1 o:p hydrogen mixture.

further questions. In particular, an important issue to address in the future is the effect of an immobilization strategy on the PHIP observed, as this significantly affects the mobility of the reaction intermediates and thus governs polarization losses via relaxation processes. Another point to be addressed is the optimization of the properties of a solid support, including its morphology, size and shape.

## 7.4
## PHIP Using Supported Metal Catalysts

In contrast to the immobilized catalysts described in Section 7.3, in the case of supported metal catalysts the primary issue in the context of PHIP is that of the reaction mechanism. When a hydrogenation reaction is carried out on the surface of a supported metal cluster (e.g. Pt or Pd dispersed on $Al_2O_3$) [52–55], $H_2$ is known to dissociate rapidly upon chemisorption on the metal surface, producing very mobile surface-bound H atoms. These atoms can then migrate rapidly over the metal surface, spill over to the support, and even dissolve into the metal lattice. As a result, it can be expected that the two H atoms rapidly lose each other, and

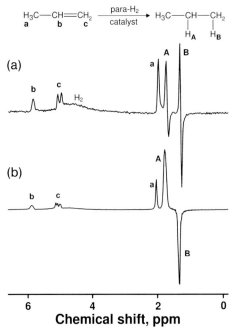

**Figure 7.2** PASADENA (a) and ALTADENA (b) polarization patterns observed in the hydrogenation of propylene gas using the immobilized rhodium complex, RhCl(PPh$_3$)$_2$PPh$_2$(CH$_2$)$_2$-SiO$_2$, as the catalyst. The two polarized hydrogens in the product are labeled A and B; the residual NMR signals of propylene are labeled a–c. A broad signal labeled H$_2$ belongs to the unreacted o-H$_2$.

the hydrogenation reaction involves random H atoms either from the surface pool or those emerging to the surface from the metal bulk. Therefore, the reaction mechanism is apparently inconsistent with the pairwise addition requirement, and supported metal catalysts have never been considered in the context of PHIP experiments.

Although the heterogeneous hydrogenation of unsaturated compounds has been studied for decades, the reaction mechanism remains a controversial matter, even for simple substrates such as ethylene and ethyne (acetylene) [56]. Therefore, in light of what has been achieved with PHIP in homogeneous catalysis, it would be useful to have a PHIP-based tool for studying also heterogeneous reactions. At the same time, supported metal catalysts could represent a viable alternative to immobilized complexes in producing polarized fluids for various NMR/MRI applications. These two issues were the primary motivation for the studies described in the following sections. It appears that the possible contribution of a pairwise addition route for supported metal catalysts has never been addressed within the context of catalytic studies. Moreover, there is probably no other way to address this issue but to use spin isomers of H$_2$. At the same time, in homogeneous catalytic reactions involving hydrogen, the observation of PHIP effects is considered as direct evidence for the pairwise addition mechanism. The point of interest was, therefore, to determine whether pairwise addition was indeed an impossibility on supported metals.

The first experiments employed $Pd/Al_2O_3$ catalysts which had been used previously in the MRI studies of an operating catalytic reactor [57, 58]. The hydrogenation of propylene on these catalysts has demonstrated weak, but unmistakable, ALTADENA effects. Interestingly, although the original experiments were performed with catalysts doped with paramagnetic Mn ions [57, 58], polarization was observed successfully. In later experiments, Mn-free $Pd/Al_2O_3$ catalysts were used [59], but the polarization level was similar. In subsequent studies, a series of $Pt/Al_2O_3$ catalysts with four different metal particle sizes (0.6, 1.1, 3.5 and 8.5 nm) were used. The results demonstrate [59] that all four catalysts yield a pronounced polarization of propane NMR lines (both PASADENA and ALTADENA) when propylene is hydrogenated with $p\text{-}H_2$ over these catalysts. There is an indication that larger polarizations are observed for those catalysts with smaller metal particles; in particular, the catalyst with 0.6 nm Pt particles (Figure 7.3) exhibited the strongest polarization in both PASADENA and ALTADENA experiments when compared to the other three Pt catalysts, but displayed a somewhat lower activity in terms of product formation.

The results obtained seem to prove beyond doubt that a pairwise addition route is involved in hydrogenation over the supported metal catalysts, even though it is probably not the dominating process (see below). A tentative explanation of these results is based on the fact that metal surface during a catalytic process is often covered with a variety of species (coke or carbonaceous deposits, reactant molecules, reaction intermediates, side products). Therefore, the metal surface is partitioned into smaller areas of clean metal [60], and this can in turn force an $H_2$ molecule to be adsorbed and activated in the vicinity of a substrate molecule and to follow the pairwise addition route. Smaller metal particles should be easier to partition into such localized reaction sites, which is in a qualitative agreement with the experimentally observed trends [59]. Further studies are needed to quantify the contribution of the pairwise hydrogen addition route. Nonetheless, it is possible at this point to estimate the minimum contribution which would be required to provide the observed signal enhancement. The maximum signal enhancement that can be produced in a PASADENA experiment at 4.7 T, 300 K and for a $\pi/4$ detection pulse is $31240f$, where $f$ is the uncompensated fraction of $p\text{-}H_2$ in the hydrogen mixture used ($f = 1/3$ for 1:1 o:p ratio) [61]. The experimental signal enhancement was evaluated by comparing the $CH_2$ multiplets in the NMR spectra of propane (labeled A in Figure 7.3a) produced with the use of $n\text{-}H_2$ and $p\text{-}H_2$. The spectral resolution was not high enough for the individual components of the $CH_2$ multiplets to be resolved, and therefore the integral of the entire $CH_2$ multiplet in the case of $n\text{-}H_2$ was evaluated, while the positive half of the antiphase multiplet was integrated in the case of $p\text{-}H_2$. The expected pattern of line intensities for the $CH_2$ group coupled to six methyl protons in propane is (1,6,15,20,15,6,1) for the case of thermal equilibrium and (1,4,5,0,−5,−4,−1) for the PASADENA pattern.

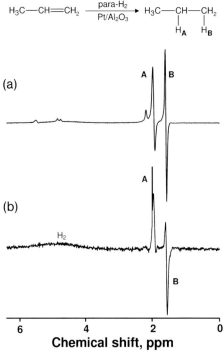

**Figure 7.3** PASADENA (a) and ALTADENA (b) polarization patterns observed in the hydrogenation of propylene gas using the supported metal catalyst, Pt/Al₂O₃, with 0.6 nm Pt clusters. The two polarized hydrogens in the product are labeled A and B. The number of NMR signal accumulations was eight for spectrum (a) and one for spectrum (b).

Therefore, the ratio of the corresponding integrals is 6.4. Taking into account that only one H atom of the two in the $CH_2$ group is derived from p-$H_2$, and that the experiments were performed at 7.05 T $(7.05/4.7 = 1.5)$, the maximum possible ratio of the integrals reduces to approximately $31\,240/(3 \times 6.4 \times 2 \times 1.5) = 542$. The ratio of the two integrals evaluated from the experimental results is approximately 15. Thus, the contribution of the pairwise addition route could be estimated at about 3%. These calculations do not take into account the relaxation-induced losses of polarization in the reaction intermediates and in the product interacting with the solid support. Therefore, the estimated 3% contribution is in fact a lower bound, while the actual value should be larger.

At this early stage of research, the results obtained raise more questions than they provide answers, and further systematic studies are required to understand the PHIP effects produced with supported metal catalysts, and to develop their practical applications. At the same time, the results show that this phenomenon will likely apply to many catalysts. In particular, other supported metals (Au, and possibly Rh) were also shown to produce PHIP effects. Although the contribution of the pairwise addition route observed in these studies is moderate, it may nevertheless be important for the theory and practice of catalysis, as it can measurably affect some of the important characteristics of the overall reaction, such as selectivity. Furthermore, these first demonstrations of PHIP in heterogeneous reactions catalyzed by supported metals are very important in the context of NMR/MRI applications of PHIP in general, and for producing polarized liquids and gases in particular. A few per cent contribution of the pairwise addition route observed in these initial experiments is in fact encouraging, since neither a wide search of the most efficient catalysts nor a systematic optimization of the existing ones have been performed so far.

## 7.5
## PHIP-Assisted Gas-Phase Imaging and Studies of Hydrogenation Selectivity

Some potential practical applications of PHIP produced in heterogeneous reactions can be demonstrated at this early stage of research. In particular, hyperpolarized propane was used to image void spaces in a tube with a cross-shaped partition or with a bundle of capillaries [51]. In both cases, detection of the NMR signal of a thermally polarized gas produced no meaningful images, while the use of p-$H_2$-induced polarization made it possible to observe the corresponding structures. Figure 7.4 shows the results obtained for the cross-shaped model object as an example; here, the observed signal enhancement of the order of 300 was sufficient

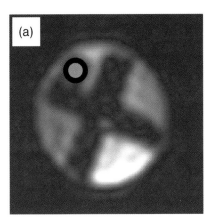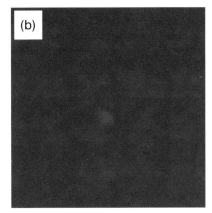

**Figure 7.4** Magnetic resonance images of a 10 mm NMR tube with a cross-shaped Teflon insert. The images were obtained using the NMR signals of propane polarized in an ALTADENA-type experiment (a) or thermally polarized propylene (b). The circle in (a) shows the position of a capillary which delivers the gas to the sample. Polarization was produced using the immobilized rhodium complex, $RhCl(PPh_3)_2PPh_2(CH_2)_2-SiO_2$, as the catalyst. The in-plane spatial resolution is $0.57 \times 0.57$ mm, the slice thickness is 10 mm. Adapted from Ref. [51].

to obtain MR images of gases with a resolution approaching that of liquid-phase MRI experiments.

In a more recent study, a microreactor loaded with catalyst was positioned inside the NMR probe and imaged while the reaction was in progress [62]. In this way it was possible to visualize those regions of the catalyst bed where the hydrogenation reaction was taking place. Furthermore, owing to the signal enhancement provided by p-H$_2$, it was possible for the first time to detect the velocity map for a hydrocarbon gas flowing in a porous medium. These experiments have revealed that neither the catalyst packing nor the flow field were spatially uniform. One potentially promising direction is to combine PHIP with the specific properties of the long-lived spin states [19–21, 23]. In particular, it was shown [62] that the application of an isotropic mixing pulse sequence can suppress relaxation of the polarization in the product molecule and thus allow the polarized molecule to escape from a tightly packed catalyst bed. This can be used not only to produce polarized fluids for the enhanced MRI studies of various objects but also to monitor further chemical transformations of the polarized molecules.

In terms of practical applications the hydrogenation of alkynes is very important, especially the selective hydrogenation of alkynes in the presence of alkenes. For an initial study, we have chosen propyne (methylacetylene). The asymmetric substitution is essential for PHIP observation [63] and also allows us to study the stereoselectivity of the hydrogenation reaction. In the case of propyne, although the *cis*- and *trans*-hydrogenations lead to the same product (propylene), they should result in different polarization patterns as the two CH$_2$ protons have different chemical shifts. The results are shown in Figure 7.5; here, the polarization of the

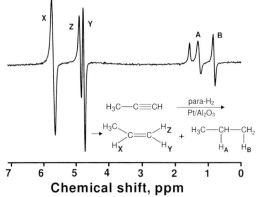

**Figure 7.5** PASADENA polarization patterns observed in the hydrogenation of propyne gas using the supported metal catalyst, Pt/Al$_2$O$_3$, with 0.6 nm Pt clusters. The polarized lines correspond to propylene generated in either cis- (X and Y) or trans- (X and Z) hydrogenation. Subsequent hydrogenation of propylene to propane is also observed (signals labeled A and B).

CH$_2$ signal for the product of trans-hydrogenation (labeled H$_Z$) is somewhat weaker than that for the cis-hydrogenation (H$_Y$). However, this cannot be taken as an immediate indication that cis hydrogenation prevails, since nuclear spin relaxation and cross-relaxation effects must also be taken into account. At the same time, it is clear that both cis- and trans-hydrogenations take place to a measurable extent. Whilst heterogeneous hydrogenation catalysts are usually not stereoselective in hydrogenation processes, it should be stressed that in PHIP experiments only those products generated via the pairwise hydrogen addition route are observed. In contrast, metal complexes in solution usually catalyze cis-hydrogenation selectively, while trans-hydrogenation with binuclear metal complexes serves as an indication that the reaction proceeds on a catalytic center which involves more than one metal atom [64]. The results presented in Figure 7.5 also show that, under experimental conditions, some of the propylene produced is further hydrogenated into propane (H$_A$, H$_B$).

## 7.6
## Conclusions

Parahydrogen-induced polarization is becoming an important tool for signal enhancement in MRI studies, both in biomedical and in technical applications. The combination of PHIP with heterogeneous hydrogenation processes appears to be a promising route towards novel approaches in the production of hyperpolarized, catalyst-free liquids and gases. Furthermore, PHIP has the potential to become a useful tool for studying industrially important heterogeneous catalytic processes such as hydrogenation, hydroformylation and hydrodesulfurization. At present, neither immobilized metal complexes nor supported metal catalysts are optimal for producing hyperpolarized fluids: the former have the necessary reaction mechanism but give low conversion of reactants into products, while the latter exhibit an almost 100% conversion but with a low contribution of the pairwise hydrogen addition. Nevertheless, the successful demonstration of PHIP effects in heterogeneous hydrogenation processes leads us to believe that further studies in this field will identify heterogeneous catalysts with much higher polarization yields.

## Acknowledgments

These studies were partly supported by grants from CRDF (RUC1-2581-NO04), RFBR (07-03-12147 and 08-03-00661), SB RAS (integration grant 11) and RAS (5.1.1 and 5.2.3). I.V.K. thanks the Russian Science Support Foundation for financial support. We also thank our colleagues from UC Berkeley (A. Pines, S.R. Burt, L.-S. Bouchard), Texas A&M (Christian Hilty), UC Santa Barbara (Song-I Han), LUMS, Pakistan (M.S. Anwar) and BIC SB RAS, Novosibirsk, Russia (V.I. Bukhtiyarov, I.E. Beck) for their contributions to the studies discussed above.

**References**

1 Bowers, C.R. and Weitekamp, D.P. (1987) *Journal of the American Chemical Society*, **109**, 5541–2.

2 Eisenschmid, T.C., Kirss, R.U., Deutsch, P.P., Hommeltoft, S.I., Eisenberg, R., Bargon, J., Lawler, R.G. and Balch, A.L. (1987) *Journal of the American Chemical Society*, **109**, 8089–91.

3 Blazina, D., Duckett, S.B., Dunne, J.P. and Godard, C. (2004) *Dalton Transactions*, 2601–9.

4 Duckett, S.B. and Sleigh, C.J. (1999) *Progress in Nuclear Magnetic Resonance Spectroscopy*, **34**, 71–92.

5 Natterer, J. and Bargon, J. (1997) *Progress in Nuclear Magnetic Resonance Spectroscopy*, **31**, 293–315.

6 Eisenberg, R. (1991) *Accounts of Chemical Research*, **24**, 110–16.

7 Bargon, J. (2007) *The Handbook of Homogeneous Hydrogenation* (eds J.G. de Vries and C.J. Elsevier), Wiley-VCH Verlag GmbH, Weinheim, p. 313.

8 Duckett, S.B. and Colebrooke, S.A. (2002) *Encyclopedia of Nuclear Magnetic Resonance*, Vol. 9 (eds D.M. Grant and R.K. Harris), John Wiley & Sons, Ltd, Chichester, p. 598.

9 Golman, K., Axelsson, O., Johannesson, H., Mansson, S., Olofsson, C. and Petersson, J.S. (2001) *Magnetic Resonance in Medicine*, **46**, 1–5.

10 Bhattacharya, P., Harris, K., Lin, A.P., Mansson, M., Norton, V.A., Perman, W.H., Weitekamp, D.P. and Ross, B.D. (2005) *Magma (New York, N.Y.)*, **18**, 245–56.

11 Mansson, S., Johansson, E., Magnusson, P., Chai, C.-M., Hansson, G., Petersson, J.S., Stahlberg, F. and Golman, K. (2006) *European Journal of Radiology*, **16**, 57–67.

12 Bhattacharya, P., Chekmenev, E.Y., Perman, W.H., Harris, K.C., Lin, A.P., Norton, V.A., Tan, C.T., Ross, B.D. and Weitekamp, D.P. (2007) *Journal of Magnetic Resonance*, **186**, 150–5.

13 Golman, K., in't Zandt, R. and Thaning, M. (2006) *Proceedings of the National Academy of Sciences of the United States of America*, **103**, 11270–5.

14 Hubler, P., Bargon, J. and Glaser, S.J. (2000) *Journal of Chemical Physics*, **113**, 2056–9.

15 Anwar, M.S., Blazina, D., Carteret, H.A., Duckett, S.B. and Jones, J.A. (2004) *Chemical Physics Letters*, **400**, 94–7.

16 Bargon, J., Kandels, J. and Woelk, K. (1993) *Zeitschrift für Physikalische Chemie*, **180**, 65–93.

17 Bargon, J., Kandels, J. and Woelk, K. (1990) *Angewandte Chemie – International Edition*, **29**, 58–9.

18 Armstrong, R.L. and Kalechstein, W. (1978) *Chemical Physics*, **28**, 125–7.

19 Carravetta, M. and Levitt, M.H. (2004) *Journal of the American Chemical Society*, **126**, 6228–9.

20 Carravetta, M. and Levitt, M.H. (2005) *Journal of Chemical Physics*, **122**, 214505.

21 Pileio, G. and Levitt, M.H. (2007) *Journal of Magnetic Resonance*, **187**, 141–5.

22 Sarkar, R., Vasos, P.R. and Bodenhausen, G. (2007) *Journal of the American Chemical Society*, **129**, 328–34.

23 Gopalakrishnan, K. and Bodenhausen, G. (2006) *Journal of Magnetic Resonance*, **182**, 254–9.

24 Buntkowsky, G., Walaszek, B., Adamczyk, A., Xu, Y., Limbach, H.-H. and Chaudret, B. (2006) *Physical Chemistry Chemical Physics*, **8**, 1929–39.

25 Aroulanda, C., Starovoytova, L. and Canet, D. (2007) *The Journal of Physical Chemistry A*, **111**, 10615–24.

26 De Vries, J.G. and Elsevier, C.J. (eds) (2007) *The Handbook of Homogeneous Hydrogenation*, Wiley-VCH Verlag GmbH, Weinheim.

27 Bowers, C.R. and Weitekamp, D.P. (1986) *Physical Review Letters*, **57**, 2645–8.

28 Pravica, M.G. and Weitekamp, D.P. (1988) *Chemical Physics Letters*, **145**, 255–8.

29 Barkemeyer, J., Haake, M. and Bargon, J. (1995) *Journal of the American Chemical Society*, **117**, 2927–8.

30 Aime, S., Gobetto, R., Reineri, F. and Canet, D. (2006) *Journal of Magnetic Resonance*, **178**, 184–92.

31 Natterer, J., Schedletzky, O., Barkemeyer, J., Bargon, J. and Glaser, S.J. (1998) *Journal of Magnetic Resonance*, **133**, 92–7.

**32** Stephan, M., Kohlmann, O., Niessen, H.G., Eichhorn, A. and Bargon, J. (2002) *Magnetic Resonance in Chemistry*, **40**, 157–60.

**33** Johannesson, H., Axelsson, O. and Karlsson, M. (2004) *Comptes Rendus Physique*, **5**, 315–24.

**34** Goldman, M., Johannesson, H., Axelsson, O. and Karlsson, M. (2006) *Comptes Rendus Chimie*, **9**, 357–63.

**35** Goldman, M. and Johannesson, H. (2005) *Comptes Rendus Physique*, **6**, 575–81.

**36** Haake, M., Natterer, J. and Bargon, J. (1996) *Journal of the American Chemical Society*, **118**, 8688–91.

**37** Barkemeyer, J., Bargon, J., Sengstschmid, H. and Freeman, R. (1996) *Journal of Magnetic Resonance A*, **120**, 129–32.

**38** Natterer, J., Barkemeyer, J. and Bargon, J. (1996) *Journal of Magnetic Resonance A*, **123**, 253–6.

**39** Aime, S., Gobetto, R., Reineri, F. and Canet, D. (2003) *Journal of Chemical Physics*, **119**, 8890–6.

**40** Kuhn, L.T., Bommerich, U. and Bargon, J. (2006) *The Journal of Physical Chemistry A*, **110**, 3521–6.

**41** Eisenschmid, T.S., MacDonald, J., Eisenberg, R. and Lawler, R.G. (1989) *Journal of the American Chemical Society*, **111**, 7267–9.

**42** Permin, A.B. and Eisenberg, R. (2002) *Journal of the American Chemical Society*, **124**, 12406–7.

**43** Godard, C., Duckett, S.B., Henry, C., Polas, S., Toose, R. and Whitwood, A.C. (2004) *Chemical Communications*, 1826–7.

**44** Ertl, G. Knözinger, H. Schüth, F. and Weitkamp, J. (2008) *Handbook of Heterogeneous Catalysis*, Wiley-VCH Verlag GmbH, Weinheim.

**45** Eichhorn, A., Koch, A. and Bargon, J. (2001) *Journal of Molecular Catalysis A – Chemical*, **174**, 293–5.

**46** Cole-Hamilton, D.J. and Tooze, R.P. (eds) (2006) *Catalysis by Metal Complexes, Vol. 30, Catalyst Separation, Recovery and Recycling. Chemistry and Process Design*, Springer, Dordrecht.

**47** Barbaro, P. and Bianchini, C. (2002) *Topics in Catalysis*, **19**, 17–32.

**48** Merckle, C. and Blumel, J. (2005) *Topics in Catalysis*, **34**, 5–15.

**49** Carson, P.J., Bowers, C.R. and Weitekamp, D.P. (2001) *Journal of the American Chemical Society*, **123**, 11821–2.

**50** Koptyug, I.V., Kovtunov, K.V., Burt, S.R., Anwar, M.S., Hilty, C., Han, S., Pines, A. and Sagdeev, R.Z. (2007) *Journal of the American Chemical Society*, **129**, 5580–6.

**51** Bouchard, L.-S., Kovtunov, K.V., Burt, S.R., Anwar, M.S., Koptyug, I.V., Sagdeev, R.Z. and Pines, A. (2007) *Angewandte Chemie – International Edition*, **46**, 4064–8.

**52** Wasylenko, W. and Frei, H. (2007) *The Journal of Physical Chemistry C*, **111**, 9884–90.

**53** Wasylenko, W. and Frei, H. (2005) *The Journal of Physical Chemistry B*, **109**, 16873–8.

**54** Cremer, P.S., Su, X., Shen, Y.R. and Somorjai, G.A. (1996) *The Journal of Physical Chemistry*, **100**, 16302–9.

**55** Cremer, P.S., Su, X., Shen, Y.R. and Somorjai, G.A. (1996) *Journal of the American Chemical Society*, **118**, 2942–9.

**56** Bond, G.C. (1997) *Applied Catalysis A: General*, **149**, 3–25.

**57** Koptyug, I.V., Lysova, A.A., Sagdeev, R.Z., Kirillov, V.A., Kulikov, A.V. and Parmon, V.N. (2005) *Catalysis Today*, **105**, 464–8.

**58** Lysova, A.A., Koptyug, I.V., Kulikov, A.V., Kirillov, V.A., Sagdeev, R.Z. and Parmon, V.N. (2007) *Chemistry Engineering Journal*, **130**, 101–9.

**59** Kovtunov, K.V., Beck, I.E., Bukhtiyarov, V.I. and Koptyug, I.V. (2008) *Angewandte Chemie*, **47**, 1492–5.

**60** Somorjai, G.A. and Zaera, F. (1982) *The Journal of Physical Chemistry*, **86**, 3070–8.

**61** Bowers, C.R. (2002) *Encyclopedia of Nuclear Magnetic Resonance*, Vol. **9** (eds D.M. Grant and R.K. Harris), John Wiley & Sons, Ltd, Chichester, p. 750.

**62** Bouchard, L.-S., Burt, S.R., Anwar, M.S., Kovtunov, K.V., Koptyug, I.V. and Pines, A. (2008) *Science*, **319**, 442–5.

**63** Haake, M., Barkemeyer, J. and Bargon, J. (1995) *The Journal of Physical Chemistry*, **99**, 17539–43.

**64** Schleyer, D., Niessen, H.G. and Bargon, J. (2001) *New Journal of Chemistry*, **25**, 423–6.

# 8
# Towards Posture-Dependent Human Pulmonary Oxygen Mapping Using Hyperpolarized Helium and an Open-Access MRI System

*Ross W. Mair, Leo L. Tsai, Chih-Hao Li, Michael J. Barlow, Rachel N. Scheidegger, Matthew S. Rosen, Samuel Patz and Ronald L. Walsworth*

## 8.1
## Introduction

In recent years, the MRI of inhaled, hyperpolarized $^3$He gas [1, 2] has emerged as a powerful method for studying lung structure and function [3, 4]. This technique has been used to make quantitative maps of human ventilation [5], to obtain regional acinar structural information via measurements of the $^3$He apparent diffusion coefficient (ADC) [6], and to monitor the regional $O_2$ concentration ($p_AO_2$) via the $^3$He spin-relaxation rate [7, 8]. These techniques have applications to basic pulmonary physiology [9] as well as to lung diseases such as asthma and emphysema [9, 10].

Despite its clinical utility, all subjects undergoing lung imaging in a traditional clinical MRI scanner must lie horizontally, this being a consequence of the magnet design. Similar restrictions are imposed on subjects being studied by other clinical imaging modalities such as computed tomography (CT) and positron emission tomography (PET). However, body orientation and postural changes have a significant effect on pulmonary function – indeed, much more so than any other body organ – and hence studies of pulmonary function at variable postures could have significant impact on both pulmonary physiology and medicine. For example, the regional distribution of pulmonary perfusion and ventilation have been a source of renewed interest in recent years [11–14], due to significant questions relating to the care and survival of patients with severe lung diseases such as acute respiratory distress syndrome (ARDS) [13].

Some initial studies with hyperpolarized $^3$He have shown that posture changes, even while horizontal, affect the lung structure modestly in a way that can nonetheless be clearly probed by $^3$He MRI [15]. To enable complete posture-dependent lung imaging, we have developed an open-access MRI system based on a simple electromagnet design that operates at a field strength ~200 times lower than a traditional clinical MRI scanner [16, 17]. To perform MRI at such a field strength, we exploit the practicality of hyperpolarized $^3$He MRI at magnetic fields <10 mT [18–21]. $^3$He hyperpolarized to 30–60% can be created by one of two laser-based

*Magnetic Resonance Microscopy.* Edited by Sarah L. Codd and Joseph D. Seymour
Copyright © 2009 WILEY-VCH Verlag GmbH & Co. KGaA, Weinheim
ISBN: 978-3-527-32008-0

optical pumping processes [1, 2] prior to the MRI procedure; high-resolution gas-space imaging can then be performed without the need of a large applied magnetic field. Such high-spin polarization gives $^3$He gas a magnetization density similar to that of water in ~10 T fields, despite the drastically lower spin density of the gas.

We have previously described in detail the clinical motivation for posture-dependent pulmonary studies [16], and the design and operation of our imager [17]. In this chapter, we report progress towards posture-dependent functional human lung imaging: that is, mapping regional pulmonary $O_2$ partial pressure. Such experiments require higher $^3$He hyperpolarization than that required for simple spin-density imaging. We outline recent improvements to our $^3$He hyper-polarization apparatus, quantitative testing of the oxygen-mapping procedure using phantoms, and initial human $p_AO_2$ maps using our open-access MRI system.

## 8.2
## Experimental

### 8.2.1
### Imager Design

A detailed description of the design and operation of our open-access human MRI system is presented elsewhere [17]. Here, we provide a brief overview of the system.

The imager operates at an applied static magnetic field, $B_0 = 6.5$ mT (65 G). This field is created by a four-coil, biplanar magnet design [22] with pairs of coils measuring 2 m and 0.55 m in diameter, separated by ~80 cm. All four coils are powered by a single dc power supply with 42.2 A of current to reach the desired field of 6.5 mT, allowing $^3$He MRI at a frequency of 210 kHz. After shimming, the $B_0$ field exhibits a total variation of less than ~5 µT (0.05 G) across the volume of a human chest. The resulting $^3$He NMR signals from such volumes exhibit spectral FWHM (full width at half maximum) line-widths of ~30 Hz.

Planar gradient coils were built for the imager, thereby eliminating another restrictive cylindrical geometry found in clinical MRI scanners. The coils were designed to allow the acquisition of $256 \times 256$ $^3$He images across a 40 cm field of view (FOV) with an imaging bandwidth of 10 kHz, while avoiding noticeable con-comitant field effects [23]. The coils are coplanar with each $B_0$ magnet coil, maintaining the ~80 cm spacing for subject access. The gradient coils are powered by Techron 8607 gradient amplifiers and, at maximum current, the three gradients each provide ~0.07 G cm$^{-1}$ gradient strength.

Radiofrequency (RF) and gradient control is accomplished using a Tecmag Apollo MRI console, which operates at a frequency of 210 kHz without hardware modification. RF pulses are fed to a Communications Power Corp. Inc. NMR Plus 5LF300S amplifier, which provides up to 300 W of RF power. A single RF coil is used for $B_1$ transmission and detection, in conjunction with a probe interface-T/R switch optimized for 200 kHz operation. The RF coil is a large solenoid ~50 cm in

(a)

(b)

**Figure 8.1** (a) A photograph of the open-access human MRI system. The 2 m $B_0$ coil is denoted by an arrow at the top left; the lower arrow denotes the 0.55 m $B_0$ coil. The other circular and grid patterns are the gradient coils, located parallel to the $B_0$ coils. The entire design is mirrored on the other side. The gap between the two pairs of coils is 80 cm; (b) A human subject positioned in the imager for vertical orientation imaging.

diameter and length, that accommodates the subject's shoulders and arms, and completely covers their chest. Being a solenoid, the coil has very high $B_1$ homogeneity [17], and can be rotated along with the subject in the imaging plane, while remaining perpendicular to the direction of $B_0$. The coil has a quality factor $Q \sim 30$, implying operating bandwidths of ~10 kHz at the $^3$He Larmor frequency of 210 kHz.

In order to improve the signal-to-noise ratio (SNR), the $B_0$, gradient and $B_1$ coils were housed inside an RF-shielded room (Lindgren RF Enclosures) that attenuates environmental RF interference above 10 kHz by up to 100 dB. Power lines for the $B_0$ magnet, preamplifier and RF coil connections all pass through commercial filters that shield out noise above 10 kHz. The gradient lines pass into the shielded room via three sets of custom, high-current passive line filters that produce ~25 dB attenuation at 100 kHz. The complete imager system, demonstrating subject access, is shown in Figure 8.1.

### 8.2.2
### Hyperpolarized $^3$He Production and Delivery

Hyperpolarized $^3$He gas is produced via the spin-exchange optical pumping (SEOP) technique using vaporized Rb as an intermediate [1]. The modular $^3$He polarization apparatus [20] is located adjacent to, but outside, the RF-shielded room. A magnetic field of 2.3 mT provides a quantization axis for optical pumping and prevents rapid polarization decay. It also provides a homogeneous $B_0$ field for *in situ* polarization monitoring using a bench-top Magritek Aurora spectrometer. For each experiment, a ~80 cm$^3$ Pyrex-glass polarization cell is filled with ~5–6 bar of $^3$He and 0.1 bar of N$_2$. The cell is heated to >170 °C, and ~30 W of circularly polarized light at 794.7 nm is applied. After optical pumping for ~8–10 h, the $^3$He nuclear spin polarization reaches ~20–30%. The polarized gas is then expanded from the optical pumping cell into an evacuated glass chamber with a Teflon piston, which enables delivery of the $^3$He via Teflon tubing through a feedthrough in the RF-shielded room, either directly to a phantom or to a delivery manifold adjacent to the human subject.

One significant limitation of the SEOP method is the slow production rate of hyperpolarized $^3$He. The demands for higher $^3$He polarization levels required for $pO_2$ mapping experiments, and for higher volumes of hyperpolarized $^3$He for administration to human subjects in a timely manner, has necessitated a number of modifications to the polarizer hardware, which are described below.

The slow production rate of hyperpolarized $^3$He is the result of a small spin-exchange collision coefficient between Rb and $^3$He, ~6.8 × 10$^{-20}$ cm$^3$ s$^{-1}$. Operating the SEOP cell at higher temperatures dramatically increases the vaporized Rb density in the cell, and hence the Rb–$^3$He spin-exchange rate. However, the laser must provide sufficient resonant light in order to fully polarize the vaporized Rb at this higher density; otherwise, the $^3$He polarization level will actually decrease as unpolarized Rb atoms collide with, and depolarize, $^3$He atoms. Although the current generation of laser diode arrays (LDA) commonly used for SEOP may be rated at 60 W or higher, their optical spectra are broad, with the desired wavelength of 794.7 nm generally appearing as a broad peak of ~3 nm width (see Figure 8.2).

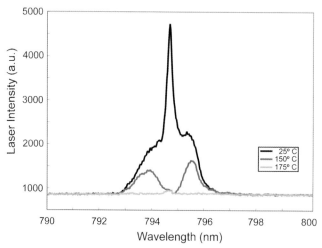

**Figure 8.2** Optical spectra from the broadband Coherent and VHG-line-narrowed Comet™ laser, after transmission through the $^3$He SEOP cell, at ambient and operating temperatures. The locked output from the Comet™ laser appears as an intense, narrow peak at 794.7 nm, while the output from the Coherent laser is the broad peak. At 150 °C, the Rb vapor absorbs all the laser light from Comet™, but not the Coherent laser.

As a result, relatively little of the rated laser power is actually resonant for the Rb polarization transition, making higher $^3$He spin polarization levels or production rates difficult to obtain without lasers of enormous power.

A concerted effort has taken place in recent years to attempt to narrow the laser spectral output, with the aim of making the most of the laser light resonant for Rb polarization. Many of these investigations initially involved the use of significant amounts of additional optical hardware and an optical table (external cavity approach) [24]. A much simpler option – and one that can be easily integrated with standard LDAs – has emerged during the past year with the advent of volume holographic gratings (VHGs) [25]. These gratings can be used in an optical mount in front of a bare laser diode bar, or incorporated directly into the LDA modules by laser manufacturers, resulting in a resonant line of ~0.2–0.3 nm FWHM. The VHGs self-seed the laser to provide a narrow resonant line output with only minimal laser power reduction. To date, we have trialed two Comet™ prototype commercial, fiber-coupled, VHG-narrowed 30 W LDA modules, a product line under development by Spectra-Physics (Newport Corp., Tucson, AZ). The Comet module is also the first LDA system with power >10 W that has employed a polarization-preserving fiber. Consequently, the laser output emerges from the fiber retaining >90% of its linear polarization, in a single beam that can be easily matched to the diameter of the polarization cell, and aligned parallel to the cell. This development further increases the amount of resonant light on the optical pumping cell over the traditional fiber-coupled LDAs that required additional optical polarization hardware, and resulted in two nonparallel beams incident on the optical pumping cell [1, 3]. Relative polarization measurements obtained using a standard 30 W fiber-coupled LDA system (Coherent FAP), and from the VHG-narrowed Comet module employing the polarization-preserving fiber, are listed in Table 8.1. Optical spectra of the output from the Coherent FAP and Comet as a function of temperature are shown in Figure 8.2.

**Table 8.1** A comparison of $^3$He NMR signal measurements (arbitrary units) obtained with a broadband 30 W laser diode array (Coherent FAP) and a VHG-narrowed 30 W LDA module (Spectra-Physics Comet™), under a variety of configurations at a cell temperature of 180 °C. The FAP laser employed an optical polarizing beam splitter, which resulted in one or two beams of polarized light, each of ~14 W, available for the cell [26]. The Comet™ used a polarization preserving fiber. Measurements were made with a cell with $^3$He $T_1$ of ~12 h. The performance of the Comet™ relative to the FAP laser will be greater in cells with longer $T_1$. Additionally, at higher operating temperatures, which favor faster polarization rates due to a higher density of vaporized Rb, the Comet™ will be more likely to achieve full polarization of the Rb vapor, and so be even more effective than the FAP than under current conditions.

| Laser | FAP: 1 beam | FAP: 2 beams | Comet: 1 beam |
|---|---|---|---|
| $^3$He polarization | $36 \pm 5$ | $40 \pm 1$ | $47 \pm 1$ |

We have also begun to modify the polarizer hardware to enable the production of multiple batches of hyperpolarized $^3$He. With a modified cleaning/high-temperature/vacuum treatment procedure of the glass [27] prior to fabrication of the optical pumping cells, we have been able to achieve $^3$He polarization lifetimes ($T_1$) in valved Pyrex cells of ~60 h, up from ~10–15 h previously. This result not only leads to a ~30% improvement in the attainable polarization during the SEOP procedure, but potentially allows the gas to be stored for later use while additional batches are polarized. Storage also requires an extremely homogeneous holding field to reduce the effects of polarization loss via diffusion through field gradients. For the 2.3 mT magnetic field in the $^3$He polarizer and a $^3$He diffusion coefficient, $D = 1.84\,\text{cm}^2\text{s}^{-1}$ at 1 bar, transverse field gradients must be minimized to $<2.5 \times 10^{-3}\,\text{mT}\,\text{cm}^{-1}$ in order to achieve a relaxation rate $>(100\,\text{h})^{-1}$ [28]. A simple pair of Helmholtz coils of reasonable size does not provide a large enough region of uniform holding field to maintain four storage cells without significant polarization loss. We therefore added an additional coil at the center of the polarizer, in order to increase the region of high-field homogeneity approximately sixfold [29] – large enough to accommodate the SEOP cell and four storage cells. This modification will allow the production of up to five batches of hyperpolarized $^3$He over a ~50 h period prior to a single human imaging session, enabling posture-dependent studies on the same day, or ventilation imaging from successive breaths of $^3$He.

## 8.2.3
## MRI Techniques

Oxygen partial-pressure ($pO_2$) mapping using $^3$He MRI is accomplished using repeated two-dimensional (2-D) gradient-recalled echo FLASH (fast low angle single shot) images, and observing the time-decay of the $^3$He MRI signal [7, 8]. For phantom and human studies, 2-D images were acquired without slice selection, using an excitation flip angle of ~3°, a dataset size of 128 × 32, and a 50 × 50 cm FOV in ~2 s. All imaging acquisitions used the following parameters: bandwidth = 4.6 kHz, 300 μs square RF pulse, TE/TR ~29/86 ms, number of experiments (NEX) = 1. The datasets were zero-filled to 128 × 64 points before fast-Fourier transformation. To follow the time course of the $^3$He magnetization, three to eight images were acquired successively, with 5 s inter-image delays, resulting in experiments ranging from ~15 to 50 s in duration. Flip angle maps were acquired with an identical procedure, except that the inter-image delay was set to 100 ms.

$pO_2$ determination is possible due to the effect of molecular $O_2$ in greatly reducing $^3$He $T_1$ via an intermolecular dipolar relaxation process. The $O_2$-induced relaxation follows the relationship: $T_1^{\text{ox}} = 1800\,(\text{torr.s})/pO_2\,(\text{torr})$ [7]. As the typical alveolar oxygen partial pressure, $p_AO_2$ is ~90–120 torr in the human lung, $T_1^{\text{ox}}$ is calculated as ~10–15 s. As the inherent spin-lattice $T_1$ of $^3$He is of the order of hours, and the surface relaxation of minutes, the observed $T_1$ of $^3$He in the lung is dominated by oxygen relaxation [7, 8]. The method for image-based $pO_2$ determination is similar to measurements of hyperpolarized noble gas flip-angle or $T_1$, derived from the relationship between the NMR signal from the $n$th pulse in a series of pulses, $\ln(S_n)$, with the pulse flip angle $\theta$, and inter-pulse time $\tau$ [19]. However, instead of acquiring a series of free induction decays (FIDs), we obtain a series of $m$ images with $n$ RF pulses per image, and a time $\tau$ between each image acquisition (inter-image delay + image acquisition time). The MRI signal relationship can then be written [30]:

$$\ln(S_m/S_0) = nm\ln(\cos\theta) - m\frac{\tau}{T_1^{ox}} = nm\ln(\cos\theta) - m\frac{\tau.pO_2}{1800} \tag{8.1}$$

Plotting $\ln(S_m/S_0)$ versus $m\tau$ on a pixel-by-pixel basis yields lines with slope = $(n\ln(\cos\theta))/\tau - pO_2/1800$, from which $pO_2$ can be calculated if $\theta$ is known. Image-based flip-angle calibrations were performed prior to $pO_2$ measurements, in order to calibrate $\theta$. The image analysis was identical, except that in the absence of $O_2$, the slope of $\ln(S_m/S_0)$ versus $m\tau$ reduces to $(n\ln(\cos\theta))/\tau$.

### 8.2.4
### Human Imaging Protocol

After expiration to their lung functional residual capacity (~2.5–3 l for a healthy adult), the subject inhales ~500 cm$^3$ of hyperpolarized $^3$He gas, followed by a small breath of air to wash the helium out of the large airways. The MRI sequence begins immediately after inhalation, and proceeds during breath-hold for ~30–40 s. All human experiments are performed according to a protocol approved by the Partners Human Research Committee at Brigham and Women's Hospital, under an inter-institutional IRB agreement with the Harvard University Committee for the Use of Human Subjects in Research.

### 8.3
### Results and Discussion

Figure 8.3 shows an image-based flip-angle calibration conducted using ~500 cm$^3$ of hyperpolarized helium expanded into a 2 l Tedlar plastic bag which was previously filled with ~1 l of N$_2$. The bag measures ~15 × 15 cm. Figure 8.3 a–d shows four of the eight spin-density images acquired during the time-course experiment. The spin-density images vary due to the nonuniform mixing of the hyperpolarized $^3$He in the N$_2$, and distortions of the partially filled bag. As the image acquisition time is finite, the time $t$ represents the 'average' time-point during the experiment at which the image was acquired (i.e. half the image acquisition time of 1.9 s). For this experiment, with an inter-image delay of 100 ms, $\tau = 2.0$ s. Figure 8.3e shows the flip-angle map, and Figure 8.3f shows a semi-log plot of signal attenuation versus total experiment time ($m\tau$) from a pixel near the center of the image. Despite the nonuniformity of the spin-density images, the flip-angle is very uniform across the dimensions of the bag (~3.9 ± 0.2°). This ~5% variation matches precise spectroscopy measurements taken at various positions in this RF

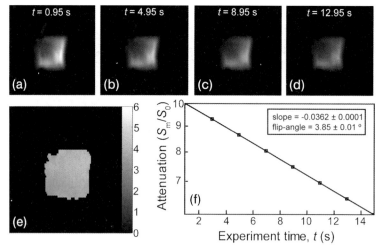

**Figure 8.3** (a–d) Time series of $^3$He MR images obtained from a plastic bag phantom containing only $^3$He and $N_2$, in order to determine the excitation flip-angle. The plotted FOV is 50 cm. Every second image from the time series is shown; (e) Flip-angle map obtained from analysis of signal attenuation. The scale bar values have units of degrees; flip-angle = ~3.9 ± 0.2°; (f) Signal attenuation analysis for a single pixel in the middle of the bag.

coil, and agrees with the standard equations for a RF field in a solenoid coil [17]. Importantly, at the frequency of 210 kHz – below the sample-noise-dominated regime for human MRI – the RF coil loading is negligible and thus the above calibration can be used reliably for human experiments, vastly simplifying human $p_AO_2$ measurements in comparison to those performed at high-field with clinical scanners [7, 8].

The $pO_2$ mapping protocol was tested on Tedlar plastic bags with $pO_2$ ranging from 50 to 160 torr. Figure 8.4a shows a $pO_2$ map derived from one such experiment. The bag was filled carefully by syringe to contain $pO_2 = 68 \pm 5$ torr after the addition of 500 cm$^3$ of hyperpolarized $^3$He. We had previously calibrated θ, using the procedure shown in Figure 8.3, to be 2.3° in this instance, and τ ~7 s (inter-image delay = 5 s). There is some slight variation in $pO_2$ throughout the bag, due to imperfect mixing of the gases prior to image acquisition; however, the average value obtained across the bag is $pO_2 = 60.5 \pm 7$ torr. Individual pixels exhibited $pO_2$ ranging from 56–70 torr.

An initial human $p_AO_2$ map obtained using our open-access imager is shown in Figure 8.4b. For this trial, the subject was lying horizontally, and the image obtained in the sagittal orientation – that is, a transverse cross-section of the lungs. Images acquired in this orientation show the lungs in a gravity-dependent state; that is, gravitational force is acting down in the plane of the image. However, the lung dimensions over which gravity is acting are minimized, and the lungs are constrained by the chest wall. Therefore, a minimal $p_AO_2$ gradient is observed with gravity in this orientation. The average measured $p_AO_2$ value throughout the lungs

(a)

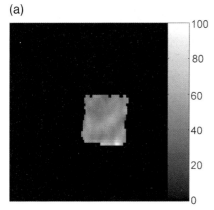

(b)

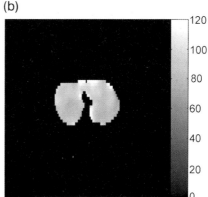

**Figure 8.4** (a) $pO_2$ map obtained from a plastic bag phantom containing $^3$He and $68 \pm 5$ torr of $O_2$. The MRI method yields an average value throughout the bag of $60.5 \pm 7$ torr; (b) $p_AO_2$ map from a human volunteer, lying horizontally in the open-access imager, after inhalation of $500\,cm^3$ of hyperpolarized $^3$He. In this sagittal orientation, there is little variation in $p_AO_2$ across the lungs. The average value is $95 \pm 6$ torr, close to the physiologically expected value of $\sim$90–120 torr. The scale bar values for both images have units of torr.

was $95 \pm 6$ torr, within the expected physiological range of 90–120 torr. Our image acquisition protocol of 2 s image time and 5 s inter-image delays allowed the acquisition of five to six images during the maximum 30–40 s breath-hold that is tolerable for most subjects. By reducing the number of phase encodes and acquisition points, we can decrease the overall spatial resolution but boost the SNR in the later images, at which time the $^3$He polarization has undergone significant $T_1$ decay. The resulting spatial resolution of $\sim$0.5 × 2 cm is adequate for an initial $pO_2$ mapping of the lung and the detection of gross changes in $pO_2$ as a function of posture. In these initial studies, we do not calculate the effect of oxygen uptake from the lungs during the breath-hold. This value is known to be $\sim$2 torr/s [7], and in our calculation its minor effect is included in the error quoted on $pO_2$.

It should be noted that these images were acquired without slice selection. $p_AO_2$ mapping relies on accurately measuring MRI signal attenuation as a function of time; however, the use of narrow slice-selective imaging methods can lead to reduced accuracy in quantitative data as out-of-slice magnetization diffuses into the image slice during pulse-sequence and inter-image delays. The use of 2-D projection methods reduces this effect.

## 8.4
## Conclusions

We have demonstrated initial regional measurements of the partial pressure of oxygen, $pO_2$, using hyperpolarized $^3$He and an open-access, very-low-field MRI system operating at 6.5 mT (210 kHz for $^3$He). Image-based flip-angle measurements have confirmed that the solenoid RF coil exhibits a very homogeneous $B_1$ field and therefore uniform RF excitation flip-angles. At the operating frequency used, coil loading is negligible, allowing the use of a single, calibrated flip-angle for human $p_AO_2$ measurements without the need to perform a calibration with each measurement and subject. Initial $pO_2$ mapping experiments with phantoms validate this procedure, and an initial human $p_AO_2$ map yields physiologically expected values. The maps have a resolution of ~0.5 × 2 cm. Recently, we have also made significant hardware improvements to our helium polarizer in order to increase the maximum polarization attainable and the production rate. These experiments indicate that the open-access imager will enable posture-dependent pulmonary functional imaging and thereby serve as a valuable tool for the study of critical pulmonary diseases and questions relating to posture-dependence on pulmonary function.

## Acknowledgments

We are indebted to Kenneth Tsai, MD, who acted as observing physician for the human imaging trials, and George Topulos, MD, who devised the human protocols. We are grateful to Rick Frost and the Spectra-Physics division of Newport Corp. for assistance with and trials of novel Comet VHG-narrowed laser modules. Support is acknowledged from NASA grant NAG9-1489 and NIH grant R21 EB006475-01A1. This material is also based upon work supported by the National Science Foundation under Grant PHY-0618891. Any opinions, findings, and conclusions or recommendations expressed in this material are those of the author(s) and do not necessarily reflect the views of the National Science Foundation.

## References

1 Walker, T.G. and Happer, W. (1997) *Reviews of Modern Physics*, **69**, 629–42.

2 Nacher, P.J. and Leduc, M. (1985) *Journal de Physique*, **46**, 2057–73.

3 Leawoods, J.C., Yablonskiy, D.A., Saam, B., Gierada, D.S. and Conradi, M.S. (2001) *Concepts in Magnetic Resonance*, **13**, 277–93.

4 Moller, H.E., Chen, X.J., Saam, B., Hagspiel, K.D., Johnson, G.A., Altes, T.A., de Lange, E.E. and Kauczor, H.-U. (2002) *Magnetic Resonance in Medicine*, **47**, 1029–51.

5 Wild, J.M., Paley, M.N.J., Kasuboski, L., Swift, A., Fichele, S., Woodhouse, N., Griffiths, P.D. and van Beek, E.J.R.

(2003) *Magnetic Resonance in Medicine*, **49**, 991–7.

6 Salerno, M., de Lange, E.E., Altes, T.A., Truwit, J.D., Brookeman, J.R. and Mugler, J.P., III (2002) *Radiology*, **222**, 252–60.

7 Deninger, A.J., Eberle, B., Ebert, M., Grossmann, T., Hanisch, G., Heil, W., Kauczor, H.-U., Markstaller, K., Otten, E., Schreiber, W., Surkau, R. and Weiler, N. (2000) *NMR in Biomedicine*, **13**, 194–201.

8 Rizi, R.R., Baumgardner, J.E., Ishii, M., Spector, Z.Z., Edvinsson, J.M., Jalali, A., Yu, J., Itkin, M., Lipson, D.A. and Gefter, W. (2004) *Magnetic Resonance in Medicine*, **52**, 65–72.

9 Mills, G.H., Wild, J.M., Eberle, B. and Van Beek, E.J. (2003) *British Journal of Anaesthesia*, **91**, 16–30.

10 Samee, S., Altes, T.A., Powers, P., de Lange, E.E., Knight-Scott, J., Rakes, G., Mugler, J.P., Ciambotti, J.M., Alford, B.A., Brookeman, J.R. and Platts-Mills, T.A.E. (2003) *Journal of Allergy and Clinical Immunology*, **111**, 1205–11.

11 Mure, M., Domino, K.B., Lindahl, S.G.E., Hlastala, M.P., Altemeier, W.A. and Glenny, R.W. (2000) *Journal of Applied Physiology*, **88**, 1076–83.

12 Mure, M. and Lindahl, S.G.E. (2001) *Acta Anaesthesiologica Scandinavica*, **45**, 150–9.

13 Gattinoni, L., Tognoni, G., Pesenti, A., Taccone, P., Mascheroni, D., Labarta, V., Malacrida, R., Di Giulio, P., Fumagalli, R., Pelosi, P., Brazzi, L. and Latini, R. (2001) *New England Journal of Medicine*, **345**, 568–73.

14 Musch, G., Layfield, J.D., Harris, R.S., Melo, M.F., Winkler, T., Callahan, R.J., Fischman, A.J. and Venegas, J.G. (2002) *Journal of Applied Physiology*, **93**, 1841–51.

15 Fichele, S., Woodhouse, N., Swift, A.J., Said, Z., Paley, M.N.J., Kasuboski, L., Mills, G.H., van Beek, E.J.R. and Wild, J.M. (2004) *Journal of Magnetic Resonance Imaging*, **20**, 331–5.

16 Tsai, L.L., Mair, R.W., Li, C.-H., Rosen, M.S., Patz, S. and Walsworth, R.L. (2008) *Academic Radiology*, **15**, 728–39.

17 Tsai, L.L., Mair, R.W., Rosen, M.S., Patz, S. and Walsworth, R.L. (2008) *Journal of Magnetic Resonance*, **193**, 274–85.

18 Tseng, C.-H., Wong, G.P., Pomeroy, V.R., Mair, R.W., Hinton, D.P., Hoffmann, D., Stoner, R.E., Hersman, F.W., Cory, D.G. and Walsworth, R.L. (1998) *Physical Review Letters*, **81**, 3785–8.

19 Wong, G.P., Tseng, C.H., Pomeroy, V.R., Mair, R.W., Hinton, D.P., Hoffmann, D., Stoner, R.E., Hersman, F.W., Cory, D.G. and Walsworth, R.L. (1999) *Journal of Magnetic Resonance*, **141**, 217–27.

20 Mair, R.W., Hrovat, M.I., Patz, S., Rosen, M.S., Ruset, I.C., Topulos, G.P., Tsai, L.L., Butler, J.P., Hersman, F.W. and Walsworth, R.L. (2005) *Magnetic Resonance in Medicine*, **53**, 745–9.

21 Ruset, I.C., Tsai, L.L., Mair, R.W., Patz, S., Hrovat, M.I., Rosen, M.S., Muradian, I., Ng, J., Topulos, G.P., Butler, J.P., Walsworth, R.L. and Hersman, F.W. (2006) *Concepts in Magnetic Resonance Part B: Magnetic Resonance Engineering*, **29**B, 210–21.

22 Morgan, P.S., Conolly, S. and Mazovski, A. (1997) Proceedings, 5th Scientific Meeting of ISMRM, Vancouver, Canada, 5, 1447.

23 Yablonskiy, D.A., Sukstanskii, A.L. and Ackerman, J.J.H. (2005) *Journal of Magnetic Resonance*, **174**, 279–86.

24 Babcock, E., Chann, B., Nelson, I.A. and Walker, T.G. (2005) *Applied Optics*, **44**, 3098–104.

25 Havermeyer, F., Liu, W., Moser, C., Psaltis, D. and Steckman, G.J. (2004) *Optical Engineering*, **43**, 2017–21.

26 Zook, A.L., Adhyaru, B.B. and Bowers, C.R. (2002) *Journal of Magnetic Resonance*, **159**, 175–82.

27 Jacob, R.E., Morgan, S.W. and Saam, B. (2002) *Journal of Applied Physics*, **92**, 1588–97.

28 Schearer, L.D. and Walters, G.K. (1965) *Physical Review*, **139**, 1398–402.

29 Wang, J., She, X.S. and Zhang, S.J. (2002) *Review of Scientific Instruments*, **73**, 2175–9.

30 Tsai, L.L. (2006) Development of a low-field $^3$He MRI system to study posture-dependence of pulmonary function, PhD thesis, Harvard University.

# 9
# Hyperpolarized $^{83}$Kr MRI

*Galina E. Pavalovskaya and Thomas Meersmann*

## 9.1
## Introduction

The use of hyperpolarized (hp) noble gas isotopes allows for nuclear magnetic resonance (NMR) signal enhancement of many orders of magnitude compared to thermally polarized NMR spectroscopy at ambient temperatures [1, 2]. The availability of hp $^3$He and hp $^{129}$Xe (both isotopes having a nuclear spin $I = 1/2$) has enabled a wide range of novel applications over the past 15 years [3–5], including magnetic resonance imaging (MRI) of the gas space of the lungs [6]. Pulmonary $^3$He MRI has been accomplished even in very weak magnetic fields with $^3$He resonance frequencies around 200 kHz [7]. There are, however, three other stable noble gas isotopes available that are also NMR-active. The lightest of these is $^{21}$Ne, which has a nuclear spin $I = 3/2$ but, because of its very low natural abundance, no MRI experiments using this isotope have been reported to date. The two other isotopes, $^{83}$Kr and $^{131}$Xe, can be used for MRI experiments without isotopic enrichment (see Table 9.1).

Noble gas isotopes with spin $I > 1/2$ possess a nuclear quadrupolar moment that is a probe for distortions of the noble gas electron shell symmetry. In the dissolved phase, the electric quadrupole-driven relaxation is usually the dominating relaxation mechanism [9, 10]. Noble gases dissolved in ordered liquid crystals or other systems with other macroscopic anisotropy can display a quadrupolar splitting in their NMR spectra [11–14]. The nuclear quadrupole moment is also a sensor for surfaces that are in contact with the gas because of its coupling with the electric field gradient (EFG) generated in the noble gas electron cloud during surface adsorption. Information about the void space in porous materials can be extracted directly from the NMR spectra [15–18], or indirectly through multiple quantum filtered (MQF) experiments [19, 20]. Quadrupolar splitting and relaxation effects have been studied even in macroscopic glass containers with centimeter-sized dimensions using optically detected magnetic resonance experiments with hyperpolarized spin $I > 1/2$ noble gases [21–24] and with NMR spectroscopy of thermally polarized noble gases [19, 25, 26]. An excellent treatment of trans-

*Magnetic Resonance Microscopy.* Edited by Sarah L. Codd and Joseph D. Seymour
Copyright © 2009 WILEY-VCH Verlag GmbH & Co. KGaA, Weinheim
ISBN: 978-3-527-32008-0

**Table 9.1** Spin, gyromagnetic ratio γ of the noble gas isotope (in %) of that of $^1$H, isotopic natural abundance of the five stable (i.e. nonradioactive) NMR active isotopes of the noble gas group. The nuclear electric quadrupole moment of the three spin $I > 1/2$ isotopes is also listed [8].

| Parameter | Noble gas isotope | | | | |
| --- | --- | --- | --- | --- | --- |
| | $^3$He | $^{21}$Ne | $^{83}$Kr | $^{129}$Xe | $^{131}$Xe |
| Spin $I$ | 1/2 | 3/2 | 9/2 | 1/2 | 3/2 |
| $\left\lvert\dfrac{\gamma}{\gamma_{^1H}}\right\rvert \cdot 100\%$ (%) | 76.2 | 7.89 | 3.85 | 27.8 | 8.24 |
| Natural abundance (%) | $^a$ | 0.27 | 11.5 | 26.4 | 21.2 |
| Quadrupole moment ($Q/fm^2$) | – | 10.2 | 25.9 | – | (–)11.4 |

a   $^3$He obtained from tritium decay.

verse relaxation on the surface of a macroscopic glass container is provided in Ref. [23].

Although the quadrupole moment of $^{83}$Kr is about twice that of $^{131}$Xe (see Table 9.1), quadrupole interactions are usually less intense for krypton compared to xenon because of the higher spin and because of the smaller and much less polarizable electron cloud of krypton that leads to smaller electric field gradients. As a result, $^{83}$Kr is the isotope of choice for hp MRI with spin $I > 1/2$ isotopes as its quadrupolar-driven relaxation is slow enough to allow for hp $^{83}$Kr transfer into a variety of porous media. As experiments with $^{131}$Xe can be used to illustrate the concept of surface-sensitive contrast, MRI with thermally polarized $^{131}$Xe will be described in Section 9.2, while Section 9.3 will introduce hp $^{83}$Kr MRI.

## 9.2
### Surface-Sensitive Contrast in Porous Media using Thermally Polarized $^{131}$Xe MRI

Magnetic resonance imaging of aerogels with thermally polarized $^{131}$Xe has been used to demonstrate the concept of surface-sensitive imaging contrast in porous media [27]. Sufficient signal intensity was generated through the use of liquefied xenon at high magnetic field strength (14 T), without the concerns regarding depolarization that are typical for studies with nonequilibrium MRI using hyperpolarized spin systems. To date, studies with liquefied $^{131}$Xe have produced the only transverse relaxation-weighted MRI contrast obtained with a spin $I > 1/2$ noble gas isotope.

Figure 9.1 shows two $^{131}$Xe NMR spectra of aerogel fragments immersed in liquefied xenon. Both spectra were recorded using the same aluminum silicate aerogel material with 0.1 g cm$^{-3}$ density, but the two samples had different levels

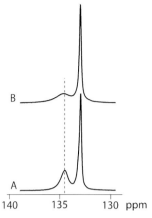

**Figure 9.1** $^{131}$Xe NMR spectra of 0.1 g cm$^{-3}$ aerogel sample immersed in pressurized, liquefied xenon at 283 K. Spectrum A = untreated 0.1 g cm$^{-3}$ aerogel; spectrum B = 0.1 g cm$^{-3}$ aerogel with water removed from the surface. The 'bulk-phase' liquefied xenon outside the aerogel fragments displays a chemical shift of 133 ppm, the xenon within the porous aerogel resonates at 134.5 ppm (A) and 134.7 ppm (B), where 0 ppm refers to the predicted shift of xenon gas at 273 K and zero pressure. Note the significant broadening of an aerogel peak in the dried sample (B). Adapted with permission from Ref. [27].

of surface hydration. The $^{131}$Xe NMR peak at around 134.5 ppm originates from xenon inside the porous material and shows a strong increase in line width for the dehydrated aerogel compared to the hydrated aerogel sample. The change in line broadening was found to be completely reversible through multiple hydration and dehydration cycles.

The observed $^{131}$Xe line width at 134.5 ppm is caused predominantly by quadrupolar interactions in $^{131}$Xe atoms during brief periods of surface adsorption, when the quadrupolar relaxation is strongly accelerated compared to relaxation in the gas or liquid phase. The aerogel exhibits a large surface area due to the fractal aggregation of 1–2 nm-sized aluminosilicate particles, and allows for a high water uptake of up to 30% of the aerogel dry weight. A substantial change in the surface morphology upon hydration and dehydration was the most likely cause for the observed changes in line width. The most significant contribution to the line width is transverse quadrupolar relaxation. Even without a detailed exploration of the exact mechanism or mechanisms that cause the observed line width, it is clear that surface hydration-dependent line broadening can be used as a source of contrast in spin–echo weighted MRI experiments, as shown in Figure 9.2. Note here that the assignment of $T_2$ or $T_1$ values is questionable for the dehydrated sample in light of the results from MQF experiments [20]; however, a crude monoexponential fitting leads to a much shorter $T_1 = 3.6$ ms for the dehydrated aerogel compared to $T_1 = 11.2$ ms for the hydrated aerogel.

## 9.3
## Feasibility of Hyperpolarized $^{83}$Kr MRI

High-spin density in the condensed phase was needed to generate sufficient signal intensity for the MR images described in Section 9.2. Xenon liquefaction was needed despite the fast quadrupolar-driven $T_1$ relaxation of $^{131}$Xe in

Hydrated aerogel          Dehydrated aerogel

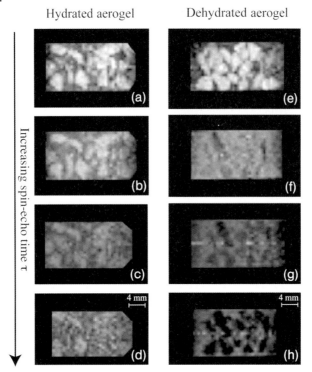

**Figure 9.2** Contrast as a function of echo time (TE) in $^{131}$Xe MR images of hydrated 0.1 g cm$^{-3}$ aerogel fragments (a, b, c, d) and of dehydrated aerogel pieces (e, f, g, h). The echo times used were (a, e): TE = 3.6 ms, (b, f): TE = 7 ms, (c, g): TE = 9 ms, and (d, h) TE = 12 ms. For short echo times both aerogel samples appear brighter than the background. Note that the background is caused by incomplete suppression of the very intense signal arising from liquid xenon surrounding the aerogel fragments. Partial suppression of the background has been accomplished through saturation and $T_1$ weighting. The location of dehydrated fragments in the 10 mm single crystal sapphire tube is shifted compared to the hydrated sample due to the dehydration procedure. For larger echo times, the signal from the dehydrated sample decreases rapidly and cannot be distinguished from the background at TE = 7 ms (f) and appears as 'dark' fragments at TE = 9 ms (g), where the corresponding signal from the nontreated sample is still above the background intensity (c). Adapted with permission from Ref. [27].

microporous aerogels that allows for rapid signal averaging. This fact illustrates the limitations of MRI with thermally polarized spin $I > 1/2$ noble gas isotopes. Even liquefaction (or the use of a high-density supercritical state) will not always provide sufficient MRI signal intensity, in particular in macroporous media with micron- to millimeter-sized pores where the relaxation rates of all noble gas isotopes are too slow. It should be noted that mammalian lungs have typical alveolar diameters ranging from tens of microns to hundreds of microns, and the resulting slow longitudinal relaxation makes the application of thermally polarized spin $I > 1/2$ noble gas MRI impossible, particularly when taking into consideration the required ambient pressure and temperature conditions. Fortunately, the relatively slow relaxation allows for the use of hyperpolarized spin $I > 1/2$ noble gas MRI.

Spin-exchange optical pumping (SEOP) of $I > 1/2$ noble gas isotopes, including $^{83}$Kr, has been described previously [22, 24, 28], but without any attempt to separate the spin $I > 1/2$ noble gas from the highly reactive alkali metal vapor. A further concern is that relaxation during SEOP will limit the obtained hyperpolarization below the required level for MRI. This problem is best illustrated by comparing the spin $I = 1/2$ isotope $^{129}$Xe that has a gas-phase relaxation $T_1$ time on the order of 2 h with the hypothetical relaxation time of $T_1 = 25$ s for gas-phase $^{131}$Xe at 100 kPa [29]. In any practical application, the $^{131}$Xe $T_1$ relaxation time is further reduced due to the presence of surfaces such as container walls [19, 25].

The depolarization problem was solved by choosing $^{83}$Kr because it exhibits relaxation rates that are typically one order of magnitude slower than those found in $^{131}$Xe under similar conditions [9]. The hypothetical longitudinal relaxation time for $^{83}$Kr that would be measured in the absence of a container wall at 300 K, 100 kPa and 2.1 T was reported as $T_1 = 470$ s [30]. Although the presence of the glass surface greatly accelerates longitudinal $^{83}$Kr relaxation, the relaxation times found are long enough to permit short-term storage and transfer of hp $^{83}$Kr gas between various containers at high magnetic field strengths. The $^{83}$Kr $T_1$ relaxation time at 175 kPa pressure in 4–5 cm-long glass cylinders with 11–13.5 mm inner diameters maintained at 289 K and at 9.4 T was found to be reduced to $T_1 = 90$–202 s, depending on the inner diameter and other conditions such as surface hydration and hp gas mixture composition [31, 32]. At low magnetic field strength (i.e. 0.05 T), the relaxation times measured by remotely detected relaxometry in a typical SEOP cell (without alkali metal) at 433 K (i.e. the temperature used for the SEOP process) was $T_1 = 220$ s, which was long enough to allow for the SEOP of $^{83}$Kr [33].

Figure 9.3a shows a continuous-flow hp $^{83}$Kr MR image of a phantom using krypton in bulk gas phase between glass structures. The hp $^{83}$Kr was generated in continuous-flow mode with polarization enhancement factors of about 27 times the thermal equilibrium value [34], and signal averaging led to a total duration of 2.1 h for the MRI experiment. A much higher hp $^{83}$Kr signal enhancement of more than three orders of magnitude (i.e. 1200 times the thermal polarization) compared to thermally polarized $^{83}$Kr at 9.4 T field strength can be obtained in stopped-flow type SEOP experiments [33]. This improvement is enough to allow for MR images with similar resolution, as shown in Figure 9.3, without the need for signal averaging (see Section 9.4). It should be noted that rubidium metal was used for all SEOP experiments described in this chapter.

Although the feasibility of alkali metal-free hp $^{83}$Kr void space imaging of bulk gas phase is demonstrated in Figure 9.3a, the transfer of hp $^{83}$Kr into porous materials may cause a significant reduction in image quality due to the vastly increased surface area that leads to accelerated quadrupolar relaxation in these media. When using hp $^{83}$Kr, the longitudinal relaxation times in various porous

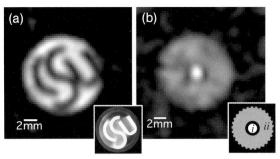

**Figure 9.3** (a) Hyperpolarized $^{83}$Kr void space MR image of a glass phantom (photograph of the phantom shown in the inset figure at the lower right-hand corner of (a). Gas flow was held constant at 125 cm$^3$ min$^{-1}$; (b) Hyperpolarized $^{83}$Kr MR image of a porous polyethylene sample with 70 μm average pore size. The inset sketch (lower right-hand portion of (b)) is of the phantom used for the MRI. The center of the sample (*i*) is a 1.65 mm diameter void space surrounded by a 0.76 mm PFA (Teflon® perfluoroalkoxy copolymer) wall and an 11 mm wide area of a porous polymer (*ii*). The measurement obtained under continuous flow (100 cm$^3$ min$^{-1}$) conditions of hp $^{83}$Kr at 9.4 T took about 2.1 h and led to a raw data resolution of 650 × 650 μm. Adapted with permission from Ref. [34].

polyethylene samples with average pore sizes ranging from 70 to 250 μm, and with different chemical surface compositions, were determined to range between $T_1 = 2.5$–5.9 s [31]. Figure 9.3b demonstrates the feasibility of hp $^{83}$Kr MRI in a porous polymer with a 70 μm average pore diameter, which is well below the average alveolar diameter of 225 μm in the adult human lung. For comparison, an average alveolar diameter of 94 μm is reported for rat lung, while 58 μm-diameter alveoli are typically found in mice [35]. It should be noted that that the hydrophobic polymer surfaces presumably cause faster $^{83}$Kr relaxation than the lung surfaces at similar surface-to-volume ratios.

Another estimate of $T_1$ time that could be expected for pulmonary MRI with hp $^{83}$Kr was obtained with a desiccated canine lung tissue sample that was structurally similar to *in vivo* samples [34]. A spin-lattice relaxation time of $T_1 = 10.5$ s was determined for hp $^{83}$Kr at 9.4 T and 289 K in this sample. Recently, *ex vivo* hp $^{83}$Kr measurements in freshly excised rat lungs found $T_1$ times on the order of few seconds (Z.I. Cleveland *et al.*, 2008, unpublished results).

## 9.4
### Hyperpolarized $^{83}$Kr as a Contrast Agent for MRI

In Section 9.3 the feasibility of using hp $^{83}$Kr for gas-space MRI of porous materials, including desiccated lungs was outlined but the currently obtained hyperpolarization of $^{83}$Kr is still substantially below that obtained with spin $I = 1/2$ noble gas isotopes. Nonetheless, hp $^{83}$Kr is still regarded as a very useful novel contrast agent because it provides information that is complementary to that obtained with hp $^3$He and hp $^{129}$Xe. Interactions of the $^{83}$Kr quadrupole moment with the electronic environment are modulated by surface adsorption processes, and are reflected in the $^{83}$Kr nuclear spin relaxation. In macroporous materials, where the $^{129}$Xe chemical shift is typically of little diagnostic value, the quadrupolar-driven relaxation of $^{83}$Kr is a sensitive probe of the surfaces. The $T_1$ time of hp $^{83}$Kr is affected by the surface-to-volume ratio, chemical composition [31], and the temperature [33] of the surfaces that are in contact with the krypton gas.

## 9.4.1
### Effect of the Surface-to-Volume Ratio on $^{83}$Kr Relaxation

The effect of the surface-to-volume ratio on $^{83}$Kr $T_1$ relaxation is best understood by considering the long $T_1$ relaxation time of $^{83}$Kr gas in the absence of surfaces (see Section 9.3) that can last for hundreds of seconds, depending on the gas pressure. Therefore, the dominant contribution to the longitudinal relaxation of $^{83}$Kr in porous materials is caused by quadrupolar interactions during brief periods of surface adsorption that strongly reduce the apparent gas-phase relaxation time in the vicinity of the surface. An increased surface-to-volume ratio leads to a higher number of surface adsorption events per krypton atom, and thus to a faster relaxation. In very high-surface-area materials such as zeolites, the longitudinal relaxation time can range from milliseconds to tens of milliseconds [18].

Generally, the $^{83}$Kr relaxation rates increase with decreasing pore size because of the increasing surface-to-volume ratio. The $^{83}$Kr $T_1$ relaxation dependence on surface-to-volume ratio in macroscopic systems was demonstrated using a series of borosilicate glass bead samples with differing bead diameters ranging from 0.1 to 2.5 mm. The ideal packing of equally sized spherical beads leads, in theory, to a total pore volume that is 26% of the entire sample, independent of the chosen bead diameter [36]. The geometry of the pores (i.e. tetrahedral and octahedral holes) is also unaffected by the bead diameter, and simple geometric considerations show that the surface-to-volume ratio is inversely proportional to the bead radius [31]. Figure 9.4 depicts experimentally determined hp $^{83}$Kr relaxation rates ($1/T_1$) as a function of inverse bead radius $1/r$ for different surfaces. The $^{83}$Kr relaxation rates were found to increase with decreasing bead radius (i.e. increasing surface-to-volume ratio), as expected, for all surfaces studied. A linear dependence of relaxation rates on the inverse radii is predicted if quadrupolar interactions on the surface are the only source for the observed relaxation and the beads of perfect spherical shape are used. The deviation from the linear behavior apparent in Figure 9.4 was attributed to variations in bead diameter, nonideal packing and nonuniform bead surfaces with complex topology.

## 9.4.2
### Effect of Surface Chemical Composition on $^{83}$Kr Relaxation

The effect of surface chemistry on $^{83}$Kr $T_1$ relaxation was demonstrated through surface-chemistry modifications of the beads, and is also shown in Figure 9.4. As

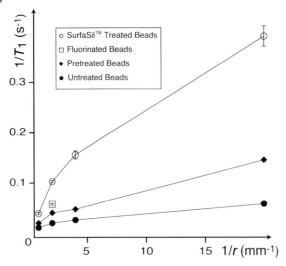

**Figure 9.4** ⁸³Kr relaxation rates $1/T_1$ versus inverse bead radii $1/r$ using hp ⁸³Kr (95% Kr, 5% N₂) and closest packed glass beads with various surface treatments, as indicated. The untreated beads were used as obtained from the supplier. The 'pretreated' beads were exposed to an NH₄OH/H₂O₂ wash followed by an HCl/H₂O₂ wash. Similar treatments are known to increase the hydrophobicity of glass surfaces slightly [37]. The surfaces with higher hydrophobicity were coated by chemical reaction with a fluorosilane siliconizing agent or with siloxane using Surfrasil™. The standard deviations resulting from at least four replicate measurements are represented by the error bars; the connecting lines are intended only as guidance for the eyes. Reproduced with permission from Ref. [31].

the surface hydrophobicity at constant surface-to-volume ratio is increased, the relaxation rate is also increased. The nonpolar, but highly polarizable, krypton electron cloud leads to higher surface adsorption enthalpies for nonpolar, hydrophobic surfaces than for polar, hydrophilic surfaces. The observed $T_1$ in the untreated 0.1 mm mean diameter borosilicate glass bead sample (hydrophilic) shown in Figure 9.4 is approximately sixfold that of the siliconized sample (strongly hydrophobic) with the same bead diameter. This effect is based solely on the surface chemistry of the porous materials. The approximately fourfold difference in ⁸³Kr relaxation (from $T_1 = 35.3$ s to $T_1 = 9.0$ s) by siliconizing the untreated glass surface observed in the 1.0 mm beads was used to generate the $T_1$ contrast shown in Figure 9.5.

A variety of hydrophobic substances in tobacco smoke condensate accelerate the ⁸³Kr $T_1$ relaxation, and the effect can be used to generate hp ⁸³Kr MRI contrast [38]. This effect may be helpful for mapping the regions of cigarette smoke particulate deposition within the lung, and thus may serve for the refinement of mathematical models of particulate deposition in the respiratory tract [39, 40]. Figure 9.6 demonstrates a clear contrast between smoke-treated and untreated surfaces of glass capillaries in variable flip angle fast low-angle shot (FLASH) hp ⁸³Kr MRI. Of potential biomedical relevance is that a smoke-induced acceleration of the ⁸³Kr

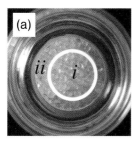

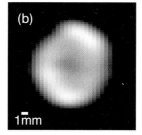

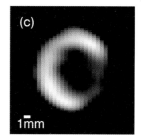

**Figure 9.5** (a) Photograph of surface-siliconized 1.0 mm hydrophobic glass beads located in the center area (*i*) that are surrounded by an outer region (*ii*) of untreated 1.0 mm glass beads. Both regions are separated by an untreated glass ring (white); (b) Hyperpolarized ⁸³Kr MRI of the glass bead sample, taken approximately 3 s after filling the sample with hp krypton (note, the glass wall is not resolved as the resolution is 424 × 864 µm); (c) Same as (b) except that the hp ⁸³Kr MRI is recorded after a 9 s waiting period. A clear contrast between the hydrophobic inner sample region ($T_1 = 9$ s) and the hydrophilic outer region ($T_1 = 35$ s) appears. Adapted with permission from Ref. [34].

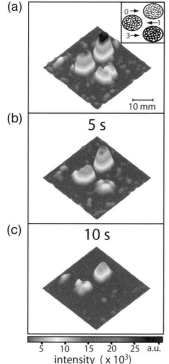

**Figure 9.6** Three $T_1$-weighted hp ⁸³Kr MR images obtained without signal averaging by a variable angle FLASH sequence. The scale at the bottom of the figure indicates the signal intensity in arbitrary units.
(a) FLASH MRI acquisition immediately after stopped-flow transfer of the hp ⁸³Kr into the sample; (b) Image acquired 5 s after hp ⁸³Kr transfer; (c) Image acquired 10 s after hp ⁸³Kr transfer. The inset in the top right is a sketch of the phantom used to produce the images, where the labels 1–3 indicate the number of cigarettes combusted for smoke deposition in a given tube. Reproduced with permission from Ref. [38].

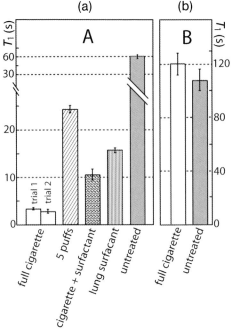

**Figure 9.7** (a) Hp $^{83}$Kr $T_1$ values for a 95% Kr and 5% N$_2$ gas mixture in contact with 1 mm-diameter borosilicate glass beads. Where indicated, the beads were exposed to cigarette smoke of one full cigarette or 5 puffs only; (b) Hp $^{129}$Xe $T_1$ values for a 20% Xe, 75% He and 5% N$_2$ gas mixture for comparison. The label 'lung surfactant' describes borosilicate glass beads that were coated with the commercially available bovine pulmonary extract Survanta®. Reproduced with permission from Ref. [38].

self-relaxation was also observed using glass surfaces coated with bovine lung surfactant extract (see Figure 9.7 for details), which suggested that hp $^{83}$Kr might represent a promising MRI contrast agent to explore the spatial distribution of particle deposition and the persistence of particle deposits in the lung.

### 9.4.3
### Effect of Paramagnetic Impurities and Molecular Oxygen on $^{83}$Kr Relaxation

A similar increase in longitudinal relaxation to that of hp $^{83}$Kr interacting with hydrophobic surfaces has also been observed for hp $^{131}$Xe ($I = 3/2$) on surface-treated optical pumping cell walls [21]. Remarkably, the opposite effect on a much longer time scale has been observed with $^{129}$Xe and $^3$He NMR because the surface coating insulates the noble gas atoms from paramagnetic sites in the glass surface [41–43]. In the case of $^{83}$Kr and $^{131}$Xe, the separation from paramagnetic sites in the glass surface is outweighed by increased quadrupolar relaxation on the hydrophobic surface.

There is however another factor, namely the low $^{83}$Kr gyromagnetic ratio (see Section 9.1), that causes the $^{83}$Kr nuclear spin to be relatively little affected by paramagnetism because of the $T_1 \propto \gamma_{nuc}^{-2}$ dependence of paramagnetic relaxation. For example, the ratio $(\gamma_{129\,Xe}/\gamma_{83\,Kr})^2 \approx 50$ leads in theory to a ~50-fold lower sensitivity of the $^{83}$Kr longitudinal relaxation to paramagnetic species compared to that of $^{129}$Xe [32, 34]. This effect is not offset by the distance $r$ dependence of paramagnetic relaxation ($T_1 \propto r^6$) because the 1.9 Å van der Waals radius of krypton is only slightly smaller than the 2.2 Å radius of xenon. In fact, the smaller krypton radius leads to a lower polarizability of the krypton electron cloud that reduces its surface-adsorption affinity and thus decreases the contact with the paramagnetic sites. The comparison of $^{83}$Kr relaxation to that of $^{129}$Xe, as shown in Figure 9.7, has been used to rule out paramagnetic species within the deposited tobacco smoke as the cause for the MRI contrast shown in Figure 9.6.

The insensitivity of $^{83}$Kr to paramagnetic substances can be advantageous for future *in vivo* MRI of lungs where a breathable mixture of hp $^{83}$Kr with molecular oxygen will be required. A significant acceleration of the $^{83}$Kr relaxation by paramagnetic oxygen would not only lead to a faster depolarization of hp $^{83}$Kr but would also obscure the contrast caused by quadrupolar-driven surface-sensitive relaxation. However, the presence of 20% $O_2$ was previously found to reduce the $^{83}$Kr $T_1$ time in desiccated canine lung tissue by only 18%, from 10.5 to 8.6 s [34]. In comparison, the high gyromagnetic ratio of $^3$He causes the $T_1$ time that can be hundreds of hours in the absence of paramagnetic species [44, 45] to be reduced to 10–20 s in lungs containing a breathable oxygen mixture [46, 47].

### 9.4.4
### Effect of Surface Temperature on $^{83}$Kr Relaxation

Using remotely detected relaxometry, a $^{83}$Kr spin-lattice relaxation of $T_1 = 85$ s was measured at 0.05 T and 297 K in a 24 mm internal diameter and 53 mm-long Pyrex storage cell [33]. The observed relaxation time rose by a factor of approximately 2.6 to $T_1 = 220$ s when the temperature of the storage cell was increased by a factor of 1.5 to 433 K. This effect is caused by surface-dominated relaxation that is reduced at higher temperature due to decreased surface residence times leading to an expected exponential dependence of $T_1$ with the temperature.

## 9.5
## Hyperpolarized $^{83}$Kr MRI Practicalities

### 9.5.1
### The Effect of Gas Mixture on $^{83}$Kr Polarization and on $^{83}$Kr Relaxation

A $^{83}$Kr signal enhancement of 1200 times the thermal equilibrium signal at 9.4 T field strength was reported in Section 9.3, obtained through SEOP with rubidium (RbSEOP). The RbSEOP mixture used for this experiment consisted of 95% Kr and 5% N$_2$, and is denoted as mixture-I for the remainder of this chapter. A different mixture (mixture-II) consisting of 25% Kr, 70% He and 5% N$_2$ yielded polarization enhancements up to 4500 times that of thermal equilibrium [32]. The effect of increasing polarization with decreasing xenon concentration is well known for $^{129}$Xe SEOP, and is probably due to reduced Rb electron spin relaxation and slower $^{83}$Kr self-relaxation during the RbSEOP process. Unfortunately, the improved spin polarization in mixture-II does not lead to a higher signal intensity because of the reduced spin density of the dilute krypton mixture. In contrast to SEOP with $^{129}$Xe [2], there is currently no concentration process available that would preserve $^{83}$Kr in its hyperpolarized state. However, the effect of krypton concentration on polarization is still useful as the concentration can be varied over a fairly broad range, without causing substantial changes in the magnitude of the obtained signal. Therefore, the hp $^{83}$Kr relaxation that depends on the composition of the gas mixture (see Figure 9.8) can be adjusted (or 'tuned') to some extent to meet experimental needs. At ambient pressure (85.0 kPa) and 289 K, the $^{83}$Kr relaxation from gas mixture-I in untreated dehydrated borosilicate glass beads is

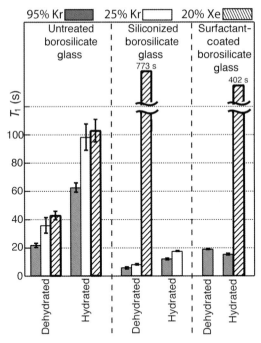

**Figure 9.8** $T_1$ values of hp $^{83}$Kr and hp $^{129}$Xe in contact with hydrated and dehydrated 1 mm-diameter borosilicate glass bead samples with various surfaces. The error bars represent standard deviations resulting from at least four replicate measurements. All gas mixtures contained 5% nitrogen and were balanced with helium (if applicable). The dehydrated samples were obtained through prolonged exposure to room temperature vacuum. The hydrated samples were exposed to the hp gases without prior evacuation. The lung surfactant used is the same as in Figure 9.7. Reproduced with permission from Ref. [32].

prolonged by 63%, from $T_1 = 22.1\,s$ to $T_1 = 36\,s$, for mixture-II. Similar changes can be seen for the other samples described in Figure 9.8. The cause of this effect is unknown and will require further investigation. It should be noted that the $^{83}$Kr relaxation measurements in porous materials are relatively little affected by changes in the total gas pressure, leading to reliably reproducible data for a given surface at constant temperature and gas composition.

### 9.5.2
### Effect of Surface Hydration on $^{83}$Kr Relaxation

Reminiscent of the line-width effect observed with $^{131}$Xe in aerogels, as described in Section 9.2, surface hydration leads also to prolonged $^{83}$Kr longitudinal relaxation time in glass bead samples (see Figure 9.8) with surface-to-volume ratios that are orders of magnitude smaller than those of aerogels [32]. The explanation for this effect is similar to that for the reduced relaxation in dilute krypton gas mixtures. Competitive surface adsorption of water molecules reduces the $^{83}$Kr surface adsorption. Amazingly, a strong effect of water vapor on $^{83}$Kr relaxation is also observed for siliconized (hydrophobic) surfaces. Of note, a special apparatus had to be developed to allow for the measurement of surfaces under near-ambient conditions without the need to pre-evacuate the sample for the hp gas transfer, as described in Ref. [32].

### 9.5.3
### The Effect of Magnetic Field Strength on $^{83}$Kr Relaxation

Measurements in macroscopic glass containers and in desiccated canine lung tissue at field strengths between 0.05 and 3 T using remotely detected hp $^{83}$Kr NMR spectroscopy have revealed that the longitudinal relaxation dramatically accelerates as the magnetic field strength decreases [33]. The relaxation time of krypton in lung tissue was found to be $T_1 = 10.5\,s$ at 9.4 T using 95% krypton gas with no oxygen present (see Section 9.3). The relaxation time in this tissue dropped to $T_1 = 7\,s$ at 3 T magnetic field strength and to $T_1 = 5\,s$ at 1.5 T magnetic field. The relaxation was slow enough to suggest that *in vivo* hp $^{83}$Kr MRI of human lungs at 1.5 T may be feasible, in particular considering the relaxation-slowing effects of surface hydration and the possibility of decreasing the $^{83}$Kr relaxation rate through dilution of the krypton concentration, as described above.

### 9.6
### Conclusions

Hyperpolarized $^{83}$Kr MRI is a technology that is still in its infancy, and further research is required to determine its practical value. However, preliminary results obtained with hp $^{83}$Kr point towards two potential paradigm shifts for MRI with hyperpolarized spin systems. Thus far, relaxation has been considered primarily as a limiting factor for hp MRI applications as it is destructive for the hyperpolarized spin state. A noticeable exception is the use of depolarization through paramagnetic relaxation to determine oxygen partial pressures in lungs [47]. However, it is worth embracing the idea of using relaxation also as a source of information about surfaces in hp MRI, perhaps even with spin $I = 1/2$ systems. In the case of hp $^{83}$Kr, the $T_1$ relaxation has been shown to be sensitive to surface-to-volume ratios, as well as to the chemical composition of surfaces, surface hydration and tobacco smoke deposition. The $^{83}$Kr $T_1$ relaxation time can be 'tuned' by choosing different optical pumping mixtures and is also affected by the applied magnetic field strength. The comparison of hp $^{83}$Kr relaxation with that of hp $^{129}$Xe can provide further information about surfaces. Additional surface-sensitive NMR parameters such as $^{83}$Kr $T_2$ and $T_{1\text{rho}}$ relaxation, as well as parameters obtained through multiple quantum filtered hp $^{83}$Kr MRI, are still unexplored to date.

The second conceptually new idea is that a low gyromagnetic ratio can actually be beneficial for MRI because of reduced paramagnetic relaxation. This is despite the inherently low signal intensities associated with low gyromagnetic ratios, provided that a hyperpolarized spin state can at least partially offset this problem. The effect of hyperpolarization on $^{83}$Kr signal intensity is complemented by the high-spin $I = 9/2$ and perhaps by reduced inductive losses compared to other spin systems with higher resonance frequency [48]. The SEOP process thus far has only generated a 0.3% $^{83}$Kr spin polarization in a high-krypton concentration gas mixture. Further polarization enhancements with improved SEOP and laser technology are one focal point of current research, although an increase in signal-to-noise ratio of almost an order of magnitude is possible without further technological advancements by switching to isotopically enriched gas. Natural-abundance krypton gas can be obtained from the atmosphere at costs that are currently 10- to 20-fold lower than those for natural-abundance xenon. This low cost of krypton gas may result in currently prohibitively expensive, isotopically enriched $^{83}$Kr becoming more affordable in the future. As a final note, krypton – unlike xenon – does not possess any anesthetic properties at ambient pressure [49].

## Acknowledgments

The studies described in this chapter were largely supported by the National Science Foundation under Grant numbers CHE-0135082 and CHE-0719423.

## References

1 Walker, T.G. and Happer, W. (1997) *Reviews of Modern Physics*, **69**, 629–42.

2 Raftery, D., Long, H., Meersmann, T., Grandinetti, P.J., Reven, L. and Pines, A. (1991) *Physical Review Letters*, **66**, 584–7.

3 Goodson, B.M. (2002) *Journal of Magnetic Resonance*, **155**, 157–216.

4 Raftery, D. (2006) *Annual Reports on NMR Spectroscopy*, **57**, 205–7.

5 Schroder, L., Lowery, T.J., Hilty, C., Wemmer, D.E. and Pines, A. (2006) *Science*, **314**, 446–9.

6 Albert, M.S., Cates, G.D., Driehuys, B., Happer, W., Saam, B., Springer, C.S. and Wishnia, A. (1994) *Nature*, **370**, 199–201.

7 Mair, R.W., Hrovat, M.I., Patz, S., Rosen, M.S., Ruset, I.C., Topulos, G.P., Tsai, L.L., Butler, J.P., Hersman, F.W. and Walsworth, R.L. (2005) *Magnetic Resonance in Medicine*, **53**, 745–9.

8 Harris, R.K., Becker, E.D., De Menezes, S.M.C., Goodfellow, R. and Granger, P. (2001) *Pure and Applied Chemistry*, **73**, 1795–818.

9 Holz, M., Haselmeier, R., Klein, A. and Mazitov, R.K. (1995) *Applied Magnetic Resonance*, **8**, 501–19.

10 Luhmer, M. and Reisse, J. (1998) *Progress in Nuclear Magnetic Resonance Spectroscopy*, **33**, 57–76.

11 Ingman, P., Jokisaari, J. and Diehl, P. (1991) *Journal of Magnetic Resonance*, **92**, 163–9.

12 Jokisaari, J., Ingman, P., Lounila, J., Pukkinen, O., Diehl, P. and Muenster, O. (1993) *Molecular Physics*, **78**, 41–54.

13 Long, H.W., Luzar, M., Gaede, H.C., Larsen, R.G., Kritzenberger, J., Pines, A. and Crawford, G.P. (1995) *Journal of Physical Chemistry*, **99**, 11989–93.

14 Li, X.X., Newberry, C., Saha, I., Nikolaou, P., Whiting, N. and Goodson, B.M. (2006) *Chemical Physics Letters*, **419**, 233–9.

15 Moudrakovski, I.L., Ratcliffe, C.I. and Ripmeester, J.A. (2001) *Journal of the American Chemical Society*, **123**, 2066–7.

16 Millot, Y., Man, P.P., Springuel-Huet, M.A. and Fraissard, J. (2001) *Comptes Rendus de l'Académie des Sciences. Série II. C*, **4**, 815–18.

17 Clewett, C.F.M. and Pietrass, T. (2005) *Journal of Physical Chemistry B*, **109**, 17907–12.

18 Horton-Garcia, C.F., Pavlovskaya, G.E. and Meersmann, T. (2005) *Journal of the American Chemical Society*, **127**, 1958–62.

19 Meersmann, T., Smith, S.A. and Bodenhausen, G. (1998) *Physical Review Letters*, **80**, 1398–401.

20 Meersmann, T., Deschamps, M. and Bodenhausen, G. (2001) *Journal of the American Chemical Society*, **123**, 941–5.

21 Wu, Z., Happer, W., Kitano, M. and Daniels, J. (1990) *Physical Review A: General Physics*, **42**, 2774–84.

22 Raftery, D., Long, H.W., Shykind, D., Grandinetti, P.J. and Pines, A. (1994) *Physical Review A: General Physics*, **50**, 567–74.

23 Butscher, R., Wäckerle, G. and Mehring, M. (1994) *Journal of Chemical Physics*, **100**, 6923–33.

24 Butscher, R., Wäckerle, G. and Mehring, M. (1996) *Chemical Physics Letters*, **249**, 444–50.

25 Meersmann, T. and Haake, M. (1998) *Physical Review Letters*, **81**, 1211–4.

26 Deschamps, M., Burghardt, I., Derouet, C., Bodenhausen, G. and Belkic, D. (2000) *Journal of Chemical Physics*, **113**, 1630–40.

27 Pavlovskaya, G., Blue, A.K., Gibbs, S.J., Haake, M., Cros, F., Malier, L. and Meersmann, T. (1999) *Journal of Magnetic Resonance*, **137**, 258–64.

28 Schaefer, S.R., Cates, G.D. and Happer, W. (1990) *Physical Review A*, **41**, 6063–70.

29 Brinkmann, D., Brun, E. and Staub, H.H. (1962) *Helvetica Physica Acta*, **35**, 431–6.

30 Brinkmann, D. and Kuhn, D. (1980) *Physical Review A*, **21**, 163–7.

31 Stupic, K.F., Cleveland, Z.I., Pavlovskaya, G.E. and Meersmann, T. (2006) *Solid State Nuclear Magnetic Resonance*, **29**, 79–84.

32 Cleveland, Z.I., Stupic, K.F., Pavlovskaya, G.E., Repine, J.E., Wooten, J.B. and Meersmann, T. (2007) *Journal of the American Chemical Society*, **129**, 1784–92.

33 Cleveland, Z.I., Pavlovskaya, G.E., Stupic, K.F., LeNoir, C.F. and Meersmann, T. (2006) *Journal of Chemical Physics*, **124**, 044312.

34 Pavlovskaya, G.E., Cleveland, Z.I., Stupic, K.F. and Meersmann, T. (2005) *Proceedings of the National Academy of Sciences of the United States of America*, **102**, 18275–9.

35 Ochs, M., Nyengaard, L.R., Jung, A., Knudsen, L., Voigt, M., Wahlers, T., Richter, J. and Gundersen, H.J.G. (2004) *American Journal of Respiratory and Critical Care Medicine*, **169**, 120–4.

36 Housecroft, C.E. and Sharpe, A.G. (2005) *Inorganic Chemistry*, 2nd edn, Pearson, Prentice Hall, Harlow.

37 Cras, J.J., Rowe-Taitt, C.A., Nivens, D.A. and Ligler, F.S. (1999) *Biosensors & Bioelectronics*, **14**, 683–8.

38 Cleveland, Z.I., Pavlovskaya, G.E., Stupic, K.F., Wooten, J.B., Repine, J.E. and Meersmann, T. (2008) *Magnetic Resonance Imaging*, 26.

39 Nazaroff, W.W., Hung, W.Y., Sasse, G.B.M. and Gadgil, J. (1993) *Aerosol Science and Technology*, **19**, 243–54.

40 Yeh, H.C. and Schum, G.M. (1980) *Bulletin of Mathematical Biology*, **42**, 461–80.

41 Driehuys, B., Cates, G.D. and Happer, W. (1995) *Physical Review Letters*, **74**, 4943–6.

42 Breeze, S.R., Lang, S., Moudrakovski, I., Ratcliffe, C.I., Ripmeester, J.A., Santyr, G., Simard, B. and Zuger, I. (2000) *Journal of Applied Physiology*, **87**, 8013–7.

43 Jacob, R.E., Driehuys, B. and Saam, B. (2003) *Chemical Physics Letters*, **370**, 261–7.

44 Hsu, M.F., Cates, G.D., Kominis, I., Aksay, I.A. and Dabbs, D.M. (2000) *Applied Physics Letters*, **77**, 2069–71.

45 Babcock, E., Chann, B., Walker, T.G., Chen, W.C. and Gentile, T.R. (2006) *Physical Review Letters*, **96**, 083003.

46 Moller, H.E., Chen, X.J., Saam, B., Hagspiel, K.D., Johnson, G.A., Altes, T.A., de Lange, E.E. and Kauczor, H.U. (2002) *Magnetic Resonance in Medicine*, **47**, 1029–51.

47 Fischer, M.C., Kadlecek, S., Yu, J.S., Ishii, M., Emami, K., Vahdat, V., Lipson, D.A. and Rizi, R.R. (2005) *Academic Radiology*, **12**, 1430–9.

48 Hoult, D.I., Chen, C.N. and Sank, V.J. (1986) *Magnetic Resonance in Medicine*, **3**, 730–46.

49 Cullen, S.C. and Gross, E.G. (1951) *Science*, **113**, 580–2.

# 10
# Fast Gradient-Assisted 2-D Spectroscopy in Conjunction with Dynamic Nuclear Polarization

*Rafal Panek, James Leggett, Nikolas S. Andersen, Josef Granwehr,*
*Angel J. Perez Linde and Walter Köckenberger*

## 10.1
## Introduction

Nuclear magnetic resonance (NMR) is a relatively insensitive spectroscopic technique due to the weak magnetic moments of the nuclear spins and the correspondingly small nuclear Zeeman energy splitting, which result in only small levels of nuclear polarization at thermal equilibrium. From the earliest stages of NMR research, much effort has been directed into finding a general strategy to overcome this limitation. One possibility is to enhance nuclear polarization via optical pumping [1]. This technique depends on the hyperfine interaction between spin-polarized electrons produced by optical excitation and nuclear spins of noble gases such as $^3$He and $^{129}$Xe. The implementation of this approach has led to rapid progress in *in vivo* lung imaging [2] and a range of other interesting applications, including the characterization of porous media and material surfaces [3]. However, attempts to transfer the polarization of $^{129}$Xe nuclei to other chemical compounds via thermal mixing in the solid state has so far proved to be difficult and has resulted in only small enhancements [4]. The use of parahydrogen provides another means of generating large polarization of nuclear spins [5]. This approach requires the pair-wise transfer of parahydrogen nuclei onto a suitable target molecule in a hydrogenation reaction. However, whilst the hydrogenation technique with parahydrogen can generate large spin polarization (in principle 100%), its general use has been limited by the requirements for target molecules with suitable double or triple bonds that can be functionalized with parahydrogen. Hence, to date mainly inorganic molecules in organic solvents have been hyperpolarized using this technique, and in particular this technique has proved to be useful in organometallic catalysis research to monitor the intermediates and products of hydrogenation reactions [6].

A third approach, known as dynamic nuclear polarization (DNP), which was described in 1958 by Abragam *et al.* [7] for insulators, can significantly enhance nuclear polarization by transferring the large electron polarization to the nuclear

*Magnetic Resonance Microscopy.* Edited by Sarah L. Codd and Joseph D. Seymour
Copyright © 2009 WILEY-VCH Verlag GmbH & Co. KGaA, Weinheim
ISBN: 978-3-527-32008-0

spin system in solids. The maximum possible DNP enhancement is proportional to the ratio of the electron and nuclear gyromagnetic ratios and, in principle, enhancements of three orders of magnitude relative to thermal polarization can be obtained (theoretical enhancement for $^1$H: ~660, for $^{13}$C ~2600). The polarization transfer from the unpaired electron containing free radical molecules to the nuclear spin system is driven by microwave (mw) irradiation at or near the electron Larmor frequency. The overall DNP enhancement depends on the details of the polarization transfer pathways and the properties of the participating molecules. This strategy has been used recently to enhance the polarization of nuclear spins in solid-state NMR in conjunction with magic angle spinning at a temperature of 100 K [8]. Furthermore, it was demonstrated that the nuclear spin polarization generated at low temperature (~1.5 K) using DNP and a purpose-designed triaryl radical with narrow electron line width can be preserved in a fast temperature jump which involved the fast dissolution of the sample in a hot solvent [9]. Another strategy is based on melting of the sample using the heat energy provided by a laser [10]. The solution containing highly polarized spin systems can then be used in *in vitro* liquid-state spectroscopy experiments, or in *in vivo* magnetic resonance imaging (MRI) experiments which require the rapid injection of the solution into an animal system [11]. The fast change of temperature results in an additional enhancement in comparison to thermal polarization at ambient temperature which is due to the Bolzmann factor. The total enhancement $\varepsilon^\dagger = \varepsilon \times (T_{detect}/T_{DNP})$, where $\varepsilon$ is the enhancement due to DNP, $T_{detect}$ is the temperature at which the signal is detected, and $T_{DNP}$ is the temperature at which the polarization was built up using DNP [10].

In liquid state the nonthermal polarization generated in a dissolution or melting DNP experiment is decaying with the longitudinal relaxation time constant $T_1$ which limits the time period during which the signal can be detected with an appreciable enhancement (after $5 \times T_1$ only 0.7% of the signal is left). Furthermore, a repetition of the solution NMR experiment with spins polarized to the same high degree can only be performed after the polarization-enhancement step using DNP has been repeated. These limitations of liquid-state NMR experiments with hyperpolarized spin systems make it necessary to use fast and efficient spectroscopy techniques to acquire the maximum information in a time short enough so that a high level of spin polarization is still available during signal acquisition.

Many multidimensional NMR spectroscopy techniques that can be used to measure connectivity, distances and exchange in a molecular spin system require repeated excitation of the spin ensemble in conjunction with incrementing evolution delays to obtain the desired modulation of the NMR signal due to chemical shift or spin–spin interactions [12]. However, it was recently demonstrated by Frydman *et al.* [13] that it is possible to create a spatial dependence of these evolution delays by using a special excitation scheme based on frequency-swept radiofrequency pulses (chirp pulses) [14]. This strategy can be used to acquire a set of data with a single excitation of the sample from which a multidimensional spectrum can be reconstructed. Schemes based on this principle are ideally suited to maximize the information that can be obtained in single fast acquisition from a spin system that was highly polarized by DNP. A first demonstration of DNP-enhanced fast 2-D spectroscopy was published recently by Frydman and Blazina using a commercial polarizer and methanol as a solvent [15]. In the following, we report on the implementation of these techniques in conjunction with the design and construction of a home-built DNP polarizer. Furthermore, we show that it is possible to use water as a solvent in such experiments, thus demonstrating the potential use of this strategy for applications to molecular systems with biological relevance.

## 10.2
## The DNP Stand-Alone-Polarizer

The design of our stand-alone polarizer (Figure 10.1) is based on the prototype developed by Wolber *et al.* [16]. The 3.35 T magnet and variable temperature insert (VTI) are components of the commercial Hypersense system (Oxford Instruments Molecular Biotools, Abington, UK). A needle valve controls the flow of liquid He

(a)                                                               (b)

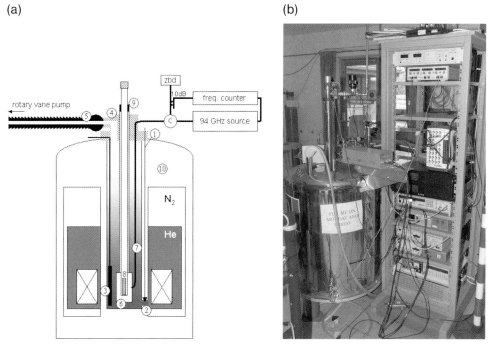

**Figure 10.1** (a) Schematic diagram of the stand-alone polarizer. 1, needle valve handle; 2, needle valve; 3, He level gauge; 4, commercial variable temperature insert; 5, vacuum line with cut-off valve; 6, microwave shield; 7, waveguide; 8, sample cup with sample holder; 9, vacuum-tight seal; 10, 3.35 T magnet; (b) Photograph of the experimental set-up showing the polarizer and control unit.

between the main Dewar within the magnet and the space of the inner bore. The VTI has a pump port to which a strong rotary vane pump (model Duo 65 m; Pfeiffer Vacuum, Asslar, Germany) is connected through a 50 mm-diameter vacuum line and a zeolite filter. The filter is important to prevent the diffusion of oil vapor into the polarizer under steady-state pumping conditions. This arrangement makes it possible to generate a temperature of 1.5 K by pumping on the He bath, whilst liquid He can be continuously supplied from the main magnet dewar. The He level in the inner bore space is monitored by a capacitance gauge consisting of two concentric stainless steel tubes [17] fitted in the lower part of the VTI next to the mw shield and connected to a capacitance-to-digital converter (AD7746; Analog Devices, Norwood, MA, USA) which directly outputs the capacitance. The capacitance variations are proportional to the He level in the inner bore space [17, 18]. The temperature in the VTI is measured by two different techniques. An Allan–Bradley resistor is used at the lower end of the VTI that accommodates the sample. A pressure gauge (model 720; Setra Systems Inc., Boxborough, MA, USA) mounted on the top panel of the VTI makes it possible to calculate the temperature through conversion of the He pressure measured in the system. He boil-off in the polarizer can be monitored using a He flow meter (model F-111B-5K0-AAD-22-V; Bronkhorst, Ruurlo, Holland) attached to the outlet of the rotary vane pump. By measuring the He boil-off rate and the base temperature it is possible to monitor the performance of the polarizer and to determine when the inner sample space of the VTI needs to be cleaned from contamination admitted during the sample loading process.

The VTI carries at its lower end a brass cylinder which acts as a structure for the confinement of the microwaves. The microwaves are transmitted through a circular waveguide from the top end of the VTI down to the lower end, and then fed through the sidewall of the brass cylinder. Mylar foil is used between the waveguide coming from the polarizer and the circulator to seal the waveguide at the outside of the magnet and to assure vacuum tightness of the inner bore space. The waveguide is connected, using a circular to WR10 transition piece, to a circulator which is fed by a 94 GHz (W-band) mw source (Model VCOM-10; ELVA-1, St. Petersburg, Russia) (Figure 10.2). The second port of the circulator is connected to a bidirectional coupler and a zero-biased diode (ELVA-1) that can be used to measure the reflected mw power. The stability of the 94 GHz mw source is improved by using a frequency counter (model 578B, equipped with frequency extension to 110 GHz; Phase Matrix Inc., San Jose, CA, USA) that measures the frequency continuously via the attenuated output (−10 dB) port of the bidirectional coupler. The frequency counter directly adjusts the mw source via its frequency control input.

The stand-alone polarizer is controlled by a LabView program (version 7.1; National Instruments, Austin, TX, USA). By using an external analog-digital/digital-analog device (NI DAQPad-6016; National Instruments), this program continuously monitors the signals from the He level gauge, the pressure sensor and the Allen-Bradley resistor, and makes it possible to change the frequency of the mw source using a GPIB connection to the frequency counter. Furthermore,

**Figure 10.2** Details of the experimental set-up showing the top view of the polarizer with sample holder inserted into the VTI. The ELVA-1 94 GHz mw source can be seen in the background. The vacuum pump line is located behind the He chase gas control panel and can be shut off by a vacuum valve. 1, chase gas control panel; 2, pressure sensor; 3, vacuum control valve; 4, radiofrequency (RF) tuning box; 5, VTI; 6, sample holder; 7, mw source; 8, capacitance meter.

this program can communicate with the NMR console (DSX 400; Bruker, Coventry, UK) using trigger signals through the TTL input and output of the spectrometer.

The sample (~2–5 µl) is inserted into the polarizer in a small polyetheretherketone (PEEK) sample cup (diameter: 3.5 mm) that is kept in position at the bottom end of a long polytetrafluoroethylene (PTFE) tube. During irradiation with mw the sample is positioned at the center of the brass mw shield located at the lower end of the VTI. In a typical dissolution DNP experiment the sample is first continuously irradiated by mw close to the electron paramagnetic resonance (EPR) frequency. The exact frequency depends on the radical type used in the experiment. A saddle coil (diameter 14 mm) placed around the sample cup inside the mw cylinder makes it possible to monitor the build-up of the nuclear polarization. This coil is connected via a transmission line to a vacuum-tight SMA connector in the top panel of the VTI. Tuning and matching of the coil is achieved through a simple capacitor network fitted outside the VTI [19]. For signal excitation and acquisition the high-power channel and the fast ADC of the NMR console was used. Small flip angle pulses were used to minimize the loss of polarization when observing the increase in the solid-state signal, and a bandwidth of 1 MHz was used for signal acquisition.

In addition to the solid-state NMR detection capabilities we have also implemented a strategy to observe the EPR signal arising from unpaired electrons in the radical molecules under DNP conditions. This strategy is based on detection of the longitudinal component of the electron magnetization using a burst of low flip angle mw pulses and a Helmholz coil with its axis aligned parallel to the axis of the static magnetic field [20]. This technique enables us to measure a range of EPR parameters such as the line shape and the electron $T_1$ relaxation time.

## 10.3
## Dissolution and Transfer

For the dissolution step the vacuum is first released using He gas and the sample is lifted with the PTFE tube to a position above the He bath. A stainless steel tube with two fluorinated ethylene propylene (FEP) capillary tubes inside and a shaped PEEK connection piece that forms a tight fit with the sample cup is rapidly pushed through the PTFE tube and onto the sample cup. One of the two capillary tubes is connected to a small, electrically heated pressure-cooker that is mounted on the top end of the stainless steel tube. Hot solvent is injected through this tube into the sample cup. The other capillary tube serves as an outlet for the solution and is connected to a 4 m-long transfer PTFE tube that is used to shuttle the sample from the stand-alone polarizer to the high-resolution 9.4 T magnet. This tube terminates in an adaptor fitted onto a standard 5 mm-diameter NMR sample glass tube. When the stainless steel tube is moved rapidly into the polarizer and pushed onto the sample cup, an optical detector generates a trigger signal that opens a pneumatic valve (model HM20-4VSC-E1W4; Ham-Let, Lindfield, UK) at the outlet of the pressure-cooker and a small amount of superheated water (~2–3 ml, ~120–150 °C, corresponding to 5 bar pressure) is injected via the capillary tube into the sample cup. The sample is rapidly dissolved and the solution is driven into the transfer tube. Another valve that controls influx of He gas into the pressure cooker is opened to pneumatically transfer the solution over the full distance between the polarizer and the high-field magnet and into the glass NMR tube. The transfer time depends on the pressure in the pressure-cooker generated by heating the solvent, and the pressure of the He chase gas. In our set-up these parameters can be adjusted, as can the time delay between opening the pneumatic valve to release the hot solvent and the opening of the valve which controls the He chase gas. Furthermore, the transfer tube and the NMR tube can be evacuated using a small vacuum pump to decrease the transfer time further. The transfer tube between the polarizer and the high field magnet is double-walled, which makes it possible to adjust the temperature of the solution in the inner tube during the transfer using a temperature-controlled water flow in the surrounding tube.

## 10.4
## DNP Mechanisms

Various DNP mechanisms have been described in the literature (for reviews, see Atsarkin [21] Abragam and Goldman [22] and Hu [23]). Three important mechanisms are the Overhauser effect, the solid effect and thermal mixing. The

Overhauser effect, which is based on cross-relaxation, requires a time-dependent modulation of the electron–nuclear interaction on the EPR time scale, and therefore can be used only in liquid-state experiments or in metals [24]. Both, the solid effect and thermal mixing lead to DNP in systems with time-independent electron–nuclear spin interactions, such as frozen solutions forming a glassy state or solids. The solid effect depends on the mixing of nuclear eigenstates by nonsecular terms of the electron–nuclear hyperfine interaction [25]. It can only be resolved when the electron linewidth is smaller than the nuclear Larmor frequency. In contrast, thermal mixing can take place when the electron–electron dipolar reservoir, cooled by mw irradiation, is in thermal contact with the nuclear Zeeman reservoir [22]. The prerequisite of the thermal mixing pathway is that the electron linewidth is larger than the nuclear Larmor frequency. Irrespective of the pathways in a particular sample, it is crucial to achieve the best possible saturation of the selected electron frequency to maximize the efficiency of the DNP process. Furthermore, the maximal enhancement depends on the generation of high electron polarization via the use of low temperatures and high magnetic fields. At 1.5 K and 3.35 T, corresponding to an electron Larmor frequency of 94 GHz, the electrons are polarized to about 99%. Both conditions can be achieved relatively easily, while the generation of lower temperatures or the use of higher frequencies becomes increasingly technically demanding.

The optimal mw frequency for generating maximal enhancement of the polarization of a particular spin system can be obtained in an experiment in which the mw frequency is incremented and the polarization build-up is measured for each individual frequency. Figure 10.3 shows a representative response of $^{13}$C polarization to mw irradiation in a 1 : 1 v/v $H_2O$-glycerol glass doped either with 18 m$M$ TEMPO or 15 m$M$ Oxo63 and containing 2 $M$ $^{13}$C-labeled formic acid.

## 10.5
## Principle of Fast Gradient-Assisted 2D Spectroscopy

Gradient-assisted ultra-fast spectroscopy techniques consist of two different parts – the excitation module and the acquisition module [13]. In the excitation module, frequency swept pulses (chirp pulses) in conjunction with gradient pulses are used to generate evolution periods with durations depending on a spatial coordinate [14, 26]. This in turn leads to a modulation of the magnetization arising from distinct chemical groups in the form of magnetization helices with pitches given by their offset frequencies. In the acquisition module, these helices are unwound by the application of a magnetic field gradient and separate echoes for each chemical group are generated. These echoes can be continuously dephased and refocused using an oscillating gradient train similar to that used in echo planar imaging experiments (Figure 10.4). In the echo train the echo amplitudes are modulated by the chemical shift and coupling Hamiltonian.

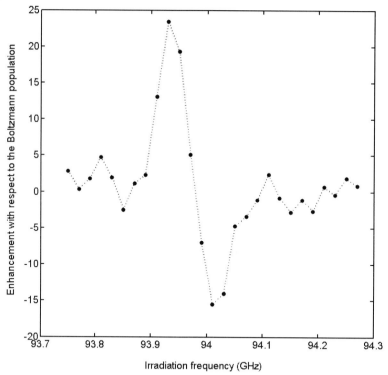

**Figure 10.3** $^{13}C$ Polarization build-up relative to the thermal polarization at 1.5 K and 3.4 T versus mw frequency. The thermal polarization was acquired for the fully relaxed $^{13}C$ spin system, and for each datum point the spin system was fully saturated by a comb of hard pulses before the build-up was measured over 3 h and extrapolated to a steady-state value.

The simplest building block that can be used in the excitation module is a pair of chirp $\pi/2$ pulses that are applied during a pair of gradient pulses. During the frequency sweep of the chirp pulse in the presence of a gradient, the magnetization is progressively rotated in the sample in one spatial dimension. The spatial dependence of the evolution period is generated by either changing the polarity of the second gradient pulse with respect to the first or by changing the sweep direction of the second chirp pulse with respect to the first. By using one of these strategies, the sample magnetization at the position that is initially affected by the first chirp pulse is affected at the end of the second chirp pulse. Conversely, the sample magnetization at a position that is rotated towards the end of the first chirp pulse is rotated first by the second chirp pulse. In this way the spin system evolves between the two chirp pulses with evolution periods that depend on a spatial coordinate determined by the direction in which the magnetic field gradient was applied (Figure 10.5). A

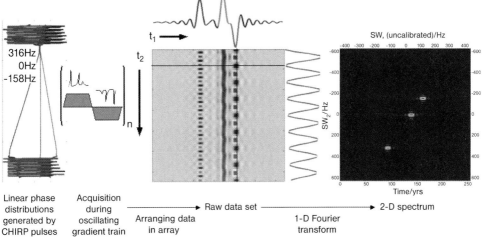

Linear phase distributions generated by CHIRP pulses → Acquisition during oscillating gradient train → Arranging data in array → Raw data set → 1-D Fourier transform → 2-D spectrum

**Figure 10.4** Simulation of a fast gradient-assisted 2-D spectroscopy experiment for the case of nuclei in three different chemical groups with offsets of −158 Hz, 0 Hz and 316 Hz. Spatially dependent phase distributions are generated by the application of chirp pulses for each chemical group. The phase modulations are refocused by magnetic field gradients, resulting in distinct echoes for each chemical group. The echoes are dephased and refocused by an oscillating gradient, and their amplitudes modulated by the offset frequency. Fourier transformation of this modulation results in a 2-D spectrum with the three peaks on the diagonal representing the three chemical groups. No interactions between the nuclear spins from different groups were assumed in this simulation.

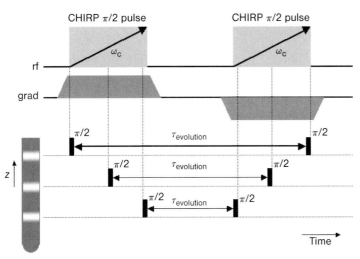

**Figure 10.5** Schematic diagram explaining the generation of evolution periods with different durations depending on a spatial coordinate in the sample. Three representative slices in the sample are shown. The magnetization in the first slice is rotated at the start of the first chirp pulse before the magnetization in the other two slices is affected. However, the magnetization in the first slice is rotated at the end of the second chirp pulse, thus generating the longest evolution period τ in the sample for this position.

single chirp pulse generates a quadratic phase modulation. However, this quadratic phase modulation can be canceled by using two or more chirp pulses consecutively [27].

A set of variants of this basic principle has been published by Frydman *et al.* [14], Pelupessy [28] and Andersen and Köckenberger [26], which depend on whether chirp pulses are used as excitation and/or refocusing pulses, and whether the modulation of the magnetization which is generated during the evolution period between the two chirp pulses is stored as an amplitude modulation of the longitudinal magnetization or used as a phase modulation of the transverse magnetization.

The spectral width $SW_1$ in the indirectly detected dimension is set by the sweep rate of the chirp pulse and the amplitudes of both the gradients applied during excitation and acquisition:

$$SW_1 = \frac{R\gamma_a G_a T_a}{2\gamma_e G_e} = \frac{\Delta O_{sweep}\gamma_a G_a T_a}{2\tau_p \gamma_e G_e} \tag{10.1}$$

where $R = \dfrac{\Delta O_{sweep}}{\tau_p}$ is the sweep rate and $T_a = n_1 t_{DW}$ is the acquisition time during a single gradient lobe, $\Delta O_{sweep}$ is the sweep width of the chirp pulse, $\tau_p$ is the duration of the chirp pulse, $\gamma$ and $G$ are the gyromagnetic ratios and gradient amplitudes during acquisition and excitation, respectively as indicated by $a$ and $e$ subscripts. As this technique can be adapted for fast heteronuclear single quantum coherence spectroscopy during which coherence is transferred between $^{13}C$ and $^1H$ nuclei, the nuclear species can be different for acquisition and detection. Note that $\Delta O_{sweep} \le L\gamma_e G_e$, where $L$ is the length of the sample, or the effective length over which a uniform excitation can be achieved.

Equation 10.1 can be understood by breaking it up into two parts which represent an opposing action on the sample magnetization. $\dfrac{2\tau_p \gamma_e G_e}{\Delta O_{sweep}}$ is a measure of how tightly wound the space-dependent helical phase distribution of the magnetization is after excitation, and $\gamma_a G_a T_a$ is a measure of how large a pitch can be unwound to form echoes during the acquisition. Thus, the desired spectral width in the indirect dimension is obtained by adjusting the corresponding parameters accordingly.

The achievable spectral width of the directly detected dimension is limited by the frequency of the gradient oscillation. Since it is convenient to process echoes from positive and negative gradient lobes separately, the spectral width of the directly detected dimension is given by

$$2(T_a + 2T_{ramp}) \le \frac{1}{SW_2} \tag{10.2}$$

where $T_{ramp}$ is the time required to switch and stabilize the acquisition gradients.

## 10.6
## Fast HSQC Spectroscopy with Hyperpolarized $^{13}$C Spin Systems

Since transfer of the solution containing the hyperpolarized compounds between the stand-alone polarizer and the high-field magnet takes several seconds (between 2 and 5 s depending on the pressure settings used and the length of transfer tubing), and since an additional settling delay $t_s$ (~2–3 s) needs to be included to allow any motion within the solution after injection into the NMR glass tube to decay away, it is advantageous for initial demonstrations to use heteronuclei with long longitudinal relaxation times $T_1$ such as $^{13}$C nuclei in a carbonyl group. The polarization of these nuclei can be used in HSQC (heteronuclear single quantum coherence [29]) experiments in which the carbon chemical shift is spatially encoded using two chirp π pulses followed by polarization transfer using a conventional refocused INEPT (Insensitive Nuclei Enhanced by Polarization Transfer [30]) sequence. The two chirp pulses generate a spatially dependent phase modulation with a frequency that depends on the offset frequency of the $^{13}$C nuclei with respect to the carrier frequency of the π/2 excitation pulse (this frequency can be considered as the $^{13}$C rotating-frame frequency when analyzing the signal evolution during the HSQC sequence). This phase modulation can be envisaged as a spatial modulation of the $^{13}$C magnetization arising from each individual chemical group in the form of different helices. These phase modulations can be unwound in the $^1$H single quantum coherences using suitable magnetic field gradients after coherence transfer between $^{13}$C and $^1$H nuclei using the INEPT module. In this way, each $^{13}$C nucleus with its individual chemical shift generates a separate echo during application of the magnetic field gradient. The echoes can be repeatedly dephased and refocused using an oscillating gradient train (Figure 10.6).

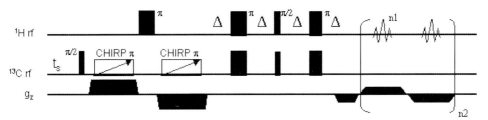

**Figure 10.6** Pulse sequence for HSQC experiment with hyperpolarized $^{13}$C spin systems. The settling delay $t_s$ was 3 s, and chirp pulse duration was 1 ms with a sweep $\Delta O = 40$ kHz. Each echo was acquired using 64 points (n1) with a total duration of 190 μs. Sixty-four echoes during positive gradient lobes and 64 echoes during negative gradient lobes were acquired (n2). $\Delta = 1289$ μs (1/4J), transfer time from polarizer was 3.5 s. $G_e = 2.7$ kHz mm$^{-1}$, $G_a = 8.8$ kHz mm$^{-1}$, $L = 14.8$ mm for the experiment using $^{13}$C-labeled formic acid.

**10.7**
**Signal Post-Processing**

Figure 10.7 shows the raw data obtained from an HSQC experiment using 1 m$M$ $^{13}$C-labeled formic acid after enhancing the polarization of the $^{13}$C nuclei using DNP at 1.5 K and 3.35 T in conjunction with a trityl radical (Ox063; Oxford Instruments Biomolecular Tools). Hot water was used for the fast dissolution. The echo train contains 64 echoes generated by gradients with positive amplitudes, and 64 echoes generated by gradients with negative amplitudes. Clearly, the signal amplitude has not decayed away at the end of the oscillating gradient train and could be used to generate further echoes. However, the number of echoes in an acquisition is currently limited by the available memory of our spectrometer.

Separating echoes from positive and negative gradient lobes and reordering them in 2-D arrays yields two complex data arrays (Figure 10.8).

The echoes are shifted due to a small mismatch between the positive and negative gradient amplitudes. This artifact leads to a broadening of the peaks in the final HSQC spectrum in the $^{13}$C chemical shift dimension. Therefore, the echoes are shifted in post-processing by applying a shift function that is linearly increasing with the echo number. A 1-D Fourier transformation along the directly detected dimension provides the final HSQC spectrum showing the connectivity of $^{13}$C nuclei with bonded $^1$H nuclei.

Since no decoupling was used during acquisition, a doublet can be resolved in the $^1$H chemical shift dimension. The intensities of the two peaks in the doublet are not equal since, after DNP, the difference between the two possible states of the $^{13}$C spin is now appreciable which directly affects the population differences in the four-level system of two coupled spins. Therefore, the observed asymmetry is due to the strong polarization of the $^{13}$C nuclei in the formic acid molecules after DNP; however, it cannot be used quantitatively to

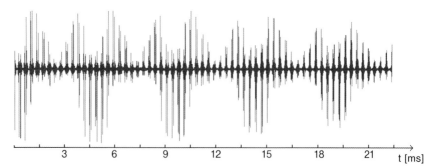

**Figure 10.7** Echo train acquired in a fast 2-D spectroscopic experiment. Echoes are generated using alternating positive and negative gradient lobes.

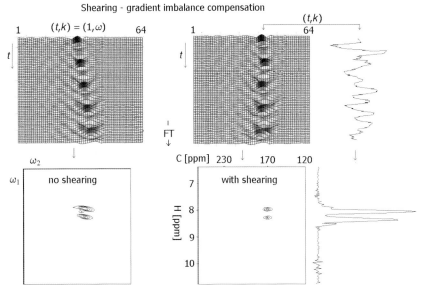

**Figure 10.8** Post-processing of data obtained by a fast 2-D spectroscopy experiment. Only one of the two complex data arrays is shown. The echoes in the left data matrix are shifted due to a small imbalance of the positive and negative gradient amplitudes that generate a broadening of the correlation peaks after Fourier transformation. The echoes in the right matrix were shifted by a linear correction function before Fourier transformation. Traces on the right-hand side correspond to the real part of the time signal at $\omega_1 = 180\,\mathrm{ppm}$ and the real part of the signal intensity after Fourier transformation.

calculate the $^{13}$C polarization due to cross-relaxation processes during the transfer [31].

In this experiment the enhancement measured in comparison to a thermally polarized liquid-state sample was 8330. As the solid-state enhancement factor $\varepsilon^\dagger$ is usually bigger than 50, at least 20% of the polarization is lost during dissolution and transfer. The polarization loss depends strongly on the $T_1$ of the type of nucleus under observation, and the chemical environment. The resonance signals of $^{13}$C nuclei in different chemical groups can be attenuated to different extents due to relaxation during sample transfer, which makes quantitative use of the DNP enhancement technique difficult.

## 10.8
## Conclusions

Fast 'single-shot', gradient-assisted 2-D spectroscopy schemes are ideally suited for maximizing the information that can be obtained in experiments with hyperpolarized molecules using DNP. As the transfer times between the polarizer and the high-field magnet are currently several seconds, these techniques function best for heteronuclei with long $T_1$ relaxation time constants. Whilst a number of interesting applications using such molecules are already available, it will be important to develop alternative hardware solutions to reduce the transfer time and the concomitant loss of polarization due to relaxation during the transfer. This would broaden the array of possible applications for DNP and would, in particular, make it possible to use hyperpolarized $^1$H spin systems. The transfer in solid state of the sample between two separate magnets, thus avoiding the shorter relaxation time constants in liquid state or frozen solids kept at only liquid nitrogen temperature, remains a challenge as this requires the temperature of the sample to be kept at a very low level. Another strategy is the combination of a low-field magnet (3.35 T) with a high-field magnet in the same housing. In this configuration, it may be possible to transfer the sample between the two isocenters in solid state and at a relatively low temperature. As this strategy requires only the transfer of the liquid sample over a very short distance after dissolution, it may be also feasible to reduce the settling time in the NMR detection cell.

With currently existing technology it should be possible to use highly polarized small molecules to investigate the binding centers of proteins in more detail. The fast 2-D spectroscopy techniques will play an important role in studying conversion using EXSY-type implementations or through-space correlation between highly polarized labels within the small molecules and the binding center using nuclear Overhauser enhancement spectroscopy (NOESY)-type experimental protocols. A reduction of the settling time will be required to make the observation of early events in such dynamic processes possible.

Fast spectroscopy in conjunction with DNP may also be of interest when studying molecular dynamics, such as the folding of small proteins. However, in this respect it is important first to investigate the effect of low temperature and rapid warming of the sample on the protein. The current time resolution of spectroscopic experiments using the fast single excitation strategy for 2-D spectra outlined in this chapter is in the order of a few tens of milliseconds.

The advent of dissolution DNP and the availability of fast, multidimensional spectroscopy techniques will undoubtedly open a new direction for NMR technology that will require major hardware developments for optimal experimental protocols.

### Acknowledgments

The authors gratefully acknowledge their many discussions with L. Frydman, Rehovot, Israel. R.P. thanks the Marie Curie program within the EU framework 6 for a studentship, and J.G. acknowledges support from the Mansfield Research Fellow program of the University of Nottingham. These research studies were supported by an instrument development grant awarded by EPSRC to W.K., and also industrial sponsorship from Oxford Instrument Molecular Biotools Ltd.

# References

1 Leawoods, J.L., Yablonsky, D.A., Saam, B., Gierada, D.S. and Conradi, M.S. (2001) *Concepts in Magnetic Resonance*, **13**(13), 277–93.

2 Moller, H.E., Chen, X.J., Saam, B., Hagspiel, K.D., Johnson, G.A., Alter, T.A., de Lange, E.E. and Kauczor, H.U. (2002) *Magnetic Resonance in Medicine*, **47**, 1029–51.

3 Goodson, B.M. (2002) *Journal of Magnetic Resonance*, **155** (2), 157–216.

4 Cherubini, A., Payne, G.S., Leach, M.O. and Bifone, A. (2003) *Chemical Physics Letters*, **371**, 640–4.

5 Natterer, J. and Bargon, J. (1997) *Progress in Nuclear Magnetic Resonance Spectroscopy*, **31**, 293–315.

6 Duckett, S.B. and Blazina, D. (2003) *European Journal of Inorganic Chemistry*, **16**, 2901–12.

7 Abragam, A., Landesman, A. and Winter, J.M. (1958) *Comptes Rendus de l'Académie des Sciences*, **247**, 1852–3.

8 Hall, D.A., Maus, D.C., Gerfen, G.J., Inati, S.J., Becerra, L.R., Dahlquist, F.W. and Griffin, R.G. (1997) *Science*, **276**, 930–2.

9 Ardenkjaer-Larsen, J.H., Fridlund, B., Gram, A., Hansson, G., Hansson, L., Lerche, M.H., Servin, R., Thaning, M. and Golman, K. (2003) *Proceedings of the National Academy of Sciences of the United States of America*, **100**, 10158–63.

10 Joo, C.G., Hu, K.N., Bryant, J. A. and Griffin, R.G. (2006) *Journal of the American Chemical Society*, **128**, 9428–32.

11 Golman, K., Ardenkjaer-Larsen, J.H., Petersson, J.S., Månsson, S. and Leunbach, I. (2003) *Proceedings of the National Academy of Sciences of the United States of America*, **100**, 10435–9.

12 Ernst, R.R., Bodenhausen, G. and Wokaun, A. (1987) *Principles of Nuclear Magnetic Resonance in One and Two Dimensions*, Oxford University Press, Oxford.

13 Frydman, L., Scherf, T. and Lupulescu, A. (2002) *Proceedings of the National Academy of Sciences of the United States of America*, **99**, 15858–62.

14 Shrot, Y., Shapira, B. and Frydman, L. (2004) *Journal of Magnetic Resonance*, **171**, 163–70.

15 Frydman, L. and Blazina, D. (2007) *Nature Physics*, **3**, 415–19.

16 Wolber, J., Ellner, F., Fridlund, B., Gram, A., Johannesson, H., Hansson, G., Hansson, L.H., Lerche, M.H., Mansson, S., Servin, R., Thaning, M., Golman, K. and Ardenkjaer-Larsen, J.H. (2004) *Nuclear Instruments and Methods in Physics Research. Section A*, **526**, 173–81.

17 Velichkov, I.V. and Drobin, V.M. (1990) *Cryogenics*, **30**, 538–44.

18 Celik, D., Hilton, D.K., Zhang, T. and Van Sciver, S.W. (2001) *Cryogenics*, **41**, 355–66.

19 Conradi, M.S. (1993) *Concepts in Magnetic Resonance*, **5**, 243–62.

20 Granwehr, J., Leggett, J. and Köckenberger, W. (2007) *Journal of Magnetic Resonance*, **187**, 266–76.

21 Atsarkin, V.A. (1978) *Soviet Physics Uspekhi*, **21**, 725–45.

22 Abragam, A. and Goldman, M. (1982) *Nuclear Magnetism: Order and Disorder*, Oxford University Press, Oxford.

23 Hu, K.N. (2006) Polarizing Agents for High-Frequency Dynamic Nuclear Polarization – Development and Application, Thesis, Massachusetts Institute of Technology.

24 Müller-Warmuth, W. and Meise-Gresch, K. (1983) *Advances in Magnetic Resonance*, **11**, 1–45.

25 Duijvestijn, M.J., Wind, R.A. and Smidt, J. (1986) *Physica*, **138B**, 147–70.

26 andersen, N.S. and Köckenberger, W. (2005) *Magnetic Resonance in Chemistry*, **43**, 795–7.

27 Cano, K.E., Smith, M.A. and Shaka, A.J. (2002) *Journal of Magnetic Resonance*, **155**, 131–9.

28 Pelupessy, P. (2003) *Journal of the American Chemical Society*, **125**, 12345–50.

29 Bodenhausen, G. and Ruben, D.J. (1980) *Chemical Physics Letters*, **69**, 185–9.

30 Morris, G.A. and Freeman, R. (1979) *Journal of the American Chemical Society*, **101**, 760–2.

31 Merritt, M.E., Harrison, C., Mander, W., Malloy, C.R. and Sherry, A.D. (2007) *Journal of Magnetic Resonance*, **190**, 115–20.

# 11
# Dynamic Nuclear Polarization-Enhanced Magnetic Resonance Analysis at X-Band Using Amplified $^1$H Water Signal

*Songi Han, Evan R. McCarney, Brandon D. Armstrong and Mark D. Lingwood*

## 11.1
## Motivation and Background

The *in vitro* and *in vivo* analysis of biological samples relies heavily on noninvasive spectroscopic techniques, inert probe molecules, and the ability to perform measurements at ambient temperature in aqueous solutions. According to these criteria, NMR is a superior tool because it provides detailed molecular signatures and images utilizing very low-energy radiofrequency (RF) irradiation (10–900 MHz) and endogenous probes (e.g. $^1$H) of biological samples that sustain sufficiently long coherence times to allow room-temperature analysis. In the human body, MRI detection is achieved by employing the protons of the most abundant molecule in biology, water, as its probe. However, both NMR and MRI suffer from low sensitivity and signal overlap of abundant endogenous protons. Electron spin resonance (ESR), a sister technique to NMR, utilizes stable nitroxide radicals which are either embedded or site-specifically attached to the molecule or material of interest as probes that have a 660-fold higher resonance frequency than that of proton nuclei in the same magnetic field. Accordingly, ESR utilizes higher polarizations and frequencies, and thus is more sensitive. On the other hand, ESR is limited by the larger heat absorption in aqueous media compared to NMR, and that no direct signal from the host molecule or solvent is detected. The DNP amplification of $^1$H NMR signal via polarization transfer from spin labels to water adds extra capabilities to ESR because information on solvent interaction and translational dynamics can be obtained [1–6]. DNP can also enhance MRI contrast as the highly amplified $^1$H NMR signal provides a new mechanism to differentiate between flowing and stationary or bulk and localized water [1, 3, 4, 7–10]. The key feature of our approach is to obtain high DNP enhancement at room temperature upon ESR irradiation, which uniquely allows DNP to be used for monitoring dynamic events.

The main aim of this chapter is to provide research groups interested in applying DNP analysis to their own investigations with information on quantitative DNP studies using a standard continuous-wave (cw) ESR spectrometer with minimal

*Magnetic Resonance Microscopy.* Edited by Sarah L. Codd and Joseph D. Seymour
Copyright © 2009 WILEY-VCH Verlag GmbH & Co. KGaA, Weinheim
ISBN: 978-3-527-32008-0

technical modifications. Should there be a desire to further pursue DNP analysis with higher performance, then additional microwave amplification modules can be added, or a home-built device used to approach the theoretical maximum NMR signal amplification [11].

Although an understanding of the basic DNP mechanism is essential for the effective application of this technique, at this point we will only briefly describe the basic principle in order that the data obtained can be interpreted and the rationale for further improvements made comprehensive. In a system of two interacting spins – in this case a proton of water and the free electron of a nitroxide radical – proton polarization will depend on the population distribution of the electron spin energy levels [12–14]. Saturation of the equilibrium electron spin population with an oscillating magnetic field can cause a net nonequilibrium proton spin polarization if there is dipolar or scalar coupling between the two spins. For nitroxide radicals and water, the coupling has been shown to be primarily dipolar [13]. If the cross-relaxation between the electron and proton spins is mediated by molecular motion (rotation, translation), then the DNP mechanism is known as the Overhauser effect [14], which is more effective in liquid samples and lower magnetic fields (<1 T, depending on the correlation time of motion). In liquids, the DNP-induced NMR signal enhancement, $E$, depends on interaction between the electron and proton as:

$$E = 1 - \rho f s \frac{|\gamma_S|}{\gamma_I} \tag{11.1}$$

where $\rho$ is the coupling factor that describes the dipolar coupling between the electron and proton, $f$ is the leakage factor that describes the electrons' ability to relax the proton, $s$ is the saturation factor that describes the saturation of the electron Zeeman transition, $\gamma_S$ is the magnetogyric ratio of the electron, and $\gamma_I$ is the magnetogyric ratio of the proton, where $\gamma_S/\gamma_I = 658$ [12]. The negative enhancement of the Overhauser effect for protons originates from the dipolar nature of the motional coupling to the electrons, and manifests itself as an amplified NMR signal that is 180° out of phase compared to the equilibrium NMR signal. For dipolar coupling, $\rho$ lies between 0 and 0.5, and depends on the distance of closest approach and the diffusion coefficient of the radical and the proton-containing molecule; thus, the physical interaction between the radical and solvent water molecules plays a key role. The leakage factor depends on the type of radical and its concentration, and can vary between 0 and 1. The saturation factor depends on the irradiation power on-resonant with the ESR transition and the efficiency of the mixing of the nitroxide radical's hyperfine state, and also varies between 0 and 1 [5]. Maximum possible DNP enhancement is achieved if $f \sim 1$, $s \sim 1$ and $\rho$ approaches 0.5 in the dipolar limit, and thus is equal to $(1 - 0.5 \times \gamma_S/\gamma_I)$. For nitroxide radicals dissolved in water, $\rho$ was measured at about 0.2 at 0.35 T [5]. At radical concentrations >10 m$M$ we determined $f \sim 1$ and $s \sim 1$; therefore, a $^1$H water enhancement of ~130-fold is expected when using a high-power microwave amplifier, and this was recently confirmed experimentally [11].

## 11.2
## Unique Application of ¹H DNP at X-Band

### 11.2.1
### Hyperpolarized Water as a Contrast Agent

Previously, we have shown that water with a highly amplified ¹H NMR signal can be used as an authentic contrast agent for imaging the macroscopic flow dispersion of water [7]. Hyper-polarization represents an ideal contrast mechanism to highlight the ubiquitous and specific function of water in physiology, biology and materials because the physiological, chemical and macroscopic function of water is not altered by the degree of magnetization. This technique, which is referred to as remotely enhanced liquids for image contrast (RELIC), utilizes the ¹H signal of continuously flowing water that is enhanced outside the sample and immediately delivered to the sample to obtain maximum contrast between entering and bulk fluids. Figure 11.1 shows a sketch of the set-up used, which is based on an electromagnet with a field set to 0.35 T. Agarose-based gel beads that are functionalized with effectively 10 m*M* nitroxide radical concentration when suspended in water [15] are loaded into a quartz column through which water is continuously flowing.

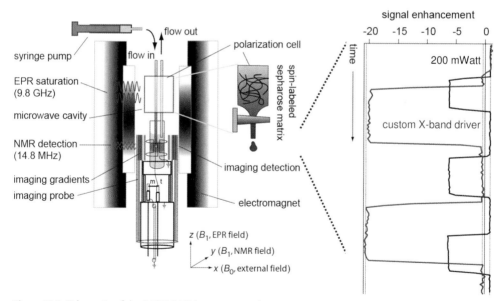

**Figure 11.1** Schematic of the 0.35 T DNP instrument and measured enhancement of ¹H NMR signal of water in continuous flow. The right-hand panel shows that signal contrast can be turned on and off repeatedly by turning on and off the 9.8 GHz microwave power that saturates the ESR signal of the spin labels attached to agarose-based gel beads.

This spin-labeled gel column is placed inside an X-band ESR cavity ($TE_{102}$; Bruker Biospin) that remains tuned and matched during the experiment. When water flows through the gel column, it interacts with the spin labels to efficiently drive dipolar cross-relaxation, and the $^1$H NMR signal is amplified within the time scale of $T_1$ (~140 ms). As the spin labels are bound to the gel via stable amide bonds, no measurable radical concentration leaks into the water, so the outflowing and hyperpolarized water is perfectly free of residual radicals. Figure 11.1 shows that the water-based contrast agent is flowing into an NMR imaging probe, where the flow images are detected at 14.8 MHz proton NMR frequency. A simple test experiment was performed of repeatedly turning on and off the microwave source on-resonant with the ESR transition of the nitroxide spin labels while continuously monitoring the $^1$H NMR signal amplification, as shown on the right-hand side in Figure 11.1. It can be seen that the signal contrast is also turned on and off repeatedly. The smaller-amplitude step function is measured with the commercial ESR set-up with up to 200 mW output power, while the larger-amplitude step functions were measured using our custom X-band driver with higher-output power of up to 6 W, thus providing better DNP efficiency. A signal enhancement factor of only about −20-fold was achieved in continuous-flow, despite sufficient microwave power, partly because the spin labeled gel material cannot yet provide comparable performance to spin labels freely dissolved in water, and partly because some $^1$H polarization is lost due to $T_1$ decay at the time of detection. As discussed recently [15], it has been found that the DNP efficiency of immobilized spin labels can be maximized by ensuring high water mobility and access to the spin labels. Thus, highly solvated and swollen hydrogel-type gel materials with good tortuosity would appear to serve as ideal polarization matrices.

We have shown previously that the continuous polarization of radical-free, flowing water allows the distinctive and direct visualization of vortices and mechanical flow dispersion in a molecular sieve bead pack [7]. Currently, these efforts are being continued to enhance $^1$H NMR image contrast for flow dispersion, with similar goals as described before – that is, to study the type of contrast and sensitivity that can be achieved and also to optimize the flow-mode DNP efficiency. Some of the preliminary flow images using RELIC are shown in Figure 11.2 and Figure 11.3. A 5 mm glass tube was packed with 1 mm diameter glass spheres, and perfused with hyperpolarized water via a small glass capillary into the water-saturated structure (Figure 11.2). The top-left image in Figure 11.2 is a standard 2-D spin warp image, while the three images to its right are RELIC images obtained at different flow rates. The point of entrance of the hyperpolarized water can be clearly seen, with a strong mechanical dispersion around the beads, as well as some back-flow that is created due to local pressure build-up. A different and also more distinct contrast is achieved by creating phase maps that only display negative versus positive signal amplitude from the corresponding RELIC images. The voxels originating from the inverted polarization contain a 180° phase-shifted NMR signal when detected as transverse magnetization. This phase shift of 180° is displayed in a phase map, and demonstrates the spatial occupation of the negatively polarized water as it enters the cell; this is shown in

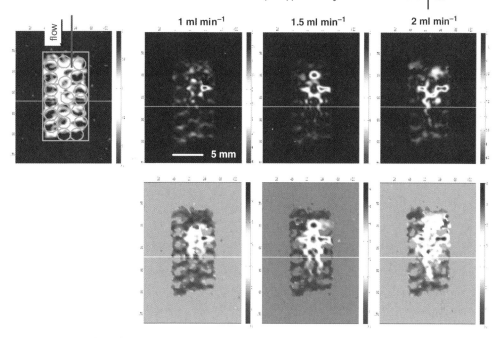

**Figure 11.2** RELIC through a bead pack of 1 mm-diameter glass beads inside a 5 mm column. The top-left image is a 2-D spin warp image without RELIC; the three images to the right on the top row are with RELIC turned on, while images in the bottom row are phase maps created from the RELIC images above.

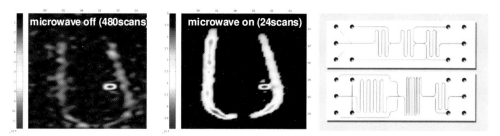

**Figure 11.3** RELIC through a U-channel of 1 mm inner diameter. The left-hand image is a 2-D spin warp image without RELIC; the center image is with RELIC turned on; the right-hand sketch shows which applications may become possible through $^1$H DNP amplification of water signal. Because $^1$H polarization comparable to the thermal polarization achievable in an imaginary 45 T magnet may become available in continuous flowthrough RELIC, flow visualization in very small structures such as microfluidic systems can be greatly improved.

the three lower images in Figure 11.2. Such a phase map can be used to interpret flow contrast with less ambiguity, for example, whether intensity near zero originates from low fluid density or from flowing water with negative polarization that has decayed almost to zero. The phase maps in Figure 11.2 show that the mechanical flow dispersion is visible for a considerable dispersion displacement along approximately five bead diameters at a flow rate of 2 ml min$^{-1}$. Based on our studies of nitroxide radicals dissolved free in solution, an optimized polarization matrix would be capable of providing enhancements of $-100$-fold instead of the $-10$- to $-20$-fold currently achieved.

For some applications, RELIC can also enhance sensitivity and image resolution, as well as providing unique flow contrast. In flow dispersion experiments, hyperpolarized water normally enters an already water-saturated system, where the bulk water has a greater volume compared to the entering fluid to be tracked; thus, there will be a significant contribution from the background signal. However, for microfluidic applications the entire fluid volume entering the fluidic system can be hyperpolarized and monitored. Although Figure 11.3 does not represent a microfluidic system (as the U-channel has a relatively large, 1 mm inner diameter), it does demonstrate that barely visible fluids can be clearly visualized by providing access to a significantly higher nuclear spin polarization. Due to the small size and fast total flow transport time within these devices, the hyperpolarization is able to survive over the entire flow path.

Whilst the fundamental disadvantage of RELIC is that the hyperpolarization decays with the relatively short $T_1$ of water protons, its greatest advantage is that the hyperpolarization can be driven in continuous flow. Given that hyperpolarized water is radical-free and therefore has a $T_1$ of ~2.5 s, the $^1$H signal of water when enhanced by a factor of $-10$ and $-100$ provides an observation time of 6 and 11 s, respectively. It should be noted that this observation time is not to be confused with the total image acquisition time. As hyperpolarization is driven continuously, conventional imaging sequences can be carried out with minimal adjustment, and hence total imaging times far exceeding the observation time can be used. However, for each unit of hyperpolarized volume, the tracking 'depth' is determined by the observation time. For example, the total blood circulation time in a mouse is on the order of 3–4 s, whereas the flow time from an intravenous injection to the right side in a human heart is ~4 s. Thus, RELIC offers appropriate time scales for observation under optimized conditions for biomedical blood perfusion MRI on small animals and also for some human applications. For *in vivo* perfusion applications, one important detail to note is that the $^1$H signal of distilled water, saline solution or plasma fluids can be amplified to a similarly high level, as has been studied by others [16] and experimentally verified by us, as long as the sample's heat absorption is minimized by the correct adjustment of fluid volume and geometry inside the microwave cavity. One likely scenario for the successful application of RELIC *in vivo* is to employ an open MRI system that operates at 0.3–0.4 T (Figure 11.4), and where DNP can be performed immediately adjacent to the subject, inside the same magnetic field that is used for MRI acquisition. Because the lower-field (0.3–0.5 T) MRI systems are mostly open and have completely acces-

**Figure 11.4** Today, numerous 0.3–0.4 T open MRI systems are available. Examples include the IGNA OVATION EXCITE from GE Healthcare, the Magnetom C! by Siemens, and the AIRIS by Hitachi Medical Systems (shown here), all of which based on permanent resistive magnets and operate at 0.3 or 0.35 T fields. These illustrations show a trip to the Pacifica Hospital (access and support generously provided by Dr. Sam Tokuyama) in Burbank, California, where the purpose of our team (M. Lingwood, P. Bluemler, S. Tokuyama, S. Han) was to verify that $^1$H DNP enhancement of static as well as flowing water can be achieved in a 0.3 T open MRI system using our fully portable DNP set-up.

sible geometry, the installation of DNP equipment and intravenous injection can be performed in relatively straightforward fashion. Overall, the implications of RELIC are immediate for both chemical engineering or biomedical applications, in that either hyperpolarized solvents or physiological fluids can be used to visualize mass transport and perfusion, with high and authentic MRI contrast originating from the water itself rather than from foreign contrast agents.

## 11.2.2
## Adding an NMR Dimension to ESR Analysis

Here, we demonstrate a completely different use of $^1$H DNP amplification of water NMR signal under ambient conditions, namely as a complement to ESR to characterize fluid dynamics in soft molecular assemblies such as micelles, vesicles and lipid bilayer membranes, or in protein aggregation processes. The DNP enhancement originates from spin labels that are covalently attached to specific molecular sites. DNP amplification of the local water around the spin labels can be interpreted in terms of the distance of closest approach between the spin labels and protons and the diffusion coefficient of the local water; the effects of both parameters are contained in the coupling factor, $\rho$ [12]. Localization of the spin label to specific sites of molecules that incorporate themselves into assembly structures provides information on local fluid dynamics through the interpretation of $^1$H DNP enhancement effects [1, 3, 4, 12]. Here, we summarize our combined DNP and ESR studies on model oleate molecular assemblies that form micelle or vesicle structures [6]. The dynamics of water in these interfacial systems is of special interest because they play a key role in membrane assembly, solute transfer and membrane dynamics. For these studies, the spin probes 5- and 16-doxyl stearic acid (5-DS, 16-DS) were used, where the 5-position is at the hydrophobic surfactant

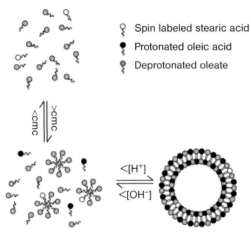

○ Spin labeled stearic acid

● Protonated oleic acid

○ Deprotonated oleate

**Figure 11.5** Oleate surfactants form micelles at pH > 10.5, vesicle structures at pH < 8, and mixed structures in between. Spin-labeled stearic acid at the 5- and 16-positions was employed as the probe species, which have been demonstrated to incorporate themselves into the micelle or vesicles structures without forming a separate phase [17].

tail, but closer to the charged hydrophilic head group, and the 16-position is at the end of the hydrophobic tail of the doxyl surfactant (Figure 11.5). It has been shown previously that these doxyl spin probes incorporate well into micelle or vesicle assemblies formed by oleate surfactants [17]. The $^1$H NMR detection coil is implemented inside the ESR cavity such that independent ESR and DNP measurements can be performed on the same sample.

Solutions were prepared containing 300 μ$M$ 5-DS spin probe in 200 μ$M$ (<<critical micelle concentration, CMC) and 25 m$M$ (>>CMC) sodium oleate, respectively. The same solutions were also prepared using the 16-DS spin probe. Below oleate concentrations of 1 m$M$, the surfactants were completely dispersed in solution, whilst at oleate concentrations of 25 m$M$ and at pH 11, both types of spin probe were completely incorporated into the oleate micelles [17]. At pH >10.5 micelle structures were predominantly formed, between pH 10.5 and 8 a mixture of micelles and vesicles were formed, and below pH 8 the surfactants were predominantly incorporated into the oleate vesicular structures. A series of cw ESR spectra were measured and DNP characterization was performed for both 300 μ$M$ 5-DS and 16-DS spin probes dispersed in surfactant solution (<<CMC) and incorporated into 25 m$M$ oleate solutions at pH values ranging from 7 to 12 (Figure 11.6).

The ESR spectra of 5-DS and 16-DS in dispersed surfactant solutions are almost identical (Figure 11.6, top row), and provide similarly short rotational correlation times ($\tau \approx 9 \times 10^{-11}$ s) for both spin probes, which means that the spin-labeled surfactants are freely moving in solution, and the different labeling position on the surfactant chain does not alter the rotational freedom. When the spin probes

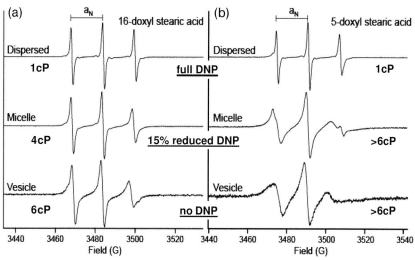

**Figure 11.6** ESR spectra of 16-DS (A) and 5-DS (B) dispersed in solution below the CMC (top), incorporated into micelles at pH > 11 (middle) and vesicles at pH < 7 (bottom). The ¹H DNP enhancement through the spin labels is equal for 16-DS versus 5-DS as long as they are part of the same structure [6].

are incorporated into the oleate micelles, the ESR spectrum broadens for both 5-DS and 16-DS, although the broadening on the 5-DS probe is more dramatic (Figure 11.6, middle row). The rotational correlation time slows for 16-DS from $\tau = 9 \times 10^{-11}$ s in dispersed surfactant solution to about $\tau = 3 \times 10^{-10}$ s in micelle solutions due to some motional restriction of the spin probe inside the micelle structures. The reason why the ESR spectra of 5-DS shows significantly greater line broadening and anisotropic features compared to 16-DS, even though both probes are incorporated into the same micelle structures, is because the 5-DS is situated closer to the charged head (but still inside the micelle core) of the stearate molecule – that is, it is closer to the more rigid interfacial layer. So, while the overall motion of both spin probes slows down due to the increase in local fluid viscosity in the micelle structures compared to bulk water, 5-DS is additionally restricted by its interaction with the micelle–solution interface. Here, we made the interesting observation that the measured DNP enhancements of the two different spin labels with very different rotational motion were equal, within error. The ESR spectra of 5-DS and 16-DS incorporated into vesicle structures (25 mM oleate at pH < 8) show a 30–50% increase in rotational correlation times compared to those of micelle structures, but the general features remain unchanged (Figure 11.6, bottom row). The hyperfine coupling constant $a_N$, which measures the ESR splitting due to the neighboring nitrogen nuclear spin, gradually decreases with increasing vesicle fraction, from a pure micelle solution to a pure vesicular solution. This decrease in $a_N$ can be attributed to a decrease in polarity or water removal around the spin label, as both affect the electron density distribution of the radical, although the two effects cannot be differentiated [17].

The $^1$H DNP enhancement factor of water was highest in the dispersed surfactant solution, and slightly decreased (by 15%) in micelle solutions. When steadily increasing the vesicle fraction from pure micelles to pure vesicles, by gradually lowering the pH, the enhancement factors also gradually decreased until no enhancement could be observed in the vesicle solution. The key observation here is that the DNP-induced $^1$H signal enhancement measured with the different spin probes, 5-DS and 16-DS, resulted in equal DNP enhancement factors in the different systems (dispersions, micelles or vesicles), although the ESR spectra between 5-DS and 16-DS were markedly different in micelle and vesicle structures. This experimental observation is in agreement with the expectation that $^1$H DNP measures the content and/or translational dynamics of the fluid water in the local soft structures through spin labels that are probing the local volume, and is not primarily sensitive to the dynamics of the spin probe as part of the surfactants.

Thus, we propose three simplified scenarios that occur in the oleate micelle vesicle system:

- The radical is freely moving in a solution-like environment
- The radical is trapped inside a semipermeable container where molecular diffusion is hindered
- The radical is sequestered into a dehydrated local environment.

The first situation is applicable to the 200 μM oleate samples where 16-DS and 5-DS are dispersed free in solution, and therefore does not reveal any difference in ESR or DNP features. The second situation applies to the 25 mM oleate pH > 11 samples, where the micelles are the dominant species. Water protons in the oleate micelles show a decreased DNP enhancement compared to water in free solution, which we attribute to a decrease in the coupling factor. The observation of significant DNP enhancements inside the micelle volume supports the discussions reported elsewhere that the interior of the oleate micelle is hydrated [17]. The third situation applies to the 25 mM oleate pH < 8, where vesicles composed of hydrophobic bilayer structures are the dominant species. When the radicals are sequestered in a hydrophobic environment or desolvated, they no longer interact with the water protons, and consequently the DNP effect vanishes as the distance of closest approach ($d$) becomes very large [5, 12]. This discussion highlights the different types of information that can be gathered by a combined ESR and DNP analysis of soft matter. ESR detects changes in the rotational dynamics of the spin probe and therefore shows a difference between the 16-DS and 5-DS probes. In contrast, $^1$H DNP of water is a measure of the content or translational dynamics of water around the probe, and therefore displays the same average DNP effect through spin labels that are incorporated into the same samples. One application that we are currently pursuing, and which is supported by these results, is the dynamic monitoring of protein aggregation through a combination of ESR and DNP analysis, where the formation of hydrophobic sites and gradual water exclusion play key roles.

## 11.3
## Instrumentation for $^1$H DNP at X-Band

### 11.3.1
### DNP with Commercial X-Band ESR Equipment

Figure 11.7 shows a commercial cw X-band ESR set-up with DNP capability composed of a rectangular $TE_{102}$ resonant cavity at ~9.8 GHz and a microwave bridge (Klystron source; Bruker Biospin) in a 0–1.5 T electromagnet, that is controlled by an EMX ESR spectrometer (Bruker Biospin). For NMR detection, an Avance 300 NMR spectrometer (Bruker Biospin) as well as an inexpensive and portable *Kea* NMR spectrometer (Magritek Limited) were used. The resonant cavity needs to remain highly tuned ($Q > 2000$) while the RF coil is placed inside, as the lower than ideal irradiation output power of a commercial ESR spectrometer needs to be best exploited. We were able to maintain a $Q$ of 2000–2500 when using a sample holder made of a quartz capillary and a small piece of chlorinated Teflon which holds a silica capillary (which contains the sample) and is pierced with channels to thread through thin silver wires that form a double-U RF NMR coil. Loosely wrapped two- to four-turn solenoid coils also have been employed for our DNP experiments when a higher NMR detection sensitivity was needed. However, the double-U coil allows the resonant cavity to tune to a higher $Q$ (quality factor). A custom resonant LC circuit tuned to ~14.8 MHz with a tuning range of about 5 MHz was built using 120 pF variable capacitors and with the double U-coil (or

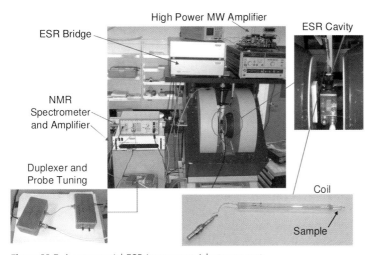

**Figure 11.7** A commercial ESR instrument (electromagnet, ESR bridge, ESR cavity) is shown that is equipped with DNP capability, by incorporating a tuned RF U-coil (14.8 MHz) into a tuned ESR cavity. The RF coil is connected via a tuning box and duplexer to a portable, single-channel, NMR spectrometer and amplifier.

solenoid coil) serving as the inductor. This set-up allows for ESR, NMR and DNP measurements of aqueous solutions without changing the set-up and sample. The Bruker EMX ESR spectrometer can be operated unleveled, giving 800 mW of output power.

When using this commercial Bruker microwave source and ESR cavity, we have measured $^1$H signal enhancements of water (all enhancements discussed here are for $^1$H of water) of $-73 \pm 2$ using 15 mM $^{15}$N-substituted 4-oxo-2,2,6,6-tetramethyl-piperidine-1-oxyl (4-oxo-TEMPO) while operating the microwave source in the unleveled mode with a maximum output power of 800 mW. However, we have measured enhancements as high as 55-fold while operating the bridge at only 200 mW in leveled mode. Although these enhancements are well below the theoretical maximum, they are adequate for the characterization and determination of key DNP parameters such as the coupling factor, $\rho$, maximum saturation factor, and maximum DNP enhancement factor, $E_{max}$, which is the DNP enhancement extrapolated to maximum microwave power.

### 11.3.2
#### DNP with a Custom Microwave Source, Commercial Resonator and Electromagnet

If achieving the largest possible signal enhancements is the goal, or the sample has broad ESR lines which are difficult to saturate, then more than 800 mW of output power is needed. In order to achieve this, we built a custom X-band transmission device using commercially available parts [11]. The transmission device uses a phase-locked YIG synthesizer (Microlambda Wireless) as the frequency source, which is tunable between 8 and 10 GHz. The output is sent through two step attenuators for power control, then split and directed through four solid-state power amplifiers (Advanced Microwave Technologies). The amplified signal is then recombined and directed into a SMA output. An SMA-to-wave guide connector is used to couple the microwave output to the Bruker TE$_{102}$ cavity. Approximately up to 23 W of microwave power leaves the source to be coupled to the cavity. As we used the resonant cavity unmatched and there are multiple connections and parts between the source and the sample, a substantial amount of this initial power is lost due to reflections and the creation of standing waves; the actual power present at the sample is estimated at $\sim$6 W at full power output. Continuous air flow and attenuation of the power down to $\sim$2 W at the sample was required to avoid substantial heating (see Ref. [11] for more details).

Our custom microwave source, together with the Bruker resonance cavity and electromagnet, provided $^1$H signal enhancements of water with TEMPO derivatives of $-112 \pm 4$ with the predicted maximum enhancement of $-130$. Additionally, we measured near-equal enhancements for $^{14}$N and $^{15}$N versions of 4-oxo-TEMPO and 4-amino-TEMPO, and observed that at high radical concentrations the maximum enhancement is independent of whether natural-abundance $^{14}$N or isotope-enriched $^{15}$N 4-oxo-TEMPO is used. This is because there is a rapid Heisenberg exchange between the different ESR transitions at these concentrations, causing $s_{max}$ ($s$ at infinite microwave power) $\rightarrow 1$, and therefore negating the advantage of using a radical with fewer ESR lines [5]. At lower concentrations, however, there is a difference in $E_{max}$ between $^{14}N$ and $^{15}N$ 4-oxo-TEMPO, which is due to Heisenberg exchange not mixing the hyperfine electron states rapidly enough compared to the intrinsic electron spin relaxation rate. Our experiments have also shown there to be little dependence of $E_{max}$ on the functional group of the TEMPO derivatives as similar concentrations of 4-oxo- and 4-amino-TEMPO approach the same enhancements at full power.

## 11.3.3
## Portable DNP Instrument

The advantage of using a resonant cavity in an electromagnet is that the magnetic field can be easily adjusted to precisely meet the ESR resonance condition of the given radical so that efficient $B_{1e}$ transmission to the sample is ensured. If the given sample and NMR probe device can be accommodated into the cavity and portability is not an issue, this set-up seems most versatile to easily achieve near-optimum DNP enhancement.

If portability of the set-up is desired, a more lightweight permanent magnet must be employed. Thus, the large electromagnet was replaced by two different types of permanent magnets. One magnet is a fixed-field, commercially available, permanent magnet with a 35 mm gap, where the $B_0$ field is normal to the pole surface [11]. The second is a field-adjustable Halbach magnet [18]; this has a cylindrical bore with 100 mm inner diameter and with the $B_0$ field perpendicular to the cylinder axis. Both magnets have sufficiently large gaps, so that the commercial Bruker $TE_{102}$ resonant cavity fits inside [11].

It is important to be able either to tune the microwave frequency or to adjust the magnetic field in order to precisely match the ESR condition. As the frequency bandwidth of X-band ESR cavities is usually very narrow, the cavity must be made frequency-tunable for the former scenario, while the latter case is necessary for a fixed-frequency cavity. These methods have advantages over adding resistive coils to a fixed-field permanent magnet in that another power supply is not necessary, and the field of the permanent magnet is not affected by the heating of the resistive coils.

Our commercial $TE_{102}$ resonant cavity was modified so that the resonant frequency could be made tunable for its use in the fixed-field permanent magnet. This was realized by making the wall opposite along which the microwave is coupled in via the wave guide (the long axis of the resonator) adjustable with a copper plate attached to a nonmagnetic screw. A nonmagnetic spring was placed between the moveable wall and a fixed plate was attached to the cavity; by turning the screw the copper plate can move further in or out of the resonator, thus varying the length of the cavity and resonant frequency. To test the range over which significant DNP enhancements could be measured in the modified cavity, it was placed in the electromagnet with a sample of 15 m$M$ $^{15}N$ 4-oxo-TEMPO. The frequency at which the cavity was resonating was then monitored using the Bruker EMX spectrometer and EIP frequency counter. Our custom microwave source was then coupled to the cavity, set to the resonant frequency of the cavity, and the DNP enhancement measured. Enhancements were measured over a range of 9.9025 GHz to 9.5285 GHz, corresponding to a change in field from 3522 G to 3389 G, which were the mechanical limits of our modified cavity. Over this range, large enhancements were measured with the maximum enhancement being −98 fold. An enhancement of −112 was measured for the same sample in the unmodified commercial cavity. We attribute this small decrease due to a lower $Q$ in the modified cavity of ~1400, while the commercial cavity has a loaded $Q$ of ~2200, and thus more power reaching the sample. This experiment shows that we are able to reach a wide and useful tuning range by a small modification of the commercial Bruker $TE_{102}$ cavity. This modified cavity can be readily employed with a fixed-field magnet and tunable microwave source to perform DNP experiments.

The modified cavity was then placed inside the portable, fixed-field, permanent magnet. The entire DNP set-up is readily portable (Figure 11.8, left) and has been transported to different buildings on campus and also used for demonstration purposes (Figure 11.4, left; Figure 11.8, right). In order to optimize the DNP efficiency, the length of the cavity was adjusted while the proton enhancement was monitored. So far, the maximum measured enhancement in this fully portable DNP set-up, under conditions where minimal sample heating occurs, was −65 ± 9 for a 15 mM $^{15}N$ 4-oxo-TEMPO aqueous sample. This was approximately 25% less than that measured in the electromagnet with the modified cavity, and was attributed to the difficulty in tuning the cavity to precisely meet ESR conditions with sufficiently high accuracy. With increased microwave power and sample heating, a higher enhancement of −92 ± 11 was measured. Further design optimizations, such as decreasing the amount of movement of the movable wall of the cavity per turn of the screw to allow for finer adjustments, should greatly improve these results.

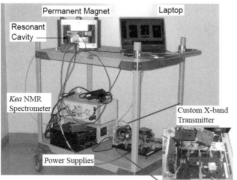 

**Figure 11.8** Left: A fully portable $^1H$ DNP system based on X-band ESR frequencies [11]. The set-up comprises a permanent magnet, the field of which is fixed at 0.35 T, a square ESR cavity, a portable NMR spectrometer (Magritek Limited) and amplifier, a custom X-band driver and a laptop computer. Right: The system can be easily transported across the campus for demonstration purposes.

## 11.4
## Conclusions

We have demonstrated new and useful applications of DNP-enhanced $^1$H NMR analysis at X-band (8–12 GHz), which is the most commonly used frequency band for ESR. As both ESR and DNP analysis of solution-state samples depend critically on the dynamics of the spin label as well as the molecule which NMR nuclei is enhanced, DNP at X-band together with ESR experiments can provide valuable information about the dynamics of the spin labels and the local solvents. A commercial ESR spectrometer can easily be modified to perform DNP experiments, provided that a simple NMR spectrometer operating at 14–15 MHz is available. With this set-up, significant signal enhancements as well as the quantification of key DNP parameters can be achieved. For experiments where large signal enhancements are preferred or required, such as in a RELIC experiment for perfusion contrast imaging, a relatively inexpensive custom microwave transmitter device is capable of reaching near-full saturation of broad ESR lines. $^1$H enhancements at 0.35 T of up to −112 fold have been measured when using the custom transmitter along with a commercial electromagnet and resonant cavity. Our goal to build a portable DNP polarizer has been achieved with $^1$H enhancements up to −92 fold thus far. The custom microwave source can be coupled into a resonant cavity and placed inside a portable permanent magnet. Indeed, the entire portable system fits onto a small trolley and has been transported to different places for demonstration experiments (Figure 11.4, left and Figure 11.8, right).

## Acknowledgments

These studies were partially supported by the Materials Research Laboratory program of the National Science Foundation (DMR00-80034), the Petroleum Research Funds (PRF#45861-G9) of the American Chemical Society and the Faculty Early CAREER Award (20070057) of the National Science Foundation.

## References

1 Nicholson, I., Lurie, D.J. and Robb, F.J.L. (1994) The application of proton-electron double-resonance imaging techniques to proton mobility studies. *Journal of Magnetic Resonance*, **104**, 250–5.

2 Guiberteau, T. and Grucker, D. (1993) Dynamic nuclear polarization of water protons by saturation of σ and π EPR transitions of nitroxides. *Journal of Magnetic Resonance*, **105**, 98–103.

3 Barros, W. and Engelsberg, M. (2002) Ionic transport, reaction kinetics and gel formation: a low-field Overhauser magnetic resonance imaging study. *The Journal of Physical Chemistry A*, **106**(32), 7251–5.

**4** Barros, W. and Engelsberg, M. (2007) Enhanced Overhauser contrast in proton-electron double resonance imaging of the formation of an alginate hydrogel. *Journal of Magnetic Resonance*, **184**(1), 101–7.

**5** Armstrong, B.D. and Han, S. (2007) A new model for Overhauser enhanced nuclear magnetic resonance using nitroxide radicals. *The Journal of Chemical Physics*, **127**, 104508–10.

**6** McCarney, E.R. and Han, S. (2008) Dynamic nuclear polarization enhanced magnetic resonance analysis of local water dynamics in soft molecular assemblies at 9.8 GHz using amplified $^1$H water signal. *Langmuir*, **24**, in press.

**7** McCarney, E.R., Armstrong, B.D., Lingwood, M.D. and Han, S. (2007) Hyperpolarized water as an authentic magnetic resonance imaging contrast agent. *Proceedings of the National Academy of Sciences of the United States of America*, **104**, 1754–9.

**8** Stevenson, S., Glass, T. and Dorn, H.C. (1998) Dynamic nuclear polarization: an alternative detector for recycled-flow NMR experiments. *Analytical Chemistry*, **70**, 2623–8.

**9** Dorn, H.C., Glass, T.E., Gitti, R. and Tsai, K.H. (1991) Transfer of 1 H and 13 C dynamic nuclear polarization from immobilized nitroxide radicals to flowing liquids. *Applied Magnetic Resonance*, **2**, 9–27.

**10** Gitti, R., Wild, C., Tsiao, C., Zimmer, K., Glass, T.E. and Dorn, H.C. (1988) Solid/liquid intermolecular transfer of dynamic nuclear polarization. Enhanced flowing fluid proton NMR signals via immobilized spin labels. *Journal of the American Chemical Society*, **110**, 2294.

**11** Armstrong, B.D., Lingwood, M.D., McCarney, E.R., Brown, E.R., Blümler, P. and Han, S. (2008) Portable X-band system for solution state dynamic nuclear polarization. *Journal of Magnetic Resonance*, **191**, 273–81.

**12** (a) Hausser, K.H. and Stehlik, D. (1968) Dynamic nuclear polarization in liquids. *Advances in Magnetic Resonance*, **3**, 79–139. (b) Muller-Warmuth, W. and Meise-Gresch, K. (1983) Molecular motions and interactions as studied by dynamic nuclear polarization (DNP) in free radical solutions. *Advances in Magnetic Resonance*, **11**, 1–45.

**13** Benial, A.M.F., Ichikawa, K., Murugesan, R., Yamada, K. and Utsumi, H. (2006) Dynamic nuclear polarization properties of nitroxyl radicals used in Overhauser-enhanced MRI for simultaneous molecular imaging. *Journal of Magnetic Resonance*, **182**, 273–82.

**14** Overhauser, A.W. (1953) Polarization of nuclei in metals. *Physical Review*, **92**, 411–15.

**15** McCarney, E.R. and Han, S. (2008) Spin-labeled Gel for the production of radical-free dynamic nuclear polarization enhanced molecules for NMR spectroscopy and imaging. *Journal of Magnetic Resonance*, **190**, 307–15.

**16** Ardenkjaer-Larsen, J.H., Laursen, I., Leunbach, I., Ehnholm, G., Wistrand, L.-G., Petersson, J.S. and Golman, K. (1998) EPR and DNP properties of certain novel single electron contrast agents intended for oximetric imaging. *Journal of Magnetic Resonance*, **133**, 1–12.

**17** Fukuda, H., Goto, A., Yoshioka, H., Goto, R., Morigaki, K. and Walde, P. (2001) Electron spin resonance study of the pH-induced transformation of micelles to vesicles in an aqueous oleic acid/oleate system. *Langmuir*, **17**, 4223–31.

**18** Raich, H. and Blümler, P. (2004) Design and construction of a dipolar Halbach array with an homogeneous field from identical bar-magnets – NMR-mandhalas. *Concepts in Magnetic Resonance Part B*, **23**B(1), 16–25.

# Part Three    Transport Phenomena

# 12
# Localized NMR Velocity Measurements in Small Channels

*L. Guy Raguin and Luisa Ciobanu*

## 12.1
## Introduction

During the past decade, the technology associated with biomechanical and elec-
tromechanical devices has become increasingly miniaturized, resulting in micro-
total-analysis-systems (μTAS) and lab-on-a-chip concepts. At present, the
investigation of fluid transport in such miniaturized devices is primarily performed using optical imaging techniques [1], such as microscopic particle image
velocimetry (μPIV) [2], defocusing digital particle image velocimetry (μDDPIV) [3]
and microscopic molecular tagging (μMTV) [4]. These methods require not only
optically accessible media (transparent flow cells, mixers or reactors with matched
index of refraction), but also microscopic or nanoscopic particles [2, 3] or dissolved
dyes [4] to trace flow paths and obtain quantitative velocity measurements [1].
Tracer particles and dyes modify the hydrodynamic properties of the fluid, which
can highly impact the flow characteristics, especially at short length scales [5]. This
is mitigated experimentally by using low seeding concentrations and small-scale
particles (with diameters equal to 100 to 300 nm for μPIV [5] and 2 μm for μDDPIV
[3], for instance) for particle-based methods, and low dye concentration for methods
using dyes. However, the low particle or dye concentration and the velocity errors
induced by Brownian motion of the small particles or dye molecules then require
spatial and/or temporal averaging in order to obtain accurate and highly resolved
spatial distributions of the velocity field [1–5].

Nuclear magnetic resonance (NMR) imaging techniques constitute valid alternatives for flow studies, given that NMR is noninvasive and compatible with both
opaque and transparent media. It is important to be aware of the limitations,
advantages and disadvantages of the different velocimetry methods that are
available to fully understand the flow within microchannels in order to design and
improve flow cells, micromixers, reactors and other microfluidic devices [6]. In
this chapter, we review several established NMR velocimetry protocols (spin
tagging, flow-compensated phase contrast, and pulsed gradient spin echo), as well
as newer techniques such as multiple modulation multiple echo velocimetry

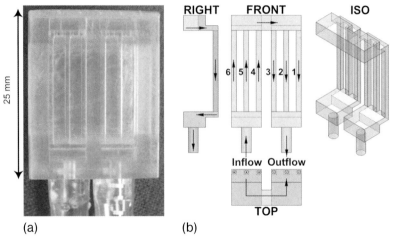

(a)  (b)

**Figure 12.1** Stereolithographic resin microchannel network.
(a) Digital photograph. The channels have rectangular cross-sections of about 0.615 mm$^2$ and are 18 mm long. (b) Front, right, top, and isometric (ISO) views showing the flow paths. The positions of the voxels in which the velocities are measured are shown. Adapted from Ref. [7] Copyright (2007), with permission from Elsevier.

(MMMEV). For quantitative purposes, we implement them for the measurement of velocity fields in a microchannel network and compare the experimental results.

All of the experiments described in this chapter were performed using a Varian Unity/INOVA 600 MHz NMR spectrometer equipped with gradients with a maximum strength of 90 G cm$^{-1}$ and a 30 mm-diameter radiofrequency (RF) probe. The phantom used in these studies is a six-microchannel network built using stereolithography (Figure 12.1a). Each channel is 18 mm long with a rectangular cross-section of about 0.615 mm$^2$. The inlet and outlet of the microchannel network are connected to Teflon tubing. The inlet tubing is attached to a syringe pump (Harvard Apparatus, Holliston, MA, USA). Water doped with copper sulfate ($T_1 = 0.25$ s, $T_2 = 0.22$ s) flows in one direction inside three channels and returns through the other three (Figure 12.1b). By design, the flow rate in the three inflow channels is equal to the flow rate in the three outflow channels, resulting in a zero net flow rate in a cross-section. This design allows for a quick assessment of the accuracy of the velocity measurements.

## 12.2
### NMR Velocimetry Methods

There are two main classes of NMR velocimetry methods: spin tagging and phase-encoding techniques. Spin tagging (ST) methods involve tagging and tracking a material volume of fluid, while phase-contrast (PC) techniques use magnetic field gradients to encode velocity information into the phase of the NMR signal. In this section, we briefly describe several of these methods, and discuss their advantages and disadvantages.

## 12.2.1
## Spin Tagging

Spin tagging consists of presaturating the magnetization in a particular group of spins in a fluid flow and subsequently imaging the displacement of those spins. The tagging can be in the form of a series of parallel planes or two sets of parallel planes that intersect perpendicularly. This is achieved by a combination of RF pulses and magnetic field gradients. Images are taken in the direction perpendicular to the tagged planes, such that parallel tagged lines (1-D combs) or 2-D grids appear in the images [8]. Originally, two different methods were developed to generate 1-D combs or 2-D grids: spatial modulation of magnetization (SPAMM) [9], and delays alternating with nutations for tailored excitation (DANTE) [10]. For the saturation of each set of parallel planes, SPAMM uses a train of RF pulses of modulated amplitude interleaved with blipped magnetic field gradients of constant strength, while DANTE consists of short and intense RF pulses superimposed on a constant magnetic field gradient. In both cases, the separation between the tagged planes is given by

$$\Delta x = \frac{2\pi}{\gamma \int G_t dt} \qquad (12.1)$$

where $\gamma$ is the gyromagnetic ratio, $G_t$ the tagging gradient strength, and the integral is performed over the time delay between two consecutive RF pulses in the RF pulse train.

In a ST pulse sequence, the grid generation module is followed by a time delay (flow evolution time, $\Delta$) to allow the tagged planes to evolve in the flow. Finally, the pulse sequence ends by an imaging sequence: a spin echo sequence for the study of steady flows, or a fast imaging sequence for unsteady flows, such as the snapshot FLASH (Fast Low Angle SHot) [11], an ultra-fast multiple-shot magnetic resonance imaging (MRI) technique [12]. The flow evolution time, $\Delta$, needs to be chosen appropriately depending on several factors. First, based on the range of expected velocities in the flow, $\Delta$ must be long enough such that displacements of the original tagged lines are detectable. At the same time, $\Delta$ should be short enough to satisfy several criteria. The tagged spins must remain within the RF coil over the total experiment time. Moreover, their detectability depends on the contrast between the tagged and untagged fluid, which decays with the $T_1$ relaxation time of the fluid, and the smearing caused by diffusion and shear present in the flow. Finally, since velocity fields (as well as displacement and strain fields) are inferred from the grid deformation over the flow evolution time, $\Delta$ should also be short enough such that the approximation that the instantaneous Eulerian velocity is given by the Lagrangian displacement of the tags within $\Delta$ is valid.

The main advantage of ST methods is their visual appeal, as they allow for an intuitive visualization of both complex fluidic and granular flows, as well as the deformation of soft materials and tissues. The disadvantages are their relatively coarse resolution and the extensive post-processing steps required to extract quantitative information [13]. A large number of studies have focused on the reconstruction of 2-D velocity fields in single or multiple planes. Then, for 3-D flows, additional algorithms are required to extract the third velocity component [14]. 3-D multistripe/multiplane tagging imaging is possible by tagging three sets of orthogonal planes and using a 3-D imaging sequence, which Weis *et al.* [15] employed to provide a time-resolved illustration of thermal convection patterns. However, the contrast to the grid stripes was not good enough for quantitative measurements and a phase encoding velocimetry method was used for quantitative characterization of the 3-D flow field.

For the present study, a spin echo pulse sequence with a SPAMM module is used, as shown in Figure 12.2, in order to produce a 1-D comb along the direction of the channels (read-out direction). The SPAMM module consists of a Gaussian-modulated RF pulse train with eight RF pulses of duration 200 μs and a cumulative effect equivalent to an 180° RF pulse, interleaved with seven blipped tagging gradients of constant strength $G_t = 10 \, G \, cm^{-1}$ and duration 100 μs. The repetition and echo times are set to 1 s and 8 ms, respectively. The field of view (FOV) is

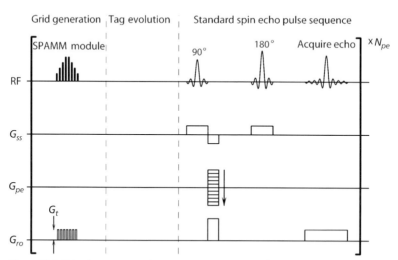

**Figure 12.2** ST pulse sequence. A SPAMM module (eight RF pulses of duration 200 μs interleaved with seven blipped tagging gradients of constant strength and duration 100 μm) is inserted as a preparation module prior to a standard spin echo imaging pulse sequence, thus resulting in a set of tagged planes perpendicular to the read-out direction. Notations: $G_t$ tagging gradient, $G_{ss}$ slice selection gradient, $G_{pe}$ phase-encoding gradient, $G_{ro}$ read-out gradient.

2 cm × 1.6 cm (and 1.5 mm slice thickness) with a matrix size of $200 × 160$ complex points in the read and phase directions, respectively, yielding an isotropic in-plane resolution of 100 μm.

## 12.2.2
### Phase-Encoding Methods

The principle of phase-encoding methods relies on the behavior of the phase $\phi(t)$ of the spin system in the presence of magnetic field gradients. For example, one can encode the $z$-component of the velocity vector into the phase of the signal, by inserting a time-varying flow-encoding gradient along the $z$-direction, $G_z(t)$, in an imaging pulse sequence. Spins located at $z(t)$ and experiencing the magnetic field gradient $G_z(t)$ precess at the Larmor frequency: $\omega(t) = \gamma [B_0 + G_z(t) \, z(t)]$. The phase of the spins, $\phi$, which is accumulated during the time duration $T_{PE}$ of the phase-encoding gradient $G_z(t)$, is given by:

$$\phi = \int_0^{T_{PE}} \gamma [B_0 + G_z(t) z(t)] dt. \tag{12.2}$$

Both the position $z$ of nonstationary spins and the gradient $G_z$ are functions of time. Using a Taylor series expansion about an arbitrarily chosen point in time $t = \tau$, Equation 12.2 becomes:

$$\phi = \gamma B_0 T_{PE} + \int_0^{T_{PE}} \gamma G_z(t) \left[ z_0 + (t - \tau) \left( \frac{dz}{dt} \right)_{t=\tau} + \frac{(t - \tau)^2}{2} \left( \frac{d^2z}{dt^2} \right)_{t=\tau} + O(t^3) \right] dt. \tag{12.3}$$

The first term in the right-hand-side of this equation is a constant for all spins in the magnetic field and is easily subtracted from the phase of the signal. Neglecting the higher-order terms, Equation 12.3 can be rewritten as:

$$\phi = \gamma m_0(\tau) z(\tau) + \gamma m_1(\tau) v(\tau) + \gamma m_2(\tau) a(\tau), \tag{12.4}$$

with

$$m_n(\tau) = \int_0^{T_{PE}} G_z(t) \frac{(t - \tau)^n}{n!} dt, \tag{12.5}$$

where $z(\tau)$, $v(\tau)$ and $a(\tau)$ are the position, velocity and acceleration components along the direction of the flow-encoding gradients at $t = \tau$, $m_n$ represents the $n$-th moment of the gradient $G_z(t)$. Equation 12.4 suggests that quantities such as position, velocity and acceleration can be encoded into the phase of the signal by the appropriate selection of the gradient $G_z(t)$ and its moments. In particular, when $m_0(\tau)$ and $m_2(\tau)$ are zero, one can encode the velocity:

$$\phi = \gamma m_1(\tau) v(\tau) \tag{12.6}$$

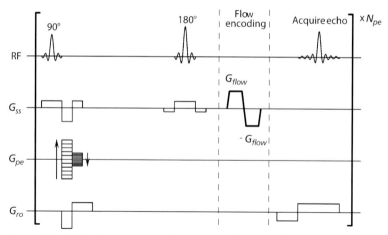

**Figure 12.3** PCSE pulse sequence. A flow-encoding bipolar gradient is inserted in a flow-compensated spin echo imaging pulse sequence to encode one velocity component (here, the one in the slice-selection direction) into the phase of the image. Notations: $G_{flow}$ flow-encoding gradient, $G_{ss}$ slice selection gradient, $G_{pe}$ phase-encoding gradient, $G_{ro}$ read-out gradient. Compared to a standard spin echo pulse sequence, additional gradient pulses are introduced after the 90° pulse in order to null the first gradient moment $m_1$ in all three directions.

### 12.2.2.1 Phase-Contrast Velocimetry

A typical flow-compensated phase-contrast spin echo (PCSE) pulse sequence is shown in Figure 12.3. Flow compensation means that the first moments $m_1$ due to the imaging magnetic gradients in the read-out, phase-encoding and slice direction are zero, so that velocity contribution to the phase is only due to the flow-encoding gradient $G_z(t)$ [16]. To that end, moment-nulling gradients are added after the 90° pulse. A complete discussion on possible choices for moment-nulling is available in Ref. [16]. For simplicity reasons, a bipolar flow-encoding gradient is often used ($m_0 = 0$ and $m_1$ is independent of $\tau$). If the acceleration is small, the last term in Equation 12.4 can be neglected. Hence, the fluid velocity is directly related to the phase:

$$\phi = \gamma m_1(0) v(0). \tag{12.7}$$

It is often assumed that the velocity is constant over each repetition time of the pulse sequence, so that $v(0)$ corresponds to the instantaneous velocity, which is valid for most laminar flows.

In order to obtain accurate velocity data, four images are necessary: one pair of images with flow (subscript 'flow') and another pair at rest (subscript 'static'). Each pair of images consists of one image with the bipolar gradients with one polarity (e.g. positive lobe first, and then negative lobe, labeled with a superscript '+'), and a second image with the opposite polarity (superscript '−'). This process is necessary to compensate for the spurious phase contributions caused by the eddy currents generated by the flow-encoding and imaging gradients, as well as imperfections in the homogeneity of the static magnetic field $B_0$, such that zero phase corresponds to zero velocity [16, 17]. Phase differences are then taken, and the point-wise velocity is given by:

$$v = \frac{\left(\phi_{flow}^+ - \phi_{flow}^-\right) - \left(\phi_{static}^+ - \phi_{static}^-\right)}{2\gamma m_1(0)} \quad (12.8)$$

with

$$m_1(0) = G_{flow}\left(t_{flow}^2 + 3t_{flow}t_{rise} + 2t_{rise}^2\right) \quad (12.9)$$

The PC technique provides full-field velocity data for steady or periodic flows, with significantly higher spatial resolution than the ST method. The strength and duration of the flow-encoding gradients control the velocity resolution, as well as the maximum velocity, $V_{enc}$, that can be encoded without causing phase wrapping ($V_{enc} = \pi/(2\gamma m_1)$) [16]. Phase unwrapping algorithms are available [18], such that $V_{enc}$ is not the maximum velocity that can be measured, although excessive phase wrapping is not desirable. The minimum velocity, $V_{min}$, that can be measured is based on the incoherent Brownian motion, which is quantified by the spin self-diffusion coefficient, $D$, resulting in $V_{min} \approx (D/T_2)^{1/2}$ for spin echo imaging and $V_{min} \approx (D/T_1)^{1/2}$ for stimulated echo imaging [16].

For the study of steady flows, a spin echo imaging pulse sequence is typically used. For the investigation of time-dependent flows, fast acquisition can be obtained via echo planar imaging pulse sequences. For time-periodic flows, interleaved pulse sequences (PC cine MRI, or 4-D PC MRI) have been developed which allow the acquisition of time-resolved 3-D velocity fields in 2-D slices or 3-D volumes [19, 20]. Additionally, Moser *et al.* [21] reported on the use of a synchronized PC echo planar imaging (EPI) protocol in a mixing reactor that still uses four images in order to obtain accurate velocity measurements. The clear advantage of PC velocimetry over ST is its capability to directly provide voxel-wise measurements of 3-D velocity components in 3-D space. While ST is intuitive and relatively unambiguous in its interpretation of the displaced tags, PC methods require additional phase correction steps or subtraction methods, and often produce large measurement errors near the flow boundaries due to susceptibility artifacts. These errors can be corrected a posteriori by invoking the no-slip boundary condition in case of stationary boundaries (see Section 12.2.3).

For the present study, the flow-compensated PCSE pulse sequence shown in Figure 12.3 is used. The flow-encoding bipolar gradient is characterized by the gradient strength at the plateaus, $G_{flow} = 30\,\text{G}\,\text{cm}^{-1}$, the duration of the plateaus, $t_{flow} = 2\,\text{ms}$, and the ramp time, $t_{rise} = 100\,\mu\text{s}$. The repetition and echo times are set to 1 s and 21 ms, respectively. The FOV is 1.6 cm × 0.3 cm in the read and phase directions, respectively, with a 1.5 mm slice thickness. The isotropic in-plane resolution is varied from 200 μm, 100 μm, down to 64 μm.

### 12.2.2.2 Pulsed-Field Gradient Velocimetry

Single-step phase-encoding methods, such as PC velocimetry, provide average velocity components for each pixel in the FOV. This might not be desirable when the signal in one pixel arises from spins with different velocities. In the case of molecular ensembles even at very high spatial resolutions, one can have a distribution of velocities. For example, there may be stationary and moving spins in the same pixel, and therefore both Brownian and flow components should be considered. Neglecting the Brownian motion in the microscopic regime can induce errors in the velocity measurements for flows with small Reynolds numbers (*Re*) [8, 22]. In order to be able to analyze such nonhomogeneous behavior, it is necessary to use a dynamic (also known as *q*-space) imaging technique where the average propagator is measured across the image. The most common *q*-space experiment is a pulsed-field gradient spin echo (PGSE) sequence shown in Figure 12.4 [23–26], which consists of a spin echo imaging sequence with a pair of gradient pulses inserted, one on each side of the 180° RF pulse.

The echo signal, *E*, in the PGSE experiment is given by:

$$E_\Delta(q) \propto \left(e^{j2\pi q_v v}\right)\left(e^{-4\pi^2 q_v^2 D/\Delta}\right) \tag{12.10}$$

where $j^2 = -1$, $\Delta$ is the time between the two gradient pulses, *D* is the diffusion coefficient, and $q_v$ is defined as:

$$q_v = \frac{\gamma}{2\pi} G\delta\Delta \tag{12.11}$$

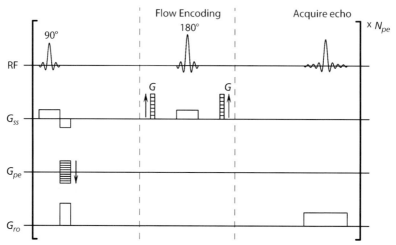

**Figure 12.4** PGSE pulse sequence. The velocity is encoded using a pair of gradient pulses with duration $\delta$ and separation $\Delta$. The strength of both gradient pulses (*G*) is varied in 10 steps from 0 to 54 G cm$^{-1}$. Notations: $G_{ss}$ slice selection gradient, $G_{pe}$ phase-encoding gradient, $G_{ro}$ read-out gradient.

where δ is the width of the gradient pulses of strength $G$. We are in the narrow-pulse approximation, $\delta \ll \Delta$, and therefore neglect the motion of the spins over the duration of the gradients. We recognize the first term in the right-hand-side of Equation 12.10 as the phase of the signal. The second term represents the attenuation of the signal amplitude. In a PGSE experiment a series of images are obtained with gradient amplitude values evenly spaced between zero and a maximum value. The dynamic displacement profile is then calculated by taking the inverse Fourier transform in $q$-space for each pixel in the image. This results in a Gaussian-shaped function, $\bar{P}_s$, which is centered at a value $k_v$:

$$k_v = \frac{\gamma}{2\pi} \frac{N}{n_D} G_{inc} v \delta \Delta \tag{12.12}$$

where $n_D$ is the number of increments, $N$ is the digital array size of the $q$-space transformation, and $G_{inc}$ is the gradient amplitude increment [23]. As a result, the velocity range that can be encoded without phase wrapping is given by $2\pi/(\gamma G_{inc}\delta\Delta)$, and the velocity resolution is $2\pi/[\gamma(n_D-1)G_{inc}\delta\Delta]$.

Compared to the ST and PC methods, the $q$-space method described here provides higher accuracy and the ability to measure velocity and diffusion simultaneously. The major drawback of this method is that, depending on the number of flow-encoding measurements in $q$-space, the acquisition time can be much longer.

For the present study, the PGSE pulse sequence shown in Figure 12.4 is used. A set of 10 flow images is acquired with equal spacing in $q$-space, with a maximum strength of the flow-encoding gradient of $54\,G\,cm^{-1}$ (resulting in $G_{inc} = 6\,G\,cm^{-1}$), a flow-encoding gradient duration of $\delta = 1.8\,ms$, and $\Delta = 18\,ms$. For post-processing, $N = 128$. The slices are oriented perpendicularly to the long axis of the channels with a FOV of $1.6\,cm \times 0.3\,cm$ in the read and phase directions, respectively. The repetition and echo times are set to $1\,s$ and $25\,ms$, respectively. The isotropic in-plane resolution is varied from $200\,\mu m$, $100\,\mu m$, down to $64\,\mu m$, with a slice thickness of $1.5\,mm$.

### 12.2.2.3 Multiple Modulation Multiple Echo Velocimetry

The two phase-encoding methods presented so far are typically spin echo methods in which only one coherence pathway is selected and therefore only one modulation is used for the phase determination. Several groups have introduced a new class of methods for flow and diffusion measurements called multiple modulation multiple echo (MMME) methods, in which multiple echoes with different flow-induced phase-shift coefficients are produced using several RF pulses in the presence of a constant field gradient [27–29].

The MMME pulse sequence consists of a train of $N$ pulses, which generate $[3^{(N-1)} + 1]/2$ echoes, with each echo corresponding to a different coherence pathway. Each coherence pathway is characterized by $N + 1$ numbers, $Q = (q_0, q_1, \ldots, q_N)$, where $q_j$ corresponds to the magnetization state for the time interval $j$ (e.g. $q_0 = 0$

is the magnetization state before the first pulse, $M = M_0$). We can now generalize Equation 12.6 and write the phase of the $j$-th echo as:

$$\phi_j = \gamma m_{1,j}(\tau)v(\tau) = \gamma \int_0^{T_j} q(t-\tau)G(t-\tau)v(t-\tau)dt = c_Q(j)v \tag{12.13}$$

where $T_j$ is the time at which the $j$-th echo appears and

$$c_Q(j) = \gamma \int_0^{T_j} q(t-\tau)G(t-\tau)(t-\tau)dt. \tag{12.14}$$

It follows that the velocity $v$ can be experimentally determined, in one single scan, as the slope of the measured phase at the $j$-th echo, $\phi_j$, as a function of $c_Q(j)$. Here, voxel-selective gradients can be used in order to localize the velocity measurement to a certain region. Our group has recently introduced such a voxel-resolved MMMEV [7].

Compared to conventional velocimetry experiments (ST, PC, PGSE), MMMEV is much faster, with acquisition times of several tens of milliseconds. Moreover, in a recent publication, Cho et al. [30] showed that the MMME technique could be extended to the measurement of an average flow velocity along an arbitrary direction in one scan (a conventional method would require at least three separate experiments with orthogonal gradient directions).

In the present study, we used a pulse sequence consisting of four rectangular RF pulses, which generated 13 echoes. The RF pulse flip angles are $\alpha_1/\alpha_2/\alpha_3/\alpha_4 = 51°/77°/77°/110°$, and had been shown previously to result in echoes of similar magnitudes [27]. The phases of the RF pulses are all zero. The magnetic field gradients are applied between, during and after the pulses. Along the direction of the velocity component of interest (e.g. along the $z$-direction), a constant field gradient $G_v$ is applied between and after the RF pulses to encode the velocity component. Gradients along the $x$-, $y$- and $z$-directions are applied during the RF pulses in order to select only the voxel of interest. In addition, along both the $x$- and $y$-directions, refocusing gradient lobes equal in magnitude to $G_x$ and $G_y$ are applied, as shown in Figure 12.5. The delays between the RF pulses are $\tau_1 = 900\,\mu s$, $\tau_2 = 3\,\tau_1$ and $\tau_3 = 9\,\tau_1$, respectively, and all RF pulses are $200\,\mu s$ long. The durations for the refocusing gradients are: $\delta_1 = 109.972\,\mu s$, $\delta_2 = 110.064\,\mu s$, $\delta_3 = 109.992\,\mu s$, and $\delta_4 = 110.02\,\mu s$. Data points are collected over a period of $20\,ms$ (np = 400, SW = $20\,kHz$, nt = 1), resulting in a total acquisition time per velocity component of $32\,ms$. The dimensions of the selected voxels are $4 \times 1.5 \times 1.5\,mm^3$. Measurements for flow rates ranging from 2 to $10\,ml\,h^{-1}$ are performed, and the magnitude of the flow-encoding gradient $G_v$ is varied between 6 and $12\,G\,cm^{-1}$, depending on the flow rate.

## 12.2.3
### Additional Post-Processing Steps for PCSE and PGSE

The data acquisition protocol for the PC and PGSE experiments involves acquiring a high-resolution reference image (25 μm isotropic in-plane resolution) to charac-

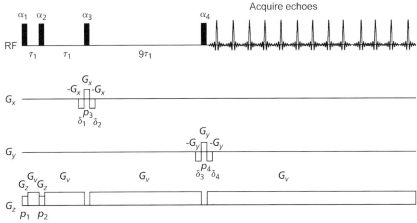

**Figure 12.5** MMMEV pulse sequence. The four RF pulses generate 13 echoes, each of which contains phase information. The $G_v$ gradient acts as the flow-encoding gradient, while $G_x$, $G_y$ and $G_z$ select the voxel of interest. Adapted from Ref. [7] Copyright (2007), with permission from Elsevier.

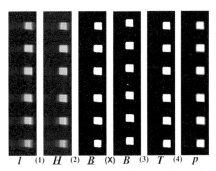

$I$    (1) $H$   (2) $\underline{B}$   (X) $B$   (3) $T$   (4) $p$

**Figure 12.6** Four-step algorithm to obtain a low-resolution partial volume mask image $p$ starting from a low-resolution image $I$ and a high-resolution reference image $H$. $\underline{H}$ is the image obtained by zero-filled up-sampling in Fourier domain of $I$, $\underline{B}$ and $B$ are binary masks obtained from $\underline{H}$ and $H$, respectively, and $T$ are images obtained by maximizing the cross-correlation coefficient between $\underline{B}$ and $B$.

terize the flow phantom geometry (channel size and locations), while the velocity-encoded images are acquired at lower resolutions (200, 100, and 64 µm). The velocity data needs to be coregistered with the high-resolution geometry image using a data fusion algorithm, and a partial volume mask needs to be created to accurately estimate the flow rates within the individual small channels.

To that end, four steps are implemented, as illustrated in Figure 12.6. Step (1) consists of performing a zero-filled up-sampling in Fourier domain [31] of the low-resolution image $l$ to obtain an image $\underline{H}$, with the same matrix size as the high-resolution image $H$. In step (2), $\underline{H}$ is binarized by using iterative thresholding such that the total flow area computed as the sum of the intensity of the resulting binary image $\underline{B}$ becomes equal to that for the binary mask $B$ obtained from the high-resolution image, $H$. Step (3) consists of aligning $B$ with $\underline{B}$ by maximizing the cross-correlation coefficient between the two images to determine the spatial offsets, resulting in a set of translated images $T$ from $B$. Finally, in step (4), $T$ is spatially averaged to decrease the resolution back to the original low-resolution image. This procedure yields a low-resolution partial volume mask image $p$ that is aligned with the velocity data, and possesses the same total flow area as the high-resolution image. This procedure makes it possible to account for partial volume effects and improve the estimation of flow rates in the individual small channels.

Finally, for PC, the velocity measurements in the pixels closest to the boundaries are the most likely to be polluted by susceptibility artifacts, and should be close to zero because of the no-slip boundary condition. Thus, if the measured velocity closest to the boundaries exceeds 15% of the maximum velocity measured in the channel, the measurement is corrected by linear interpolation between zero and the nearest neighbor in the direction normal to the boundary.

## 12.3
### Experimental Results and Discussion

The ST velocity measurement results are shown in Figure 12.7. The tags were digitally enhanced by coloring in black the pixel with the lowest intensity within the tagged lines. This allows for the quantitative determination of the 1-D profiles of the velocity averaged over the depth of the microchannels at several positions along the network. The cumulative flow rates for the three inflow channels and the three outflow channels (channels 4 to 6, and 1 to 3 in Figure 12.1, respectively) for ST, PC and PGSE velocimetry techniques are shown in Table 12.1 for each in-plane resolution used and an imposed flow rate of $14 \, \text{ml} \, \text{h}^{-1}$. Both inflow and outflow rates should be equal to the nominal flow rate. Moreover, our flow phantom is such that the net flow rate (outflow rate subtracted from the inflow rate) should be equal to zero by mass conservation, and thus can be used to assess the consistency of the velocimetry technique. For all methods and for all resolutions tested, both the average measurement error and the consistency error are found to be less than 6%.

PCSE and PGSE were used to obtain 2-D velocity profiles in all six microchannels for low ($200 \, \mu\text{m}$), medium ($100 \, \mu\text{m}$) and high ($64 \, \mu\text{m}$) in-plane resolution with a slice thickness of $1.5 \, \text{mm}$. Figures 12.8 and 12.9 show the experimental velocity contours obtained for channel #5 for a flow rate of $14 \, \text{ml} \, \text{h}^{-1}$, and their difference with the predicted contours calculated from the analytical solution of steady fully developed laminar flow within a rectangular duct. It can be noticed that, for both methods, the flow contours agree better with the theoretical predictions for higher

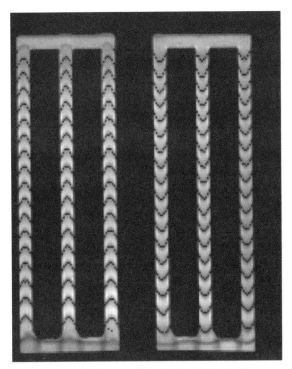

**Figure 12.7** Axial velocity profiles obtained with the ST pulse sequence presented in Figure 12.2. The tags were digitally enhanced by coloring in black the tagged pixels with the lowest intensity.

**Table 12.1** Flow rates in the inflow (4 to 6 in Figure 12.1) and outflow (1 to 3) channels obtained via PC and PGSE velocimetry for a nominal flow rate of $14\,ml\,h^{-1}$.

| Method | ST | | PCSE | | PGSE | | |
|---|---|---|---|---|---|---|---|
| Resolution (µm) | 100 | 200 | 100 | 64 | 200 | 100 | 64 |
| Inflow rate ($ml\,h^{-1}$) | 14.40 | 14.01 | 14.39 | 14.70 | 13.43 | 13.28 | 13.24 |
| Outflow rate ($ml\,h^{-1}$) | 14.53 | 13.36 | 14.15 | 14.31 | 14.22 | 14.02 | 14.00 |
| Net flow rate ($ml\,h^{-1}$) | 0.13 | 0.65 | 0.24 | 0.39 | −0.79 | −0.74 | −0.77 |

resolution, as expected. Phase-contrast velocity data at the highest resolution (64 µm in-plane resolution) exhibit more noise than both PGSE velocity data at this resolution and PC data at lower resolutions. This indicates that the lower signal amplitude caused by the smaller voxel size going from 100 µm to 64 µm in-plane resolution makes the phase measurements in PC velocimetry more prone to noise.

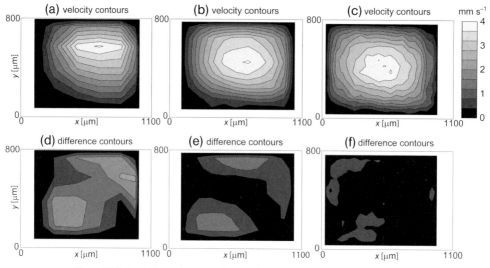

**Figure 12.8** (a–c) 2-D velocity profiles in channel #5 obtained via PCSE for $Re = 3$ ($Q = 14\,\mathrm{ml\,h^{-1}}$), for three different in-plane resolutions 64, 100 and 200 µm, respectively; (d–f) difference between experimental and predicted contours.

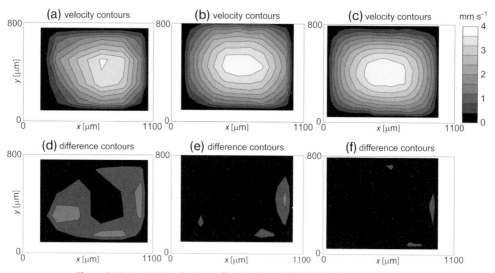

**Figure 12.9** (a–c) 2-D velocity profiles in channel #5 obtained via PGSE for $Re = 3$ ($Q = 14\,\mathrm{ml\,h^{-1}}$), for three different in-plane resolutions 64, 100 and 200 µm, respectively; (d–f) difference between experimental and predicted contours.

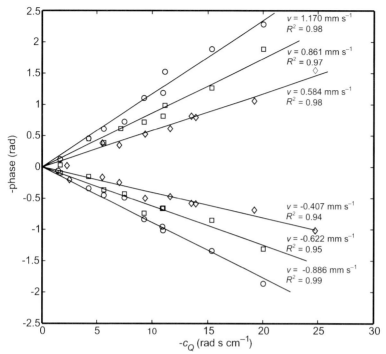

**Figure 12.10** Flow-induced phase shift versus $c_Q$ for three different flow rates in microchannels #1 (negative velocities) and # 6 (positive velocities) obtained using MMMEV. The data points for each flow rate were fitted with straight lines, the slope giving the corresponding mean flow velocities. Symbols: $\bigcirc$, 8 ml h$^{-1}$; $\square$, 6 ml h$^{-1}$; $\diamond$, 4 ml h$^{-1}$. Reprinted from Ref. [7] Copyright (2007), with permission from Elsevier.

MMMEV data sets are acquired for each of the six microchannels, first with no flow, and then with flow. The static data sets are used as phase references for the flow measurements. In terms of numerical data processing, for each flow rate we compute the phase differences $\phi_{1,j} = \phi_{1,j,\text{flow}} - \phi_{1,j,\text{static}}$, where $j = \{1, \ldots, 13\}$ is the echo index. The coefficients $c_{Q,z}(j)$ are calculated for each echo. The times at which the echoes occur are determined experimentally (these times can be calculated theoretically using the formalism developed by Sodickson and Cory [32]). As representative examples of the velocity measurements, in Figure 12.10 the phase shift $\phi_1(j)$ is plotted as a function of $c_Q(j)$ for two voxels, corresponding to microchannels #1 and #6, for imposed flow rates of 4, 6 and 8 ml h$^{-1}$. The flow velocities are extracted from the data points as the slope of zero-intercept linear fits.

The results for individual microchannels for all flow rates are presented in Table 12.2, where the standard deviation is calculated from 10 measurements to quantify any degree of uncertainty. This uncertainty corresponds to 5–10% of the measured

**Table 12.2** The average velocities in each individual channel.
The values are obtained from 10 measurements.

| Channel | Average channel velocity (mm s$^{-1}$) | | | | |
|---|---|---|---|---|---|
| | $Q = 10\,\mathrm{ml\,h^{-1}}$ | $Q = 8\,\mathrm{ml\,h^{-1}}$ | $Q = 6\,\mathrm{ml\,h^{-1}}$ | $Q = 4\,\mathrm{ml\,h^{-1}}$ | $Q = 2\,\mathrm{ml\,h^{-1}}$ |
| 1 | $-1.08 \pm 0.06$ | $-0.88 \pm 0.07$ | $-0.60 \pm 0.03$ | $-0.39 \pm 0.03$ | $-0.18 \pm 0.02$ |
| 2 | $-1.41 \pm 0.08$ | $-1.14 \pm 0.03$ | $-0.90 \pm 0.04$ | $-0.68 \pm 0.04$ | $-0.35 \pm 0.02$ |
| 3 | $-1.64 \pm 0.04$ | $-1.35 \pm 0.08$ | $-1.00 \pm 0.03$ | $-0.67 \pm 0.02$ | $-0.34 \pm 0.02$ |
| 4 | $1.46 \pm 0.04$ | $1.19 \pm 0.06$ | $0.95 \pm 0.04$ | $0.68 \pm 0.04$ | $0.36 \pm 0.03$ |
| 5 | $1.57 \pm 0.09$ | $1.32 \pm 0.11$ | $0.83 \pm 0.05$ | $0.51 \pm 0.03$ | $0.27 \pm 0.02$ |
| 6 | $1.53 \pm 0.05$ | $1.21 \pm 0.05$ | $0.90 \pm 0.04$ | $0.57 \pm 0.04$ | $0.28 \pm 0.05$ |

**Table 12.3** Comparison of experimental and predicted average inflow and outflow velocities.

| $Q$ (ml h$^{-1}$) | Average velocity (mm s$^{-1}$) | | | |
|---|---|---|---|---|
| | Experimental | | Predicted | |
| | Inflow | Outflow | Inflow | Outflow |
| 10 | $1.52 \pm 0.04$ | $-1.38 \pm 0.04$ | 1.520 | $-1.491$ |
| 8 | $1.24 \pm 0.06$ | $-1.12 \pm 0.03$ | 1.216 | $-1.193$ |
| 6 | $0.89 \pm 0.03$ | $-0.83 \pm 0.03$ | 0.912 | $-0.895$ |
| 4 | $0.59 \pm 0.02$ | $-0.58 \pm 0.03$ | 0.608 | $-0.597$ |
| 2 | $0.32 \pm 0.02$ | $-0.29 \pm 0.02$ | 0.304 | $-0.298$ |

velocity. The cross-sectional area for each channel was measured via the same
high-resolution spin echo image (25 μm in-plane resolution) that was used for the
additional post-processing of the PC and PGSE data sets. Both the experimental
and predicted average inflow and outflow velocities can then be computed; the
results are shown in Table 12.3. As illustrated in Table 12.3, the measured average
inflow and outflow velocities agree with each other to within 8%, and also match
the imposed flow rate to within 8%.

## 12.4
## Conclusions

MR velocimetry experiments were conducted in a microchannel network to
compare different MRI protocols (ST, PC, PGSE, MMMEV) velocity measurements at different resolutions. For in-plane velocimetry, ST provides good qualitative and quantitative results, but lacks in resolution. For out-of-plane velocimetry,

the velocity measurements via PC and, in particular, PGSE are in very good agreement with the predicted velocities. At higher resolutions, PC velocimetry is more susceptible than PGSE to spurious errors caused by the lower signal-to-noise ratio (SNR) per voxel. Finally, MMMEV is a fast and localized velocimetry technique which produces average velocities consistent with the imposed flow rate and measured cross-sectional areas. Thus, the slice-selective gradients used to localize the microchannel of interest were successfully decoupled from the velocity-encoding gradient. Additionally, MMMEV obviates the resolution issues involved with ST, PC and PGSE when it is necessary to measure average velocities.

The measurement of flow rates, which is critical to the design and quality control of microfluidic devices, involves the measurement of both velocities and cross-sectional areas. While the focus of this chapter is on MR velocimetry techniques, the problem of accurately measuring flow rates also relies heavily on the accurate measurement (or a priori knowledge) of cross-sectional areas. For traditional engineering applications, cross-sectional areas are well characterized, since the parts are machined to specified tolerances. However, for microfluidics the use of soft polymers such as polydimethylsiloxane (PDMS) for their microfabrication [33, 34] introduces some variability, in particular for pressure-driven flows [35]. Either MRI or another measurement technique (e.g. optical microscopy, scanning electron microscopy, profilometry, atomic force microscopy) would then be required to characterize the flow domain of these devices.

## Acknowledgments

These studies were supported by the Biomedical Imaging Center, Beckman Institute for Advanced Science and Technology, University of Illinois at Urbana-Champaign, the Richard W. Kritzer Foundation, and the NSF. Science and Technology Center of Advanced Materials for Purification of Water with Systems (Water-CAMPWS, Grant CTS-0120978). The authors thank Prof. John Georgiadis for providing the multichannel network.

## References

1 Sinton, D. (2004) *Microfluidics and Nanofluidics*, **1**, 2–21.
2 Santiago, J.G., Wereley, S.T., Meinhart, C.D., Beebe, D.J. and Adrian, R.J. (1998) *Experimental Fluids*, **25**, 316–19.
3 Pereira, F., Lu, J., Castaño-Graff, E. and Gharib, M. (2007) *Experimental Fluids*, **42**, 589–99.
4 Maynes, D. and Webb, A.R. (2002) *Experimental Fluids*, **32**, 3–15.
5 Meinhart, C.D., Wereley, S.T. and Santiago, J.G. (1999) *Experimental Fluids*, **27**, 414–19.
6 Stone, H.A., Stroock, A.D. and Ajdari, A. (2004) *Annual Review of Fluid Mechanics*, **36**, 381–411.
7 Raguin, L.G. and Ciobanu, L. (2007) *Journal of Magnetic Resonance*, **184**, 337–43.
8 Fukushima, E. (1999) *Annual Review of Fluid Mechanics*, **31**, 95–123.

**9** Axel, L. and Dougherty, L. (1989) *Radiology*, **172**, 349–50.

**10** Mosher, T.J. and Smith, M.B. (1990) *Magnetic Resonance in Medicine*, **15**, 334–9.

**11** Moser, K.W., Raguin, L.G., Harris, A., Morris, H.D., Georgiadis, J.G., Shannon, M. and Philpott, M. (2000) *Magnetic Resonance Imaging*, **18**, 199–207.

**12** Haase, A. (1990) *Magnetic Resonance in Medicine*, **13**, 77–89.

**13** Ozturk, C. and McVeigh, E.R. (2000) *Physics in Medicine and Biology*, **45**, 1683–702.

**14** Raguin, L.G. and Georgiadis, J.G. (2004) *Journal of Fluid Mechanics*, **516**, 125–54.

**15** Weis, J., Kimmich, R. and Müller, H.P. (1996) *Magnetic Resonance Imaging*, **14**, 319–27.

**16** Pope, J.M. and Yao, S. (1993) *Concepts in Magnetic Resonance*, **5**, 281–302.

**17** Moser, K.W., Georgiadis, J.G. and Buckius, R.O. (2000) *Magnetic Resonance Imaging*, **18**, 1115–23.

**18** Ghiglia, D.C. and Pritt, M.D. (1998) *Two-Dimensional Phase Unwrapping: Theory Algorithms and Software*, John Wiley & Sons, Inc., New York.

**19** Pelc, N.J., Herfkens, R.J., Shimakawa, A. and Enzmann, D.R. (1991) *Magnetic Resonance Quarterly*, **7**, 229–54.

**20** Markl, M., Chan, F.P., Alley, M.T., Wedding, K.L., Draney, M.T., Elkins, C. J., Parker, D.W., Wicker, R., Taylor, C.A., Herfkens, R.J. and Pelc, N.J. (2003) *Journal of Magnetic Resonance Imaging*, **17**, 499–506.

**21** Moser, K.W., Raguin, L.G. and Georgiadis, J.G. (2003) *Magnetic Resonance Imaging*, **21**, 127–33.

**22** Callaghan, P. T. (1999) *Reports on Progress in Physics*, **62**, 599–670.

**23** Callaghan, P.T., Eccles, C.D. and Xia, Y. (1988) *Journal of Physics E: Scientific Instruments*, **21**, 820–2.

**24** Callaghan, P.T., MacGowan, D., Packer, K.J. and Zelaya, F.O. (1990) *Journal of Magnetic Resonance*, **90**, 177–82.

**25** Callaghan, P.T. and Xia, Y. (1991) *Journal of Magnetic Resonance*, **91**, 326–52.

**26** Stejskal, E.O. and Tanner, J.E. (1965) *The Journal of Chemical Physics*, **42**, 288–92.

**27** Song, Y.-Q. and Tang, X.P. (2004) *Journal of Magnetic Resonance*, **170**, 136–48.

**28** Song, Y.-Q. and Scheven, U.M. (2005) *Journal of Magnetic Resonance*, **172**, 31–5.

**29** Sigmund, E.E. and Song, Y.-Q. (2006) *Magnetic Resonance Imaging*, **24**, 7–18.

**30** Cho, H., Ren, X.-H., Sigmund, E.E. and Song, Y.-Q. (2007) *Journal of Magnetic Resonance*, **186**, 11–16.

**31** Seppä, M. (2007) *Medical Image Analysis*, **11**, 346–60.

**32** Sodickson, A. and Cory, D.G. (1998) *Progress in Nuclear Magnetic Resonance Spectroscopy*, **33**, 77–108.

**33** Duffy, D.C., McDonald, J.C., Schueller, O.J.A. and Whitesides, G.M. (1998) *Analytical Chemistry*, **70**, 4974–84.

**34** Jo, B.H., Van Lerberghe, L.M., Motsegood, K.M. and Beebe, D.J. (2000) *Journal of Microelectromechanical Systems*, **9**, 76–81.

**35** Holden, M.A., Kumar, S., Beskok, A. and Cremer, P.S. (2003) *Journal of Micromechanics and Microengineering*, **13**, 412–18.

# 13

# Electro-Osmotic Flow and Dispersion in Microsystem Channel Networks and Porous Glasses: Comparison with Pressure-Induced Transport and Ionic Currents

*Bogdan Buhai, Yujie Li and Rainer Kimmich*

## 13.1
### Introduction

Nuclear magnetic resonance (NMR) techniques, both with and without spatial resolution, permit the quantitative measurement of transport quantities such as the flow velocity, the electric current density of ionic currents, and incoherent molecular displacements due to hydrodynamic dispersion [1]. Investigations of this type are of interest for fluids in complex systems where the spatial distribution of transport obstacles and confining pore spaces give rise to heterogeneous transport fields [2]. The question is, therefore, are the transport patterns for the quantities mentioned all the same in such systems?

Microsystem devices such as the so-called 'lab-on-a-chip' are promising tools as chemical sensors and control elements in chemical and biochemical engineering. Transport phenomena of interest in this sort of system may be ionic currents, electro-osmotic flow (EOF) [3] and pressure-induced hydrodynamic flow (PIF). In this chapter, maps of transport quantities such as ionic current density (ICD) and flow velocity due to electro-osmosis and hydrodynamic pressure gradients under the same geometric conditions, and a comparison of patterns of transport pathways, will be presented. Test objects of varying complexity were examined. In this context, random site percolation clusters [4] may be considered as paradigms for systems of particularly high complexity, implying the competitive action of thin and thick channels or bottlenecks connected in parallel [5, 6]. The ultimate degree of complexity is certainly reached in random porous media such as porous glasses. The flow of liquids through such samples is subject to hydrodynamic dispersion [7, 8] as a further transport phenomenon, irrespective of whether it is induced by a pressure gradient or electro-osmosis.

All of the techniques required for this type of experiment are available in the huge arsenal of NMR methods for the measuring and mapping of hydrodynamic and electric quantities. Apart from the porous glasses, all of objects studied were fabricated – that is, milled in solid matrix materials – using computer-generated templates [9]. The use of such model objects has three important advantages:

*Magnetic Resonance Microscopy.* Edited by Sarah L. Codd and Joseph D. Seymour
Copyright © 2009 WILEY-VCH Verlag GmbH & Co. KGaA, Weinheim
ISBN: 978-3-527-32008-0

- Channel networks of any complexity can be made at will in the frame of the mechanical resolution of the milling tools.

- The fact that the channel structure is known in terms of the computer-generated coordinates permits simulations using finite element and finite volume methods [5, 6, 9]–that is, the experimental and simulated data refer to identical channel networks.

- There is a free choice of the matrix material. For example, congruent channel networks can be milled in a nonpolar plastic material on the one hand, and in polar ceramics on the other hand. If electric surface properties determine the phenomenon under investigation, then comparisons will unambiguously permit their identification [6].

As a unique feature of NMR imaging, all forms of transport quantities mentioned so far can be mapped in the same object, using the same apparatus. Different transport phenomena can be related on this basis; for example, hydrodynamic flow is subject to viscosity and, in principle, also to inertia. This is in contrast to electric currents where no such effects matter. The resistivities of hydrodynamic flow and electric currents also scale with different exponents with the channel width [5, 6, 10].

## 13.2
## Principles of Electro-Osmotic Flow

Electro-osmotic flow occurs in liquid electrolyte solutions confined between the polar surfaces of a solid matrix in the presence of an electric field [3, 11]. In the simplest case, this scenario may be represented by the scheme shown in Figure 13.1. An electric double layer (EDL) is formed along the polar walls, and the electrostatic potential, $\Phi_{EDL}$, due to this charge distribution obeys the Poisson equation

$$\nabla^2 \Phi_{EDL} = -\frac{1}{\varepsilon_0 \varepsilon} \rho_q, \tag{13.1}$$

where $\varepsilon$ is the dielectric constant of the electrolyte solution, and $\varepsilon_0$ is the electric field constant. The charge density, $\rho_q$, is distributed in the electrostatic potential according to a Boltzmann factor:

$$\rho_q = e \sum_i c_{i0} z_i \exp\left\{-\frac{z_i e \Phi_{EDL}}{k_B T}\right\}, \tag{13.2}$$

where $e$ is the positive elementary charge, $c_{i0}$ and $z_i$ are the bulk concentration and the valence of the $i$-th ionic species, respectively, $k_B$ is Boltzmann's constant, and $T$ is the absolute temperature.

In the Debye–Hückel or high-temperature approximation, $|z_i e \Phi_{EDL}| \ll k_B T$, the right-hand side of Equation 13.2 can be approximated linearly. The boundary

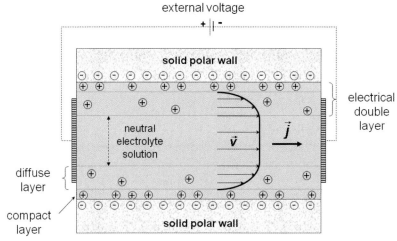

**Figure 13.1** Schematic illustration of the principle of electro-osmotic flow. An electrolyte solution in contact with polar walls produces an electrical double layer at the surfaces. All effects described in this study refer to the flow of the neutral electrolyte solution; the electric double layer is too thin to be resolved by these techniques.

condition for Equation 13.2 is given by the so-called zeta potential, $\zeta$. Next to the surface, we have $\Phi_{EDL} = \zeta$, while $\Phi_{EDL} = 0$ at infinity [3, 11]. The Debye–Hückel parameter is defined by

$$k^2 = \frac{2z_i^2 e^2 n_{i,\infty}}{\varepsilon \varepsilon_0 k_B T} \tag{13.3}$$

where $n_{i,\infty}$ is the bulk concentration of ion species $i$. The width of the electric double layer is of the order $1/k$.

The flow velocity field, $\vec{v} = \vec{v}(\vec{r})$, can be described by the Navier–Stokes equation under steady-state conditions for external electrostatic forces:

$$\rho_m(\vec{v}\cdot\vec{\nabla})\vec{v} + \vec{\nabla}p - \eta\nabla^2\vec{v} = \rho_q\vec{E}_{ext} \tag{13.4}$$

with the condition $\vec{\nabla}\cdot\vec{v} = 0$. Here, $p$ is the pressure in the fluid, the quantities $\rho_m$ and $\rho_q$ are the mass and charge densities, respectively, $\eta$ is the dynamic viscosity, and $\vec{E}_{ext}$ is the external electrical field strength. At the walls, the implicit 'no-slip' boundary condition should be valid, that is, $v_{wall} = 0$. According to the Helmholtz–Smoluchowski equation, the mean EOF velocity is predicted to obey

$$\langle v_{EOF} \rangle \approx \frac{\varepsilon}{\eta}\zeta E_{ext} \tag{13.5}$$

The EOF velocity is expected to increase linearly with the voltage applied to the electrodes, provided that the system is open and no dynamic counterpressure is established.

## 13.3
## Experimental Set-Ups

### 13.3.1
### Microsystem Model Objects

Test objects have been fabricated using silica-based ceramics (MACOR; Schröder Spezialglastechnik (http://www.schroederglas.com/download/ MacorProduktbeschreibung)) and polystyrene plates (BASF (https://www. plasticsportal.net/wa/EU/Catalog/ePlastics/doc/BASF/prodline/polystyrol/)) as polar and nonpolar materials, respectively. The porosity of these materials is negligible. The objects under consideration are represented in Figure 13.2 as computer-generated templates (left column) and spin density maps of the water-filled pore space (right-hand column). The fabrication method is described in detail in Ref. [9].

The three model objects represented in Figure 13.2 are of increasing complexity from the top of the figure to the bottom. The simplest object (Figure 13.2a and b) consists of seven parallel channels of varying thickness. The structure of the channels shown in Figure 13.2c and d is characterized by edges and steps. Figure 13.2e and f represent a random-site percolating cluster with a site occupation probability of 0.6 and a base matrix $100 \times 100$ [4]. The in-plane size is $6\,cm \times 6\,cm$ (or $7.5\,cm \times 6\,cm$ if the in- and outflow and electrode compartments are included) in the case of the first two objects, and $4\,cm \times 4\,cm$ (or $6\,cm \times 4\,cm$ with the in- and outflow and electrode compartments included) for the percolation cluster. The structures were milled $1\,mm$ deep in $2\,mm$-thick ceramic or polystyrene plates. Seven to ten identical, but independently micro-machined, plates were stacked and studied at a time for sensitivity reasons and to ensure the quasi 2-D character of the experiments. In this way, any possible imperfections such as surface defects are averaged out. In the case of the channel structure with edges (Figure 13.2c), the quasi-two-dimensionality of the EOF was further approached by choosing polar ceramics material only for the in-plane matrix material, whereas the top and bottom covers were nonpolar.

For pressure-induced flow, a hydrostatic pressure gradient was generated by a reservoir which was situated $2\,m$ above the sample and kept permanently filled to a constant level with the aid of a peristaltic pump.

In all cases, the electrodes used for EOF and ICD mapping experiments were gold-plated to ensure an uncorroded galvanic contact with the electrolyte solution during the measuring time. An interdigitated structure of two electrode combs prevented the generation of eddy currents when the field gradient pulses were applied.

The electric conductivity of the electrolyte solution is given by the empirical Kohlrausch law [12]

Templates

Water spin density
maps

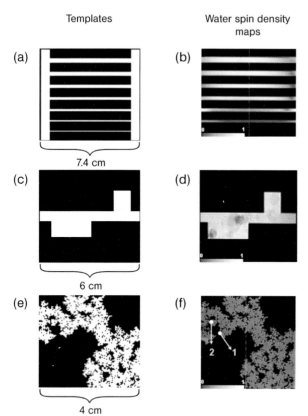

(a)

(b)

7.4 cm

(c)

(d)

6 cm

(e)

(f)

4 cm

**Figure 13.2** Typical structures of quasi 2-D model objects investigated in this study. The computer-generated templates in the left column are used to fabricate quasi-2-D model objects. The experimental spin density maps (256 × 256 pixels) shown in the right-hand column refer to water filled into the pore spaces. (a,b) Object with seven parallel channels of various widths (0.8 mm, 1.2 mm, 1.6 mm, 2.0 mm, 2.4 mm, 2.8 mm, 3.2 mm). The channel length is 6 cm plus 2 × 0.7 cm for the input and output compartments where the electrodes are placed; (c,d) Channel of varying widths with edges. The top and bottom covers of this object consist of nonpolar polystyrene plates, whereas the in-plane matrix material is cut in polar ceramics; (e,f) Random site percolation cluster as a paradigm for more complex channel networks. Isolated clusters were omitted. The size of the quadratic base lattice was 100 × 100. With a resolution of 400 μm, this corresponds to an object size of 4 cm × 4 cm without input and output compartments (not shown at the left and right-hand side, respectively). The regions marked by 1 and 2 refer to the detail plots to be discussed.

$$\sigma(c) = \Lambda_0 c - k_0 c^{3/2} \tag{13.6}$$

where $\sigma$ represents the conductivity, $\Lambda_0$ is the molecular conductance characteristic for the salt dissolved, and $c$ is the salt concentration. $k_0$ is a constant depending on the salt valence. On this basis, the experimental parameters for EOF velocity mapping were estimated prior to the measurements.

For the EOF experiments, the model objects were filled with $1\,mM$ $CaCl_2$ in distilled water. A conductivity of $2.44 \times 10^{-4}\,S\,m^{-1}$ at room temperature was measured in the bulk solution with a conductivity meter (HANNA Instruments HI 8633). The voltage applied across the 7.5 cm distance between the electrodes was in the range 300 to 900 V and was generated by a KEPCO BOP 1000 M power supply controlled by a home-made electronic device. The available range of current strength was between 10 mA and 1 A; however, the maximum current used practically was only 0.1 A. Under such conditions any joule heating during the electro-osmosis experiments could safely be neglected. An estimation based on Equation 55 of Ref. [3] (see Section 4-3.7) suggests a temperature increase smaller than $10^{-3}\,°C$. All experiments were carried out at room temperature.

The objects were closed so that no inflow or outflow of the electrolyte solution was possible. The ICD mapping experiments were carried out again with an aqueous $CaCl_2$ solution, but with a 10-fold higher concentration and conductivity in order to optimize the ICD encoding efficiency. It should be noted that the value of the conductivity is needed as a parameter for the computer simulations carried out in parallel to the experiments [5, 6].

## 13.3.2
### Porous Glasses

For the hydrodynamic dispersion studies, porous VitraPor glasses #1, #C, #4, and #5 with nominal pore sizes of 100–160 µm, 40–60 µm, 10–16 µm and 1–1.6 µm were investigated. Some typical scanning electron micrographs, recorded with a Zeiss DSM 962 scanning electron microscope, are shown in Figure 13.3. The VitraPor samples were of a cylindrical shape with diameter 6 mm and length 40 mm, and were contained in polychlorotrifluoroethylene (PCTFE) sample holders. In order to avoid any bypassing of the water to be pressed through the pore spaces, the sample holders were thermally shrunk onto the sample cylinders by first heating them to 197 °C for about 20 min. When cold, the sample could be dropped into the sample holder. Following thermal equilibration, the sample was tightly embraced by the PCTFE cylinder, from which it could no longer be removed without causing mechanical damage.

Pressure-induced flow was produced at constant flow rates of 0.1 to 0.9 ml min$^{-1}$ using degassed, distilled water. The experimental set-up is shown schematically in Figure 13.4a, where the arrows indicate the direction of flow.

To measure the EOF, a voltage in the range of 300 to 900 V was applied to the VitraPor samples, which were saturated with an aqueous solution of NaCl (1 mM). The current pulses were again generated using the KEPCO power supply (see above). The conductivity of the bulk solution was measured in the range 1.22 to $1.25 \times 10^{-4}\,S\,m^{-1}$ at room temperature; the length of the electrolyte solution along the U-shaped sample container was 9 cm from electrode to electrode.

Potential problems arising due to joule heating of the electrolyte solution [3, 13] and oxidation and reduction processes on the electrode surfaces [13] were avoided

500 µm
(a)
500 µm
(b)
50 µm
(c)
10 µm
(d)

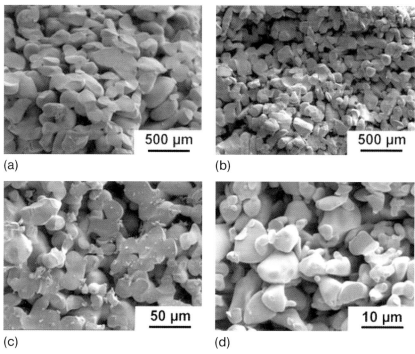

**Figure 13.3** Typical scanning electron microscopy (SEM) images of (a) VitraPor #1, (b) VitraPor #C, (c) VitraPor #4 and (d) VitraPor #5 samples with nominal pore sizes of 100–160 µm, 40–60 µm, 10–16 µm and 1–1.6 µm, respectively.

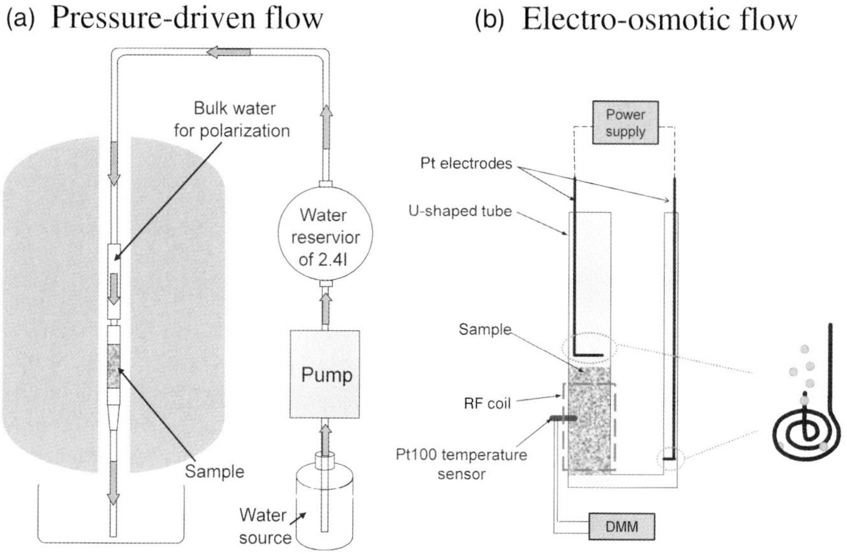

(a) Pressure-driven flow
(b) Electro-osmotic flow
Bulk water
for polarization
Water
reservior
of 2.4l
Pump
Sample
Water
source
Power
supply
Pt electrodes
U-shaped tube
Sample
RF coil
Pt100 temperature
sensor
DMM

**Figure 13.4** Schematic representation of the sample set-ups for (a) PIF and (b) EOF hydrodynamic dispersion experiments. The enlarged detail sector on the right-hand side of (b) shows the geometry of the electrodes.

by a special design of the electro-osmotic cell (Figure 13.4b). A U-shaped sample container and a spiral form of the electrodes prevented distortions by gas bubbles. The electrodes were made from 1 mm-thick platinum wires in the spiral form shown in Figure 13.4b. This also ensured that no perceptible eddy currents arose during the field gradient pulses. A Pt100 resistive temperature sensor was embedded inside the sample to measure the sample temperature *in situ*. The sample temperature was carefully controlled within the range of 22.1–22.3 °C, and the duty cycle of the current pulses was chosen as sufficiently low such that joule heating did not influence the results.

### 13.3.3
### Instruments

The hydrodynamic dispersion experiments were performed using a Bruker DSX 400 NMR spectrometer equipped with a vertical 9.4 T magnet with a 89 mm-wide room temperature bore. The maximum gradient strength was $1.0 \, T \, m^{-1}$. A variant of the pulsed gradient spin echo (PGSE) technique was applied with a velocity-compensated pulse sequence to measure the effective dispersion coefficient as a function of the effective dispersion time [8].

The much larger model objects (see Figure 13.2) were examined using an NMR tomograph equipped with a home-made radiofrequency (RF) console and a Bruker 4.7 T magnet with a 40 cm horizontal bore and a birdcage resonator with an inner diameter of 195 mm. The maximum field gradient was $50 \, mT \, m^{-1}$ in all three space directions. The echo time was typically 28 ms, and the repetition time 4 s. The spatial resolution of the experimental maps was in the range 180 to 300 μm, and the velocities were resolved with an accuracy of $0.03 \, mm \, s^{-1}$. The experimental error of the ICD measurements was estimated as $\pm 2 \, \mu A \, mm^{-2}$.

### 13.4
### Pulse Sequences

### 13.4.1
### Mapping of Electro-Osmotic and Pressure-Induced Velocity

Typical NMR techniques for the mapping of the EOF and PIF velocity are described elsewhere [1, 6, 14]. 'Mapping'–in contrast to the more unspecific 'imaging'–means that the spatial distribution of the selected quantity of interest is quantitatively rendered. Figure 13.5 shows schemes of the diverse RF, field gradient and current pulse sequences. Depending on the quantity to be mapped, the standard spin echo RF pulse sequence is combined with various magnetic flux density gradient and electric current pulses.

The field gradient pulses in Figure 13.5a serve 2-D imaging of an *x/y* slice of the object. It should be noted that the spin echo selectively originates from the

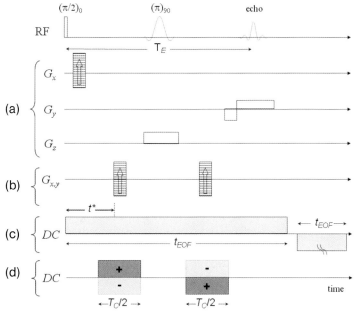

**Figure 13.5** Radiofrequency (RF), magnetic field gradient ($G_x$, $G_y$, $G_z$), and ionic current (DC) pulse sequences used in the NMR mapping experiments in this study. A standard Hahn echo with the echo time $T_E$ is acquired as the NMR signal of the liquid filled into the objects. Spin density mapping: (a) The combination with the gradient pulses in the schemes (a), provides 2-D images (see Refs [1, 15]) of a slice normal to the z direction, $\rho(x, y)$. The arrow indicates the direction of the increments in subsequent transients. The ramps of the gradient pulses are not shown for simplicity but are taken into account in the evaluations. Mapping of the PIF velocity: (a,b) Maps of the flow velocity components in the sensitive slice can be recorded by adding the gradient pulse pairs shown in (b) to the imaging part given in (a) (see Refs [1, 10], for instance) . That is, two 3-D experiments, $x,y,v_x$ and $x,y,v_y$, are performed resulting in maps of the magnitude $v_{PIF}(x, y)$. Mapping of the EOF velocity: (a–c) In the absence of hydrostatic pressure gradients, flow can be generated by electro-osmosis arising in polar matrices by ionic currents applied perpendicular to the z direction as shown in scheme (c). The results are maps of the magnitude $v_{EOF}(x, y)$. The interval $t^*$ represents the time necessary to achieve the steady state and is given in Equation 13.7. $t_{EOF} > T_E$ is the total length of the current pulse. The current acts during the whole echo time, so that no phase shifts directly induced by the current arise. The only phase shifts relevant in EOF velocity mapping are those due to spatial and velocity encoding. In order to reverse electrode reactions, identical current pulses of opposite polarity are applied during the recycle delay as indicated on the right-hand side of row (c). Mapping of ICD: (a,d) [5, 6, 10, 16, 17] In samples invariant along the z axis, 2-D current density maps of the x/y plane, $j(x, y)$, can be recorded based on the phase shifts produced by sequence (d). $T_c$ is the total application time of ionic currents. Note that, in contrast to the EOF pulse scheme (c), the phase shifts induced by the current pulses before and after the 180° RF pulse do not cancel each other owing to the bipolarity. Apart from this, the polarity is alternated in subsequent scans while subtracting the current densities from each other according to $j_{x,y} = (j_{x,y\pm} - j_{x,y\mp})/2$. In this way, undesired offsets are eliminated. Note that all maps mentioned can be recorded in the same object just by choosing the appropriate pulse sequence.

liquid in the objects, whereas the solid matrices do not contribute to the echo signal, even if they contain protons. If the spin density and relaxation times are distributed homogeneously in the fluid, then maps of liquid spin density can be rendered in this way without taking relaxation losses into account. The spatial resolution achieved in all three space directions was around 300 μm with our hardware configuration, so that the pore space of the model objects could be well resolved in all details.

Adding two more Fourier domains in the form of gradient pulse pairs along the $x$ and $y$ directions to be incremented in subsequent scans permits phase encoding of the velocity components in $x$ and $y$ directions, respectively, for the evaluation of velocity maps (Figure 13.5b) [1, 14, 15]. The maximum phase-encoding gradient determines the velocity resolution, while the increment gives the maximum range of velocities (the velocity 'field of view', so to speak). The velocity magnitudes rendered in the maps refer to the mode of the velocity distribution in the voxels. The intricacies of extremely broad variations of the flow velocity, as they are expected in complex channel systems such as percolation clusters, are discussed in Ref. [18], where a measuring protocol is provided to avoid artifacts on these grounds.

In order to acquire maps of the EOF velocity, phase encoding of the velocity must be synchronized with a current pulse, as shown in Figure 13.5c. Prior to the proper phase-encoding experiment, the current pulse must remain on for long enough to ensure steady-state conditions. A characteristic time to reach steady state in porous materials can be calculated on the basis of Equation 90 in Ref. [3]:

$$t^* = \frac{\rho_m a^2}{\eta \lambda_1^2} \tag{13.7}$$

where $\lambda_1 = 2.405$ is determined by the first null of the zero-order Bessel function, $J_0(\lambda_1) = 0$, and $a$ is the channel radius under consideration (or the mean pore size in the case of porous materials). An estimate leads to $t^* \approx 27$ ms. The echo time was varied in the range from 10 to 50 ms with no effect on the EOF velocity maps, and consequently steady-state conditions were considered to be well established. The EOF velocity is phase-encoded in the same way as the PIF velocity. In this context, it should be noted that the magnetic field gradients produced by the currents do not interfere with the velocity-encoding gradients as they are not incremented.

### 13.4.2
### Echo Attenuation by Hydrodynamic Dispersion

The instantaneous velocity $\vec{v}$ can be analyzed in two terms according to

$$\vec{v}(t) = \vec{u}(t) + \vec{V}, \tag{13.8}$$

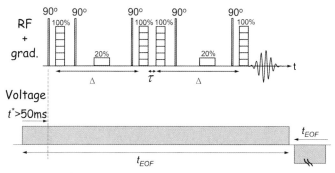

**Figure 13.6** Coherent-velocity compensated RF and field gradient pulse sequence practically used for NMR experiments probing echo attenuation by hydrodynamic dispersion due to PIF and EOF. The bottom row shows the electrical field pulses needed for electro-osmosis. See also the legend of Figure 13.5.

where $\vec{V} \equiv \lim_{t \to \infty} \langle \vec{v} \rangle$ is the average velocity, and $\vec{u}(t)$ is the fluctuation of the Lagrangian velocity. The principle of the NMR echo attenuation pulse sequence in Figure 13.6 is to compensate the phase shift effect by $\vec{V}$ and to examine the attenuation by $\vec{u}(t)$ [8]. Since all position and velocity symbols to be used in this context refer to components along the gradient direction, we will omit the vector sign in the following.

The trajectory of a nucleus during the time $t$ can be expressed as

$$r(t) = r_0 + \int_0^t v(t')dt' = r_0 + \int_0^t [V + u(t')]dt' = r_0 + Vt + \int_0^t u(t')dt', \qquad (13.9)$$

where $r_0$ is the initial position. A field gradient pulse of an arbitrary shape,

$$G_z = \begin{cases} G(t) & \text{for } 0 \le t \le T \\ 0 & \text{otherwise} \end{cases} \qquad (13.10)$$

produces the cumulative phase shift

$$\phi(T) = \gamma \int_0^T G(t)r(t)dt$$

$$= \gamma \left[ r_0 \int_0^T G(t)dt + V \int_0^T G(t)t\,dt + \int_0^T G(t) \int_0^t u(t')dt'\,dt \right]$$

$$= \phi_0(T) + \phi_1(T) + \phi_2(T), \qquad (13.11)$$

where $\gamma$ is the gyromagnetic ratio. The phase shifts by the position-dependent term $\Phi_0(T)$, and the coherent velocity-dependent term $\Phi_1(T)$ can be compensated by using a bipolar gradient pulse of the form [14]:

$$G_z(t) = \begin{cases} G & \text{for } 0 \leq t \leq \tau \\ -G & \text{for } \tau \leq t \leq 3\tau \\ G & \text{for } 3\tau \leq t \leq 4\tau \\ 0 & \text{otherwise} \end{cases} \tag{13.12}$$

Both, $\Phi_0(4\tau)$ and $\Phi_1(4\tau)$ vanish after this gradient pulse. Merely the incoherent fluctuation term $\phi_2(4\tau) = \gamma G \left[ \int_0^\tau u(t)dt - \int_\tau^{3\tau} u(t)dt + \int_{3\tau}^{4\tau} u(t)dt \right]$ still contributes and gives rise to attenuation of the spin echo amplitude.

From a practical viewpoint it is more convenient and reliable to use a combination of 180° RF pulses and unipolar gradient pulses instead of a bipolar version as suggested by Equation 13.12. For technical reasons, it is moreover favorable to split the 180° pulses into two pulses each of 90°, and to compose the gradient pulses of identical pulses of length $\delta$ and strength G. This ensures that the gradient amplifier produces identical pulses of well-defined 'area' $G\delta$ (Figure 13.6). It should be noted that a combination of two 90° RF pulses effectively inverts the effect of the gradient pulses with respect to the sign. The spoiling gradient pulses in the $\Delta$ intervals spoil all spin coherences. The use of 90° pulse pairs instead of 180° pulses reduces the echo amplitude by a factor of 1/4 on the one hand, but avoids excessive transverse relaxation losses on the other hand. The relevant dispersion time is given by the combined interval $2\Delta$.

For studies of hydrodynamic dispersion due to EOF, voltage pulses are applied during the spin echo sequence (Figure 13.6) in analogy to Figure 13.5c. The steady-state establishment time $t^*$ was varied from 50 ms to 150 ms without any perceptible influence on the hydrodynamic dispersion results.

The effective (time-dependent) dispersion coefficient, $D_{eff} = D_{eff}(2\Delta)$, defined by mean-squared (incoherent) displacement

$$\langle Z^2 \rangle = 4D_{eff}\Delta \tag{13.13}$$

was evaluated from echo attenuation curves by fitting the limiting expression

$$E(q, 2\Delta)_{q \to 0} \propto \exp\{-2q^2 D_{eff}\Delta\} \tag{13.14}$$

to the attenuation data for small values $2q^2\Delta$ as an approach. The wavenumber $q$ is defined by $q = \gamma \delta G$. Note that this evaluation formalism anticipates that the displacement propagator can be approximated by a Gaussian function [14].

During the course of the RF and field gradient pulse trains, the current pulses were cycled in forward and backward directions (see Figure 13.6) in order to avoid problems caused by gas production and chemical reactions on the electrode surfaces. Asymmetric anode and cathode reactions may cause a displacement of the whole water column in the U-shaped sample tube (see Figure 13.4b) and alter the concentration of the solution. With unidirectional flow, gas produced at the anode tends to become trapped as bubbles in the sample rather than escape from the upper part of the U-shaped sample tube. Therefore, a second electrical field pulse of identical length, $t_{EOF}$ (see Figure 13.6), but of opposite polarity, was applied in the recycle delay.

### 13.4.3
### Mapping of the Ionic Current Density (ICD)

In contrast to the pulse sequence for mapping of the EOF velocity (Figure 13.5a, b and c), the ionic current density is mapped with the aid of bipolar current pulses in the free evolution intervals of the spin-echo pulse sequence (Figure 13.5a and d). Phase shifts arising from the magnetic fields of the current distribution are intentionally not refocused but rather serve as the source of the information on the local current density.

The principle of ICD mapping is based on Maxwell's fourth equation for stationary electromagnetic fields:

$$\vec{j}(\vec{r}) = \frac{1}{\mu_0} \vec{\nabla} \times \vec{B}(\vec{r}),$$

(13.15)

where $\vec{B}(\vec{r})$ is the magnetic flux density produced by the electric current density $\vec{j}(\vec{r})$ at position $\vec{r}$, and $\mu_0$ is the magnetic field constant. The curl operation depends on all three spatial derivatives of the flux density with respect to the spatial Cartesian coordinates. The model objects examined here were designed in such a way that a quasi 2-D current distribution practically invariant along the $z$ direction exists. In this case, only two components and two derivatives in the transverse plane remain finite:

$$j_z(\vec{r}) = 0,$$
$$j_x(\vec{r}) = \frac{1}{\mu_0} \frac{\partial B_z}{\partial y},$$
$$j_y(\vec{r}) = -\frac{1}{\mu_0} \frac{\partial B_z}{\partial x}.$$

(13.16)

These derivatives can be determined experimentally from maps of the phase shifts of the local Larmor precession by evaluating the numerical derivatives along the $x$ and $y$ directions after 'unwrapping' with the aid of Goldstein's 2-D algorithm [19]. Detailed descriptions of the method can be found in Refs [1, 10, 15–17]. According to Ref. [17], the optimal value for the total current pulse length, $T_C$, depends on the transverse relaxation time, $T_2$, the diffusion coefficient of the ions, $D$, the field of view, $x_{fov}$, and the number of pixels, $N$:

$$\frac{1}{T_C} = \frac{1}{T_2} + 4D \left( \frac{4\pi N}{5 x_{fov}} \right)^2.$$

(13.17)

Practically, the current pulses in row (d) of Figure 13.5 are characterized by a current strength of 40 mA and a length of 4 ms. The voltage across the model objects (6 cm) was 300 V. The mean magnitude of the current density is estimated to be 36 A m$^{-2}$. Artifacts due to eddy currents induced by the field gradient pulses can safely be ruled out, as simulations of the current distribution coincide with the experimental data [5, 6, 10].

## 13.5
## Results

### 13.5.1
### NMR Mapping and Simulation of EOF, PIF and ICD in Microsystem Model Objects

In the following section NMR experiments performed with quasi 2-D objects, as well as simulations of the same pore space geometry of varying complexity, will be reported. The phenomena to be considered are EOF in open and closed systems, the distribution of ICD in the presence and absence of EOF, and PIF. Details can be found in previously published reports [5, 6, 9, 10].

In a system with closed electrodes, the electro-osmotic driving force causes hydrodynamic backflow so that stationary flow patterns with cyclic streamlines arise. Assuming a negative zeta potential, electro-osmosis is expected to drive the electrolyte solution from the anode to the cathode. However, the electro-osmotic pressure in thin channels is stronger than in thicker channels because the electro-osmotically active surface area is larger relative to the cross-sectional area. Hence, thin channels are more efficient and backward flow is expected in the thicker channels. This is shown in Figure 13.7 for the (quasi) 2-D object shown in Figure 13.2a and b. Both, the (cyclic) flow directions and the width-dependent flow rates predicted by simulations were verified in these experiments.

Percolation clusters are taken as a paradigm for complex channel networks as they are pertinent for microsystem technology. Typical experimental EOF and PIF velocity maps and ICD in the presence and absence of EOF are provided in Figure 13.8 for the percolation cluster shown in Figure 13.2f.

Figure 13.8b shows a typical experimental ICD map derived from an unwrapped phase map recorded in the ceramic object – that is, in the presence of EOF. This is to be compared with the ICD map (Figure 13.8c) recorded in the object made from polystyrene – that is, in the absence of EOF. The detailed vector representations in the right-hand columns of Figure 13.8 indicate that there must be a strong interference between flow and ionic current, as the ionic current distributions measured with and without background EOF are so different.

The different flow and current patterns with and without electro-osmotic flow are remarkable, as illustrated in Figure 13.9. In the presence of EOF, both the velocity and ICD maps show recirculation patterns. However, the positions where these phenomena were observed are different. At position 1 (see Figure 13.2f), no flow vortex could be identified whereas the electric current density clearly reveals

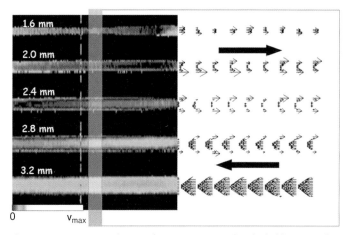

**Figure 13.7** Experimental EOF velocity map for the five widest channels of the seven-channel object shown in Figure 13.2a and b. The flow velocity field is rendered on the left-hand side by a gray-scale code, and by velocity vectors on the right-hand side. The latter refer to the shaded box area shown in the center of the gray-scale representation. The main flow directions are indicated by bold arrows. The mean electric field strength was 150 V cm⁻¹. The vertical white and partially broken line on the left is an artifact.

a sort of eddy. At position 2, on the other hand, the findings are opposite: there is no ICD vortex but a clear $\vec{v}_{EOF}$ eddy appears. In the absence of EOF, no complete ICD vortex pattern was found at any position (see Figure 13.8c, for instance). Recirculation patterns can be observed only if flow is present; however, these flow-related phenomena do not occur necessarily at the same positions.

Flow eddies attached to obstacles are well known also for very low Reynolds numbers. On the other hand, the fact that electric current vortices have been observed appears to be quite remarkable in view of the known property $\vec{\nabla} \times \vec{E} = 0$ of the electrostatic field strength. This relationship is obviously valid for the electric current density maps in the absence of EOF, but if EOF is superimposed then pronounced recirculation patterns occur. The conclusion is that the combination of hydro- and electrodynamic phenomena leads to new effects not predicted by the theory. It appears that there is no similitude between these two types of transport [20], and that the Helmholtz–Smoluchowski equation (see Equation 13.5) which relates the flow velocity with the electric potential gradient is not satisfied in the case of closed electrodes.

In order to elucidate the origin of recirculation patterns of EOF and the differences to PIF, we have studied a channel structure with edges (see Figure 13.2c and d). Figure 13.10 shows the simulated flow velocity fields for 2-D conditions. Assuming a uniform zeta potential distribution along the surfaces, irrespective of the surface topology, leads to an EOF velocity vector field without any recirculation (Figure 13.10a). However, if the zeta potential is more realistically modified at the edges (positions 1–4 in Figure 13.10b) by reducing the zeta potential to $\zeta = -10$ V instead of $\zeta = -100$ mV at the flat parts of the surface, the velocity fields Figure 13.10b

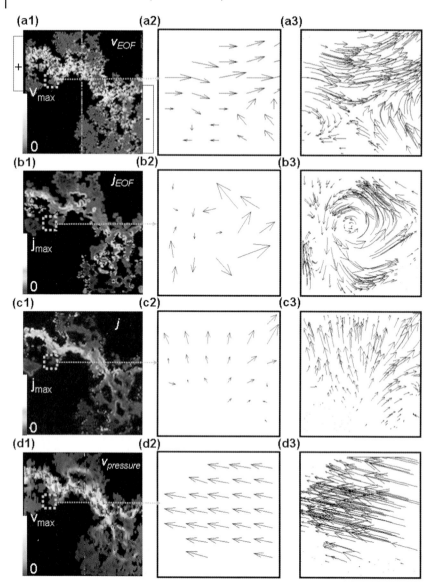

**Figure 13.8** Experimental maps of the magnitudes of different transport quantities measured in quasi 2-D percolation model objects (Figure 13.2e and f). The FOV is 4 cm × 4 cm. The matrix is rendered in black. The marked areas of 0.8 mm × 0.8 mm are detailed on the right-hand side in vector and in trajectory representation. The latter means that the velocity vector arrows are bent along the local streamlines, and the length of the curvilinear vectors is proportional to the mean magnitude along the arrow. (a) EOF velocity, $v_{EOF}$, in the polar matrix. The electric field strength was 83 V cm$^{-1}$; (b) Electric current density in the presence of EOF, $j_{EOF}$ (polar matrix); (c) Electric current density in the absence of EOF, $j$ (non-polar matrix); (d) Flow velocity caused by an external pressure gradient in the open system, $v_{PIF}$. Typical maximum values for the velocity and electric current density are $v_{max} = 4.5$ mm s$^{-1}$ and $j_{max} = 5$ mA mm$^{-2}$, respectively.

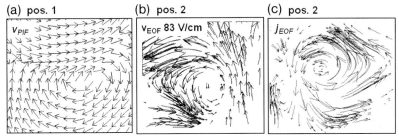

**Figure 13.9** Vortices of different transport quantities in 0.8 mm × 0.8 mm areas at positions 1 or 2 of the percolation cluster shown in Figure 13.2f. (a) 2-D simulation of the flow velocity, $\vec{v}_{PIF}$, driven by an external pressure gradient at position 1 in vector representation; (b) Experimental EOF velocity, $v_{EOF}$, at position 2 in trajectory representation. (b) Experimental electric current density, $j_{EOF}$, in the presence of EOF at position 2 again in trajectory representation.

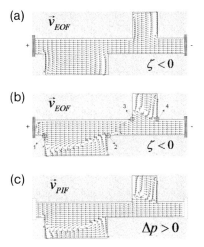

**Figure 13.10** Simulated velocity maps in vector representation in a 2-D channel with edges (Figure 13.2c and d). (a) EOF velocity for uniform zeta potential, $\zeta = -100$ mV, along the matrix surfaces. The voltage between the electrodes was assumed to be 600 V; (b) EOF velocity for a zeta potential distribution: $\zeta = -100$ mV at the flat parts of the surface and $\zeta = -10$ V at the edge points 1–4; (c) Pressure-induced flow.

and Figure 13.11b and d show such patterns in close similarity to PIF. This finding is in accordance with experimental maps for the EOF velocity components $v_x$ and $v_y$, as shown in Figure 13.11a and c. In order to mimic the two-dimensionality of the simulations, the model objects used in the experiments were made in such a way that the fluid space was bounded at the top and the bottom by nonpolar polystyrene covers, while the thin, in-plane matrix material consisted of polar ceramics.

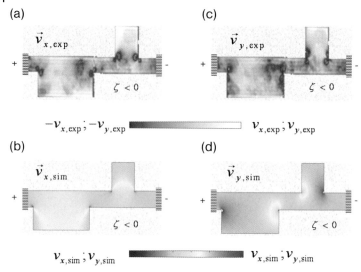

**Figure 13.11** Comparison of experimental and simulated maps of electro-osmotic velocity components in a channel with edges (Figure 13.2c and d). For the simulations, the same distribution for the zeta potential was assumed as in Figure 13.10b. In the experiments, the voltage between the electrodes was 600 V in a distance of 6 cm. The electrolyte solution was 1 mM CaCl$_2$ with a bulk conductivity of 2.44 × 10$^{-4}$ S m$^{-1}$. (a,b) Maps of the measured and simulated x-components of the EOF velocity. The maximum velocity was $| v_x| = 1.86$ mm s$^{-1}$. At the edges, the recirculation patterns appear as in Figure 13.10b. (c,d) As before but for the y-component.

## 13.5.2
### Hydrodynamic Dispersion by EOF and PIF in Porous Glasses

Aqueous solutions filled in porous silica glasses (VitraPor) of different nominal pore sizes are considered as disordered clusters percolating through the pore space. The pore space may be characterized by the nominal pore diameter, $a$, and the correlation length, $\xi$. In the so-called scaling window, $a < \sqrt{\langle Z^2 \rangle} < \xi$, a time-dependent diffusion coefficient is expected for flow through random media at large Péclet numbers, Pe >> 1 [21].

For PIF the flow rate can be adjusted by the pressure gradient. Figures 13.12a and b show plots of mean-squared incoherent displacement for water in VitraPor #1 (100–160 μm pore size) and VitraPor #C (40–60 μm) at different flow rates as a function of the dispersion time, $2\Delta$ [8]. For both low and high flow rates, the data can be described fairly well by power laws of the form

$$\langle Z^2 \rangle \propto (2\Delta)^\alpha \tag{13.18}$$

in the frame of the experimental accuracy. This applies in particular to the flow rates 0 and 0.7–0.9 ml min$^{-1}$. Clearly, a crossover between the subdiffusive limit in the absence of flow to a superdiffusive case in the presence of strong enough flow

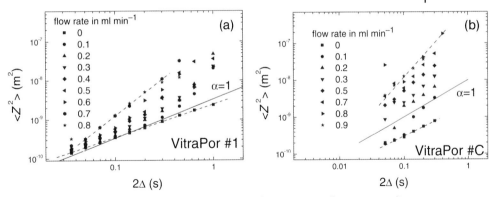

**Figure 13.12** Mean-squared (incoherent) displacement for pressure-induced hydrodynamic dispersion as a function of the effective dispersion time, $2\Delta$, at different flow rates for (a) VitraPor #1 and (b) VitraPor #C. The solid line represents the proportionality $<Z^2> \sim 2\Delta$. The lower and upper dashed lines represent fits of Equation 13.18 to the data. The fitted exponents are $\alpha = 0.84$ and $\alpha = 1.95$, respectively.

occurs [22]. The fitted exponents are $\alpha \approx 0.84$ and $\alpha \approx 1.95$, respectively, for the time window probed in the experiments.

The root mean-squared displacements range from 10 to 200 µm; this is the length scale of the pore space topology of the examined porous glasses – that is, the transport properties refer to the scaling window where a power law behavior can be expected. In the absence of flow, transport is governed by molecular diffusion obstructed by the confining geometry. The consequence is a subdiffusive mean-squared displacement law. Above a flow rate of about 0.7 ml min$^{-1}$, the exponent $\alpha \approx 1.95$ indicates a super-diffusive law approaching the 'ballistic' case ($\alpha = 2$). At this limit the particles are displaced in all directions with the same mean velocity. For hydrodynamic dispersion in disordered porous media, this is the case when pure mechanical mixing under the influence of the local geometry is relevant, as demonstrated with computer simulations by Duplay and Sen [23]. A detailed discussion of this phenomenon can be found in Ref. [8].

Figure 13.13a, c, and e show the effective dispersion coefficient $D_{eff}$ measured as a function of the dispersion time $2\Delta$ at different voltages in VitraPor #5, #4 and #1, while Figure 13.13b, d and f display the corresponding mean-squared displacement $<Z^2>$ (see Equation 13.13). For VitraPor #5 (pore size 1–1.6 µm), the EOF data for $D_{eff}$ first decrease sharply and then increase slightly with $2\Delta$ (Figure 13.13a), whereas with VitraPor #1 (pore size 100–160 µm) this tendency is reversed. In both cases, a minimum value of $D_{eff}$ could be identified at a certain characteristic dispersion time between 130 and 170 ms. Beyond this characteristic dispersion time a crossover from subdiffusive to normal or superdiffusive mean-square displacement laws occurs for small and large pores (Figure 13.13b and f). The time dependences of $D_{eff}$ and $<Z^2>$ for the 10–16 µm pores plotted in Figure 13.13c and d lie between these values.

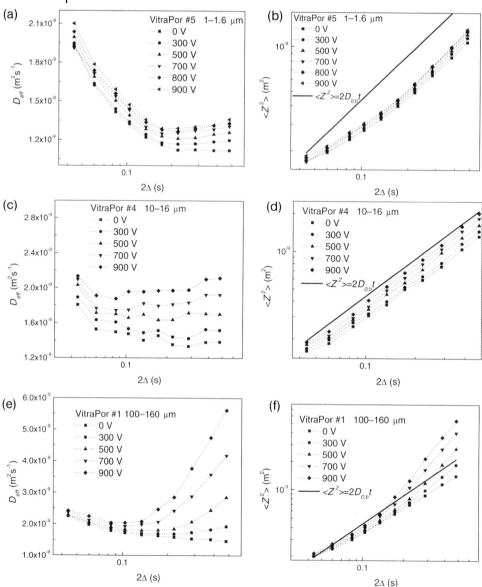

**Figure 13.13** Hydrodynamic dispersion for EOF. (a,c,e) Effective dispersion coefficient, $D_{eff}$, measured as a function of the dispersion time, $2\Delta$, at different voltages for VitraPor #5, VitraPor #4 and VitraPor #1, respectively. (b,d,f) Mean-squared incoherent displacement, $<Z^2>$, as a function of $2\Delta$ at different voltages for VitraPor #5, VitraPor #4 and VitraPor #1, respectively. The solid line represents $<Z^2> = 2D_{0,b}t$, where $D_{0,b}$ represents the diffusion coefficient in the bulk solution at zero voltage. The dashed lines between data points serve as guides for the eye.

The fact that the dispersion coefficient is enhanced with increasing external electric field strength demonstrates the existence of EOF even in networks of only micrometer-thick pores. Of course, larger pore sizes facilitate hydrodynamic flow, and the electro-osmotically induced hydrodynamic dispersion becomes more efficient. In the case of VitraPor #1 at dispersion times longer than the characteristic value of 130 ms, the exponent $\alpha$ (see Equation 13.18) increases from about 0.88 in the absence of an electrical field to about 1.85 at the strongest electrical field. On the other hand, at shorter effective dispersion times the experimental data can be fitted only with a subdiffusive displacement law.

The existence of a minimum value for $D_{eff}$ may indicate the competitive action of pore space restrictions at short times ($D_{eff}$ decreases with time) and mechanical dispersion at long times ($D_{eff}$ increases with time). In a plot of $D_{eff}$ versus $2\Delta$ for PIF in VitraPor #1 (not shown here), a tendency towards a minimum value of $D_{eff}$ was also found in the same range of dispersion time at a flow rate of less than $0.3\,\text{ml min}^{-1}$ [8].

## 13.6
## Discussion

In these studies we have investigated fluid transport in model pore and channel network systems of different degrees of complexity. The flow velocity induced by pressure gradients or electro-osmosis and the ionic current density have been mapped using NMR techniques and simulated with computational fluid dynamics algorithms. In porous glasses, where the pore size is too small to be resolved spatially by NMR mapping methods, hydrodynamic dispersion as an indicator of pressure or electro-osmotically driven flow has been examined globally with pulsed-gradient NMR diffusometry. The total pore or channel length scale probed in this study ranges from micrometers to millimeters. In this sense, NMR provides access to microfluidics in mesoscopic and microscopic pore spaces.

Studies of this type are of interest for microsystem technology, notably as microsystem instruments often consist of complex channel networks that imply features of our model objects. Cyclic electro-osmotic pumps without moving elements can be implied in such devices. The intentional production of recirculation in such systems is essential for the function of microsystem mixers.

One remarkable result has been the difference between the experimental ICD maps of the percolation cluster recorded with and without EOF. A substantial interference of hydrodynamic flow and electric current occurs. Eddies of the electric current density were observed in the presence of EOF, but not in the hydrodynamically static case, and this may be considered as a combined hydrodynamic and electrodynamic phenomenon. An ion driven by hydrodynamic flow through an inhomogeneous electrostatic field effectively experiences a time-dependent electric field strength, $\vec{E} = \vec{E}(t)$, despite stationary flow and electrostatic conditions. There is a certain analogy to the 'material' acceleration $\vec{a}_2 = (\vec{v} \cdot \vec{\nabla})\vec{v}$ known from the Navier–Stokes equation given in Equation 13.4; that is, electrostatic laws are not applicable in the presence of EOF. The similitude problem was discussed previously in Ref. [20] and more recently in Ref. [24]. By taking the results of the EOF, ICD and PIF experiments and simulations together, it can be said that recirculation patterns occur with all transport quantities if flow is present, but at different positions of the percolation cluster.

The different transport- and recirculation patterns found for PIF and EOF were shown to be due to the different influences of the surface properties. In simple channels with edges, a singular increase of flow friction at edges coincides with singularities of the zeta potential. Under such circumstances, almost identical patterns were both simulated and experimentally recorded. In more complex systems, such as the percolation cluster considered in these studies, friction and zeta potential singularities do not necessarily coincide. Therefore, it is not surprising that eddies of different flow phenomena arise at different positions of the same system.

The observation of hydrodynamic dispersion proves that microscopic flow was induced in pores as thin as one micrometer by both hydrostatic pressure gradients and electro-osmosis. EOF in even thinner pores of fixed porous bead beds has also been reported [25, 26]. The time dependence of the mean-squared displacement changes from a subdiffusive power law in the absence of these driving mechanisms to an almost ballistic superdiffusive behavior for large Péclet numbers. This is a demonstration of the anomalous displacement laws predicted by theoretical considerations [22].

## References

1 Stapf, S. and Han, S.-I. (eds) (2005) *NMR Imaging in Chemical Engineering*, Wiley-VCH Verlag GmbH, Weinheim.

2 Kimmich, R. (2002) Strange kinetics, porous media, and NMR. *Chemical Physics*, **284**, 253–85.

3 Li, D. (2004) *Electrokinetics in Microfluidics*, Elsevier, Amsterdam.

4 Stauffer, D. and Aharony, A. (1992) *Introduction to Percolation Theory*, Taylor and Francis, London.

5 Buhai, B. and Kimmich, R. (2006) Dissimilar electro-osmotic flow and ionic current recirculation patterns in porous media detected by NMR mapping experiments. *Physical Review Letters*, **96**, 17450-1–4.

6 Buhai, B., Binser, T. and Kimmich, R. (2007) Electroosmotic flow, ionic currents, and pressure-induced flow in microsystem channel networks: NMR mapping and computational fluid dynamics simulations. *Applied Magnetic Resonance*, **32**, 25–49.

7 Callaghan, P.T., Codd, S.L. and Seymour, J.D. (1999) Spatial coherence phenomena arising from translational spin motion in gradient spin echo experiments. *Concepts in Magnetic Resonance*, **11**, 181–202.

8 Li, Y., Farrher, G. and Kimmich, R. (2006) Sub- and superdiffusive molecular displacement laws in disordered porous media probed by nuclear magnetic resonance. *Physical Review E*, **74**, 066309-1–7.

9 Klemm, A., Kimmich, R. and Weber, M. (2001) Flow through percolation clusters: NMR velocity mapping and numerical simulation study. *Physical Review E*, **63**, 041514-1–8.

10 Weber, M. and Kimmich, R. (2002) Maps of electric current density and hydrodynamic flow in porous media: NMR experiments and numerical simulations. *Physical Review E*, **66**, 026306-1–9.

11 Probstein, R.F. (1994) *Physicochemical Hydrodynamics*, John Wiley & Sons, Inc., New York.

12 Moore, J.H. and Spencer, D.N. (eds) (2001) *Encyclopedia of Chemical Physics and Physical Chemistry, Volume I: Fundamentals*, IoP, Bristol.

13 Pettersson, E., Fur, I. and Stilbs, P. (2004) On experimental aspects of electrophoretic NMR. *Concepts in Magnetic Resonance A*, **22**, 61.

14 Kimmich, R. (1997) *NMR Tomography, Diffusometry, Relaxometry*, Springer, Berlin.

15 Blümich, B. (2000) *NMR Imaging of Materials*, Clarendon Press, Oxford.

16 Scott, G.C., Joy, M.L.G., Armstrong, R.L. and Henkelman, R.M. (2002) Sensitivity of magnetic-resonance current-density imaging. *Journal of Magnetic Resonance*, **97**, 235.

17 Sersa, I., Jarh, O. and Demsar, F. (1994) Magnetic resonance microscopy of electric currents. *Journal of Magnetic Resonance Series A*, **111**, 93.

18 Kossel, E. and Kimmich, R. (2005) Flow measurements below 50 µm: NMR microscopy experiments in lithographic model pore spaces. *Magnetic Resonance Imaging*, **23**, 397–400.

19 Ghiglia, D.C. and Pritt, M.D. (1998) *Two Dimensional Phase Unwrapping*, John Wiley & Sons, Inc., New York.

20 Cummings, E.B., Griffiths, S.K., Nilson, R.H. and Paul, P.H. (2000) Conditions for similitude between the fluid velocity and electric field in electroosmotic flow. *Analytical Chemistry*, **72**, 2526.

21 Sahimi, M. (1995) *Flow and Transport in Porous Media and Fractured Rock*, VCH, Berlin.

22 Metzler, R. and Klafter, J. (2000) The random walk's guide to anomalous diffusion: a fractional dynamics approach. *Physics Reports*, **339**, 1.

23 Duplay, R. and Sen, P.N. (2004) Influence of local geometry and transition to dispersive regime by mechanical mixing in porous media. *Physical Review E*, **70**, 066309.

24 Santiago, J.G. (2007) Comments on the conditions for similitude in electroosmotic flow. *Journal of Colloid and Interface Science*, **310**, 675.

25 Tallarek, U., Scheenen, T.W.J. and Van As, H. (2001) Macroscopic heterogeneities in electroosmotic and pressure-driven flow through fixed beds at low column-to-particle diameter ratio. *The Journal of Physical Chemistry*, **105**, 8591–9.

26 Tallarek, U., Rapp, E., Van As, H. and Bayer, E. (2001) Electrokinetics in fixed beds: experimental demonstration of electroosmotic perfusion. *Angewandte Chemie – International Edition*, **40**, 1684–7.

# 14
# MRI of Fluids in Strong Acoustic Fields

*Igor V. Mastikhin and Benedict Newling*

## 14.1
## Introduction

When a strong sound propagates in a fluid, it may cause several effects, the most notable being acoustic streaming (AS), which is described as the motion of liquids and gases caused by the presence of a sound field, and cavitation, which is the formation, growth and collapse of bubbles in a liquid. Both effects are very diverse [1, 2], and both are of substantial interest in the fundamental and biomedical sciences, engineering and industry – wherever strong acoustic fields are employed or encountered. The strength of the sound required to cause such effects depends both upon the frequency and the medium in which the sound is propagating.

Interest in these phenomena is longstanding, as both were first observed during the nineteenth century. Both were first rigorously analyzed by Lord Rayleigh: AS in 1884 [3] and cavitation in 1917 [4] (the authors could not resist the temptation to include the latter's opening lines: 'When reading O. Reynolds' description of the sounds emitted by water in a kettle as it comes to the boil, and their explanation as due to the partial or complete collapse of bubbles as they rise through cooler water, I proposed to myself a further consideration of the problem thus presented; but I had not gone far when I learned from Sir Parsons that he also was interested in the same question in connexion with cavitation behind screwpropellers . . .').

An impressive amount of experimental and theoretical research has been accumulated since then, but as the complexity of research increases, so too does a need for the experimental corroboration of new theories and models. There is a need for an experimental technique able to obtain information about the fluid parameters noninvasively – that is, without modifying the fluid dynamics (this is particularly important in AS in gases). Magnetic resonance imaging (MRI) satisfies that requirement. Another attractive feature of MRI is its ability to image spatially and acoustically opaque media, which is important in the case of cavitation. Based on its flexibility, MRI is also able to provide spatially resolved information on both the fluid's global parameters, such as the fluid velocity, dispersion and diffusivity, as well as local parameters such as molecular mobility, all within a reasonable amount of experimental time. In these investigations we have demonstrated the

*Magnetic Resonance Microscopy.* Edited by Sarah L. Codd and Joseph D. Seymour
Copyright © 2009 WILEY-VCH Verlag GmbH & Co. KGaA, Weinheim
ISBN: 978-3-527-32008-0

application of MRI to studies of the dynamics of cavitating liquid (Section 14.2), to comparative studies of dynamics of both liquid and dissolved gas during cavitation (Section 14.3), and also to studies of AS in gas (Section 14.4).

## 14.2
## Dynamics of a Liquid in Cavitating Fluid

Cavitation is defined as the 'formation, growth and collapse of bubbles in a liquid, the bubbles being formed from previously dissolved gas and the liquid vapor'. Cavitation is produced by a pressure drop in the liquid. If such a pressure drop is caused by a motion – for example, by a propeller rotating in water – this is a case of a hydraulic cavitation, and this is how cavitation was historically discovered. If the pressure drop is produced by a strong sound, we have an acoustic cavitation. When the cavitating bubbles collapse, shock waves, high temperatures and pressures can develop inside and near them, and this makes cavitation a highly destructive and chemically active process. Studies of cavitation represent interests in chemistry, mechanical and chemical engineering, and in industrial and biomedical applications of ultrasound.

Cavitation phenomena cover a very wide time scale, from subnanoseconds (the duration of a single flash of light from a collapsing bubble), to microseconds (a bubble oscillation period), to milliseconds, seconds, and minutes (bubble interaction, rectified diffusion, and fluid heat transfer). MRI methods can never deliver millions of frames per second, as optical methods can. Although the fastest one-dimensional (1-D) MRI acquisition cannot be made shorter than a fraction of a millisecond, MRI is able to measure processes on the time scale of milliseconds and seconds, providing statistically averaged information.

First, we need to identify the nuclear magnetic resonance (NMR) parameters that can be modulated by cavitation and so be employed in our measurements:

- Changes in spin density are not to be counted upon: the ratio of bubble to bulk liquid volume is usually much lower than 1%.

- Cavitating bubbles are filled with gas and vapor and, as such, might produce magnetic field gradients at boundaries of susceptibility discontinuity. With the very low bubble density, however, these effects will be difficult to detect except for cases of a very strong cavitation that will heat the liquid quickly.

- Changes in relaxation parameters can be masked by heating and degassing.

Initially, we decided to begin our study of cavitation with spatially resolved measurements of velocity spectra and hydrodynamic dispersion coefficient [5], for the following reasons. Acoustic streaming is a bulk advective flow caused by the attenuation of acoustic waves; the attenuation coefficient is proportional to $v^2$, where $v$ is the acoustic frequency. At MHz-frequencies, AS is an important detection mechanism of liquid-filled cavities in diagnostic ultrasound. At kHz-frequencies, the attenuation by water is negligible, but it can be amplified by the presence of absorbing agents such as bubbles. Therefore, measurements of AS can provide information on the cavitation field. (We contrast these measurements with those reported in Section 14.4, in which we study AS as a phenomenon in its own right.) A different type of streaming, microstreaming, is a microflow in the vicinity of collapsing and fragmenting bubbles. Microstreaming actively mixes the fluid and thus should contribute to the measured hydrodynamic dispersion coefficient of the fluid.

The measurements were performed during the initial stage of degassing cavitation for pure water, degassed water, water containing SDS (sodium dodecyl sulfate) as surfactant, and water containing SDS and NaCl. It is known that surfactants modify the dynamics of cavitating bubbles, thereby affecting sonochemistry and sonoluminescence in the cavitation field [6, 7]. A working hypothesis is that, as the anionic surfactant accumulates at the bubble interface, the bubbles acquire a negative charge, and the charged bubbles then tend to repel each other. The maximum effect of surfactant addition on sonochemistry was found experimentally to occur at a SDS concentration of 1–3 m$M$ [6]. An addition of NaCl was seen to cancel the SDS effects, because NaCl screens the electrostatic field, and this may permit bubble flocculation – if not coalescence. Finally, we investigated how the surfactant modifies the dynamics of cavitating fluid on the macroscale.

All measurements were performed on a 2.35 T, horizontal-bore superconducting magnet (Nalorac, Martinez) with an Apollo console (Tecmag, Houston). A quadrature birdcage radiofrequency (RF) coil (Morris, ON) was driven by a 2 kW AMT 3445 RF amplifier (Brea, CA). The 31 kHz Langevin-type ultrasonic transducer (SensorTech, ON), with a parallel beam, was situated coaxially inside a 5 cm internal diameter (i.d.) 22 cm-long cylindrical vessel at the center of the magnet (see Figure 14.1). In these experiments a transducer power of 3.48 W (measured by calorimetry) was employed. All water solutions were doped with 0.2 m$M$ manganese sulfate to decrease the spin-lattice relaxation time of the proton MRI signal to 0.58 s and, thus, reduce the imaging time. A pulsed-field gradient (PFG)

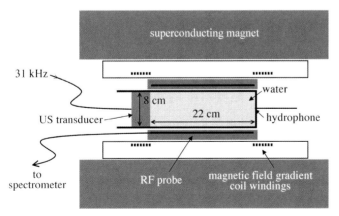

**Figure 14.1** The experimental set-up.

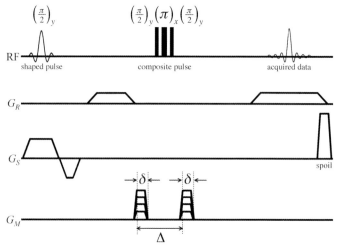

**Figure 14.2** PFG sequence timing diagram. On the RF axis are an excitation of the sample, an inversion of the magnetization, and detection of the signal. On the $G_R$ axis are applications of a motion-sensitizing magnetic field gradient (two pulses, each of duration δ) and an imaging gradient.

sequence [8] with slice selection was used (see Figure 14.2), with 32 gradient-encoding steps. A series of 32 measurements was performed, with the ultrasound turned on at #5 and turned off at #25 so that the onset of cavitation could be observed.

No AS was detected in the degassed water, and no flows or differences from free diffusion were observed. Streaming was pronounced in all measurements with nondegassed water samples, although the presence of gaseous bubbles was critical for the observed streaming. The measurements showed that, for the three water samples (pure water, water with SDS, water with SDS-NaCl), the AS patterns were different. An observed enhancement of streaming by SDS was explained by a stabilization of the bubbles and a reduction of their coalescence.

The dispersion data revealed a very interesting difference between cavitation in these three water types (see Figure 14.3): here, anisotropy of dispersion in water containing SDS (Figure 14.3b and c) is indicated by the discrepancy between the solid and dotted lines. At least two factors can increase the dispersion in cavitating fluid, namely microstreaming and bubble migration. Microstreaming is caused by cavitating bubbles, and is thought to be responsible for an active mixing of the fluid in cavitation fields [9]. The second source of dispersion in the standing wave is bubble migration between the nodes and antinodes of pressure. The origins of such migration are rectified diffusion that causes bubble growth, and the primary Bjerknes force that attracts bubbles smaller than the resonant size towards and repels larger bubbles away from the nodes. Bubble migration is expected to affect the dispersion along the direction of the sound wave propagation – that is, along

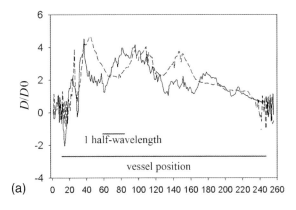

(a)

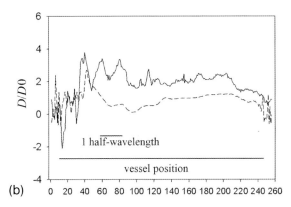

(b)

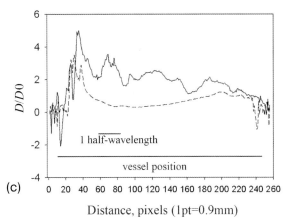

(c)

Distance, pixels (1pt=0.9mm)

**Figure 14.3** Maps of the relative dispersion coefficient ($D$) along the vessel length in the vessel containing (a) pure water, (b) water with SDS, and (c) water with SDS and NaCl. The maps were summed from the fifth to the 25th measurement (over the sonication interval). The horizontal axis shows distance along the vessel in pixels (1 pixel = 0.95 mm). The transducer is at pixel 15 and the end of the vessel is at pixel 246. The vertical axis shows the normalized dispersion coefficient. The solid line is the principal component along the vessel length of dispersion coefficient relative to the molecular self-diffusion coefficient of water at 20 °C ($2.0 \times 10^{-9} \, \text{m}^2 \, \text{s}^{-1}$); the dotted line is the principal component orthogonal to the vessel length of the relative dispersion coefficient.

the principal axis of the vessel in this case. Migration will generate an anisotropic dispersion. Microstreaming should not have any preferred direction in the bulk of the liquid, and only near the rigid boundaries will the microflows be directed towards the boundaries. This 'boundary effect' can be neglected in these experiments because most of the detected signal is derived from the bulk. Hence, microstreaming will generate an isotropic dispersion in the fluid. Accordingly, the observed anisotropy of the dispersion for water with SDS and SDS-NaCl can be explained by a reduction of microstreaming in solutions with surfactants. While the dispersion profiles of the pure water showed approximately the same locations of zones, with high dispersion for displacements both along and across the vessel, the dispersion profiles of water with surfactants showed no high-dispersion zones for displacements across the vessel. The stabilization of bubbles with surfactants caused a reduction in bubble fragmentation, decreased the microstreaming and reduced the isotropic component of dispersion.

## 14.3
## Dynamics of Gas and Liquid in the Cavitating Fluid

The defining feature of cavitation is the presence of oscillating bubbles: these grow during the rarefaction phase and shrink during the compression phase of the wave, until they collapse. The bubble volume can change by as much as a factor of one million between the two extreme states. During maximum expansion, the pressure inside is much lower than that in the surrounding liquid, which causes the dissolved gas and the vapor phase to migrate into the bubble. During maximum compression, the pressure gradient reverses such that the gas, reaction products and unreacted species are ejected from the bubble. Although information on gas dynamics during cavitation is quite important, it is unavailable by using commonly used optical and acoustical techniques. An observation of bubble evolution cannot provide information on how much of the dissolved gas participates in cavitation, how long the gas molecules stay inside the bubbles, or how the gas dynamics depends on nucleation sites. Of additional interest is the question of stability of the microbubbles that serve as nucleation sites in subsequent sonications.

The best way to obtain such information is to interrogate the gas molecules about their experiences in and around the cavitating bubbles. In this respect MRI has shown much promise to accomplish this task, as the technique is insensitive to optical and acoustical opacity, while the NMR relaxation times – which usually are much longer than the period of ultrasonic oscillation – can serve as the system 'memory'.

The aim of these studies was to measure the dynamics of both liquid and dissolved gas during cavitation at the same parameters. To achieve this, prior to the measurements a highly soluble (0.78 vol/vol) Freon-22 was dissolved in water. The $^{19}F$ NMR relaxation parameters $T_1$ and $T_2$ differed by three orders of magnitude, from several milliseconds for the free gas to several seconds for the gas dissolved in water. The same 2.35 T MRI scanner was used as in Figure 14.1, with a 19.7 kHz ultrasonic transducer attached to a cuvette filled with water (at one-wavelength standing wave condition). A standard PFG-spin-echo sequence in 32 gradient steps was used, with data being acquired at Δ-values of 28, 48 and 88 ms. The results are presented here as propagators spatially resolved along the cuvette axis, with velocity as the vertical axis, and position as the horizontal axis. The spatial resolution was 0.5 mm for the $^1$H-data, and 2.2 mm for the $^{19}$F-data. A total of 0.44 W power was absorbed by the water, as measured by calorimetry. The measurement duration was limited to 7 min in order to avoid heating and degassing of the water.

Two types of water sample were used, namely regular tap water and tap water passed through a 0.2 μm filter. (Filtration was used as the difference in the number of particles as potential nucleation sites would have a substantial effect on the cavitation intensity.)

The measurements of water showed that displacement was stronger in the unfiltered water (Figure 14.4a), with some circulation due to AS. The circulation was less developed in the filtered water (Figure 14.4b), but still qualitatively comparable to that in the unfiltered water.

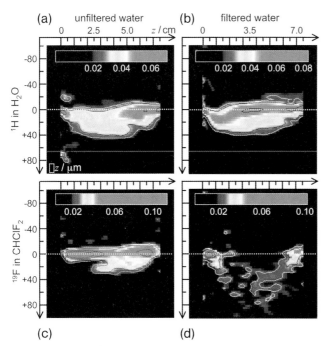

**Figure 14.4** Maps of displacement (Δz) versus position along the cuvette (z) for (a,b) the $^1$H in water and (c,d) $^{19}$F in Freon-22; the transducer is located to the left. Maps are shown for both (a,c) unfiltered tap water and (b,d) filtered tap water. The total probability is normalized to one at each z. The white contour is at a probability of 0.023

The measurements of gas showed a radically different picture (Figure 14.4c and d), with the gas dynamics clearly uncoupled from the liquid. It is well known that gaseous bubbles are present during cavitation, which explains why the gas and liquid dynamics become uncoupled. A substantial portion of gas participates in the bubble motion, which is driven by Bjerknes forces (gradients of acoustic pressure and interactions between the bubbles).

Most unexpectedly, for the filtered water, the displacements of the gas were greater than those of the water itself, at approximately 60 μm far from the transducer, at the center of the cuvette. The motion of gas in the unfiltered water, on the other hand, was localized next to the transducer where cavitation was strongest, and its displacement was smaller than that of the surrounding water.

One possible explanation for these observations for the filtered water is that the gas molecules move faster than water because their travel is mediated by the cavitating bubbles. Due to weak cavitation in the filtered water, cavitating bubbles do not interact much with each other, which in turn increases their lifetimes and makes it possible for them to travel over quite large distances. In a dense cavitation cloud – as in the unfiltered water (with many impurities) – the cavitating bubbles actively interact with each other such that they cannot move very far, and their lifetimes are also shorter. It is also well known from theory [10] that rapid changes in bubble volume limit the amount of gas involved in cavitation to that dissolved in the boundary layer of the liquid around the bubbles. If the surrounding liquid can be refreshed, for example by induced motion, then the gas exchange between the bubble and the environment is enhanced. The moving bubbles act as nucleation sites, spawning new bubbles and involving most of the dissolved gas in cavitation. This picture resembles a superheated liquid that begins to boil as soon as nucleation sites are added to the system.

At the same time, the $T_1$ and $T_2$ relaxation measurements did not show any measurable difference between relaxation parameters in quiescent and cavitating water (both filtered and unfiltered). As there is no stationary gas in filtered water during cavitation, most of it moves with the bubbles. The $T_2$-relaxation of Freon in the gaseous state is 2 ms, and therefore an average residence time for gas molecules inside the bubbles must be much less than 2 ms. By using a two-site exchange model [11], the residence time was estimated to be within 100 μs – about two oscillation periods of ultrasound – indicating a very high refreshing rate of the bubbles' gas content.

When the measurements were repeated with time delays between repeats, it was found that the propagators for gas in filtered water began to strongly resemble those in the filtered water for 2 min delays – there was no arc of displacement, and a substantial amount of the stationary gas was present (Figure 14.5a). With a 1 h delay, however, the majority of the bubbles recovered to their original shape (Figure 14.5b). Propagators for water exhibited the same trend, with motion becoming more violent in the filtered water. This is an indication of an increased number of nucleation sites leading to cavitation enhancement. As the filtered water did not contain many impurities, the only increase in nucleation sites may derive from microbubbles remaining after the previous cavitation events. Despite

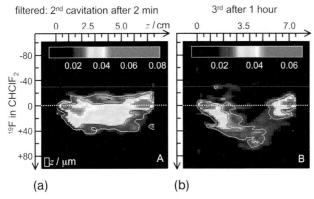

filtered: 2nd cavitation after 2 min          3rd after 1 hour

**Figure 14.5** Maps of Freon-22 displacement in filtered water
during second and third acoustic cavitation events at
(a) 2 min after the first cavitation (Figure 14.4) and
(b) 1 h after the second cavitation.

general theoretical considerations that estimate the lifetime of microbubbles to be
within the order of seconds [12], their stability proved to be remarkable.

By using MRI, we were able to show that gas and liquid dynamics during cavitation are uncoupled, with gas dynamics strongly dependent on the number of
nucleation sites. Microbubbles as remnants from previous cavitation events can
be remarkably stable, and thus influence cavitation dynamics long after the 2
minute delays between events. From these data it is possible to evaluate a 'mean
bubble path' and how much of the gas will participate in cavitation, and this may
be very useful when analyzing the efficiency of sonochemical reactors. An average
residence time for gas molecules inside cavitating bubbles is on the order of
several oscillation periods. Such measurements are made possible by the remarkable difference between NMR relaxation times in dissolved and gaseous Freon-22
but, in principle, similar measurements could be made for any gas which possessed such contrast.

## 14.4
## Acoustic Streaming in Gases

Acoustic streaming is the motion of a fluid caused by the presence of a sound
field. In the form first explained in detail by Lord Rayleigh [1], which occurs when
a standing sound wave is established in an enclosure, a pattern of circulation from
node to antinode is developed (Figure 14.6).

The phenomenon of AS has found application in cleaning, in mixing for microfluidics, and in cooling. In thermoacoustic engines, AS is usually a parasitic effect,
responsible for a reduction in efficiency. AS flows in gases are particularly delicate
and quite difficult to measure without disturbing the flow. Thus, there is a need
for noninvasive measurement methods in order to study the important applica-

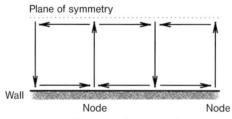

**Figure 14.6** An illustration from Rayleigh's 1884 paper on acoustic streaming. The plane of symmetry is the center line of an enclosing vessel.

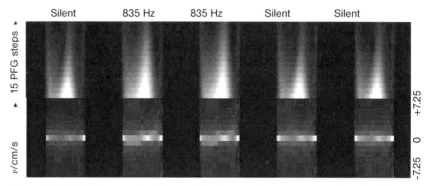

**Figure 14.7** (Top) Five sets of profiles acquired before, during and after establishment of a standing sound wave in the tube. The section of the tube imaged is 12 cm long. Each profile is motion-sensitized and 15 different profiles were acquired at each time, with increasing motion sensitization in the direction of the arrow, Fourier transformation along that dimension gives the velocity spectra (bottom), in which the horizontal axis is position along the tube, the vertical axis velocity (from −7.25 to +7.25 cm s⁻¹) and the brightness of each pixel is representative of the mass fraction of propane moving with that particular velocity at that position along the tube. Positive velocities are left to right along the tube section. The dashed vertical lines in each collection of velocity spectra mark the positions of the spectra in Figure 14.8.

tions of AS. Although laser Doppler anemometry (LDA) is one possibility, it cannot operate in optically opaque systems. Here, we demonstrate the use of MRI as another possible measurement method for the study of AS in gases [12].

The PFG sequence (see Figure 14.2) was applied at 2.35 T to a 2 m-long 20 cm-diameter cylindrical tube filled with propane gas. The MRI signal from propane is longlived for a gas (~$10^{-2}$ s). One-dimensional images (profiles) of a 12 cm section near the center of the tube are shown in the upper half of Figure 14.7. The first data set (labeled 'silent') shows 15 profiles, each acquired with a different amplitude of motion-sensitizing PFG, with the amplitude of PFG increasing in the direction of the arrow. The upper row of data shows most attenuation and is most sensitized to motion. These 15 measurements were repeated five times, at approximately 15 min intervals, before, during and after the establishment of an acoustic standing wave at 835 Hz in the tube.

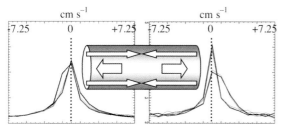

**Figure 14.8** Velocity spectra selected from Figure 14.6 at the positions indicated. Spectra from all five time points (before, during and after sound application) are overlaid. The profiles indicate a smaller mass fraction of faster counter flow (at the tube walls), as would be expected for developed Rayleigh streaming.

The pattern of attenuation changes in the presence of the sound field (indicating a change in the motion of the propane) and returns to its quiescent state when the sound is turned off. In order to quantify this change, the 15 profiles are Fourier-transformed along the dimension defined by the PFG amplitude. The resultant five images (lower half of Figure 14.7) are collections of velocity spectra.

Consider again the first ('silent') data set; the majority of propane appears in the central row and, therefore, has a velocity of $0\,cm\,s^{-1}$. The width of the spectrum represents the diffusivity (random motion) of the propane–the wider the velocity spectrum the greater the coefficient of diffusivity at that location. Before the sound field is applied, the velocity spectra along the length of the tube are identical, but in the presence of the sound field ('835 Hz') that uniformity is lost. The peak in the velocity spectrum clearly shifts towards negative velocities in the first part of the tube section, and towards positive velocities in the second part.

Figure 14.8 shows velocity spectra extracted from Figure 14.7 at the positions along the tube section which show the greatest difference from their quiescent counterparts. These positions are 7.27 cm apart, and $\lambda/4$ calculated from literature data for the speed of sound in propane [13] is 7.25 cm.

With a total measurement time of 8 min, the noninvasive acquisition of spatially resolved velocity spectra of acoustic streaming is possible. MRI measurements represent the only truly noninvasive measurement of gas dynamics, and are therefore a promising tool in the study of such delicate flows as AS. The MRI measurements are, of necessity, time averaged (on the order of minutes) and may therefore be applied to steady acoustic flows, or to measure temporal statistics in unsteady flows. The range of gas velocities measured in the MRI literature ranges from $mm\,s^{-1}$ to tens of $m\,s^{-1}$ [14].

## 14.5
## Conclusions

In this chapter we have described several applications of MRI to phenomena produced by strong sound fields in fluids, and believe that MRI–as a noninvasive measurement of great flexibility – has great promise in this very broad field.

Today, there are several likely directions for further research, including speed, stability and accessibility. As most of these problems essentially relate to fluid dynamics, the application of fast MRI methods to fluid dynamics characterization, notably in the areas of spatially resolved velocity, diffusivity and dispersion, should prove to be of great benefit. The use of a stable method in the presence of metal/charged parts of a sound emitter often proves advantageous, and this may well apply also to phase-encoding methods. Since, in general, sound emitters/transducers and sonochemical reactors cannot be conveniently positioned inside an MRI scanner, the transition from model experiment to measuring the effects produced by real-time sound fields will require the use of either embedded NMR sensors or a portable NMR instrument.

### Acknowledgments

The authors thank Duncan McLean and Scott Culligan for their help, and Murray Olive and Brian Titus for manufacturing parts of the experimental set-up. Professor B. Balcom provided much useful advice in all of our endeavors, which our NMR technician Rodney McGregor made possible. Funding from the Natural Science and Engineering Council of Canada and the Harrison McCain Foundation is gratefully acknowledged. The UNB MRI Center is supported by a NSERC Major Facilities Award.

### References

1 Young, R. (1990) *Cavitation*, CRC Press, Boca Raton.
2 Boluriaan, S. and Morris, P.J. (2003) *Aeroacoustics*, **2** (3-4), 255.
3 Rayleigh, L. (1884) *Proceedings of the Royal Society of London*, **36**, 10.
4 Rayleigh, L. (1917) *Philosophical Magazine*, **34**, 94.
5 Mastikhin, I.V. and Newling, B. (2005) *Physical Review E*, **72** (5), 056310-1–12.
6 Segebarth, N., Eulaerts, O., Reisse, J., Crum, L.A. and Matula, T.J. (2002) *The Journal of Physical Chemistry B*, **106**, 9181–90.
7 Ashokkumar, M., Hall, R., Mulvaney, P. and Grieser, F. (1997) *The Journal of Physical Chemistry B*, **101**, 10845–50.
8 Stejskal, E.O. and Tanner, J.E. (1965) *The Journal of Chemical Physics*, **42**, 288.

9 Leighton, T.G. (1994) *The Acoustic Bubble*, Academic Press, San Diego.
10 Fyrillas, M.M. and Szeri, A.J. (1994) *Journal of Fluid Mechanics*, **277**, 381–407.
11 (a) Dubois, B.W. and Evers, A.S. (1992) *Biochemistry*, **31**, 7069–76.
(b) Kabalnov, A., Klein, D., Pelura, T., Schutt, E. and Weers, J. (1998) *Ultrasound in Medicine and Biology*, **24**, 739–49.
12 Newling, B., MacLean, D.A. and Mastikhin, I.V. (2006) *Canadian Acoustics*, **34**, 48.
13 Trusler, J.P.M. and Zarari, M.P. (1996) *The Journal of Chemical Thermodynamics*, **28**, 329–35.
14 Newling, B. (2008) *Progress in Nuclear Magnetic Resonance Spectroscopy*, **52**, 31–48.

# 15
# MRI of Coherent and Incoherent Displacements in Turbulent Flow

*Benedict Newling, Zhi Yang, Mark H. Sankey, Mike L. Johns and Lynn F. Gladden*

## 15.1
### Introduction: Turbulence, and the Role of MRI in Its Measurement

Many natural and artificial flows are turbulent, and there have been attempts to describe the beautiful structures associated with turbulence since ancient times (Figure 15.1). Turbulent flows are continually changing, and must be analyzed in terms of the statistics of their descriptive parameters. As a consequence, a unifying theory of turbulence has been the stubborn subject of much study. Computational fluid dynamics (CFD) has some claim to be the 'gold standard' in predictive tools for turbulent flows, and thus there continues to be a need for measurement methods to test those predictions in applications as diverse as aeronautics, ecohydraulics and meteorology. Magnetic resonance imaging (MRI) has a place in the toolbox: hot-wire and hot-film anemometries are invasive, while laser Doppler anemometry and particle imaging velocimetry require optical transparency. MRI is a naturally three-dimensional (3-D) measurement, and has distinct time advantages over point-by-point techniques. A critical issue for all of these measurement methods is how to deal with the stochastic nature of turbulence. For a simple description, we will follow Reynolds [1] and decompose each vector component of fluid velocity into a steady average and a random fluctuation

$$v_i'(t) = v_i + V_i(t) \tag{15.1}$$

The subscript $i$ is an index of Cartesian component, such as $x$, $y$, or $z$ or 1, 2, 3, for primary, secondary and tertiary phase-encoding directions. Reynolds' other longstanding contribution to our description of turbulent flows comes in the form of the dimensionless Reynolds number, $Re = d\rho v_i/\mu$, which measures the ratio of inertial to viscous effects in a fluid flow field. Here, $\rho$ is the fluid density, $d$ is a lengthscale that is characteristic of the geometry of the flow, and $\mu$ is the viscosity of the fluid. The higher the Reynolds number, the smaller the perturbation of flow which will induce turbulence [2]. Turbulence in pipes of common materials (and common roughness) is often observed to begin around $Re = 2000$–$3000$. Most

*Magnetic Resonance Microscopy.* Edited by Sarah L. Codd and Joseph D. Seymour
Copyright © 2009 WILEY-VCH Verlag GmbH & Co. KGaA, Weinheim
ISBN: 978-3-527-32008-0

**Figure 15.1** A copy of one of Leonardo da Vinci's sketches of the vortices in turbulent, free-surface flow in the wake behind a bluff obstruction. (The Royal Collection © 2007, Her Majesty Queen Elizabeth II.)

measurement methods for turbulent flows attempt to measure an instantaneous velocity component, for example, $v_1'(t)$, as rapidly as possible. Various statistics of velocity may then be calculated; a mean velocity over time, for instance, or the correlation between velocities measured at some fixed time interval. Correlations between different velocity components are also measured, to assess the isotropicity of the fluctuations, and the possible combinations quickly multiply when one considers also correlations between velocity components measured at different points in the flow field. Rapid measurements require high signal-to-noise ratios (SNRs), and MRI has been skillfully employed in this sense by two research groups [3, 4]. Our alternative approach in these studies (we do not attempt snapshots of the flow field) has been to use the duration of the MRI measurement for intrinsic temporal averaging, which allows measurement over a greater range of samples, in particular for gaseous flows, in which the spin density is typically a factor of $10^3$ lower than in a liquid. The temporal averaging is not continuous, however, as discussed in Section 15.3.

## 15.2
### The Peculiar Advantages of Motion-Sensitized SPRITE for Measuring Turbulent Flows

Single-point imaging (SPI) techniques have, for some time, been useful in materials MRI [5, 6], where the $T_2^*$ of the samples is short (<1 ms). Our preference for efficient velocity measurements in highly turbulent, gaseous flows is for a motion-sensitized version of SPRITE (Single Point Ramped Imaging with $T_1$ enhance-

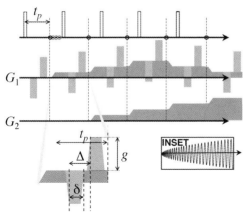

**Figure 15.2** The first few acquisition points of a centric, spiral SPRITE pulse sequence with gradient waveforms in orthogonal directions shown on the $G_1$ and $G_2$ axes. Motion sensitization is carried out by superimposition of a bipolar gradient pair, in lighter gray, in the same direction as the $G_1$ axis. The inset shows the complete gradient waveform for the $G_2$ axis as it is used to sample a span of $\vec{k}$-space.

ment [7]). The technique is purely phase-encoded. Individual $\vec{k}$-space locations are sampled in quick succession by stepping the phase-encoding gradients directly from one value to another throughout a series of many radiofrequency (RF) pulses. Figure 15.2 shows the first five RF pulses in a two-dimensional (2-D) SPRITE imaging sequence, with the accompanying position-encoding gradient waveforms shown in the darker gray shade. In materials' imaging applications, the RF pulses typically have a tip angle $<< \pi/2$, the encoding time $t_p$ is $\leq 1$ ms, and the stabilization time between gradient switching and RF application is $\approx 1$ ms (proportionally longer than shown in the diagram). The possibility exists of acquiring a single datum point after each RF pulse (black circle), as envisaged in the original development of SPI, or of acquiring multiple data points (e.g. gray circles after the first RF pulse) before the gradient is changed. Multiple-point data may be used for signal averaging or to extract relaxation information for each $\vec{k}$-space point [8, 9]. The $\vec{k}$-space raster illustrated here is centric; the first RF pulse is accompanied by no position-encoding gradient. The first datum point acquired is therefore a bulk measurement and is used as such below. The $\vec{k}$-space is typically segmented into separate leaves, which may be either spiral or piece-of-pie shapes, for example [10]. In Figure 15.2 the full gradient waveform of the secondary phase encoding gradient is shown in the inset. This waveform contains several thousand gradient steps, each of which is accompanied by an RF excitation, as a $\vec{k}$-space raster is sampled. Motion sensitization is carried out using the pulsed-field gradient (PFG) approach of Stejskal and Tanner [11, 12], by superimposing upon the position-encoding gradient a (lighter gray) displacement-encoding PFG pair.[1] This additional

---

**1)** We refer to this sequence colloquially as S′ (pronounced ess-prime, for Single Point Ramped Imaging with Motion Encoding).

gradient may be superimposed on any gradient axis in order to sensitize the measurement to displacement in that direction. In the figure, motion sensitization has been combined with the primary phase-encoding gradient $(G_1)$. The phase accumulated by a spin of position $\vec{r}(t)$ in the phase-encoding interval is

$$\phi = \gamma \int_0^{t_p} \vec{G}(t) \cdot \vec{r}(t) dt \tag{15.2}$$

in a frame of reference rotating with an angular frequency of $\gamma B_0$, where $B_0$ is the polarizing field. Three Cartesian axes may be considered separately, so that the phase accumulated due to the primary phase-encoding gradient is

$$\phi_1 = \gamma \int_0^{t_p} G_1(t) r_1(t) dt = \gamma G_1 r_1 t_p \tag{15.3}$$

for a stationary spin $(r_1(t) = r_1)$ in a constant gradient. This is, of course, exactly the same phase accumulated by a stationary spin without the PFG pair (i.e. in ordinary SPRITE). In the case of constant velocity in the primary phase-encoding direction, $v_1$, then $r_1(t) = r_1 + v_1 t_p$ and

$$\phi_1 = \gamma t_p \left( G_1 \left[ r_1 + \frac{1}{2} v_1 t_p \right] + \frac{1}{2} g v_1 \delta \right) = \gamma \Delta \left( 2 G_1 [r_1 + v_1 \Delta] + g v_1 \delta \right) \tag{15.4}$$

if $t_p = 2\Delta$. The calculation serves to illustrate that the phase of any spin will depend upon $\Delta = t_p/2$, $\delta$, $G_1$ and $g$. It is common practice to regard $\Delta$ as the flow evolution interval (see below), but with our fixed $t_p = 2\Delta$, Equation 15.4 may be expressed in terms of either time.

Constant speed is not a very good model, even for fluid elements in laminar fluid flow. The displacements of individual fluid elements (which we can think of as spin isochromats) are modified by molecular diffusion, which we may include (after Carr and Purcell [13]) as random jumps occurring every fraction of a second. The random distribution of resulting displacements gives rise to a distribution of phases (rather than the illustrative coherent change in phase in Equation 15.4), and hence leads to some cancellation of signal, and an attenuation at the center of $\vec{k}$-space $(G_i = 0)$ of the form

$$s(g) = s(0) \exp \left[ -\frac{2}{3} \gamma^2 g^2 \delta^2 D_0 \left( \frac{t_p}{2} \right)^3 \right] \tag{15.5}$$

in which $s$ is the measured MR signal, and $D_0$ is the molecular self-diffusion coefficient,[2] defined as $< [r_1(t_p) - r_1(0)]^2 > / (2 t_p)$ in the absence of any flow. This particular form of Equation 15.5 crucially assumes that the distribution of displacements due to self-diffusion is Gaussian. The measured $D_0$ can depend upon $t_p = 2\Delta$ when the diffusion is restricted in some way [11].

---

**2)** Strictly this is only true if the molecular self-diffusion is isotropic. Otherwise, the coefficient here is one of the components of the diffusion tensor, and for the pulse sequence as implemented in Figure 15.2 would be the principal component of the diffusion tensor $D_{11}$.

We may think of the random fluctuations in velocity due to turbulence as having a similar, but much larger, effect. The trajectory of each spin $r_1(t)$ will generally be rather complicated, and will quickly become very different from other spins which were initially nearby. At any point in the flow field, there will be an ensemble of spins with different histories, which have different phases as a result, leading again to a cancellation of signal in the MR image. In addition, strong, local, fluctuating accelerations may contribute to the phase of a spin (terms of order three in $t_p$, which were ignored in Equation 15.4 when we assumed constant velocity). After Kuethe [14], we may replace $D_0$ in Equation 15.5 with $D_t$, a turbulent diffusivity. $D_t$ is an indicator of the vigor of the turbulence, again due to its link with mean-squared displacement. Equation 15.5 is the limiting behavior when $t_p/2 \gg T_c$, the turbulence correlation time [15]. $T_c$ is characteristic of the time taken for a fluid element to reorient, or of two neighboring fluid elements to develop very different trajectories. In general, the measured $D_t$ will depend upon $t_p$, but is asymptotic when $t_p$ is long enough.

This attenuation modeled in Equation 15.5 with $D_t$ is responsible for signal reduction in clinical images of pathologically turbulent blood flow [16], and can be used to quantify the intensity of turbulence [14, 17]. However, in very turbulent flows and during typical echo times (TE ~ 2 ms) cancellation often leads to complete signal loss. The advantage of SPRITE techniques is that the encoding interval $t_p$ can be made very short (<1 ms without difficulty). By reducing $t_p$ to 600 μs, we have been able to regain control of the dispersion of phase in flows with Re > 200 000 [18]. Not only is signal loss not disastrous, but the phase distribution also provides propagator information [11] upon Fourier transformation of the complex signal acquired as a function of $g$, the amplitude of the PFG pair (Section 15.3). Of course, the short $t_p$ interval is responsible for the successes of SPRITE imaging in materials science, and allows the imaging of fluids with intrinsically short $T_2^*$, such as some gases. A short $T_1$ is a positive advantage in imaging with the SPRITE techniques, where a long $T_1$ can force a long TR and low flip angle in order to reduce image blurring [19]. A long $T_2^*$ can be awkward because unspoiled transverse magnetization following one pulse can interfere with the FID following the next. The combination of short $T_1$ and $T_2$ has been an advantage when measuring the flow of thermally polarized sulfur hexafluoride (SF$_6$, $T_1 = T_2 = 1.4$ s). In measuring turbulent water flow using motion-sensitized SPRITE, we will typically add a paramagnetic contrast agent to reduce both $T_2$ (and therefore $T_2^*$) and $T_1$.

## 15.3
## The Apparatus

$^{19}$F images at 2.4 T were made in a Nalorac (Martinez, USA) 2.4 T, horizontal-bore superconducting magnet using a Tecmag (Houston, USA) Apollo console. A water-cooled 7.5 cm internal diameter (i.d.) Nalorac magnetic field gradient set was driven by Techron (Elkhart, USA) 8710 amplifiers. Measurements were performed using a home-built birdcage RF coil, driven by a 2 kW AMT (Brea, USA) 3445 RF amplifier. Gas flow was driven around a closed loop of cylindrical pipe through the center of the magnet by a centrifugal pump (Ametek Lamb Electric, Kent, USA). The loop was filled with $SF_6$ and an independent measurement of mean fluid speed performed using a Venturi flow meter. $SF_6$ is approximately five times more dense than air ($5.69 \, \text{kg m}^{-3}$ at 315 K and 0.1 MPa); the viscosity is $1.6 \times 10^{-5}$ Pa·s at 315 K [20]. Bulk values of $T_1 = T_2 = 1.4$ ms were measured at 0.1 MPa [21]. A blutt obstruction was placed 44 tube diameters downstream of the Venturi meter (inside the RF coil) to allow steady-state flow development before the obstruction.

$^1$H images at 7.0T (bubbly flow) were performed on a MARAN spectrometer (Resonance Instruments Ltd, Oxford, UK) equipped with a 7T wide-bore, horizontal superconducting magnet (7T/60/AS; Magnex Scientific Ltd, Oxford, UK). A standard microimaging gradient set (SGRAD156/100/S; Magnex Scientific Ltd) was used, powered by a set of three power amplifiers (model 7782; A.E. Techron, Elkhart, USA). A home-made 62 mm i.d. probe was used with an RF power amplifier (7T100S; Communication Power Corp., New York, USA). All measurements were carried out at ambient temperature.

## 15.4
## Mapping of Turbulent Flow Fields

The preceding discussion is illustrated by the example shown in Figure 15.3. The images in Figure 15.3a and b are side elevations of a length of cylindrical tubing which is filled with $SF_6$ gas, flowing from left to right with an average speed of

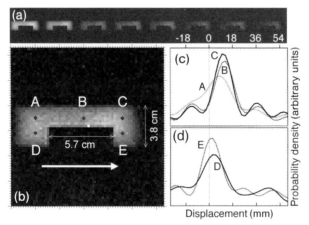

**Figure 15.3** (a) Flow of gaseous $SF_6$ gas from left to right with a mean velocity of 5.4 m s$^{-1}$ in a polarizing field $B_0 = 2.4$ T. The encoding time, $t_p$, was 600 μs, and $\delta = \Delta = t_p/2$. Eight images are shown; each was obtained with a different value of g, which increases from left to right; (b) A larger version of the first image from (a), in which five locations in the flow field are identified. The propagators at those locations are shown in (c) and (d). The resolution in the propagators is 1.80 mm per pixel.

$5.4\,\mathrm{m\,s^{-1}}$. The tube contains a bluff obstruction, not dissimilar to that which fascinated da Vinci, where the tube cross-section is suddenly halved to semicircular. Eight images are shown in Figure 15.3a: each was obtained with a different value of $g$, which increases from left to right. Eight identical images of stationary gas were also acquired to correct the signal phases for any effects of imbalance in the PFG pair. The acquisition of multiple images ensures temporal averaging of the measured flow properties. The averaging is discontinuous: consecutive acquisition points (in the single-point case) are separated in time by approximately 2 ms. However, several thousand points are collected over an interval of 4–5 min to complete an image data set, and eight images are acquired over a half-hour. Steady turbulent flow–that is, a flow for which the statistical properties do not change with time–is assumed and the duration of the measurement taken to indicate a complete averaging of the flow behavior.

In the images of Figure 15.3a, the increasing $g$ (from left to right) leads to increasing phase dispersion due to motion during the $t_p = 600\,\mathrm{\mu s}$ interval and, hence, to increasing attenuation of the signal. Fourier transformation along the $g$ dimension yields, for each pixel in the images, the propagator, such as those shown in Figure 15.3c and d. The horizontal axis is displacement, and the vertical axis is the probability density of that displacement occurring in a given time interval. The eight g-dimension data points, shown in Figure 15.3a, were zero-filled to 64 points before Fourier transformation, and some truncation in the $g$ domain accounts for the sinc-like oscillating artifacts in the baseline. Despite this effect, the propagators, which show the time-averaged motion at the positions labeled A–E in Figure 15.3b, serve to illustrate some general features. Propagator A is wider than the others, which indicates a greater variation in the phases of those spins traveling through the pixel at A than at the other locations. This we associate with a more intense turbulence (greater velocity fluctuation, $V_i(t)$) than at B or C, for example, in which locations the flow is channeled above the obstruction, and the distribution of displacements is narrowed. The peak in each propagator indicates $(v_1 \cdot \Delta)$, the steady time-averaged displacement at each pixel, because the eight images from which the propagator was constructed took approximately 30 min in total to acquire. Thus, the magnitude of $v_1$ at locations A–E may be ranked: $B > C > A \gg D > E$. We use this peak position in order to develop a map of the averaged $v_1$ for the whole flow field. Only the third term of Equation 15.4 changes with $g$, and it is this that determines the position of the peak in the propagator. (Note that a steady, mean acceleration at a given location, which would give rise to a significant term $O(t_p^3)$ in Equation 15.4, could also contribute to the measured displacement.) The other two terms in Equation 15.4 determine the spatial registration of the signal in the image: the first term gives the starting position of a fluid element; and the second term is a smearing of image caused by motion during the $t_p$ interval.

Figure 15.4 shows maps of $v_1 \equiv v_z$ (with the z-axis along the bore of a horizontal bore superconducting magnet, $B_0 = 2.4\,\mathrm{T}$) constructed as described above. In Figure 15.4a and b, we once again show the flow of $SF_6$ gas past a bluff obstruction. The mean $v_z$ is now $10.7\,\mathrm{m\,s^{-1}}$ from left to right, which gives $Re = 130\,000$.

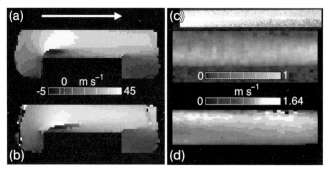

**Figure 15.4** All image data were acquired using the motion-sensitized version of SPRITE (see Figure 15.2). (a,b) Velocity maps of the flow of gaseous $SF_6$ from left to right with a mean velocity of $10.7\,\text{m s}^{-1}$ in a polarizing field $B_0 = 2.4\,\text{T}$. Maps of $v_z$ were obtained by taking the mode of the propagator defined in each pixel by eight different values of $g$. In (a) the encoding time, $t_p$, was $400\,\mu\text{s}$; in (b) the encoding time was $900\,\mu\text{s}$. In both (a) and (b) $\delta = \Delta = t_p/2$; (c) Image of the flow of bubbly water/air from left to right with a mean velocity of $1.05\,\text{m s}^{-1}$ (water) and $3.86\,\text{m s}^{-1}$ (air) in a polarizing field $B_0 = 7.0\,\text{T}$. The encoding time was $1.3\,\text{ms}$. The water was doped with $8\,\text{mM}$ GdCl$_3$, so that $T_1 = 8.4\,\text{ms}$ and $T_2 = 6.5\,\text{ms}$. The inset is a photograph of the dispersed bubbly flow; (d) A map of $v_z$ for the bubbly flow obtained by taking the mode of the propagator defined in each pixel by 32 different values of $g$. $\delta = 350\,\mu\text{s}$, $\Delta = t_p/2 = 650\,\mu\text{s}$. This separation of the PFG pair ensured their balance and removed the need for a set of images of stationary fluid.

The map clearly shows the faster distribution of gas above and recirculation (darker shades of gray) behind the obstruction, as observed in Da Vinci's sketches. In Figure 15.4a, $t_p = 2\Delta = 400\,\mu\text{s}$, while in Figure 15.4b $t_p = 2\Delta = 900\,\mu\text{s}$. The effects of the smearing term (Equation 15.4) along the z-axis can be seen to increase with $t_p$, as some of the faster spins appear spread out along the upper section of the tube. This effect is also responsible for some distortion of the boundary layer above the obstruction. Forward-moving fluid elements are misregistered into the top of the boundary layer, which therefore loses the wing-like shape that is so apparent in the $400\,\mu\text{s}$ images. Countercurrent motion in the boundary layer itself is clearly considerable, and countercurrents are clear also upstream of the obstruction. In this region of slower fluid flow, there is much less discrepancy between the $400\,\mu\text{s}$ and $900\,\mu\text{s}$ maps, although smearing is still present. Downstream of the obstruction, a large (1.7 cm-diameter) eddy is clearly visible, and its distortion in the $900\,\mu\text{s}$ image is interesting. The velocities in this region seem too low to cause significant smearing, and the eddy too large for spins to have performed a half-circulation during the $900\,\mu\text{s}$ interval (at the speeds indicated in the $400\,\mu\text{s}$ map), thereby reducing their total z displacement. The smaller displacement-per-unit-time reflected in the $900\,\mu\text{s}$ map is perhaps indicative of smaller, local eddies.

Figure 15.4c and d show SPRITE images of the flow of dispersed bubbly flow (air in water). In Figure 15.4c, the normalized image intensity is taken as illustrative of the mass fraction of liquid (0–1). The SPRITE measurement, with $t_p = 1.3\,\text{ms}$, is relatively insensitive to local field inhomogeneities caused by the magnetic

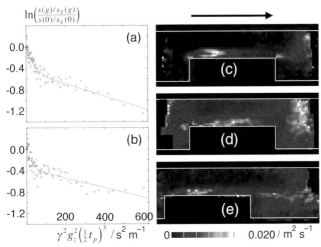

$$\ln\left(\frac{s(g)/s_S(g)}{s(0)/s_S(0)}\right)$$

(a)

0.0
-0.4
-0.8
-1.2

(c)

(b)

0.0
-0.4
-0.8
-1.2

(d)

200    400    600

(e)

$$\gamma^2 g_z^2\left(\tfrac{1}{2}t_p\right)^3 / s^2\,m^{-1}$$    0 ▬▬▬▬▬ ▮ 0.020 / m² s⁻¹

**Figure 15.5** Measurements of turbulent diffusivity, $D_t$, in gas flow past the bluff obstruction. (a) Data points from the center of $\vec{k}$-space for images acquired at eight values of $g$ and 11 values of $t_p$, normalized as described in the text. The curve is a bulk measurement of turbulent diffusivity, $D_t$, for a mean gas flow rate of 10.7 m s⁻¹. The solid line models two diffusivity coefficients (the slope according to Equation 15.5), which correspond to two different regions of more and less vigorous turbulence in the flow field. Note that the data points at shorter $t_p$ do not contribute to the shallower portion of the curve; (b) Bulk measurement of $D_t$ for a mean gas flow of 16.9 m s⁻¹. Maps of $D_t$ at a mean gas flow rate of 10.7 m s⁻¹ are shown in (c) $t_p = 400\,\mu s$ and (d) $t_p = 900\,\mu s$. A map of $D_t$ at a mean gas flow rate of 16.9 m s⁻¹ and $t_p = 600\,\mu s$ is shown in (e).

susceptibility boundaries at air–water interfaces, which makes the method uniquely suitable – among MRI methods – for the study of this type of flow. The effects of susceptibility-induced inhomogeneity can actually be quantified using multiple point data. The MR image clearly shows buoyancy-driven separation of the bubbles towards the top of the tube from left to right (in the direction of flow). This effect can also be seen in a photograph of the test section (see inset). In Figure 15.4d, the corresponding map of the horizontal component ($z$) of the water velocity shows that this separation is responsible for an evolution of the flow field from left to right. A boundary layer of slower flow begins to develop at the top of the tube, while to the left the velocity profile is quite sharply peaked at the top of the tube – an effect which is ascribed to the geometry of the mixing device. The profile becomes more pluglike (the velocity becomes more spatially homogeneous) towards the right-hand end of the tube.

Figure 15.5a and b show data from the center of $\vec{k}$-space for two different flow rates of SF₆ gas past the bluff obstruction. In Figure 15.5a, the average gas velocity is 10.7 m s⁻¹, while in Figure 15.5b it is 16.9 m s⁻¹. At the center of $\vec{k}$-space $G_i$ is zero, so these data are effectively a bulk measurement of signal attenuation with a motion-sensitizing gradient, $g$, which drives the increase in the abscissa. The data were acquired at eight different values of $t_p$, from 400 to 900 μs (in 50 μs steps), but are all combined in a single curve by normalizing twice–once with respect to the $g = 0$ data ($s(0)$) and once with respect to identical data acquired with stationary gas ($s_S(g)$, $s_S(0)$). The solid line is a model of two exponentially decaying components, like Equation 15.5, with $D_{t1}$ and $D_{t2}$. These different diffusivities correspond to different regions of the flow field (high and low turbulence intensity, see below). There is no trend in the data with $t_p$, although there is considerable variation. We have chosen to interpret the scatter as uncertainty, and to propose that $D_{t1}$ and $D_{t2}$ are both in the limit $t_p \gg 2T_c$. For the 10.7 ms$^{-1}$ gas flow in Figure 15.5a, $D_{t1} = (4.3 \pm 0.7) \times 10^{-2} \text{m}^2\text{s}^{-1}$ and $D_{t2} = (1.7 \pm 0.2) \times 10^{-3} \text{m}^2\text{s}^{-1}$, and the two regions are present in the ratio $0.37 \pm 0.03 : 0.63$. In the 16.9 ms$^{-1}$ gas flow in Figure 15.2b, $D_{t1} = (6 \pm 2) \times 10^{-2} \text{m}^2\text{s}^{-1}$ and $D_{t2} = (1.4 \pm 0.2) \times 10^{-3} \text{m}^2\text{s}^{-1}$, and the two regions are present in the ratio $0.30 \pm 0.03 : 0.70$. Within uncertainty, the turbulent diffusivities are unchanged between the two flow rates. At this stage, we cannot be certain that this is a real indication of some coherent structure [22], stable over a range of Re, downstream of the obstruction or is simply a result of the finite precision of the measurement. There is slightly less of the greater turbulent diffusivity in the faster flow (see below).

Having made the assertion $t_p \gg 2T_c$, we go on to interpret the spatially resolved signal attenuation accordingly [14, 18], to generate maps of the turbulent diffusivity in Figure 15.5c at $t_p = 400\,\mu$s and average velocity 10.7 ms$^{-1}$, Figure 15.5d at $t_p = 900\,\mu$s and average velocity 10.7 ms$^{-1}$, and in Figure 15.5e at $t_p = 600\,\mu$s and average velocity 16.9 ms$^{-1}$. Broadly, two regions of diffusivity are revealed as predicted by the bulk measurements in Figure 15.5a and b: in the boundary layer above the obstruction, and in the wake behind the obstruction, the diffusivity is high; elsewhere, and particularly in the constriction above the obstruction, the diffusivity is low. In the faster flow of Figure 15.5e, the wake region of high diffusivity is moved downstream, which may account for the changing ratios of $D_{t1} : D_{t2}$ in the graphs of Figure 15.5a and b. At short observation times, the lower turbulent diffusivity coefficient is not resolved, which accounts for the larger areas of darker shading in Figure 15.5c.

## 15.5
## Conclusions

The motion-sensitized SPRITE sequence is suitable for measuring flow in short $T_2^*$ (<1 ms) liquids and gases by virtue of the short encoding time ($t_p$). $T_2^*$ may be short because of the molecular properties of the fluids involved, or because of susceptibility heterogeneity, as in two-phase flows. The method is most suitable where the flows are fast ($\geq 1$ ms$^{-1}$), because the motion-encoding gradient amplitudes need not then be large ($\sim 10$ mT m$^{-1}$), and gradient switching demands are reduced. The flows may be staggeringly fast and at very high Re numbers in comparison with most frequency-encoding MRI methods, although the SPRITE method measures the time-averaged flow field and does not attempt to take snapshots.

## Acknowledgments

M.S., M.L.J. and L.F.G. thank the Engineering and Physical Sciences Research Council (EP-SRC) of the UK for support through a Platform Grant. Y.Z. and B.N. thank the Natural Sciences and Engineering Research Council (NSERC) of Canada for Discovery Grant support. B.N. would also like to thank the Harrison McCain Foundation for support, and the Sir Peter Mansfield Center for Magnetic Resonance at the University of Nottingham, UK – and Dr Paul Glover in particular – for providing a refuge in which to write.

## References

1 Vennard, J.K. (1966) *Elementary Fluid Mechanics*, John Wiley and Sons, Inc., New York.

2 Hof, B., Juel, A. and Mullin, T. (2003) Scaling of turbulence transition threshold in a pipe. *Physical Review Letters*, **91** (24), 244502.

3 Kose, K. (1992) Visualization of turbulent motion using echo-planar imaging with a spatial tagging sequence. *Journal of Magnetic Resonance*, **98** (3), 599–603.

4 Sederman, A.J., Mantle, M.D., Buckley, C. and Gladden, L.F. (2004) MRI technique for measurement of velocity vectors, acceleration, and autocorrelation functions in turbulent flow. *Journal of Magnetic Resonance*, **166** (2), 182–9.

5 Emid, S. and Creyghton, J.H.N. (1985) High resolution NMR imaging in solids. *Physica B and C*, **128**, 81–2.

6 Webb, A.G. (2004) Optimizing the point spread function in phase-encoded magnetic resonance microscopy. *Concepts in Magnetic Resonance Part A*, **22A** (1), 25–36.

7 Balcom, B.J., MacGregor, R.P., Beyea, S.D., Green, D.P., Armstrong, R.L., Bremner, T.W. and Logan, A. (1996) Single-point ramped imaging with $T_1$ enhancement (SPRITE). *Journal of Magnetic Resonance Series A*, **123**, 134.

8 Halse, M., Rioux, J.A., Romanzetti, S., Kaffanke, J., MacMillan, B., Mastikhin, I., Shah, N.J., Aubanel, E. and Balcom, B.J. (2004) Centric scan SPRITE magnetic resonance imaging: optimization of SNR, resolution, and relaxation time mapping. *Journal of Magnetic Resonance*, **169**, 102–17.

9 Kaffanke, J., Dierkes, T., Romanzetti, S., Halse, M., Rioux, J.A., Leach, M.O., Balcom, B.J. and Shah, N.J. (2006) Application of the chirp z-transform to MRI data. *Journal of Magnetic Resonance*, **178**, 121–8.

10 Balcom, B.J., Khrapitchev, A.A. and Newling, B. (2006) Sectoral sampling in centric-scan SPRITE magnetic resonance imaging. *Journal of Magnetic Resonance*, **178** (2), 288–96.

11 Callaghan, P.T. (1991) *Principles of Nuclear Magnetic Resonance Microscopy*, Oxford University Press, Oxford.

12 Stejskal, E.O. and Tanner, J.E. (1965) Spin diffusion measurements: Spin echoes in the presence of a time-dependent field gradient. *Journal of Chemical Physics*, **42** (1), 288–92.

13 Carr, H.Y. and Purcell, E.M. (1954) Effects of diffusion on free precession in nuclear magnetic resonance experiments. *Physical Review*, **94** (3), 630–8.

14 Kuethe, D.O. (1989) Measuring distributions of diffusivity in turbulent fluids with magnetic resonance imaging. *Physical Review A (General Physics)*, **40** (8), 4542–51.

15 Kuethe, D.O. and Gao, J.-H. (1995) NMR signal loss from turbulence: models of time dependence compared with data. *Physical Review E (Statistical Physics, Plasmas, Fluids, and Related Interdisciplinary Topics)*, **51** (4), 3252–62.

16 Gatenby, J., MacCauley, T.R. and Gore, J.C. (1993) Mechanisms of signal loss in magnetic resonance imaging of stenoses. *Medical physics*, **20** (4), 1049–57.

**17** Gatenby, J.C. and Gore, J.C. (1994) Mapping of turbulent intensity by magnetic resonance imaging. *Journal of Magnetic Resonance, Series B*, **104** (2), 119–26.

**18** Newling, B., Poirier, C.C., Zhi, Y., Rioux, J.A., Coristine, A.J., Roach, D. and Balcom, B.J. (2004) Velocity imaging of highly turbulent gas flow. *Physical Review Letters*, **93** (15), 154503.

**19** Mastikhin, I.V., Balcom, B.J., Prado, P.J. and Kennedy, C.B. (1999) SPRITE MRI with prepared magnetization and centric k-space sampling. *Journal of Magnetic Resonance (1997)*, **136**, 168.

**20** Hurly, J.J., Defibaugh, D.R. and Moldover, M.R. (2000) Thermodynamic properties of sulfur hexafluoride. *International Journal of Thermophysics*, **21**, 739–65.

**21** Prado, P.J., Balcom, B.J., Mastikhin, I.V., Cross, A.R., Armstrong, R.L. and Logan, A. (1999) Magnetic resonance imaging of gases: A single-point ramped imaging with t1 enhancement (SPRITE) study. *Journal of Magnetic Resonance (1997)*, **137**, 324–32.

**22** Bogard, D.G. and Thole, K.A. (1998) *The Handbook of Fluid Dynamics*, CRC Press, London, New York, Chapter 13.4, p. 13.40.

# 16
# Transport Properties in Small-Scale Reaction Units

*Andrea Amar, Lisandro Buljubasich, Bernhard Blümich and Siegfried Stapf*

## 16.1
## Introduction

Over the past few years, magnetic resonance microscopy has spread out from its traditional roots of medical and animal imaging, via its first thriving branch of materials science, towards an ever-growing range of applications in chemical engineering and related subjects. To name just a few, the field covers topics as diverse as reaction monitoring, food quality control, biofilm assessment and fuel cells, with boundaries to other disciplines of science frequently and deliberately being crossed. The broadening of this range of applications is supported by new developments of hardware and sensitivity-enhancement schemes that allow transparency to be brought into what only a very short time ago were regarded as 'black boxes'. Chemical engineering relies heavily on these exciting new approaches to exploit the full potential of nuclear magnetic resonance (NMR) techniques in model reactors and, eventually, in full-scale reaction apparatus.

The state of the art, in 2005, in terms of methods, hardware and data analysis concepts has been summarized [1], and a number of recent applications presented in detail. During the relatively short time that has passed since the compilation of that volume, progress in all three aspects of the toolbox has led to significant improvements in application ranges, selectivity and specificity, and in experimental speed as well as temporal and spatial resolution. In this chapter, attention will be focused on the latter point–that is, how can measurements in small objects contribute to an understanding of reaction processes and reactor performance? Likewise, what is the role for dedicated studies of high spatial resolution in the context of a proper description of industry-scale reactors?

The motivation to apply magnetic resonance imaging (MRI) or magnetic resonance microscopy (a term which appears particularly appropriate in this context) to small-scale reaction units is (at least) twofold. First, the investigation of full-scale chemical reactors used in production may be impossible due to size or cost restrictions of the imaging hardware, or may be feasible but only allow for an insufficient spatial resolution of the system, typically being on the order of one-hundredth of

*Magnetic Resonance Microscopy.* Edited by Sarah L. Codd and Joseph D. Seymour
Copyright © 2009 WILEY-VCH Verlag GmbH & Co. KGaA, Weinheim
ISBN: 978-3-527-32008-0

the resonator dimensions. For instance, the spatial heterogeneity of a fixed bed, or of the single- and multiphase fluid flows through such a bed, can well be resolved with commercial imaging equipment, with bores comparable to whole-body human imagers that currently define the ultimate limit of the accessible reactor diameter. In order to follow reactions at the level of the actual reaction sites, however, studies of a single catalyst pellet at well-defined conditions can be performed with a much higher spatial resolution, allowing the verification and discrimination of coupled diffusion/reaction models.

The second reason is more hardware-related, and exploits the superior performance of gradient and radiofrequency (RF) detectors on small scales, leading to the improved spatial and temporal resolutions that are required to understand processes which are intrinsically fast or localized to the submillimeter scale, such as transport dominated by self-diffusion, or the growth and degeneration of biofilms. Recently, the rapid growth in importance of microfluidic devices for applications as sensors or at different stages of synthesis procedures has generated a third motivation for NMR microscopy in chemical engineering on the 'real-scale', and one which may well have substantial commercial potential. Whilst a clear definition of 'high spatial resolution' does not exist, it might be agreed that a level on the order of 100 μm would be considered a suitable 'ballpark figure', with record values of 10 μm or less having been reported for pure spin-density imaging.

In time, MRI can only become a valuable tool in chemical engineering if it can be used to visualize spatial distributions, velocities, material properties and the compositions of technically relevant systems better, faster, and/or more specifically than concurring techniques:

- The quest for extremely high spatial resolution has partially been conceded to other methods due to the sensitivity limit of classical NMR, although even this boundary might be conquered by hyperpolarization and atomic force NMR approaches.

- Velocity – or, more generally, transport – measurements, on the other hand, are less straightforward to achieve with classical methods, and often require doping with tracer substances that may act as contaminants, particularly in a reaction environment. The examples described in Ref. [1], of spatially highly resolved velocity measurements, included the investigation of flow fields in dedicated rheology devices [2] and viscometers [3], the visualization of Fano flow [4] and Taylor–Couette–Poiseuille flow [5], shear flow of emulsions [6] and velocity imaging of structured media such as biofilm reactors [7] or hemodialyzer membranes [8].

- Material properties are determined by correlation with typical NMR measurement parameters, such as relaxation times and diffusion coefficients, in combination with appropriately encoded imaging schemes. Earlier examples include one-dimensional profiling of human skin and paintings [9], maps of solid hydrocarbon distribution [10] or liquid content variation [11] obtained in single catalyst pellets, pore size distribution mapping via $^{19}F$ relaxation gas imaging [12] or moisture content measurements in food [13].

- Compositional information exploits the most essential attribute of NMR, namely its chemical selectivity. Due to more severe sensitivity restrictions, the spatial resolution of such techniques may currently still be inferior, but promising polarization transfer methods are already at the application stage. For instance, reactants in fixed beds were successfully quantified using their spectral signature [14], while the strong dependence of $^{129}$Xe chemical shift on its environment was used as an indirect marker to distinguish between intraporous and extraporous spaces [15].

The continuous development in each of these topics, and their clear relevance to chemical engineering, has been demonstrated in several chapters within this volume, and together these provide an almost complete overview of the state of the art of NMR microscopy in this subdiscipline.

In this chapter, the details of two recent approaches to small-scale velocity imaging and indirect reaction monitoring are presented. Both methods focus especially on the importance of understanding the complex transport properties required to achieve optimized reactor performance: the quantification of velocity fields and transport properties inside a liquid-in-liquid droplet, and outside a single pellet during an exothermic reaction process.

## 16.2
## Single Levitated Droplets: A Model for Liquid–Liquid Extraction Columns

### 16.2.1
### Introduction and Device Presentation

Liquid–liquid extraction processes are used widely in chemical engineering, and have their most important applications in cleaning procedures where contaminants in a bulk, valuable fluid component (donator phase) are being removed by bringing it into contact with a second, disperse phase (acceptor phase). Ideally, the donator and acceptor phases are immiscible, while the contaminant (transfer phase) is soluble in both fluids. In order to provide a maximum transfer within a given amount of time, a large concentration difference of the contaminant and a large interface area between the two main phases are desired. This is often realized by dispersing the acceptor phase into a swarm of droplets and allowing it to pass through the continuous phase, thus exploiting the density differences between phases.

It is a well-known fact that the efficiency of mass transfer between the two phases is determined by convective transport made possible through circulation which occurs both inside and outside the droplets. Mass transfer can, in fact, be substantially faster than would be expected from pure diffusive transport across the drop interface. Mass transfer rates are underestimated by orders of magnitude not only by the analytical solution of Kronig and Brink [16], but also by 2-D-axisymmetric computational fluid dynamics (CFD) simulations for nondeformable droplets with

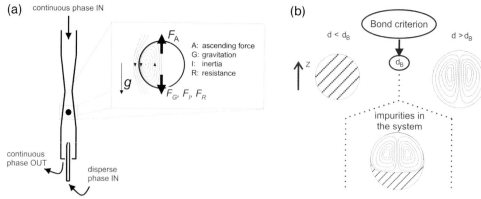

**Figure 16.1** (a) Scheme of the levitation cell. The conical measurement cell is filled with a continuous phase flowing from top to bottom, and only one drop is produced at a time. Under these conditions, and when the drop levitates (the sum of the forces vanishes), vortex patterns in the vertical plane are expected as shown here; (b) Sketch of the bond criterion. For a critical diameter $d_B$, that depends on the system chosen and the purity of the system, two regimes can be distinguished: (i) below the critical diameter rigid drops with no internal circulation are expected; (ii) above the critical diameter fully mobile surfaces and vortex patterns are to be found. When impurities are present in the system, mixed surfaces are predicted.

an ideally mobile interfacial region, which do not make use of approximated solutions of the Navier–Stokes equations [17]. The modeling of mass transfer depends, however, on a precise knowledge of the fluid dynamics inside the drop, which in turn can be understood (in theory) only by taking into account sufficiently detailed models of the boundary layer properties. The single-droplet behavior, which must be understood as a basis for extraction-column design (Figure 16.1a), is determined by mass transfer and sedimentation, which take place simultaneously and have direct influences on each other. Although in the past, several theoretical, numerical and experimental investigations on single droplets have been carried out, sedimentation velocities and mass transfer rates cannot be predicted à priori; experimental data can only be matched by additional empirical parameters. The only experimental evidence for fluid dynamics is usually delivered from integral measurements of mass transfer in an extraction column or in single-drop cells [18]. Although particle tracer methods have been used to visualize the flow pattern in drops directly, these are limited in their applicability with respect to dimensionality. In fact, they frequently monitor only motion in suitable sections within the drop due to the conditions required for appropriate lighting and optical detection. They also represent an invasive technique, which can compromise the validity of the results derived concerning the fluid flow field. For example, it is known that the fluid dynamics of a drop can be very sensitive to small concentrations of impurities in the system, that tend to accumulate at the interface.

In the present study, fast NMR imaging techniques are combined with velocity encoding in order to generate statistical and imaging information about the internal dynamics of single levitated drops inside a continuous liquid. These drops were kept in place by adjusting the countercurrent of the continuous phase in a suitably shaped device that is located inside the magnet bore, and circulation patterns as shown in Figure 16.1a are expected under these conditions. Drops of typically up to 4 mm in diameter held in this set-up had to be imaged with sufficient spatial resolution. The internal dynamics of the drops can generally be divided into different regimes. While small droplets sediment like rigid spheres, larger droplets feature pronounced internal dynamics [19], where the limit is set by the Bond Criterion (Figure 16.1b). One aim of the present studies was to discuss these limiting cases, which requires the determination of either very small or very large velocities in an otherwise identical geometry. The fact that the drops did not move as a whole allowed the application of multiple acquisition techniques, these being compromised only by the need to allow full relaxation of the spin system by introducing sufficient delays between signal encoding.

## 16.2.2
### Set-Up for the Levitated Droplet Experiment

The foremost requirement for measuring internal dynamics in levitated drops is the generation of a drop that is stable in position and shape for long periods of time, at least on scales of the desired resolution in the spatial and velocity dimensions. Furthermore, reproducibility of drop generation and position are needed for the measurements of series of drops.

Although the original motivation is based on understanding the behavior of drops in chemical extraction columns, a simplification of the apparatus was introduced to isolate the principal subject of the study–the drop. Instead of a swarm of droplets traveling through a large cell filled with a continuous liquid, only one drop at a time is generated per experiment, and this is kept in stable levitation in a cell with special geometry. Preceding studies, calibration tests and preparations are fully described in Ref. [20].

A suitable computer-controlled experimental set-up has been developed (see Figure 16.2) where a countercurrent of a surrounding continuous phase of $D_2O$ flows from the top to the bottom of the cell. The droplet is produced by a precision injector, where the volume and injection speed are chosen so as to avoid daughter-drop generation. The droplet then rises–due to the difference in density between the two phases–until vertical force equilibrium is reached in the conical part of the cell (Figure 16.1), which is positioned in the magnetic center of the apparatus. When the measurement has been made the countercurrent is switched off such that the droplet rises to the top, where it no longer affects the flow and can be collected for chemical analysis.

A set of conical glass cells was manufactured with different geometric parameters and degrees of skewness; some cells were purposely made with a bent, non-vertical flow path in order to demonstrate the influence of transverse flow components of the continuous phase on the drop behavior. One particular, 'perfectly' symmetric cell (within manufacturing errors; provided by Lehrstuhl für

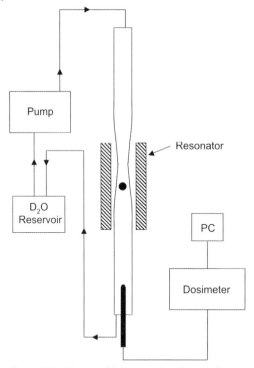

**Figure 16.2** Scheme of the experimental set-up for generating and maintaining levitated drops.

Thermische Verfahrenstechnik TVT, RWTH Aachen) allowed for optimal conditions for generating and visualizing symmetric flow conditions, and at the same
time could be fitted into the NMR microimaging device. The geometry follows the
results from a geometry optimization with the objective to maximize the stability
of droplet position [21].

The experiments were conducted as follows: first, all components of the system
in contact with either liquid were thoroughly cleaned to minimize detrimental
effects on the drop boundary mobility. This included the glass parts, Teflon pipes
and connectors, as well as the pump head itself. A schematic of the principal
components of the set-up is shown in Figure 16.2. Here, a pump maintains a
constant flow of the continuous phase, which was chosen to be $D_2O$ in order to
avoid background signal contributions as much as possible. The flow occurs from
top to bottom in the glass cell in order to compensate for the buoyancy of the drop,
which is lighter than $D_2O$. A computer-controlled dosimeter introduces one drop
of a predefined volume and an initial speed that guarantees separation from the
pipette tip. The drop rises to the point in the narrowing section of the glass cell
where all vertical forces add up to zero, and then comes to rest (this point coincides
with the center of the resonator). It takes several minutes for the drop to reach
this position with a constant pump flow rate; however, by adjusting the pump
setting manually (i.e. by allowing the drop to rise for a previously determined time

and then switching on the pump) this time can be reduced to about 20 s, making this the earliest possible starting point for the NMR acquisition. Time is a critical factor inasmuch as, for unclean systems, the accumulation of impurities – either dust particles or substances dissolved in the continuous phase – at the drop interface was found to affect its mobility within a time frame of minutes.

Before the NMR experiments were started, drops of different sizes were generated outside the magnet and checked visually. It transpired that small drops could generally be levitated with no optically detectable motion, while particularly large drops sometimes appeared to undergo a weakly damped or even undamped oscillation of their vertical position. The positional stability of the drop was verified with 1-D frequency-encoded imaging each time before performing the velocity-sensitive experiments.

The $D_2O$ (from Aldrich) had a degree of deuteration of 99.8%, while toluene (99.5% purity, density $0.867\,g\,cm^{-3}$; Aldrich) was chosen as the drop phase. Water, toluene and acetone as the transfer component is a typical test case for laboratory investigations of extraction and sedimentation processes. From this experience, it is known that an enhanced mass transfer can be observed for relatively large drops, indicating a mobile interface. To this end, toluene drops of approximately 4 mm diameter were used; larger drops were either found to be unstable in the cell nozzle or tended to touch the glass wall due to their pronounced oscillations. Before starting the experiments the $D_2O$ phase was presaturated with the disperse phase, using standard methods (by shaking a container filled with both liquids and waiting for complete phase separation).

The NMR experiments were conducted on a Bruker DSX 500 spectrometer with a 11.7 T field strength (500 MHz $^1H$ Larmor frequency). Standard Bruker microimaging hardware was used, with a birdcage resonator of 10 mm inner diameter. The microimaging units provided maximum gradient strengths of $1.0\,T\,m^{-1}$.

## 16.2.3
### Pulse Sequences

Measurements were performed using the following strategy, and with each step being carried out on different drops, starting at the earliest possible moment after drop generation. First, the positional and shape stability of the drops was verified by acquiring 1-D profiles in short succession (repetition time 200 ms) to cover the settling time and the first few minutes of the drop's life in particular. Profiles were acquired along the vertical ($z$) as well as at least one of the horizontal ($x,y$) axes. Second, 2-D and 3-D images were obtained to visualize the drop's overall shape. Third, propagator measurements were performed to monitor the velocity distribution inside the drop in the three orthogonal directions. Because the accumulation of impurities at the interface is expected to affect the maximum occurring velocities, the total acquisition time of each propagator was reduced to the smallest possible value, in this case 40 s (four scans). In the fourth and final step, velocity maps were obtained by supplementing fast-imaging sequences with a velocity-encoding module. The experiments were conducted either on individual slices inside the drop, or in an integral manner as a projection onto one plane which, despite averaging over one dimension, still contained sufficient information about the velocity field to allow a comparison with available models. Measurement of the velocity components encoded in orthogonal directions could then be combined to generate vector plots of the velocity fields.

Standard spin warp sequences were used for 2-D and 3-D imaging of the drops. Propagator measurements were performed with pulsed-gradient spin-echo (PGSE) sequences where the gradient duration and separation were set to $\delta = 1.0$ ms and $\Delta = 10$ ms, respectively.

The multiline spectrum of the toluene leads to a multiplicity of images in the read dimension. To avoid this, the isolation of one particular spectral line is required. Hence, a frequency-selective RF pulse was used, which excited only the methyl line and allowed the suppression of between 95% and 99% of the unwanted signal. Because of the higher velocities expected for the toluene drop, a single velocity encoding followed by multiple signal acquisitions was not feasible: the requirement that the initially encoded spins must not travel farther than a distance equivalent to one pixel width (typically 117 μm) was not fulfilled for a multiacquisition sequence. Therefore, a spin-echo imaging sequence was used, where each single acquisition was preceded by one velocity-encoding step, and the whole succession of $q$ and $k$ encoding was repeated in typically 128 steps, corresponding to the desired spatial resolution in the phase direction. The repetition time was chosen as 200 ms, with crusher gradients following each acquisition in order to destroy the residual transverse magnetization that persisted due to the long relaxation times of toluene (of several seconds). Because of the longer experimental duration, usually only two velocity-encoding steps were chosen, and the local velocity was determined from the difference of the phase values in each pixel according to $\phi = 2\pi(q_2 - q_1)\,v\Delta$. Due to the smaller total number of signal accumulations, the results were found to be less accurate than the propagator technique mentioned above. The total duration for encoding one velocity component with a spatial resolution of $128 \times 128$ points was 4 min (without slice selection).

### 16.2.4
### Results

When the drops had reached their equilibrium position they became stabilized, and from the 1-D projections, their time-invariant position and size could be deduced with an error of less than 1% – that is, a possible variation in position along any of the axes was less than the spatial resolution of the profiles of 20 μm [19]. The equilibrated toluene drop, however, possessed a nonspherical shape with an aspect ratio of 1.19, having an axial diameter of 3.6 mm and a sagittal diameter of 4.3 mm (34 μl volume). The results of the propagator measurements at different drop ages are presented in Figure 16.3. With an average velocity of the continuous phase near the drop position of approximately 160 mm s$^{-1}$, the highest velocities inside the drop were about 30 mm s$^{-1}$ in $z$ direction, where an asymmetry is clearly observed, compared to almost symmetric velocity distribution functions of $v_x$ and

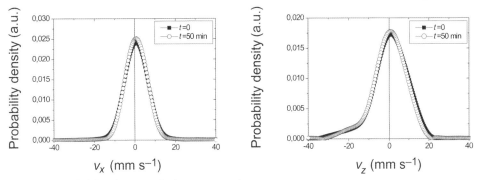

**Figure 16.3** Probability densities of velocities in x and z directions for two different drop ages of toluene drops. Motion along the vertical axis is denoted $v_z$.

$v_y$. Despite this asymmetry, the average velocity in all three directions, obtained from integrating over the functions in Figure 16.3, vanished within experimental error. Although the mobile interface of the toluene drop should be more vulnerable to accumulated impurities, a significant change in the shape and width of the propagators was not observed for experiments carried out over a total time of 8 h.

Moreover, no deviation could be determined between the velocity distributions obtained from several independent drops. Furthermore, repeated velocity images obtained for the same set-up revealed quantitatively similar (i.e. indistinguishable) velocity patterns. Such reproducibility provided a strong hint that it would be possible to generate drops with a predefined volume that possessed identical velocity patterns, so that the data acquired could be directly compared or even superposed.

Two-dimensional velocity images of the whole drop were measured, as slice-selective velocity images were much more time-consuming due to the need to use a conventional spin-echo technique. These represented integrations over the remaining coordinate axis, but were sufficient to reveal the internal vortex dynamics. The essential velocity patterns that are predicted for a drop with a mobile or partially mobile interface are summarized in Figure 16.4 [22, 23]. At the left-hand side, the different forces acting on the drop are shown. In the conical cell, the forces depend on the vertical (z) coordinate, leading to force balance at one particular position. In the drop interior, a vortex is induced which can be most simply approximated by a toroidal velocity field. The velocity distributions resulting from an integration along an axis perpendicular to the plane are drawn schematically in the remaining part of Figure 16.4, where light gray indicates positive-velocity and black indicates negative-velocity components. Flow is directed upwards in the center, and downwards around the perimeter. The fountain-like trajectories lead to a cross-shaped pattern when the transverse velocities (here, $v_x$) are considered in a projection on a vertical plane. In the lower half of Figure 16.4, the effect of a rigid cap in the drop is shown. Such a cap has been described previously as being

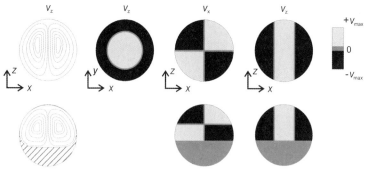

**Figure 16.4** Idealized vortex pattern in a levitated drop (top) and schematic distribution of velocities by sign (light gray = positive; dark gray = negative). In the bottom line, the equivalent velocity distribution is shown for a drop with a rigid cap.

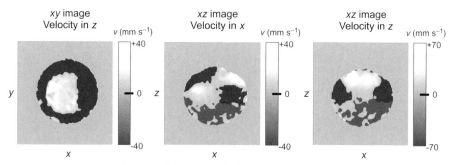

**Figure 16.5** Grayscale-coded plots of individual velocity components in projections onto different planes of the toluene drop; velocity components and plane orientations are given in the figure.

the result of an incomplete interface mobility as a consequence of impurities or agents affecting the interface tension [22, 23]. It appears at the downstream side of the drop, which coincides with the drop's lower half in this study as the flow direction of the continuous phase is from top to bottom. The rigid cap is, in principle, expected to grow with time if more impurities aggregate at the interface; therefore, a stable drop behavior is a clear indication of the low amount of impurities present in the system [24].

Figure 16.5 shows the velocity maps of the drop. Due to the absence of slice selection, the velocity value in each pixel represents an integral over the projections along the remaining dimension, which is normal to the plotted plane. The results agree qualitatively with the theoretical predictions [25], and considerably smaller velocities are visible in the bottom half of the vertical projections than in the top half (center and right). While the method applied is not expected to generate the

Velocity map in xz plane          Velocity map in xy plane

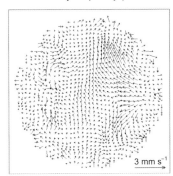

Figure 16.6 Vector plots of the internal velocities in a toluene
drop averaged over the dimension normal to the drawing
plane. Left, vertical projection; right, horizontal projection.

correct average velocity values integrated along the line of sight, the overall velocity
distributions as well as maximum values and averages were found to be in good
agreement with the propagators. In order to provide a better visualization of the
internal drop dynamics, vector plots have been reconstructed (see Figure 16.6).
The circulation (vortex) pattern is restricted to approximately the upper half of the
drop, while typically 10-fold smaller velocities occur in the bottom half. These do
not show any particular feature and appear to be random, although this might be
a consequence of the volume averaging which can mask an existing regular
pattern. The horizontal projection, on the other hand, reveals an indistinct velocity
distribution which is expected as all 'outward-flowing' and 'inward-flowing' fluid
elements would compensate when viewed from the top, with a small asymmetry
brought about by the rigid cap which deforms the toroidal velocity field. A vortex-
like pattern with small residual velocity components of below $3\,mm\,s^{-1}$ is indeed
found in the right-hand part of Figure 16.6.

To demonstrate the sensitivity of the technique, a final test was made by tilting
the whole cell deliberately. Assuming a rigid rod, the geometric constraints by the
inner magnet bore and the fixations allowed a tilt angle of less than $0.2°$, but a
possible bending of the glass tube could have led to a somewhat larger tilt angle
that could not be determined directly. Under these circumstances, the drop must
experience a superposition of internal vortex and overall rolling motions which
add up linearly in the measured velocity field. However, because of the presence
of the rigid cap that rests in place at the bottom of the drop, the internal dynamics
becomes more complicated than such a mere superposition. At the same time,
the aspherical drop is not found to tumble as a whole, which would be easily
observable by a variation of its axis length. Instead, its shape is preserved and
the asymmetric flow of the continuous phase leads to a more complex internal
circulation that can qualitatively be understood from the vector plots of Figure
16.7. While the vortex in the upper half remains the dominating feature, a circulat-
ing pattern becomes visible in the lower half (left-hand plot in Figure 16.7; the

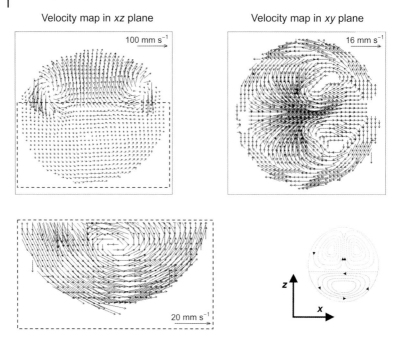

Velocity map in *xz* plane          Velocity map in *xy* plane

100 mm s$^{-1}$          16 mm s$^{-1}$

20 mm s$^{-1}$

**Figure 16.7** Vector plots of the internal velocities in a toluene drop in an asymmetric stream environment, averaged over the dimension normal to the drawing plane. Left, vertical projection; right, horizontal projection.

vector lengths increased fivefold for a better comparison in the insert at bottom). The projection onto the horizontal plane (right-hand plot) reveals a remarkable asymmetric vortex pattern containing relatively large velocity components. From the magnitude of the additional velocities observed in this drop, a rotation velocity of about 10 mm s$^{-1}$ can be estimated, corresponding to approximately one rotation per second.

In the range of stable drops within the bond criterion, several drops of different sizes, with volumes from 18 to 50 µl were measured, and both their shape and velocity distributions were studied. From Figure 16.8, the change in geometry can be directly appreciated when increasing the drop volume. As expected from theory [22], if the interfacial tension and/or viscous forces are much more important than the inertia forces, the bubbles and drops will have shapes that are close to spherical, whereas for dominating inertia forces, spheroidal or ellipsoidal shapes would be expected. It can also be observed that the rigid cap portion shrinks for larger drops. Although, quantitative computations have not yet been carried out from this set of experiments, a qualitative explanation for this behavior can be found in the surface of the rigid cap, which seems to be constant for all the drops and may be related to the constant concentration of impurities present in the system.

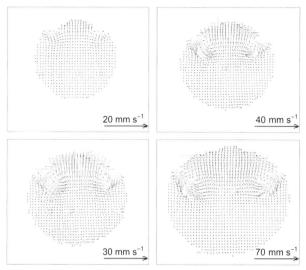

**Figure 16.8** Velocity plots for four different toluene drops, corresponding to drop volumes of 18, 27, 34 and 50 μl, respectively (label indicates vector length for given velocity). It can be appreciated how the velocities in the mobile part of the drop increase with drop volume.

Another interesting aspect highlighted by these measurements is how the velocities inside the mobile upper part of the drops are reduced with decreasing size. This is in accordance with theory which predicts that, for smaller drops, the mobility is expected to be reduced (larger rigid cap) as the Bond limit will eventually be reached and motion inside the drops is expected to be absent. On the other hand, the decrease in velocities might also be related to the presence of the rigid cap itself. It has been seen [24] that, for drops of identical size but with a growing rigid cap, the velocities in the confined mobile part of the drop were also decreasing. By combining the findings of both studies, it can be concluded that drop size and impurities both affect the magnitude of the velocities inside levitated drops and, moreover, they are actually part of the same phenomena that lead to the Bond criterion. Small drops are rigid due to the presence of surface-active substances that tend to accumulate at the drop surface. This means that the bond criterion provides a limit for which the impurities – which are always assumed present in the system – either become important to the analysis or can be neglected. In ideally pure systems, completely mobile drop surface and vortex patterns over the whole drop are expected (such as those sketched in Figure 16.4). However, when impurities are present the internal velocities of the drop will decrease. For small drops, below the bond criterion, the surface is completely covered by surface-active substances and no internal motion is expected. However, for larger drops (above the Bond criterion) two different regions can be distinguished: (i) a completely rigid region (the rigid cap), where most impurities tend to accumulate due to the continuous flow around the drop; and (ii) a partially mobile region. In the latter case, regular velocity patterns can still be seen, even though they are also affected by the impurities; hence, the magnitudes of velocities will also be influenced, by being decreased in line with the increasing amounts of surface-active substances in the system.

## 16.3
## Reaction in Single Catalytic Pellets: Model for Fixed-Bed Reactors

### 16.3.1
### Introduction

Heterogeneously catalyzed reactions mostly take place in the presence of finely dispersed catalysts (metals such as Ni, Pt, Pd, Cu, etc.); these in turn are localized in materials of large internal surfaces, that is, porous media. In the present studies pellets of $Al_2O_3$ that were several millimeters in size were employed. The reaction efficiency then depends on parameters such as internal surface area, the homogeneity of metal distribution, and the porosity and tortuosity of the pellet. In general, the pore space of catalyst pellets is described by a complex topology, having pores in the nanometer and micrometer ranges. It is also known that the presence of micrometer-scale pores has a strong influence on reaction efficiency [26, 27], since without these the reaction would predominantly take place at the outer edge of the pellet, and the core would remain mostly inactive. This, however, would require the use of smaller catalyst pellets for maximum efficiency, which in turn enhances pressure drop inside the fixed-bed reactor. A proper understanding of the processes governing mass transport to and from the pellet interior is therefore vital for an optimum design of the reactor.

In most reactions of technical interest, a gas is one of the involved components. The gas generated during the reaction – predominantly in the vicinity of the metal sites at the pore surface – is first dissolved within the surrounding liquid phase, until the maximum solubility is exceeded. The formation of a gas phase, however, depends on the interface tension and the size and tortuosity of the pore system; bubbles might therefore be generated inside large pores, or might only form at the external surface of the pellet. In general, each pore generates bubbles at a certain rate or frequency; large pores lead to large bubbles at a low frequency, and *vice versa*. For a constant reaction rate, these properties can be predicted for isolated pores [28]; in a real catalyst pellet, however, the coupling of all the pores within the interconnected pore space gives rise to a pattern of bubble generation that cannot be computed analytically [29].

Bubble formation greatly enhances not only gas transport but also fluid transport; this in turn affects the fluid both within the pellet and in the intrapellet space of a fixed-bed reactor. In consequence, material transport becomes much faster than by assuming purely diffusive processes. Hence, the reaction efficiency can increase dramatically.

In these investigations the reaction

$$H_2O_2 \text{ (liquid)} \rightarrow H_2O \text{ (liquid)} + \frac{1}{2}O_2 \text{ (gas)}$$

is studied in the presence of a cylindrical commercial porous catalyst particle of 4 mm diameter and 4 mm length, with $Al_2O_3$ as the carrier material (this carrier was doped either with Ni or Cu). Due to the fact that the reaction is strongly exothermic, and to permit a time-resolved study of the reaction strength over time, the experiments were performed in a mixture of $H_2O$ with a relatively low concentration of hydrogen peroxide.

The aim of the study was to identify the reaction speed and conversion rate, respectively, by utilizing properties that are easily accessible with NMR methods. Two such parameters are the oxygen concentration and the effective diffusion coefficient ($D_{eff}$) of the liquid in the vicinity of the pellet or in a defined closed volume. In the former case, the dependence of the liquid's relaxation time ($T_2$) on oxygen concentration allows the state of the system to be monitored during the reaction. In the second case, the production and motion of gas bubbles produces a random change in the velocities of the liquid around the pellet, which is displaced and driven by the rising bubbles. The faster the production of gas, the larger the amount of rising bubbles will be; in consequence, the increase in $D_{eff}$ will be more pronounced.

All of these experiments were performed using a single catalyst pellet immersed in an aqueous solution of initial 5% (v/v) $H_2O_2$ concentration, where a total of 1.1 ml was filled inside a 7 mm inner diameter glass tube and the pellet was fixed in the center position. The reaction was monitored for several hours, without any further supply of $H_2O_2$. To allow comparison between experiments, the particle size and fluid volume were kept constant in all cases. All experiments were performed at a Larmor frequency of 200 MHz using a Bruker DSX200 spectrometer equipped with a 4.7 T magnet.

## 16.3.2
### Monitoring the Reaction via Changes in Relaxation Time

Although the main goal of this initial study was to follow the reaction process in a semiquantitative manner, a series of preliminary experiments was carried out to provide an estimate of the dependence of $^1H$ relaxation times of the $H_2O/H_2O_2$ solutions as a function of $H_2O_2$ concentration. In the first experiment, bulk samples of different $H_2O_2$ concentration were prepared and the transverse $T_2$ was measured. As seen in Figure 16.9d, the relaxation time decreased significantly when the amount of $H_2O_2$ was increased. In order to consider all of the contributing effects for the dependence of relaxation times, the influence of the triplet state of molecular oxygen, dipolar and scalar coupling with the protons and intermolecular proton exchange between water and dissociating $H_2O_2$ molecules must be taken into account. The dependence of $T_1$ – being considerably less pronounced and less

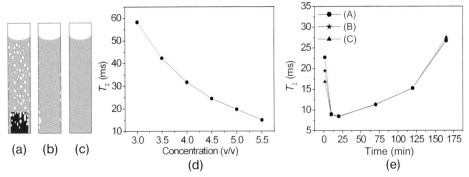

(a) (b) (c)

(d)

(e)

**Figure 16.9** General scheme and results for experiments performed to justify labeling concentrations via $T_2$. (a) $T_2$ of the liquid was measured in the presence of catalyst pellet and bubbles; (b) A second measurement of $T_2$ was performed with the bubbles adhered to the tube walls after removal of the pellet; (c) A third $T_2$ measurement was performed after removing the bubbles until reaching a completely homogeneous solution; (d) $T_2$ versus $H_2O_2$ concentration (v/v) in an aqueous solution; (e) $T_2$ versus time obtained from experiments performed as described in (a–c) (see text). Each set of three experiments was repeated by starting the reaction from the same initial conditions and waiting for different times.

suitable for reaction monitoring – was not investigated fully for the systems under study. Within the frame of this contribution, the measured values were taken as indicators for the conversion rate of $H_2O_2$ decomposition, and no attempt was made to derive absolute concentrations. A more extensive study has been conducted by L. Buljubasich *et al.* (unpublished results).

A second set of experiments was performed in order to verify the relative influence of the presence of bubbles and catalyst pellet on $T_2$. Figure 16.9a–c shows the set-up schematically: when the reaction had been started $T_2$ was measured at a fixed time with the situation shown in Figure 16.9a (i.e. during the catalytically activated $H_2O_2$ decomposition in the presence of the pellet within the measurement volume). Immediately following this first measurement (duration typically 30 s) the pellet was carefully taken out of the glass tube and $T_2$ re-measured, this time with the bubbles only adhering to the tube walls (Figure 16.9b). Finally, the tube was shaken until complete removal of the bubbles was achieved (Figure 16.9c). The procedure was repeated for different waiting times, on each occasion taking care to bring the sample to the same initial condition. The plot in Figure 16.9e shows $T_2$ versus waiting time for the three different set-ups. It is evident that the pellet itself and the presence of oxygen in a gaseous state do not affect $T_2$ significantly, with the exception of the shortest waiting time where transient phenomena might have occurred before equilibration of the reaction process over the pellet volume. It should be noted that, due to the metal content, the $^1H$ signal of fluid inside the pellet possesses a $T_2$ which is too short to be detected by the Carr–Purcell–Meiboom–Gill (CPMG) sequence used in this experiment. This set of experiments led to the conclusion that it is indeed possible to obtain a direct measure of reaction

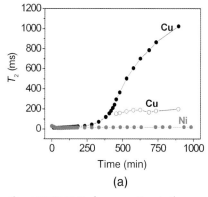 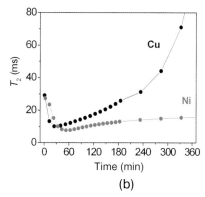

**Figure 16.10** $H_2O_2$ decomposition in the presence of two different catalyst pellets monitored via $T_2$. (a) $T_2$ versus time plot, measured every 12 min during 16 h. The decay curve for the Cu sample becomes nonexponential after about 6 h and is then fitted by a biexponential function; (b) The same plot for the first 6 h. The behavior for the first few minutes is explained in the text. Both experiments were started with an initial 5% $H_2O_2$ concentration.

progress independently of the components inside the measurement volume, by deriving information exclusively from the liquid state. A direct translation of $T_2$ values into $H_2O_2$ concentrations within a reacting system, however, requires additional independent chemical analysis due to the existence of further parameters affecting relaxation properties, such as the spatial distribution of pH values.

Nevertheless, the long-term behavior of system activity can be followed using relaxation parameters. Figure 16.10a shows the reaction monitored for both samples investigated in this study, by measuring $T_2$ every 12 min over a continuous period of 16 h. After about 6 h the signal decay became nonexponential and was tentatively fitted with a biexponential function. The same was observed in a $T_2$ measurement of pure water in the presence of a Cu-doped pellet, where the two components were found to be 1720 ms and 227 ms, respectively, with relative weights of 0.70 and 0.30. The occurrence of nonexponential relaxation decays can possibly be explained by an effect of $B_0$ inhomogeneity and partial diffusional damping due to the presence of the sample. The graph in Figure 16.10a shows that, towards the end of the observation period, both fitted values approach those determined independently for pure water. Therefore, the majority of $H_2O_2$ initially present in the sample must have decomposed after 16 h.

The situation is different in the pellet doped with Ni, which shows a much slower decomposition. In order to minimize the field-distortion effect of the ferromagnetic Ni, a relatively low concentration (5% Ni) had to be used, which is well below the amounts being conventionally employed in commercial catalysts.

The graph in Figure 16.10b shows a magnification of the same curves presented in Figure 16.10a for the first 6 h. Three observations were made: (i) the corresponding $T_2$ at the beginning of the reaction was somewhat larger than expected for the initial 5% v/v; (ii) the increase of the corresponding concentration in the first

30 min for Cu and 60 min for Ni; and (iii) the subsequent increase of $T_2$, which was much faster for the Cu-doped pellet than for the Ni pellet.

The first two findings can be explained by the presence of water inside the pellet at the start of the experiment. Initially, $H_2O_2$ is required to diffuse into the pellet from the surrounding solution, and this results in an activation period before maximum reaction efficiency in the pellet is reached. This time lag is almost absent for an initially dry pellet that is soaked with the $H_2O_2$ solution, a fact proven in corresponding experiments (data not shown). A possible interpretation of the time-dependence of the reactions (see Figure 16.10b) is the initial growth of oxygen concentration dissolved in liquid, leading to $T_2$ relaxation times slightly shorter than in bulk at equilibrium concentration. When the reaction rate has reached a sufficient level to produce bubbles, the extra oxygen is removed and $T_2$ begins to increase in agreement with the decrease in $H_2O_2$ concentration. Visual inspection of the reaction confirms that the bubbling activity builds up slowly during the activation period, and then decreases again over a period of many hours. Thus, the $T_2$ value during the first minutes must be understood as an effective number representing the combined effect of $H_2O_2$ and dissolved $O_2$ in the solution.

The third observation confirms the much lower activity of Ni at its given concentration, so that–following a similar saturation level of oxygen–the decomposition of $H_2O_2$ takes much longer and equilibrium is not reached within the continuously monitored time frame.

### 16.3.3
### Monitoring the Reaction via the Diffusion Coefficient ($D_{eff}$)

As mentioned in Section 16.3.1, $D_{eff}$ represents another useful parameter for monitoring reactions involving gas production. Starting with the same initial condition as described in Section 16.3.2, $D_{eff}$ was measured continuously every 2 min for a period of 6 h. Experiments were carried out on a Bruker MicroImaging system using a stimulated-echo sequence with gradient pulse lengths of 1 ms and separations of 100 ms. Measurements were performed alternately with gradients aligned parallel and perpendicular to the axis of gravity, the fitted results being called $D_z$ and $D_x$, respectively.

Figure 16.11a and b show $D_z$ and $D_x$ for the Cu sample, and Figure 16.11c and d for the Ni sample. In all plots, a line with the diffusion coefficient of the bulk sample with the aqueous hydrogen peroxide solution was included for comparison and to indicate the limiting value that is expected to be reached when the reaction has ceased.

The insert in Figure 16.11a further demonstrates the change of the initial decay for four different times, at 10 min, 25 min, 2.5 h and 5 h, respectively. Two principal aspects can be observed here:

- For both samples, there was a pronounced difference between diffusion coefficients in the $z$ and $x$ directions. In fact, for the Cu sample the difference was 15-fold (taking the maximum values), and for the Ni about 10-fold. In both

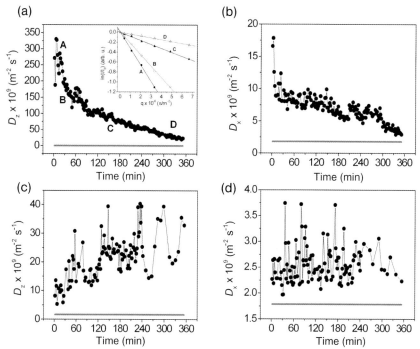

**Figure 16.11** Reaction monitored by means of the effective diffusion coefficient. (a,b) $D_{eff}$ in $z$ and $x$ direction respectively, for the Cu-doped pellet. The continuous line in both cases represents $D_{eff}$ measured with water and the pellet. The plot insert in (a) shows four individual measurements to demonstrate the quality of the experiments; (c,d) As above, but for the Ni-doped pellet. In all the cases one point was measured every 2 min for a period of 6 h.

cases, the value in $z$ direction was considerably larger. This may be easily understood from the predominantly vertical motion of the rising gas bubbles that clearly lead to a much more increased mean-squared displacement along $z$ during the evolution time of $\Delta = 100\,\text{ms}$.

- A striking difference between the two samples was also observable. Whilst in the case of the Cu pellet the maximum $D_{eff}$ in $z$ direction was about 150-fold larger than for bulk water, and changed significantly during the first 6 h, for the Ni sample $D_{eff}$ oscillated much more closely to the value found in the absence of a reaction. In the former case, the bubbles were generated and left the surface almost continuously; in the latter case, only a few bubbles appeared, grew slowly on the pellet surface and separated from it at lower frequency; this led to slow fluctuations of liquid content and motion, producing the fluctuations observed in Figure 16.11c and d.

These two characteristics together allowed it to be concluded that the rate of reaction was clearly much larger for the Cu sample, and this was in qualitative agreement with the conclusion obtained from the $T_2$ experiments. A more quantitative interpretation could be achieved by the independent determination of fluid mobility parameters in the presence of artificial bubble generation with well-defined size and frequencies.

### 16.3.4
### Monitoring the Reaction via 1-D Propagators

Following the finding of strongly different values of the $D_{eff}$ in perpendicular directions, the statistics of motion were monitored by determination of the propagators (displacement probability densities; see Refs [30, 31]) along the same axes. Using the same set-up as in the $T_2$- and $D_{eff}$-monitoring experiments, propagators were acquired every 3.5 min during 6 h in $z$ and $x$ directions, using a stimulated-echo sequence with $\delta = 1$ ms, $\Delta = 100$ ms, and 16 gradient steps per measurement. For each propagator, those velocities where the probability densities had decayed to an arbitrarily chosen value of 10% of the maximum were taken as characteristic velocities, as it was not possible to identify the maximum velocities with sufficient precision. In Figure 16.12, the velocity components in the $x$ and $z$ directions are plotted versus time for the Cu sample (Figure 16.12a) and the Ni sample (Figure 16.12b). It should be noted that in the Cu sample, $v_z$ shows a larger change than $v_x$ – approximately 4.5- and two-fold, respectively – with the latter beginning much closer to the velocity corresponding to the free diffusion case.

In the Ni sample, velocities in both directions also begin close to the diffusion case but do not change much in 6 h, in agreement with the $D_{eff}$ measurements. This observation corresponds to the different rate of bubble production between the two samples, which can be directly related with the reaction rate.

In Figure 16.12c and d, three propagators at different times are presented for $z$ and $x$ direction, respectively, to show the difference not only in velocities but also in distribution. From Figure 16.12c it becomes clear that at 5 min after the reaction has started the corresponding propagator is asymmetric, with higher velocities in the positive $z$ direction (up) due to the predominant bubble motion. At 2 and 6 h after the start of the reaction the propagators show a more symmetric shape and approach a Gaussian curve (which would be expected for an isotropic random motion such as free diffusion), with much lower velocities than at 5 min, but still considerably higher than in the case of free diffusion of water. On the other hand, in the $x$ direction (Figure 16.12d) even at the start of the reaction the propagators appear symmetric, with the highest velocities some sevenfold slower than in the former case.

### 16.3.5
### Conclusions

The decomposition of $H_2O_2$ in the presence of two different catalysts was followed, using NMR, in three different ways. The dependence of $T_2$ on $H_2O_2$ concentration allowed the decomposition to be monitored with relatively good time resolution. Whilst, initially, the majority of the $H_2O_2$ was consumed in the presence of the

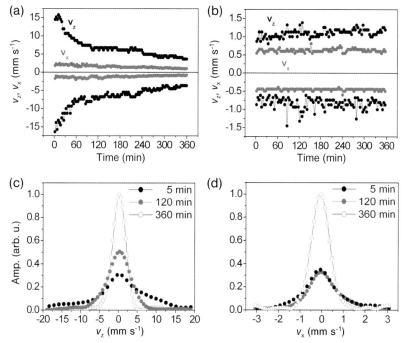

**Figure 16.12** Reaction monitored via 1-D propagators. (a,b) $v_z$ and $v_x$ versus time plotted for the Cu and Ni pellets, respectively (see text for definition of characteristic velocity). The measurements were performed every 3.5 min for a period of 6 h; (c) Propagators along $z$ and (d) along $x$ for the Cu-doped pellet at three different times after the start of the reaction.

Cu sample after 16 h, over the same time in the presence of Ni only a small fraction of the peroxide had reacted; hence, it was concluded that the rate of decomposition in the latter case is much slower than in the former. It is possible, however, to arrive at the same conclusion via the evolution of $D_{eff}$ and by means of 1-D propagators. A higher rate of decomposition – and consequently a greater production of bubbles – can be observed either by comparing the $D_{eff}$ versus time curves, or the maximum velocity plots, for both samples. In all cases higher values were observed in the vertical direction, due to the rising bubbles. It is likely that the general shape of propagators will eventually allow the proper modeling of fluid transport process in the exterior of a single pellet.

## 16.4
## Summary and Outlook

High-resolution NMR imaging has proven to be a powerful tool for the detailed *in situ* investigation of transport and reaction processes, addressing problems where either detailed spatial velocity information of semi-stationary systems is required for developing and discriminating suitable models of material transport, or where transient phenomena need to be monitored with an acceptable compromise between temporal and spatial resolutions, making use of parameters readily available by pulsed-field gradient NMR. Here, two examples have been presented, indicating the possible directions that MRI might follow in chemical engineering in the near future. In going beyond the fundamental understanding of reaction/ transport coupling, both of the suggested geometries – the levitated drop and the single catalyst pellet – show great potential as test-beds for a range of model reactions, making use of well-defined boundary conditions and optimized microimaging hardware.

## Acknowledgments

The authors are grateful to T. Oehmichen and L. Datsevich for sample preparation and fruitful discussions within the reaction part of this project. L.B. is indebted to F. Casanova and J. Perlo for their continuing support and discussion. Financial support from the Deutsche Forschungsgemeinschaft DFG (SFB 540, Sta 511/6-1) is gratefully acknowledged.

## References

1 Stapf, S. and Han, S. (eds) (2006) *NMR Imaging in Chemical Engineering*, Wiley-VCH Verlag GmbH, Weinheim.

2 Callaghan, P.T. (2006) Hardware and method development for NMR rheology, in *NMR Imaging in Chemical Engineering* (eds S. Stapf and S. Han), Wiley-VCH Verlag GmbH, Weinheim, pp. 183–205.

3 Powell, R.L. (2006) MRI viscometer, in *NMR Imaging in Chemical Engineering* (eds S. Stapf and S. Han), Wiley-VCH Verlag GmbH, Weinheim, pp. 383–404.

4 Xia, Y. and Callaghan, P.T. (2006) Imaging complex fluids in complex geometries, in *NMR Imaging in Chemical Engineering* (eds S. Stapf and S. Han), Wiley-VCH Verlag GmbH, Weinheim, pp. 404–16.

5 Georgiadis, J.G., Raguin, L.G. and Moser, K.W. (2006) Quantitative visualization of Taylor-Couette-Poiseuille flows with MRI, in *NMR Imaging in Chemical Engineering* (eds S. Stapf and S. Han), Wiley-VCH Verlag GmbH, Weinheim, pp. 416–32.

6 Shapley, N.A. and M.A. (2006) Two phase flow of emulsions, in *NMR Imaging in Chemical Engineering* (eds S. Stapf and S. Han), Wiley-VCH Verlag GmbH, Weinheim, pp. 433–56.

7 Codd, S.L., Seymour, J.D., Gjersing, E.L., Gage, J.P. and Brown, J.R. (2006) Magnetic resonance microscopy of biofilm and bioreactor transport, in *NMR Imaging in Chemical Engineering* (eds S. Stapf and S. Han), Wiley-VCH Verlag GmbH, Weinheim, pp. 509–33.

8 Han, S. and Stapf, S. (2006) Fluid flow and trans-membrane exchange in a hemodialyzer module, in *NMR Imaging in Chemical Engineering* (eds S. Stapf and S. Han), Wiley-VCH Verlag GmbH, Weinheim, pp. 457–70.

9 Casanova, J., Perlo, J. and Blümich, B. (2006) Depth profiling by single-sided NMR, in *NMR Imaging in Chemical Engineering* (eds S. Stapf and S. Han), Wiley-VCH Verlag GmbH, Weinheim, pp. 107–23.

10 Ren, X., Stapf, S. and Blümich, B. (2006) Multiscale approach to catalyst design, in

*NMR Imaging in Chemical Engineering* (eds S. Stapf and S. Han), Wiley-VCH Verlag GmbH, Weinheim, pp. 263–84.

11 Koptyug, I.V. and Lysova, A.A. (2006) In situ monitoring of multiphase catalytic reactions at elevated temperatures by MRI, in *NMR Imaging in Chemical Engineering* (eds S. Stapf and S. Han), Wiley-VCH Verlag GmbH, Weinheim, pp. 570–89.

12 Beyea, S.D., Kuethe, D.O., McDowell, A., Caprihan, A. and Glass, S.J. (2006) NMR imaging of functionalized ceramics, in *NMR Imaging in Chemical Engineering* (eds S. Stapf and S. Han), Wiley-VCH Verlag GmbH, Weinheim, pp. 304–21.

13 McCarthy, M.J., Gambhir, P.N. and Goloshevsky, A.G. (2006) NMR for food quality control, in *NMR Imaging in Chemical Engineering* (eds S. Stapf and S. Han), Wiley-VCH Verlag GmbH, Weinheim, pp. 471–90.

14 Gladden, L.F., Akpa, B.S., Mantle, M.D. and Sederman, A.J. (2006) In situ reaction imaging in fixed-bed reactors using MRI, in *NMR Imaging in Chemical Engineering* (eds S. Stapf and S. Han), Wiley-VCH Verlag GmbH, Weinheim, pp. 590–608.

15 Pavlovskaya, G.E. and Meersmann, T. (2006) Hyperpolarized $^{129}$Xe spectroscopy, MRI and dynamic NMR microscopy for the in situ monitoring of gas dynamics in opaque media including combustion processes, in *NMR Imaging in Chemical Engineering* (eds S. Stapf and S. Han), Wiley-VCH Verlag GmbH, Weinheim, pp. 551–70.

16 Kronig, R. and Brink, J.C. (1950) On the theory of extraction from falling droplets. *Applied Science Research A*, **2**, 142.

17 Waheed, M.A., Henschke, M. and Pfennig, A. (2004) Simulating sedimentation of liquid drops. *International Journal for Numerical Methods in Engineering*, **59**, 1821.

18 Henschke, M. and Pfennig, A. (1999) Mass transfer enhancement in single drop extraction experiments. *AIChE Journal*, **45**, 2079.

19 Amar, A., Groß-Hardt, E., Khrapichev, A., Stapf, S., Pfennig, A. and Blümich, B.

(2005) Visualizing flow vortices inside a single levitated drop. *Journal of Magnetic Resonance*, **1**, 177.

20 Amar, A. (2007) Fluid Dynamics of Single Levitated Drops by Fast NMR Techniques, PhD thesis, RWTH Aachen.

21 Bischof, C.H., Bücker, H.M., Rasch, A. and Slusanschi, E. (2003) Sensitivities for a single drop simulation. *Lecture Notes in Computer Science*, **2658**, 888–96.

22 Davis, J.T. (1972) *Turbulence Phenomena*, Academic Press, New York/London.

23 Clift, R., Grace, J.R. and Weber, M.E. (1978) *Bubbles, Drops and Particles*, Academic Press, New York.

24 Amar, A., Stapf, S. and Blümich, B. (2005) Internal fluid dynamics in levitated drops by fast magnetic resonance velocimetry. *Physical Review E*, **72**, 030201.

25 Gross-Hardt, E., Amar, A., Stapf, S., Pfennig, A. and Blümich, B. (2006) Flow dynamics inside a single levitated droplet. *Industrial and Engineering Chemistry Research*, **45**(1), 416–23.

26 Datsevich, L.B. (2003) Oscillations in pores of a catalyst particle in exothermic liquid (liquid-gas) reactions. Analysis of heat processes and their influence on chemical conversion, mass and heat transfer. *Applied Catalysis A: General*, **250**, 125–41.

27 Datsevich, L.B. (2003) Alternating motion of liquid in catalyst pores in a liquid/liquid–gas reaction with heat or gas production. *Catalysis Today*, **79-80**, 341–8.

28 Datsevich, L.B. (2003) Some theoretical aspects of catalyst particle at liquid (liquid-gas) reactions with gas production: oscillation motion in the catalyst pores. *Applied Catalysis A: General*, **247**, 101–11.

29 Datsevich, L.B. (2005) Oscillation theory Part 4. Some dynamics peculiarities of motion in catalyst pores. *Applied Catalysis A: General*, **294**, 22–33.

30 Kärger, J., Pfeifer, H. and Heink, W. (1988) Principles and applications of self-diffusion measurements by nuclear magnetic resonance. *Advances in Magnetic Resonance*, **12**, 2–89.

31 Callaghan, P.T. (1991) *Principles of NMR Microscopy*, Clarendon Press, Oxford.

# 17
# Solid Liquid Particulate Suspensions – Gravitational and Viscous Transport

*Sarah L. Codd, Stephen A. Altobelli, Joseph D. Seymour, Jennifer R. Brown and Einar O. Fridjonsson*

## 17.1
## Introduction

The flow of solid particulates suspended in fluids occurs throughout the natural world. The interplay of gravitational, viscous and inertial forces determines the transport behavior of the particulate and fluid phases. The modeling of biological functions such as suspension filtration feeding [1], and industrial applications such as rotary filtration [2], require an understanding of the detailed dynamics of the solid and fluid phases. Recently, significant interest has been garnered by granular flows in which the fluid phase is a gas. This is not only due to their prevalence in natural and technological systems but also because these systems serve as a statistical mechanical 'sandbox' for experiments on the complex collective dynamics of many body systems [3, 4]. In these granular systems particle interactions are typically dominated by particle–particle collisions of an elastic nature. As the viscosity of the suspending fluid increases, as with liquid suspending phases there is a transition to a regime in which the fluid viscous forces control particle interactions that are mediated through the fluid. In these regimes lubrication forces [5] and longer-range many-body hydrodynamic interactions [6] control the transport dynamics. There is a long history of the study of such systems for particle sizes ranging from colloidal (where particle sizes of the order of $\leq 10\,\mu m$ allow Brownian motion-induced collisions between the suspending fluid molecules and particulates to be of sufficient amplitude to impact particle dynamics) to larger noncolloidal particles for which these Brownian effects are negligible [7]. A common feature of many solid–fluid suspensions is their opacity, which limits the ability of light-based measurement techniques to study transport processes. This, along with the ability to measure scale-dependent dynamics, makes magnetic resonance (MR) methods ideal for noninvasively spatially resolving transport in suspensions [8, 9].

This chapter briefly summarizes the theoretical treatment of solid–fluid suspensions and indicates the regimes in which viscous forces are important in modeling transport, in order to provide context to the MR data. MR measurements of

*Magnetic Resonance Microscopy.* Edited by Sarah L. Codd and Joseph D. Seymour
Copyright © 2009 WILEY-VCH Verlag GmbH & Co. KGaA, Weinheim
ISBN: 978-3-527-32008-0

transport in solid–liquid suspensions with varying gravitational, viscous and inertial forces are presented. The focus is on elucidating the methods for resolving the
particulate and suspending fluid phases. Noncolloidal particle suspension flow is
analyzed in systems where viscous forces interplay strongly with gravity–that is,
viscous resuspension [10] in a Couette flow cell and the horizontal rotating drum
flow. MR measurement of the flow of a model hard sphere colloidal suspension
under pressure-driven flow in a capillary and a bifurcation are also studied [11].

## 17.2
## Overview of Suspension Transport Theory

Solids suspended in fluids have long been of interest, and results such as the
modification of suspension viscosity relative to the suspending fluid due to the
presence of a dilute solid phase (*e.g.* the Einstein relationship, $\mu_r = \mu_o(1 + \frac{5}{2}\phi)$) are
well known [12]. The particle volume fraction ($\phi$) dependence of material properties is a common feature of suspension models for material properties and the
transport of momentum [13]. Much of the theory deals with the integration of a
detailed description of the dynamics of particles at a microscale to predict macroscopic rheological behavior [7]. This approach involves the concept of microstructure, or the relative spatial distribution of the particle phase.

### 17.2.1
### Noncolloidal Particle Viscous Resuspension

The force balance on a noncolloidal particle in a Couette rotational flow device
indicates that the normal forces in the vertical direction are dependent upon the
particle concentration, which varies as a function of vertical position [10]. In
Couette flow with the cylinder axis oriented with gravity, the unidirectional shear
flow generates particle migration due to forces in the radial shear and axial gravitational directions. Viscous resuspension in Couette flow has received significant
theoretical and experimental treatment [10, 14]. Three important dimensionless
numbers in the characterization of noncolloidal suspension flows are the Reynolds
number (*Re*), the Stokes number, and the Shields parameter of sediment transport
theory. The particle Reynolds number (*Re*$_p$) provides the ratio of inertial to viscous
forces in suspension flows $Re_p = \rho Ua/\mu_o$ , where $\rho$ is the fluid density, $\mu_o$ the fluid
viscosity, *U* the particle velocity and *a* the particle radius. In shear flows the particle
velocity is often replaced with $U = a\dot{\gamma}$ , where $\dot{\gamma}$ is the shear rate. A low *Re* indicates that inertia in the suspending fluid is small relative to viscous forces. The
Stokes number (*St*) specifies the relative importance of particle inertia to viscous
hydrodynamic force on the particle, $St = mU/6\pi\mu_o a^2$, where *m* is the particle mass.
At low *St*, lubrication forces will dominate interparticle interactions, rather than
solid body collisions, and the particle dynamics will depend on hydrodynamic
forces [15]. For viscous resuspension phenomena in the Couette and rotating
drum flows, the balance of viscous and buoyancy forces is important. The ratio is

referred to as the Shields parameter, $\psi = \tau/\Delta\rho g a = \mu_o\dot{\gamma}/\Delta\rho g a$ , where $\tau$ is the shear stress, $\Delta\rho$ the density difference between the fluid and particle, and $g$ the gravitational acceleration [10]. The viscous resuspension phenomena is theoretically characterized for Couette flow by a flux balance between the gravitational particle flux in the vertical direction and shear-induced particle diffusion in the direction of shear [10, 14]. The theory predicts that the height $h$ of the particle resuspension in the Couette device relative to the initial particle sediment height $h_o$ scales as $h \propto A^{1/3}$ where $A = (9/2)\mu_o\dot{\gamma}/\Delta\rho g h_o$ and is independent of the particle radius [14]. Such detailed analysis for the rotating drum flow has not been developed due to the more complicated force balances.

## 17.2.2
### Colloidal Particle Flow

Dynamics in colloidal particle flows share some physical features with noncolloidal suspensions. Transport is characterized by the particle Reynolds and Stokes numbers, as discussed above. However, the presence of Brownian motion requires an additional dimensionless group, the Peclet number, $Pe = a^2\dot{\gamma}/D_{SES}$ , where $D_{SES}$ is the Stokes–Einstein–Sutherland diffusivity given by $D_{SES} = kT/6\pi\mu_o a$ [12, 16], to characterize the relative amplitude of hydrodynamic to Brownian forces. The dynamics of colloidal suspensions are well studied [7] due to their prevalence in natural systems. In biological transport the mixing and distribution of particles in capillaries and bifurcations is important for biological function and in separations processes. Of particular recent interest has been the presence of shear-induced particle migration in colloidal systems, an effect thought to be more dominant in noncolloidal systems [11, 17]. Migration of the particles is due to the presence of normal forces in the particle phase [17] or to osmotic pressure effects [18] and is an area of current research. The particle dynamics and microstructure impact the spatial distribution of particles and hence their partitioning in bifurcating flow systems.

## 17.3
### Distinguishing Particles and Suspending Fluid Using Magnetic Resonance Techniques

In suspension flow, magnetic resonance microscopy (MRM) can be used to provide information about the spatial location of either the fluid or the particle phase, or both. A variety of contrast mechanisms can be selected from, depending on the system and the desired measurements. MRM transport-encoding techniques, velocity maps and propagators can provide information regarding the suspending fluid dynamics, and if the particulates are MR active then the same information can be gleaned for the particles themselves [9]. In this chapter we review several MRM methods for distinguishing particle dynamics and distribution from that of the surrounding fluid.

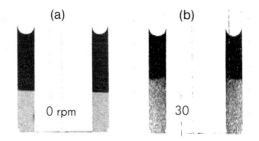

(a)  (b)

0 rpm  30

**Figure 17.1** $T_1$-weighted images for a noncolloidal suspension consisting of 100–300 µm-diameter PMMA particles in 200 cS silicone oil in a Couette (45 mm o.d. rotating, 29 mm i.d. stationary). (a) The PMMA particles are initially sedimented at the bottom; (b) The outer cylinder is rotated at 30 rpm and the particles undergo viscous resuspension. The viscous shear forces become greater as the rotation rate increases, suspending the particle phase.

### 17.3.1
### Signal Contrast Due to Volume Exclusion of Solid Particles

The simplest case is non-MR-active solid particles in a suspending liquid, where the spatial particle locations can be determined by comparing the percentage of each image voxel that is accessible to the mixture with a measurement of the total signal in the voxel. If the signal is less than that expected from 100% occupation by suspending fluid, then the remainder of the volume can be assumed to be particles. Figure 17.1a shows a mixture of 100–300 µm poly(methylmethacrylate) (PMMA) particles in 200 cS silicon oil where the mixture is in a Couette [45 mm outer diameter (o.d.) rotating, 29 mm internal diameter (i.d.)]. The PMMA particles are initially sedimented at the bottom of the cell, their location being obvious due to the lower signal in the bottom half of the Couette. The same image taken whilst the outer cylinder is rotating at 30 rpm shows clearly the particles undergoing viscous resuspension; again, their spatial distribution is evident from the image contrast.

### 17.3.2
### MR-Active Particles

Particles with a solid shell and liquid center [19] have been used to directly determine the spatial distribution of particles with MRM and, more importantly, to determine their transport characteristics distinctly from the surrounding suspending liquid. Coreshell oil-filled colloidal particles suspended in water allow measurement of the particle and liquid phase dynamics using nuclear magnetic resonance (NMR) spectral resolution. The uncharged polymer shell confers hard-sphere behavior to the particles [20]. Pulsed-field gradient spin-echo (PGSE) NMR has been previously applied to measure the oil diffusion within the particle and the particle Brownian diffusivity for 1.51 µm radius spheres as a function of particle concentration [32], to measure the nanometer scale velocity of the particle phase in Couette flow of concentrated ($\phi = 0.46$) 185 nm radius spheres [31], and to demonstrate irreversibility and particle migration for shear flow of dilute ($\phi < 0.10$) $a = 1.22 \pm 0.17\,\mu$m radius spheres in a $R = 500\,\mu$m radius capillary [11].

$T_1$ inversion selection has been applied to noncolloidal oil filled pharmaceutical pills with a hard polymer shell in the size range 1–5 mm. These particles have been used to examine viscous resuspension (Section 17.4), slurry flow (Section 17.5) and granular flows [19, 21, 22]. Semi-solid imaging techniques such as SPRITE [23] can be used to obtain the changing spatial distribution of polyethylene beads [24]. Although the $T_2$ relaxation time of the polyethylene is not long enough to allow velocity-encoding PGSE experiments, the rapid time resolution of 1-D image profiles can provide transport process information for the dynamic sedimentation or flocculation process of these particles.

## 17.3.3
### Separate Measurement of Phases

Although, as mentioned above, it is possible to create MR-active particles, it remains necessary to separate the signal from that of the surrounding fluid. If both the fluid and the particles have strong proton signals, they may appear indistinguishable to MRM. Only composite or averaged measurements will then be obtainable, unless a specific mode of differentiation is employed. Available modes include relaxation weighting ($T_2$ and $T_1$), and chemical shift.

### 17.3.3.1   Oil and Water
The specific local electronic environment within a molecule can interact with the proton spin, altering the local magnetic field and hence impacting a shift in the resonant Larmor frequency. These chemical shifts in the resonant frequency are typically of the order of a few ppm for protons. For MRM applications, a complex fluid or complex geometry can result in linewidths for each individual spectral line that make many materials indistinguishable in the MR spectrum. However, methylene and water protons are separated by a significant chemical shift of 3.5 ppm. This difference is large enough to allow selective excitation of either phase in an oil/water sample, or separate analysis of each phase via resolution of the spectral dimension, even when rather broad linewidths exist [11].

### 17.3.3.2   Contrast Agents
Gadolinium compounds are commonly used in medical scanning to reduce $T_1$ relaxation, allow rapid imaging, and highlight regions accessible by the agent. These chelated molecules can be used creatively in MRM applications to distinguish the surrounding fluid from particles in particulate flows. If the contrast agent is soluble in the suspending fluid (as are standard imaging solutions in water) but are impermeable to the particles, then the doping agent can be used to completely eliminate the signal in the suspending fluid. Any proton signal that is measured can then be assigned to the particle phase. Oil-soluble gadolinium

agents could be used to preferentially eliminate the signal from the oil contained within the discrete particle phase.

### 17.3.3.3 Deuteration

The deuteration of one phase (fluid or particle) can eliminate the proton NMR signal from that phase, but allow analysis of the deuterated phase by separate excitation at the deuteron frequency. This typically requires not only a separate experimental run but also the use of a separate or dual deuteron/proton radiofrequency (RF) coil to allow the measurement of both frequencies. The ability to resolve the deuteron dynamics is limited by the lower gyromagnetic ratio.

### 17.3.3.4 Fluorinated Signal

A $^{19}$F liquid such as polytetrafluoroethylene (PTFE) oil can be used for one of the phases. It is possible to use a dual-frequency coil, or alternatively the proton coil can be retuned to the fluorine signal. A beautiful example of the combination of solid glass (no signal) particles, solid low-density polyethylene (LDPE) (proton signal) beads and PTFE oil fluid (fluorine signal) in a three-phase sedimentation/flotation system was demonstrated by Beyea *et al.* [24] (see Figure 17.2).

### 17.3.3.5 $T_1$ Inversion

Inversion recovery is a method used to suppress signal from one specific $T_1$ relaxation component. A 180° RF pulse (the inversion pulse) is applied, disturbing the spins from equilibrium. After a set length of time, the spins are flipped into the transverse plane via a 90° RF pulse and the desired MRM sequence is run. The magnetization amplitude following a 180° RF pulse evolves in time according to the exponential growth curve:

$$M(t) = M_o[1 - 2\exp(-t/T_1)] \qquad (17.1)$$

For a spin population with a single specific $T_1$ relaxation time, the longitudinal magnetization is identically zero at a time termed the inversion time ($t_{inv} = 0.6931 T_1$).

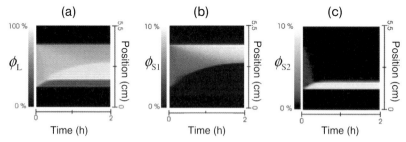

**Figure 17.2** One-dimensional images of volume fraction versus time for a three-phase sedimentation/flotation system shown for (a) PTFE liquid fraction $\phi_L$ and (b) the LDPE solid fraction $\phi_{S1}$. Volume fractions of (c) the negatively buoyant glass phase $\phi_{S2}$ were obtained by subtraction. Sixty profiles of each phase were acquired simultaneously over a 2 h period. Images are 64 points with a field-of view of 5.5 cm. Figure reproduced with permission from Beyea *et al.* [24].

If a 90° pulse is applied at that exact time following the 180° inversion pulse, then the resultant observable signal from that population will be zero. Thus, the difference in relaxation properties between different materials can be exploited to achieve signal suppression by applying a 180° pulse and then waiting a time equal to $0.6931T_1$ to minimize the undesired signal.

### 17.3.3.6 Potential Artifact Using $T_1$ Inversion Recovery in Flowing Systems

When a $T_1$ inversion recovery pulse is used to null any undesired signal in the presence of flow, an artifact due to inflow can arise. To demonstrate the signature of this artifact, propagators for water flow in a capillary are shown for four different $t_{inv}$ delay times (see Figure 17.3). The propagators collected after inversion delay times $t_1$ and $t_4$ are the expected hat function convolved with a Gaussian, since at these times there has either been no significant inflow or $T_1$ recovery ($t_1$) or the inversion time ($t_4$) is longer than $T_1$, and hence there is no memory in the spins of the inversion pulse. However, at $t_2$ the propagator is skewed towards higher velocities and at $t_3$ the propagator is skewed towards slower velocities. The spins which have been inverted have an 180° phase difference from those which have recovered, and their response to subsequent RF pulses will be 180° out of phase. For the intermediate delay times ($t_2$ and $t_3$) the signal is a combination of inverted spins, recovered spins, and inflow spins. Before the minimum $t_2 = 400\,ms$, the 'in-coil' spins are partially recovered and have a reduced signal amplitude. Spins which flow into the active region of the coil during the delay time have not been inverted and are velocity weighted, mostly higher-velocity, spins. The phase difference between the inverted and noninverted spins creates a cancellation of signal at high velocities, skewing the propagator towards lower velocities. At $t_3 = 575\,ms$, immediately following the relaxation minimum, the inverted spins have recovered

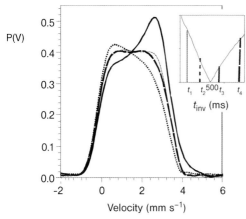

**Figure 17.3** Propagators for water flowing in a 1 mm capillary with an average velocity $V_{ave} \sim 2\,mm\,s^{-1}$, $\Delta = 30\,ms$, $TR = 4\,s$, $\delta = 2\,ms$ and $g_{max} = 1.738\,T\,m^{-1}$, corresponding to the different inversion delay times shown in the insert: $t_1 = 100\,ms$ (thin line) $t_2 = 400\,ms$ (short dashed line) $t_3 = 575\,ms$ (thick line) and $t_4 = 2000\,ms$ (long dashed line). The propagators demonstrate the impact of the inflow artifact as related to inversion delay time.

but have not regained their entire signal amplitude. In addition, the influx of refreshed spins is greater, resulting in a combined effect on the propagator of a weighting towards higher velocities which is greater than the weighting towards lower velocities that occurs at $t_3$.

Phase cycling cannot differentiate between inverted and noninverted spins which are of a single quantum coherence state [25]. This effect therefore cannot be eliminated by additional phase cycling. These impacts are significant if detailed specifics of the transport, such as the full propagator, are desired. The inversion recovery method is therefore suitable for stationary systems without significant inflow, or for velocity maps where only a mean velocity is being measured in each pixel over an appropriate short velocity observation time.

## 17.4
## Viscous Resuspension (Contrast Mechanism: $T_1$ Inversion)

### 17.4.1
### The System

A wide-gap Couette, 19 mm i.d. stationary cylinder and 45 mm outer diameter rotating cylinder was used (the rotating outer cylinder reduces Taylor-vortex instability). The Couette was filled with a 20 mm-deep bed of 1 mm oil-filled, pharmaceutical pills (Taiho Pharmaceuticals, Kyoto, Japan) that were surrounded by a 100 000 cS polymethylsiloxane (silicone) oil, such that the total fluid depth was 50 mm. The most useful silicone oils for these experiments are DOW 200 fluids (Esco Products, Houston, TX). The pills were thin spherical shells filled with vegetable oil; they were manufactured for a granular flow study [19] and ranged in size from 1 to 5 mm diameter. Although water soluble, the pill shells are known to be stable for years in silicone oil. The density of the pills is 1.045 g cm$^{-3}$; hence, they sediment quickly (on the timescale of minutes) when the system is stationary as the density of the oil is $\rho_{oil} = 0.974$ g cm$^{-3}$.

The experiments were performed in the 30 cm horizontal bore 1.89 T superconducting magnet at New Mexico Resonance, interfaced to a TECMAG imager/spectrometer and using a shielded gradient set (Magnex, Inc., Cambridge, MA) capable of producing 40 mT m$^{-1}$.

### 17.4.2
### Particle Visualization

MR images were recorded separately for both the silicone and the particle oil while the Couette cell was stationary, and then while it was rotating at between 1 and 30 rpm. An inversion pulse was used in front of a standard imaging sequence to selectively eliminate one of the oils. An inversion delay time of $t_{inv} = 740$ ms was used to eliminate the silicone oil signal, and $t_{inv} = 185$ ms to eliminate the particle oil signal. Figure 17.4a–f shows these two sets of images, where the slice thickness

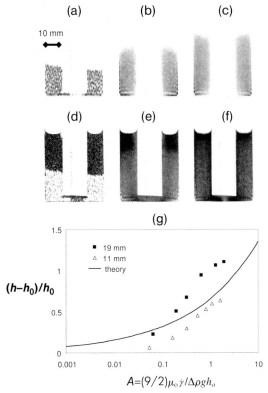

**Figure 17.4** (a–c) Images of the oil inside the pills for 0 rpm, 5 rpm ($Re_p = 2.8 \times 10^{-9}$; $Pe_p = 6.4 \times 10^{14}$; $St = 1.4 \times 10^{-4}$; and $\Psi = 3.2 \times 10^3$) and 30 rpm ($Re_p = 1.7 \times 10^{-8}$; $Pe_p = 3.8 \times 10^{15}$; $St = 18.42 \times 10^{-4}$; and $\Psi = 1.9 \times 10^4$) respectively; (d–f) Images of the surrounding silicon oil for 0, 5 and 30 rpm respectively; (g) A plot of the height of the bed against the nondimensional parameter $A = (9/2)\mu_o\dot{\gamma}/\Delta\rho gh_o$ shows the trend predicted by theory. The filled squares are for the system shown in the above images; the open triangles are for the same system but with the stationary inside cylinder replaced with a narrower 11 mm-diameter cylinder.

is 38 mm and in-plane resolution is 0.45 mm. The increase of the bed height with increasing shear rate is clearly visualized from the images of each phase.

As can be ascertained from the values of the dimensionless groups provided in the figure legend, this system is in the low *Re* viscous regime with negligible colloidal (Brownian) forces and low *St* regime with fluid lubrication force-dominated particle interactions. The trend predicted by theory for the bed height plotted against the parameter $A = (9/2)\mu_o\dot{\gamma}/\Delta\rho gh_o$ (see Figure 17.4g) gives a reasonable qualitative agreement with data, given that the Couette is nonideal. Nonideality from the wide gap of the Couette and wall effects, due to a small number of particle diameters per gap width, is significant and both are neglected by the theory [14].

(a)                                    (b)

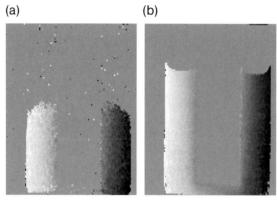

**Figure 17.5** Velocity maps of the wide-gap Couette with a 19 mm-diameter stationary inner cylinder and 45 mm-diameter outer cylinder rotating at 5 rpm and maximum velocity $V_{max} = 11.8$ mm s$^{-1}$ (white is positive velocities out of page; black are negative velocities into page). (a) Oil in the 1 mm-diameter pharmaceutical pills; (b) The surrounding 100 000 cS polymethylsiloxane (silicone) oil.

### 17.4.3
### Fluid Dynamics of Each Phase

Velocity maps were made using the same inversion pulses in front of MR velocity-encoding sequences. Velocity maps were obtained for each oil phase separately (Figure 17.5a). It should be noted that Figure 17.5a and b re-state the data of Figure 17.4b and e, respectively. A comparison of the velocity data shows that the oil beads and silicone oil move with the same velocity in the same location (as expected), and that the velocity profile of the mixture versus the pure oil at the top demonstrate the increased effective viscosity of the mixture compared to the pure fluid.

### 17.5
### Liquid–Solid Slurry in a Rotary Kiln Flow

### 17.5.1
### The System

The slurries were composed of a discrete particle phase of spherical, 1 mm-diameter oil-filled pharmaceutical pills and a continuous fluid phase of 200 cS polymethylsiloxane (silicone) oil, (DOW 200; Esco Products). The experiments were performed in the same MRI system described in Section 17.4. The slurry flow of interest here occurs in a horizontal cylinder of 6.9 cm diameter and 6 cm length. The cylinder is mounted inside the birdcage RF coil with its axis coincident with the axis of the magnet. The mounting allows stable rotation driven by an electric motor outside the Faraday cage.

17.5.2
**Granular Flow Background**

To introduce the slurry flow field in terms of its form and the types of measurement that may be of interest, a brief description of granular flow in a horizontal
rotating cylinder is given. The dry granular flow is very informationally rich and
has been studied using MRI; the results are published elsewhere [19, 21, 22, 26,
28]. Velocities, velocity fluctuations, correlation times and the axial and radial
segregation that occurs spontaneously when particles of different size or density
are combined have been measured.

Figure 17.6 shows an MRI image of a dry granular bead flow; the configuration
typical of slow dry flows is clearly visible. The discrete phase is dragged upwards,
where it first comes into contact with the rotating cylinder, and then flows downhill
in a thin shear layer, as outlined in Figure 17.6a by the white dots. Beneath the
shear layer the spheres are in rigid body rotation. In the dry case, the interstitial
fluid is air and its inertia is negligible; however, as the rotation rate increases the
surface becomes sigmoidal.

17.5.3
**The Experiments**

A $T_1$ inversion pulse was used to selectively null the signal from one phase and
measure the concentration and velocity fields separately for each phase. After a
nonselective composite inversion pulse, a delay of 185 ms nulls the signal from
the pills, while a delay of 740 ms nulls the signal from the silicone oil. A series of
1.3 cm-thick slices, perpendicular to the axis of rotation, were selected with a
shaped RF-excitation pulse. The basic sequence is velocity compensated, and a
bipolar gradient is easily added to provide velocity sensitivity in the horizontal or

(a)                                    (b)

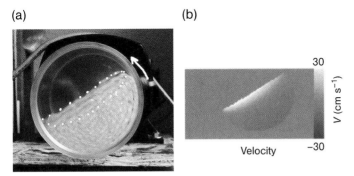

**Figure 17.6** (a) End view of a rotating kiln filled with pills in
dry granular flow and (b) the corresponding MRI velocity
map showing rotation of oil-filled pills in dry granular flow.
Reproduced with permission from Seymour and Caprihan
[22].

vertical directions. A total of nine 2-D images is used to calculate the velocity and concentration images for a particular rotation rate. With the inversion delay adjusted to image oil, three static images (balanced, $+x$ sensitivity, $+y$ sensitivity) and three flowing images are recorded. A set of three flowing images is then taken with the inversion delay adjusted to image pills. Although the configuration of the system changes between the static and flowing images, there is an oil signal from all voxels within the cylinder, and this makes use of the oil phase static images as a zero velocity reference very convenient. The in-plane resolution was 0.035 (horizontal) $\times$ 0.07 (vertical) cm, and a set of nine images required 3.5 h to acquire.

### 17.5.4
### Results

The results from the central slice at 14 rpm are shown in Figure 17.7. The cylinder rotation is counterclockwise and 14 rpm corresponds to a linear velocity of about 5 cm s$^{-1}$ at the inner surface of the cylinder. The Reynolds, Peclet and Stokes numbers were each calculated from this velocity, as the shear rate is strongly spatially varying in the rotating cylinder. The dimensionless numbers clearly indicate that the flow is in the low $Re$ viscous, high $Pe$ noncolloidal, low $St$ fluid lubrication force-mediated particle collision flow regime. In the top left (oil) image in Figure 17.7 the location of the flowing pills appears in darker shades. Clearly, the pills have displaced some oil, which in turn lowers the oil signal in proportion to the displaced volume.

The general configuration is similar to the dry granular flow of Figure 17.6. The dynamic angle of repose is reduced in the slurry, and this reduces the vertical velocity components in the lens-shaped shear layer. The shear layer is also thicker relative to the dry case. Layering of the pills is evident along the right-hand edge

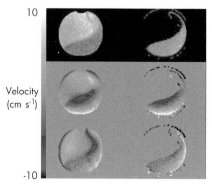

**Figure 17.7** Results from a slice through the cylinder center, rotating at 14 rpm ($Re_p = 1.25 \times 10^{-3}$, $Pe_p = 5.6 \times 10^{17}$, $St = 3 \times 10^{-1}$ and $\Psi = 5.6 \times 10^3$). The left column shows measurements from the continuous (oil) phase; the right column shows results from the discrete (pills) phase. Top row, magnitude images; middle row, horizontal velocities; bottom row, vertical velocity component.

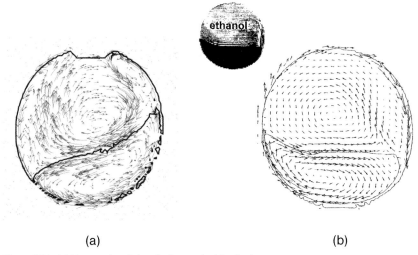

(a)                                             (b)

**Figure 17.8** (a) Vector plot of the oil-phase velocities in the central slice at 14 rpm; (b) Flow of two immiscible fluids, ethanol and oil, in a horizontal rotating cylinder shows a similar velocity pattern. Reproduced with permission from Jeong *et al.* [29].

of the cylinder, and the outermost layer adjacent to the cylinder flows upward until it reaches the small air bubble (the dark indentation) which is visible at the top of the oil image. After colliding with the air bubble, this layer of pills reverses direction and returns to the main body of pills. Although the cylinder, pills and the oil in contact with the cylinder are rotating counterclockwise, within the oil above the bed of pills there is a large cell which is slowly rotating in the opposite sense (this is shown more clearly in Figure 17.8a).

Figure 17.8a shows the horizontal and vertical velocity components combined as arrows. The pill velocities (outlined in the bed region) differ only slightly, indicating that there is no significant slip between the phases in this steady flow. The oil velocities in contact with the rotating cylinder are consistent with a no-slip boundary condition, but from the velocities along the bottom of the air bubble, it is clear that the air bubble is also rotating counterclockwise.

This general configuration, including a large counter-rotating cell, has been observed previously by others [29], and a representative result is shown in Figure 17.8b. The MRI image in the upper left of Figure 17.8b shows there is $T_1$ contrast between the lower oil layer and the ethanol layer floating above it. In this case, the cylinder was rotating clockwise. The interface between the ethanol and the oil is nearly horizontal; in this respect the slurry case of pills in silicon oil presented here is intermediate between the dry flow (see Figure 17.6) and the two-liquid case shown in Figure 17.8a.

## 17.6
## Colloidal Suspensions

### 17.6.1
### The Particles

Model 'MR-active' coreshell colloidal particles can be used to obtain spectrally resolved dynamic measurements with MRM. Coreshell particles are micron-scale particles which consist of a liquid hexadecane core and a hard outer PMMA shell [30]. By using particles which have a liquid core with a separate chemical signature from the suspending medium (water), the particles themselves can be tracked and differentiated from the fluid flow behavior of the suspending medium [11, 20, 31]. This property may be exploited in MR sequences which retain spectral information, such as those where the signal is not acquired in the presence of a magnetic field gradient.

### 17.6.2
### Particle Visualization

To differentiate between the particle oil phase and the suspending water phase, the spectral resolution of the experiment must be less than 3.5 ppm, the chemical shift between oil and water. At dilute particle concentrations, a very small fraction of the MRM signal will actually arise from the hexadecane core oil. The overwhelming signal due to the suspending water creates a broad peak which, due to the small resonant frequency difference between the two substances, overlaps with and obscures the oil peak. The excess water signal must be reduced and a paramagnetic spin-lattice relaxation ($T_1$) enhancing agent, Magnevist, added at a concentration of 0.0117 ml per milliliter of suspension is used. Due to the hard polymer shell of the coreshell particles, the particle core oil relaxation time is unaffected, but the suspending water relaxes significantly via $T_1$ relaxation during the observation time $\Delta$. The water peak is thus of the same order of magnitude as the oil signal in the acquisition window, and both peaks can be spectrally resolved. It is essential to use sufficient phase cycling to cancel contributions from the significant amount of recovered spins.

### 17.6.3
### Colloidal Suspension Flowing in a Single 1 mm Capillary

An acquisition of 128 points at 4960 Hz gave a spectral resolution of 39 Hz per point, well below the 750 Hz separation of the oil and water peaks. A Bruker DRX250 was used to acquire data from the PGSE experiment with 64 $q$-encoding steps, $\delta = 3$ ms, $g = 1.738\,\mathrm{T\,m^{-1}}$, and $\Delta = 30$ ms. A syringe pump (Harvard Apparatus) was used to flow the suspension at $V_{ave} = 0.88\,\mathrm{mm\,s^{-1}}$ ($Re_t = 0.88$; $Re_p = 3.7 \times 10^{-6}$; $Pe_t = 376$; $Pe_p = 21$) through a 1 mm-diameter capillary. The tube Reynolds ($Re_t$) and Peclet ($Pe_t$) numbers are defined as in Sections 17.2.1 and 17.2.2, with a length scale based on the tube radius rather than on the particle size.

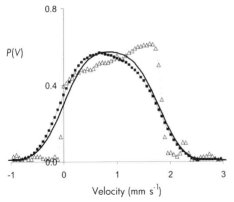

**Figure 17.9** Propagators are for water (line) and an $\phi = 0.08$ suspension (closed squares are the suspending water; open triangles are the core oil) flowing at $V_{ave} = 0.88$ mm s$^{-1}$ ($Re_t = 0.88$; $Re_p = 3.7 \times 10^{-6}$; $Pe_t = 376$; $Pe_p = 21$) over an observation time $\Delta = 30$ ms in a 1 mm-diameter capillary. Reproduced from [11] with permission.

The pure water propagator (the solid line in Figure 17.9) is the convolution of the velocity probability distribution hat function, expected for Poiseuille flow, as there is an equal probability for every velocity from 0 to $v_{max}$, and the molecular diffusion probability distribution Gaussian, which softens the edges. The spectrally resolved particle oil and suspending fluid propagators for a dilute ($\phi = 0.08$) core-shell suspension demonstrate contrasting dynamics for each phase. The suspending fluid propagator is slightly skewed towards slower velocities, while the particle oil propagator is strongly skewed to faster velocities. There is therefore a higher probability of finding oil molecules moving at faster velocities, which is an indication that shear-induced particle migration towards the center of the capillary has occurred [11]. In addition, the edges of the particle oil propagator are much sharper due to the slower self-diffusion of hexadecane $D_o = 4.6 \times 10^{-10}$ m$^2$s$^{-1}$, the restriction of the oil within the hard shell of the particle and the small Stokes–Einstein–Sutherland diffusivity for 2.49 μm-diameter particles $D_{SES} = \dfrac{kT}{6\pi\mu a} = 1.72 \times 10^{-13}$ m$^2$s$^{-1}$ [32]. The widths of the two water propagators are very similar, indicating that a similar amount of water molecular diffusion is occurring in both the pure water and the dilute suspension.

### 17.6.4
### Colloidal Suspension Flowing in a Bifurcation

Spectrally resolved propagator experiments were conducted on a bifurcation flow system (see Figure 17.10) consisting of circular capillaries laser-etched into a hard polymer with an inlet radius $1.25 \pm 0.01$ mm, bifurcating to a small radius outlet of $0.38 \pm 0.01$ mm, and a large radius outlet of $0.63 \pm 0.01$ mm, with

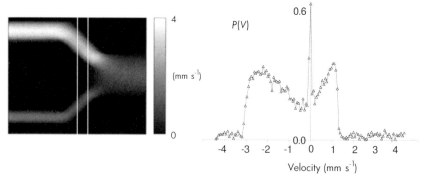

**Figure 17.10** (a) Velocity map in the axial direction for a φ = 0.22 suspension flowing at 10 ml h$^{-1}$ with ~59 × 59 μm spatial resolution. The two white lines indicate the location of the 500 μm slice used in the propagator experiment; (b) Propagator for flow in a perpendicular direction to the axial flow for a φ = 0.22 suspension flowing at 10 ml h$^{-1}$ with 128 $q$ encoding steps. Positive velocities correspond to particles being displaced in the smaller outlet channel; negative velocities correspond to particles being displaced in the larger outlet channel. The peak at zero velocity is probably due to a higher probability of stationary particles at the wall.

daughter capillaries at a 45° angle to the parent capillary. A 500 μm slice was excited across the outlet channel and propagators obtained in the direction perpendicular to the axial flow, as shown in Figure 17.10b. A PGSE experiment was used with 128 $q$ encoding steps, δ = 4 ms, $g_{max}$ = 1.738 T m$^{-1}$, and Δ = 25 ms. A KD Scientific syringe pump was used to flow the suspension through the bifurcation.

Positive and negative displacements correspond to the signal from the small and large outlet channels respectively, quantifying the amount of fluid entering each channel. Experiments were conducted for both an undoped (φ = 0.22) suspension obtaining a combined signal from the oil and water, and for a Magnevist-doped sample where the oil and water peak were spectrally resolved and a propagator for the only oil signal was obtained (see Figure 17.10b). These experiments were repeated for flow rates of 10 ml h$^{-1}$, 20 ml h$^{-1}$ and 30 ml h$^{-1}$, representing a range of tube $Re_t$ values from 1 to 8. The relative split of the oil propagator into positive and negative displacements indicates the amounts of particles entering each channel. This, in addition to the information for the combined oil and water propagators, which indicates the total fluid entering each channel, allows determination of the relative concentrations of particles entering each channel. The effect of different flow rates on particle concentrations entering each channel can then be studied. The ability to spectrally resolve the oil peak (particles) from the water provides a powerful technique for studying the fluid dynamics of suspensions in a bifurcation using dynamic MRM.

## 17.7
## Summary

The use of MR contrast mechanisms to study the dynamics of the suspending fluid and discrete particle phase in solid–liquid suspensions has been reviewed. The ability of MR to resolve the dynamics of each phase provides unique data on the transport and mixing of suspensions and slurries. These data may play a significant role in the design and control of suspension flows in processing and separation systems for applications ranging from biomedicine to composite materials processing.

## Acknowledgments

The authors acknowledge funding from the US National Science Foundation grants NSF CTS-0348076 (J.D.S.), and CBET-0642328 (S.L.C.), NIH Grant Number P20 RR16455-04 from the INBRE-BRIN Program of the National Center for Research Resources for Fellowship Support (J.R.B.). These studies were partially funded by the Department of Energy, Office of Basic Energy Sciences Division of Materials Sciences and Engineering, via Grant No. DE-FG03-93ER14316. Support was also provided by the United States Department of Energy under Contract DE-AC04-94AL85000. These sponsorships do not constitute endorsements by the funding agencies of the views expressed in this article. The authors thank Lisa Mondy for helpful discussions.

## References

1 Vogel, S. (1994) *Life in Moving Fluids: The Physical Biology of Flow*, Princeton University Press, Princeton.

2 McCabe, W.L., Smith, J.C. and Harriott, P. (2005) *Unit Operations of Chemical Engineering*, McGraw-Hill, New York.

3 Jaeger, H.M., Nagel, S.R. and Behringer, R.P. (1996) The physics of granular materials. *Physics Today*, **April**, 32.

4 Jaeger, H.M., Nagel, S.R. and Behringer, R.P. (1996) Granular solids, liquids and gases. *Reviews of Modern Physics*, **68**, 1259.

5 Batchelor, G.K. (1967) *An Introduction to Fluid Dynamics*, Cambridge University Press, Cambridge.

6 Brady, J.F. and Bossis, G. (1988) Stokesian dynamics. *Annual Review of Fluid Mechanics*, **20**, 111.

7 Russel, W.B., Saville, D.A. and Schowalter, W.R. (1989) *Colloidal*

*Dispersions*, Cambridge University Press, Cambridge.

8 Altobelli, S.A., Givler, R.C. and Fukushima, E. (1991) Velocity and concentration measurements of suspensions by nuclear magnetic resonance imaging. *Journal of Rheology*, **35**, 721.

9 Fukushima, E. (1999) Nuclear magnetic resonance as a tool to study flow. *Annual Review of Fluid Mechanics*, **31**, 95.

10 Leighton, D.T.Jr and Acrivos, A. (1986) Viscous resuspension. *Chemical Engineering Science*, **41**, 1377.

11 Brown, J.R., Seymour, J.D., Codd, S.L., Fridjonsson, E.O., Cokelet, G.R. and Nyden, M. (2007) Dynamics of the solid and liquid phases in dilute sheared Brownian suspensions: irreversibility and particle migration. *Physical Review Letters*, **99**, 240602.

**12** Einstein, A. (1957) *Investigations on the Theory of the Brownian Motion*, Dover Publications, New York.

**13** Batchelor, G.K. (1974) Transport properties of two phase materials with random structure. *Annual Review of Fluid Mechanics*, **6**, 227.

**14** Acrivos, A., Mauri, R. and Fan, X. (1993) Shear induced resuspension in a Couette device. *International Journal of Multiphase Flow*, **19**, 797.

**15** Subramanian, G. and Brady, J.F. (2006) Trajectory analysis for non-Brownian inertial suspensions in simple shear flow. *Journal of Fluid Mechanics*, **559**, 151.

**16** Sutherland, W. (1905) A dynamical theory of diffusion for non-electrolytes and the molecular mass of albumin. *Philosophical Magazine*, **9**, 781.

**17** Frank, M., Anderson, D., Weeks, E.R. and Morris, J.F. (2003) Particle migration in pressure driven flow of a Brownian suspension. *Journal of Fluid Mechanics*, **493**, 363.

**18** Brady, J.F. (1993) Brownian motion, hydrodynamics, and the osmotic pressure. *Journal of Chemical Physics*, **98**, 3335.

**19** Yamane, K., Nakagawa, M., Altobelli, S.A., Tanaka, T. and Tsuji, Y. (1998) Steady particulate flows in a horizontal rotating cylinder. *Physics of Fluids*, **10**, 1419.

**20** Wassenius, H. and Callaghan, P.T. (2005) NMR velocimetry studies of the steady shear rheology of a concentrated hard sphere colloidal system. *European Physical Journal*, **E18**, 69.

**21** Seymour, J.D., Caprihan, A., Altobelli, S.A. and Fukushima, E. (2000) Pulsed gradient spin echo nuclear magnetic resonance imaging of diffusion in granular flow. *Physical Review Letters*, **84**, 266.

**22** Caprihan, A. and Seymour, J.D. (2000) Correlation time and diffusion coefficient imaging: application to a granular flow

system. *Journal of Magnetic Resonance*, **144**, 96.

**23** Balcom, B.J., MacGregor, R.P., Beyea, S.D., Green, S.D., Armstrong, R.C. and Bremner, T.W. (1996) Single-Point Ramped Imaging with T1 Enhancement (SPRITE). *Journal of Magnetic Resonance, Series A*, **123**, 131.

**24** Beyea, S.D., Altobelli, S.A. and Mondy, L.A. (2003) Chemically selective NMR imaging of a 3-component (solid–solid–liquid) sedimenting system. *Journal of Magnetic Resonance*, **161**, 198.

**25** Hurlimann, M.D. (2001) Carr-Purcell sequences with composite pulses. *Journal of Magnetic Resonance*, **152**, 109.

**26** Nakagawa, M., Altobelli, S.A., Caprihan, A., Fukushima, E. and Jeong, E.-K. (1993) Non-invasive measurements of granular flows by magnetic resonance imaging. *Experiments in Fluids*, **16**, 54.

**27** Sanfratello, L., Caprihan, A. and Fukushima, E. (2007) Velocity depth profile of granular matter in a horizontal rotating drum. *Granular Matter*, **9**, 1.

**28** Hill, K.M., Caprihan, A. and Kakalios, J. (1997) Bulk segregation in rotated granular material measured by magnetic resonance imaging. *Physical Review Letters*, **78**, 50.

**29** Jeong, E.K., Altobelli, S.A. and Fukushima, E. (1994) NMR imaging studies of stratified flows in a horizontal rotating cylinder. *Physics of Fluids*, **6**, 2901.

**30** Loxley, A. and Vincent, B. (1998) Preparation of poly(methylmethacrylate) microcapsules with liquid cores. *Journal of Colloid and Interface Science*, **208**, 49.

**31** Wassenius, H. and Callaghan, P.T. (2004) Nanoscale NMR velocimetry by means of slowly diffusing tracer particles. *Journal of Magnetic Resonance*, **169**, 250.

**32** Wassenius, H., Nyden, M. and Vincent, B. (2003) NMR diffusion studies of translational properties of oil inside core-shell latex particles. *Journal of Colloid and Interface Science*, **264**, 538.

**Part Four    Biological Systems**

# 18
# Low-Field MR Sensors for Fruit Inspection

*Rebecca R. Milczarek and Michael J. McCarthy*

## 18.1
### Sensor Needs of the Fruit-Processing Industry

Postharvest quality assessment of fruit is increasing in importance as processing operations grow in scale and consumers demand more consistent products. Processors are striving for tighter quality tolerances despite variations in incoming fruit. These variations are a result of weather changes from season to season, natural biological diversity, and disparities in environmental and harvest conditions among growers. The 'quality assessment' of fruit is essentially the attempt to describe and quantify this variation. Every quality assessment procedure has at its core one or both of two general goals: (i) the identification and rejection of fruit that contain defects (e.g. insect, bacterial or viral damage); and (ii) the classification of sound fruit into various categories (e.g. size/color grades). Both of these goals involve the characterization of physical and chemical features of the fruit. In post-harvest operations, these features must be measured at several points: as loads of fruit reach the processing plant in trucks and train cars; as washed fruit are sent to long-term storage; as raw fruit are routed to production lines for further processing; and as whole fruit are graded before packaging and shipping. Current industry implementation of these measurement steps involves a blend of human and machine techniques. Human visual inspection is still widely used in fruit packing houses, despite its high labor costs and operator ergonomic concerns. Machine vision systems have enjoyed some success in commercial implementation, notably in the task of color sorting. However, neither human nor typical machine vision sorting procedures can assess the internal features of a fruit product. Wavelengths outside of the visible portion of the electromagnetic spectrum hold some promise for use in fruit inspection; near-infrared (NIR) imaging, in particular, has been successfully demonstrated for measuring soluble solids content and external bruising of various fruits and vegetables [1–3]. However, the penetration depth of NIR radiation into intact fruit is, at best, on the order of a few millimeters [4]. Thus, surface-scanning inspection techniques are of limited use for determining internal quality parameters.

*Magnetic Resonance Microscopy.* Edited by Sarah L. Codd and Joseph D. Seymour
Copyright © 2009 WILEY-VCH Verlag GmbH & Co. KGaA, Weinheim
ISBN: 978-3-527-32008-0

These internal quality parameters are important in the fruit-processing industry because much information about the maturity, texture and flavor of a given fruit sample cannot be determined 'with the naked eye'. For decades, statistical sampling followed by destructive testing (e.g. juicing to measure soluble solids content, cutting into cubes for texture analysis, slicing into portions for sensory analysis) has been the norm for assessing the internal features of fruit. The internal quality status of an entire batch, shipment or lot is inferred from the destructive analysis of a few representative samples. The measurement process itself may take several minutes or hours, thus delaying a decision on the fate of the batch. Processors would rather employ rapid, nondestructive techniques for 100% inspection of their products. A few such techniques have found niche applications in the fruit-processing industry. For example, impact-based testing has been demonstrated for measuring the firmness of peaches [5, 6], and X-ray inspection has been used for detecting voids in onions [7]. However, the range of measurement possibilities for each individual technique is quite narrow. Fruit processors desire a sensor system that can measure multiple internal quality attributes (both physical and chemical), is easily scalable, and produces information that is correlated to destructive measurements in a logical manner.

## 18.2
## Magnetic Resonance in Fruit Quality Assessment

Nuclear magnetic resonance (NMR) and magnetic resonance imaging (MRI) have been investigated extensively during the past two decades as an answer to the fruit processors' sensor needs. MR methods are particularly well-suited to the quality assessment of fruit because of the abundance of protons in these biological materials. Protons in fruit are most often found in water molecules; in some high-oil-content fruit, such as olive and avocado, lipid protons are also present in large numbers. Indeed, the ratio of the oil peak to the water peak in proton spectra of avocado fruit has been shown to be a good indicator of maturity of both stationary [8] and moving [9] samples. However, for most fruit, MR analysis focuses on the behavior of water protons.

In plant tissue, water protons create contrast in MR images via two mechanisms: diffusion and relaxation. Within the relaxation mechanism, three primary pathways have been identified [10]:

- Proton exchange between water and cell metabolites and macromolecules.
- Dephasing of magnetization as water molecules diffuse through internal field gradients created by susceptibility discontinuities.
- Movement of water molecules among various subcellular and extracellular compartments.

The compartmentalization of fruit tissue, in particular, creates numerous relaxation environments. The existence of multiple relaxation environments at the cellular level results in multiexponential curves for both longitudinal ($T_1$) and

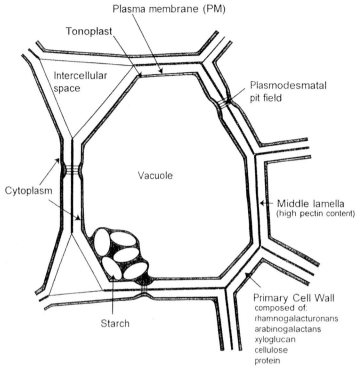

**Figure 18.1** A typical plant cell. Water may be present in any of a number of compartments, including the vacuole, cytoplasm, and intercellular space. Harker, F.R., Redgwell, R.J., Hallett, I.C. and Murray, S.H. (1997) Texture of fresh fruit, *Horticultural Reviews*, vol. 20 (ed. J. Janick), John Wiley & Sons, Inc. p. 129.

transverse ($T_2$) relaxation. This multiexponential fitting has been correlated to compartmentalization in fruit such as apple [11]. Moreover, the sizes and compositions of the relaxation environments change with time. Under most conditions, the majority of water in a plant cell resides in the vacuole, as depicted in Figure 18.1. The vacuole membrane is permeable, allowing changes in vacuole volume and internal solute concentration depending on the influx and efflux of water. In apple parenchyma, for example, the vacuole shrinks in volume and the intravacuole sugar concentration increases during maturation of the fruit [12]. Tracking of the component intensities of multiexponential MR relaxation signals has already been demonstrated for the monitoring of apple drying and freezing [13], and thus could also provide information on apple maturity. Other correlations of MR signals to the status of fruit tissue are numerous and recently have been extensively reviewed by Hills and Clark [10].

While many studies have shown such MR/quality parameter correlations, most have been performed using research-grade or medical-grade high-field magnets.

Although scanning and processing times have rarely been considered as design parameters to be optimized, these times will certainly factor prominently in commercial applications. As recently as 1999, researchers were casting doubt on the feasibility of using MR technologies for the routine inspection of fruits and vegetables [14]. Not surprisingly, the most often cited reasons for this doubt were the expense of MR equipment and difficulty of operation of a high-field magnet in a processing environment. Low-field MR sensors address both of these concerns.

Many research groups have successfully demonstrated the relaxometry of fruits and vegetables on low-field systems. The longitudinal relaxation time, commonly measured using an inversion recovery sequence, and the transverse relaxation time, commonly measured using a Carr–Purcell–Meiboom–Gill (CPMG) sequence, have been correlated to numerous quality parameters using low-field equipment. A few examples of such low-field correlations follow. The $T_2$ of navel orange segments measured at 0.23 T can be used to differentiate between normal and freeze-affected fruit [15]. For fried potato products, the dry matter content of the raw potato is an important predictor of texture. Thybo and others have obtained $T_1$ and $T_2$ measurements of potatoes at 0.47 T and correlated these with dry matter content [16]. A preliminary study on the monitoring of grape ripening using a one-sided MR sensor has been reported by Rahmatallah and coworkers [17]. Here, the sensor can be brought up to the surface of the fruit and has maximum field strength of about 0.32 T at 1.3 mm from the face of the magnet.

Whilst all of these applications show promise for the implementation of low-field MR relaxometry and imaging for the quality evaluation of individual fruit, the issue of sample motion has rarely been addressed for these sensors. In addition to the avocado experiments mentioned above, only three other reported studies have been performed on fruit samples moving on a conveyor belt through an MR sensor. Hernandez-Sanchez and coworkers have imaged citrus fruit for the detection of freeze damage [18] and presence of seeds [19] as the fruit traveled on a conveyor belt through a horizontal-bore MR system. By using a Fast Low-Angle SHot (FLASH) imaging sequence, they were able to obtain usable images at conveyor speeds of up to $0.10 \, m \, s^{-1}$. However, the experiments were carried out on a high-field (4.7 T field strength) research-grade system that, while showing promise for a laboratory setting, would be difficult to install in a typical fruit-processing plant. Chayapresert and Stroshine [20] constructed a low-field MR sensing system, including a conveyor belt, to bring their success at measuring the defect of apple internal browning [21] from an off-line to an on-line regime. Healthy and internally browned apple samples passed through a 0.13 T magnet at speeds ranging from $0 \, m \, s^{-1}$ (stationary) to $0.25 \, m \, s^{-1}$. The $T_2$ values, measured using a CPMG sequence, differed significantly between the healthy and browned apples for speeds of $0.15 \, m \, s^{-1}$ or slower. This enabled the slow-speed sorting of the apples based on the state of their internal tissue, but unfortunately the misclassification rates increased as the conveyor speed increased. Sample misalignment was also identified as a source of classification error. However, this study showed that sample motion and relatively poor homogeneity (600 ppm in this case) are not insurmountable problems for MR sensors, although ideally the line speed must be an order of magnitude higher than that demonstrated. An ideal inspection system installed on a conveyor belt would process 8–12 fruit per second, corresponding to a singulated fruit linear velocity of about $2\,\mathrm{m\,s^{-1}}$.

## 18.3
## Motion Considerations for Implementing Magnetic Resonance in Quality Sorting

The 8–12 fruit per second rate is the current standard commercial line speed. Each of the fruit are conveyed in an individual cup; hence, they are referred to as singulated. The successful integration of MR-based quality sensors will require the same throughput while maintaining the singulated identity of the fruit. This rate of sorting can be accomplished by schemes that may be organized into three categories:

- Fruit moves through the sensor at conveyor speeds
- Fruit slows through the measurement zone
- Fruit stops in the measurement zone for a specific time

A procedure involving a stop-measure-go approach has implications for both pulse sequence selection and the contrast available to detect quality factors in the image.

The magnitude of the proton signal from a fruit will depend upon the time permitted for spin-lattice relaxation. In a stop-measure-go operation mode the entire time that the fruit remains in the magnetic field will be greater than for continuously moving samples (for a set magnetic field volume). Depending on the strength and extent of the fringe field, it may be possible to approach two to three $T_1$ values which would result in significantly larger $M_o$ values compared to fruit that move continuously. Consider a linear dimension in the direction of the conveyor travel of 300 mm containing three individual fruit. If the system operates continuously at $1\,\mathrm{m\,s^{-1}}$ line speed the fruit would remain in the magnetic field for 0.3 s. If the fruit stop in the measurement zone, they may remain for approximately 0.6 s, or about twice as long as for continuous motion. This additional time permits different timing options for pulse sequences than if the samples were moving.

The influence of motion on the NMR signal is manifest through changes in the effective relaxation times and signal intensity. The influence of motion on relaxation times is expressed as [22]:

$$\frac{1}{T_1^{observed}} = \frac{1}{T_1^{static}} + \frac{1}{\tau}$$

$$\frac{1}{T_2^{observed}} = \frac{1}{T_2^{static}} + \frac{1}{\tau}$$

where the inherent relaxation times are designated with the superscript 'static', the relaxation times observed during motion are designated with the superscript 'observed', and τ is the lifetime of the nuclei in the measurement zone (0.3 s from the example above). Consider the impact on $T_1$ for flesh versus pit proton signals. In the case of a static sample the flesh has an approximate $T_1$ of 1 s and the pit 0.1 s. If the fruit is moving at 1 m s$^{-1}$ the observed $T_1$ values become ~0.23 s for the flesh and 0.075 s for the pit. The ultimate contrast obtained in the image will depend upon the value of the magnetization, the range of relaxation times, the speed of motion and the pulse sequence. Additionally, with a natural product such as fruit there will be a 'normal' range in relaxation values that may depend upon maturity, temperature, sugar content and other factors.

Incorporating or eliminating motion during the imaging of fruit for defect detection may yield different contrast between defects and sound tissue. Almost all defects have been 'identified' in images based on relaxation time differences compared to sound tissue. Depending on the relative differences in relaxation times, imaging a static or moving sample may yield the same results, or different results.

One novel solution to the conveyor velocity problem is to use the sample motion itself to create an NMR signal. This technique – known as 'motional relativity' – was recently explored by Hills and Wright [23, 24] to obtain $T_1$ measurements, $T_2$ measurements and 1-D profiles of objects moving through a prototype MR sensor. Conventional NMR spectrometers use an RF pulse of time-dependent amplitude to tip the longitudinal magnetization vector in the sample through a given angle. MR measurements based on motional relativity, in contrast, use a time-independent RF field (still transverse to the main magnetic field, $B_0$) through which the sample passes at a known velocity and acceleration. The tip angle of the magnetization vector is dependent on the strength of the RF field as a function of spatial coordinates and the velocity/acceleration of the sample. By using a low-field (0.06 T) magnet, Hills and Wright demonstrated several of the advantages of this approach, including the elimination of extensive RF pulse programming and associated eddy currents, robust performance in situations of high $B_1$ inhomogeneity, and effectiveness for samples moving at fruit processing line speeds (e.g. 1.25 m s$^{-1}$ for a $T_2$ measurement experiment). Motional relativity, however, does present some challenges that must be addressed before commercial implementation would be feasible. For samples with long $T_1$-values (including many fruits and vegetables), prepolarizing magnets used upstream of the $B_1/B_0$ unit are advantageous. The design and cost of these prepolarizers must be included when assessing the entire motional relativity system. Also, a motional relativity sensor inherently precludes the use of pulse sequences that involve time-dependent techniques such as repeated acquisition and phase cycling. This disadvantage may be somewhat overcome with creative single-shot sequence designs, but techniques such as diffusion tensor imaging remain impracticable on this type of apparatus at present. Nonetheless, motional relativity presents an intriguing alternative to conventional approaches for bringing MR sensors on-line.

Sample motion effects will likely be compounded if MR imaging is deemed a more useful alternative to MR relaxometry for a given fruit inspection task. While the literature on low-field MR relaxometry of agricultural products is vast, low-field

MR imaging has not been explored as closely. Fruit packinghouses already use high-speed imaging for the surface inspection of their products; MR imaging is a logical next step for their internal assessment needs. Certain fruit features such as bulk texture and maturity are amenable to measurement by MR relaxometry, but other features (especially spatially localized defects) require 2-D image data for identification. In the following section, we discuss a recent application of a low-field industrial-grade imaging system.

## 18.4
### Low-Field MRI Application: Seed Detection in Citrus Fruit

The state of California produces 2.5 million tons of fresh-market orange crop per season, accounting for 80% of the supply in the United States [25]. Of this crop, seedless mandarin orange varieties are experiencing an increase in both popularity among consumers and total planted acreage [26]. While mandarin cultivars have been carefully bred to eliminate the seed-producing trait, unintended pollination can result in seeded fruit. The primary vector for this pollination is the large honeybee population present in central California. Pollination by bees is required for the successful production of US$ 6 billion of other crops such as peach, almond and avocado. Conflict between seedless citrus fruit growers and beekeepers resulted in the February 2007 introduction of California Assembly Bill No. 771, which established a Seedless Mandarin and Honeybee Coexistence Working Group [27]. Until the Working Group can find a solution that satisfies both parties (and likely afterward, since 100% elimination of mandarin tree pollination is improbable), citrus growers in California will need some way to determine which of their fruit are seeded. As there is no external evidence that a fruit contains seeds, nondestructive internal inspection is required.

The feasibility of low-field MRI for such inspection was examined in March and April 2007. In this experiment, 140 samples of 'W. Murcott' mandarin oranges (*Citrus reticulata*) were hand-picked from test plots of Paramount Citrus near Delano, California. The fruit were imaged in a horizontal-bore ASPECT AI 1T industrial MR imaging magnet, as depicted in Figure 18.2. This magnet has a footprint of approximately 350 cm$^2$ and no fringe fields, enabling use in a packing line environment. A 60 mm inner diameter (i.d.) solenoid coil was used for transmission of RF pulses and reception of signal from the sample.

Each image was $128 \times 128$ pixels, with a resolution of 570 μm per pixel. Images of a transverse ('wagon-wheel') slice of the fruit were obtained using three different MR sequence types: Turbo Fast Low-Angle Shot (Turbo FLASH); Fast Spin Echo (FSE); and Gradient Recalled Echo (GRE). The Turbo FLASH imaging sequence used an inversion pulse time of 700 ms, and the FSE sequence had an effective $T_2$ of 412 ms. The observe frequency (in MHz) and attenuation (in dB, used to set the 90° and 180° pulse powers) were reset for each sample. The observe frequency shifted 18 kHz over the course of the three days of imaging, due to temperature variations of the magnet. The entire imaging system was enclosed in a tempera-

**Figure 18.2** The ASPECT AI industrial MR imaging system. Although this 1 T system was used with stationary samples for the current study, a conveyor belt can be passed through the bore for sample handling.

ture-controlled tent and kept at 25 °C to minimize this variation. Other experiments with similar fruit (data not shown) indicated that attenuation increases in a roughly linear manner with fruit mass; the attenuation values for the fruit in the current study varied from 13.4 to 16.8 dB. As the seeds are most likely to develop along the plane of the fruit halfway between the stem and blossom ends, the imaging slice used was centered on this plane. The slice thickness for the Turbo FLASH and FSE sequences was 17 mm. The GRE sequence produced a 64-slice image set, centered at the same location as the Turbo FLASH and FSE sequences. The images of the central 16 slices (1.07 mm per slice) of the 64-slice set were averaged to create a GRE image that was congruent to the Turbo FLASH and FSE images.

Within 24 h of imaging, the presence and count of seeds were verified by an expert grader through destructive testing. Each fruit was cut using the 'Volume Cut' guidelines set forth in the California Code of Regulations, 3 CCR Section 1430.9.1. The 'Volume Cut' is comprised of three transverse cuts through the fruit: one cut one-third of the distance from the blossom end to the stem end, one cut one-third of the distance from the stem end to the blossom end, and one cut halfway in between the first two cuts. The number and size (<1.5 mm or >1.5 mm) of the seeds in each of the four resulting slices was recorded. If a seed had been cut into two pieces, it was assigned to only one slice: the slice in which more than 50% of the seed was found. Note also that the four slices examined in this protocol comprise the central third of the fruit–roughly 1.5 cm in thickness for an average mandarin orange.

Before multivariate image analysis (MIA; described later in this section) could be applied to the mandarin images, MR artifacts and inherent features of the fruit samples necessitated two preprocessing steps: masking and detrending. The

(a)                    (b)                    (c)

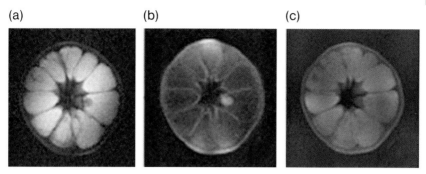

**Figure 18.3** MR images of a typical mandarin orange. Note the seed near the right-center area of the fruit. (a) Fast Spin Echo; (b) Turbo FLASH; (c) Gradient Recalled Echo.

intent of masking was to identify only those areas of the images that contained pulp and seeds. It is best to avoid performing MIA on, for example, the space outside the sample as this area does not supply any relevant information to the model. A mask was created from the FSE image for each sample in a series of steps in Matlab (R2006b; The MathWorks, Inc., Natick, MA). First, a binary image was created using a threshold that separated the pulp signal from the seed, rind, segment membrane and background air signals. The leftmost point of the resulting object was found, and 60 points were traced out clockwise along the edge of the object. A circle was then fitted to the traced-out curve. A circle with a radius 70% the size of the radius of the best fit circle (but centered at the same point) served as the base for the mask. The center of the fruit, comprised only of air for this variety, was then identified and removed from the mask. Thus, the final mask consisted of a solid circle, minus the fruit center, that covered most of the area of the pulp. The same mask was applied to each of the three images (FSE, Turbo and GRE) for a given fruit sample.

Near-coil artifacts were observed in the raw Turbo and GRE images. To reduce these artifacts, a quadratic filter was applied to each individual fruit sample image. In this way, images were 'detrended'. Examples of the three images types for a typical orange sample are shown in Figure 18.3.

Other research groups have used FLASH images to identify seeds in citrus fruit [19], but no attempt had been made thus far to optimize selection of pulse sequences for seed detection on a low-field instrument. In the current experiment, the technique of MIA was used to evaluate the three congruent images of each fruit as a step towards image optimization. Here, 'optimization' refers to maximizing the contrast between the seed and nonseed portions of the image. MIA is a natural extension of standard multivariate statistical techniques to image-based data sets. (A brief introduction to MIA will be made here; for more detailed coverage of the topic, Geladi and Grahn [28] is an excellent reference.) A multivariate image is essentially a set (or 'stack') of multiple congruent images of the same object. Each individual image may be obtained in any number of ways; in this case, images were obtained using different MR pulse sequences. By 'unwrapping' multivariate images row-by-row or column-by-column, the images can be analyzed in the same ways as regular multivariate data sets. Multivariate statistical techniques such as principal components analysis, partial least squares and cluster analysis are applied to the unwrapped data set, and the results are rewrapped to create new images that are then interpreted statistically.

In this case, partial least squares (PLS) was the multivariate technique used to interpret the data. PLS is an iterative technique that seeks to maximize the explanation of variance in both the X-block (the data that is used for prediction) and the Y-block (what is being predicted). The result of PLS is a model for creation of a prediction image. The prediction image is a linear combination of the original $k$ images, where $k$ is the number of individual images making up the multivariate image. In the present study, $k = 3$. The 140 masked, detrended FSE images were arranged end-to-end to create an image of dimensions $128 \times 17\,920$. The Turbo and GRE images were arranged in the same way and then added to the FSE image to create a $128 \times 17\,920 \times 3$ data set – the X-block.

On each masked and detrended Turbo image, pixels belonging to seeds were identified visually and marked in a binary image (1 = seed, 0 = not seed). In most cases, the number of seeds marked in this way matched the offline count. However, artifacts, damaged tissue and partial volume effects made some seeds nearly impossible to identify by eye. Only objects that were clearly seeds (corroborated by the off-line counts and digital photographs of the cut fruit) were marked in the binary image. The binary images were arranged end-to-end as above to create a $128 \times 17\,920$ image – the Y-block.

The X- and Y-blocks were input to the PLS routine of the Matlab MIA toolbox (Eigenvector Research, Wenatchee, WA). The resulting model, using one latent variable, was applied to the raw data to create a prediction image in which seeds appeared hyperintense and orange pulp appeared hypointense. Figure 18.4 shows the prediction image created from the three X-block images in Figure 18.3.

In the overall prediction image (for all 140 fruit samples), the ratio of average seed intensity to average pulp intensity was 2.90. This is a clear improvement in

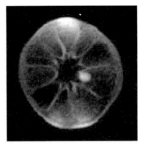

**Figure 18.4** Prediction image from the PLS model. Corresponding to the fruit shown in Figure 18.3, this is a $128 \times 128$ section of the $128 \times 17\,920$ overall prediction image for the entire data set.

seed/pulp contrast compared to that of the Turbo FLASH (1.40), FSE (1.21) and GRE (1.05) images taken separately. In the in-line implementation of the MR imaging system, the enhanced contrast of the prediction image is expected to increase the accuracy of the seed detection process, reducing both misses and false positives. However, this increase in accuracy must be weighed against the concomitant increase in scan time, since two or more images must be obtained, as opposed to just one image, in the multivariate approach. The next step in this study will be the assessment of this trade-off between time and accuracy. MIA-assisted development of hybrid scan techniques, combining the most contrast-enhancing aspects of the three different imaging sequences, is also a possibility. Although several technical challenges remain to be addressed in order to automate the imaging and seed detection process, the multivariate approach shows promise as a method for enhancing image contrast without extensive pulse sequence optimization. Thus, the multivariate analysis of MRI data obtained on a low-field instrument has been shown to be a feasible approach for quality assurance of citrus fruit.

## 18.5
## Conclusions

Low-field MR techniques, both relaxometry and imaging, can meet the fruit processors' need for rapid, nondestructive internal quality assessment of their products. Several approaches to the motion problem are currently under investigation, and the solution for each application will depend on the specifications of the fruit and the conveying system. For data analysis, a multivariate approach may compensate for the limitations encountered in the low-field regime. It should be noted that, while the techniques described here are low-field MR applications in fruit, the general approach can certainly be extended to vegetables, meat, bakery products and other foodstuffs. During the next few years, one can expect to see a full integration of low-field MR relaxometry and imaging in the industrial packing/processing lines of many agricultural products.

## References

1 Liu, Y., Chen, Y.R., Wang, C.Y., Chan, D.E. and Kim, M.S. (2006) Development of hyperspectral imaging techniques for the detection of chilling injury in cucumbers; Spectral and image analysis. *Applied Engineering in Agriculture*, **22**, 101–11.

2 Slaughter, D. (1995) Non-destructive determination of internal quality in peaches and nectarines. *Transactions of the American Society of Agricultural Engineers*, **38**, 617–23.

3 Lu, R. (2003) Detection of bruises on apples using near-infrared hyperspectral imaging. *Transactions of the American Society of Agricultural Engineers*, **46**, 523–30.

4 Lammertyn, J., Peirs, A., De Baerdemaeker, J. and Nicolai, B. (2000) Light penetration properties of NIR radiation in fruit with respect to non-

destructive quality assessment. *Postharvest Biology and Technology*, **18**, 121–32.

5 Gutierrez, A., Burgos, J.A. and Molto, E. (2007) Pre-commercial sorting line for peaches firmness assessment. *Journal of Food Engineering*, **81**, 721–7.

6 Slaughter, D.C., Cristoso, C.H., Hasey, J.K. and Thompson, J.F. (2006) Comparison of instrumental and manual inspection of clingstone peaches. *Applied Engineering in Agriculture*, **22**, 883–9.

7 Tollner, E.W., Gitaitis, R.D., Seebold, K.W. and Maw, B.W. (2005) Experiences with a food product x-ray inspection system for classifying onions. *Applied Engineering in Agriculture*, **21**, 907–12.

8 Chen, P., McCarthy, M.J., Kauten, R., Sarig, Y. and Han, S. (1993) Maturity evaluation of avocados by NMR methods. *Journal of Agricultural and Food Research*, **55**, 177–87.

9 Kim, S-M., Chen, P., McCarthy, M.J. and Zion, B. (1999) Fruit internal quality evaluation using on-line nuclear magnetic resonance sensors. *Journal of Agricultural and Food Research*, **74**, 293–301.

10 Hills, B.P. and Clark, C.J. (2003) Quality assessment of horticultural products by NMR, in *Annual Reports on NMR Spectroscopy*, vol. 50 (ed. G.A. Webb), Elsevier Science, Oxford, UK, pp. 75–120.

11 Snarr, J.E.M. and Van As, H. (1992) Probing water compartments and membrane permeability in plant cells by 1H NMR relaxation measurements. *Biophysical Journal*, **63**, 1654–8.

12 Yamaki, S. and Ino, M. (1992) Alteration of cellular compartmentation and membrane permeability to sugars in immature and mature apple fruit. *Journal of the American Society for Horticultural Science*, **117**, 951–4.

13 Hills, B.P. and Remigereau, B. (1997) NMR studies of changes in subcellular water compartmentation in parenchyma apple tissue during drying and freezing. *International Journal of Food Science & Technology*, **32**, 51–61.

14 Abbott, J.A. (1999) Quality measurement of fruits and vegetables. *Postharvest Biology and Technology*, **15**, 207–25.

15 Gambhir, P. N., Choi, Y.J., Slaughter, D.C., Thompson, J.F. and McCarthy, M.J. (2005) Proton spin-spin relaxation time of peel and flesh of navel orange varieties exposed to freezing temperature. *Journal of the Science of Food and Agriculture*, **85**, 2482–6.

16 Thybo, A.K., Andersen, H.J., Karlsson, A.H., Donstrup, S. and Stokilde-Jorgensen, H. (2003) Low-field NMR relaxation and NMR-imaging as tools in differentiation between potato sample and determination of dry matter content in potatoes. *Lebensmittel-Wissenschaft und-Technologie*, **36**, 315–22.

17 Rahmatallah, S., Li, Y., Seton, H.C., Gregory, J.S. and Aspden, R.M. (2006) Measurement of relaxation times in foodstuffs using a one-sided portable magnetic resonance probe. *European Food Research and Technology*, **222**, 298–301.

18 Hernandez-Sanchez, N., Barreiro, P. and Ruiz-Cabello, J. (2004) Detection of freeze injury in oranges by magnetic resonance imaging of moving samples. *Applied Magnetic Resonance*, **26**, 431–45.

19 Hernandez-Sanchez, N., Barreiro, P. and Ruiz-Cabello, J. (2006) On-line identification of seeds in mandarins with magnetic resonance imaging. *Biosystems Engineering*, **95**, 529–36.

20 Chayaprasert, W. and Stroshine, R. (2005) Rapid sensing of internal browning in whole apples using a low-cost, low-field proton magnetic resonance sensor. *Postharvest Biology and Technology*, **36**, 291–301.

21 Keener, K.M., Stroshine, R.L., Nyenhuis, J.A. (1999) Evaluation of low field (5.40-MHz) proton magnetic resonance measurements of Dw and T2 as methods of nondestructive quality evaluation of apples. *Journal of the American Society of Horticultural Science*, **124**, 289–95.

22 Tellier, C. and Mariette, F. (1995) On-line applications in food science, in *Annual Reports on NMR Spectroscopy*, Vol. 31 (eds G.A. Webb, P.S. Belton and M.J. McCarthy), Elsevier Science, Oxford, UK, pp. 105–22.

23 Hills, B.P. and Wright, K.M. (2005) Towards on-line NMR sensors, in *Magnetic Resonance in Food Science: The Multivariate Challenge* (eds S.B. Engelsen, P.S. Belton and H.J. Jakobsen), Royal Society of Chemistry, Cambridge, UK, pp. 175–85.

**24** Hills, B.P. and Wright, K.M. (2006)
Motional relativity and industrial NMR
sensors. *Journal of Magnetic Resonance*,
**178**, 193–205.

**25** United States Department of Agriculture
Economic Research Service (2007)
California's Citrus Industry, http://www.
ers.usda.gov/News/CAcitrus.htm
(accessed 10 July 2007).

**26** California Agricultural Statistics Service
(2004) California Citrus Acreage Report,

http://www.calcitrusgrowers.com/
200412citac.pdf (accessed 10 July 2007).

**27** California Legislature (2007) 2007–08
Regular Session, Assembly Bill No. 771
http://www.leginfo.ca.gov/pub/07-08/bill/
asm/ab_0751-0800/ab_771_bill_20070906_
amended_sen_v92.pdf (accessed 19
September 2007).

**28** Geladi, P. and Grahn, H., (1996)
*Multivariate Image Analysis*, Wiley-VCH
Verlag GmbH & Co. KGaA, New York.

# 19
# Microscopic Imaging of Structured Macromolecules in Articular Cartilage

*Yang Xia and ShaoKuan Zheng*

## 19.1
## Introduction

The degradation of articular cartilage, a thin layer of connective tissue which covers the load-bearing ends of bones in joints and absorbs shocks and distributes loads, plays a critical role in the development of osteoarthritis (OA) and related joint diseases, which affect 33% of the US population [1]. While the structure and properties of healthy and diseased cartilage have been studied extensively, our understanding of how healthy cartilage bears load, lubricates joints and progressively loses these 'normal' functions in OA remains limited. This is because degradation of the tissue is preceded 'unnoticeably' by a number of insidious processes characterized by subtle changes in tissue's fine structure and delicate functions/compositions/interactions [2]. Consequently, an accurate diagnosis of *early* OA remains elusive in clinical practice.

Articular cartilage has a specialized organization, both molecularly and structurally. Molecularly, it is composed essentially (as % wet weight) of three molecules: water (65–80%), collagen (~20%) and proteoglycan (~5%) [3–5]. The collagen in cartilage is primarily type II, and is present in the form of a rigid 3-D matrix of triple-helical fibrils. The proteoglycan (aggrecan) has a bottle-brush-like structure with a central protein core and side chains of heavily sulfated glycosaminoglycan (GAG). In healthy tissue, 100 to 200 monomers of proteoglycan bind to a hyaluronic acid via the link protein to form a massive proteoglycan aggregate, with a length up to 1 micron. In addition to these three major components, cartilage also has a number of minor molecules, including small leucine-rich proteoglycans, cartilage oligomeric matrix protein, matrilins, thrombospondin, fibronectin and other types of collagen (type I, V, VI, IX and XI). Although the function of these minor molecules remains largely unknown, some have been shown to be responsible for assisting the connection of the collagen fibril network [6–8].

On the histological scale, articular cartilage contains a scattered population of living cells (chondrocytes), which accounts for less than 1% of the tissue. Most cartilage is extracellular and highly structured [6, 9–13]. In the simplest analysis,

*Magnetic Resonance Microscopy.* Edited by Sarah L. Codd and Joseph D. Seymour
Copyright © 2009 WILEY-VCH Verlag GmbH & Co. KGaA, Weinheim
ISBN: 978-3-527-32008-0

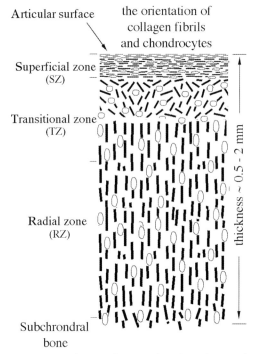

Articular surface    the orientation of
                     collagen fibrils
                     and chondrocytes

Superficial zone
(SZ)

Transitional zone
(TZ)

thickness ~ 0.5 - 2 mm

Radial zone
(RZ)

Subchrondral
bone

**Figure 19.1** Schematic diagram of cartilage showing the relative diameter and the orientation of the collagen fibers (the short lines) in different histological zones and the size and orientation of the chondrocytes (circles and ovals). Note: the collagen fibrils and chrondrocytes are not drawn to scale.

the histological structure of noncalcified cartilage has been commonly considered to comprise of three subtissue zones based on the orientation of local collagen fibrils (Figure 19.1). These three zones are: (i) the superficial zone (SZ), where the collagen is oriented parallel with the articular surface; (ii) the transitional zone (TZ), where the collagen is oriented rather randomly; and (iii) the radial zone (RZ), where the collagen is oriented mainly perpendicularly to the articular surface. The *spatial orientation* of the collagen matrix in cartilage essentially defines the depth-dependent anisotropic nature of the tissue properties as measured by any imaging technique [12].

Due to its multiscale hierarchical organization, measurements that interrogate cartilage at different length-scales and technical modalities are warranted. Due to its depth-dependent and heterogeneous structure, a thorough understanding of the tissue's response to external loading (exercising) and degradation (disease) requires microscopic imaging. Since 1992, we have been using various imaging techniques to study articular cartilage, including microscopic magnetic resonance imaging (μMRI), polarized light microscopy (PLM) and Fourier-transform infrared imaging (FTIRI). Each of these techniques examines a unique aspect of the tissue's structure and property. Collectively, the use of multidisciplinary techniques provide a better understanding of the various molecular mechanisms that govern the properties of articular cartilage as an effective load-bearing tissue in humans and animals.

## 19.2
## Imaging the Orientation of Collagen Fibrils by $T_2$ Anisotropy

Although it has been known for over four decades that the NMR properties of collagen fibrils in tendons can have a strong anisotropy due to the collagen orientation in the polarizing magnetic field $B_0$ [14–16], its implication in the MRI of articular cartilage was not fully appreciated until the documentation of the so-called 'magic angle effect' in the clinical MRI of cartilage during the early 1990s [17–19]. Since then, many studies have been conducted at high resolution to investigate the influence of this tissue anisotropy in MRI [12]. A set of examples is shown in Figure 19.2a, where the intensity of the cartilage images clearly changes when the tissue block is oriented differently with respect to $B_0$ (pointing upwards) [20, 21]. These images unmistakably demonstrate a depth-dependent motional anisotropy for the protons, due to the close interaction between the protons in water molecules and the collagen matrix [12]. The $T_2$ maps (shown in Figure 19.2b as the 1-D profiles) at these orientations show distinct anisotropy, which is depth-dependent and has three featured 'zones' for mature cartilage. Because these MRI 'zones' are statistically equivalent to the three histologic zones in cartilage [22], these results show that, just as noncalcified cartilage can be conceptually subdivided based on the orientation of the collagen fibers into three distinct structural zones in histology, a piece of cartilage can also be subdivided based on the regional characteristics of the $T_2$ relaxation in μMRI into three structural zones. The fact that the $T_2$-based MRI zones are equivalent to the collagen-based histology zones suggests that $T_2$ relaxation is a sensitive and noninvasive MRI marker to study the tissue structure and molecular interactions in articular cartilage.

For example, in one cartilage study where the tissue blocks were rotated within the full 360° angular space (Figure 19.2c and d) [21], the relative amplitudes of the $T_2$ maxima (approximately at 55°, 125°, 235° and 305°) were found to deviate somewhat from those expected from the dipolar interaction. The discrepancy between the observed $T_2$ anisotropy and the angular dependence of the dipolar interaction was explained by means of a simple model which considers the average of one isotropic and two anisotropic spin populations, the first being associated with 'free' water, and the latter two arising from collagen-associated waters. We show that, even for the 'long' $T_2$ components, which arise in multiple-compartment studies of collagen–water systems, there appears to be two subpopulations, each having the same peak value of $T_2$ but with the angular dependence of one shifted in phase by 90° relative to the other by virtue of the fact that each

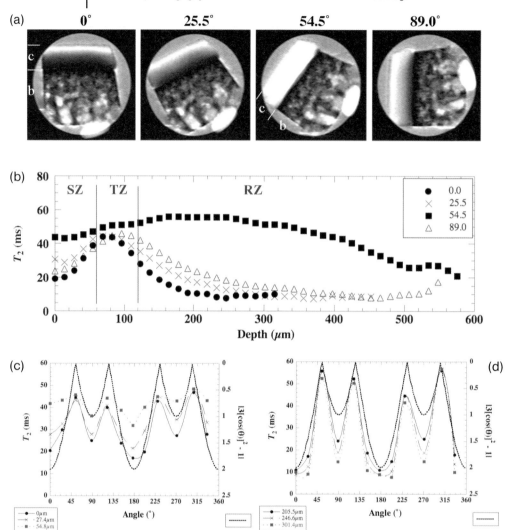

**Figure 19.2** (a) Proton images of one specimen imaged at four angles and with the echo time (TE) of 8.86 ms. At each orientation, four proton images were acquired, each with a different $T_2$ weighting. The same specimens were imaged at 12 particular orientations within the 0°–360° angular space. The $B_0$ direction was pointing vertically upwards. The cartilage-bone specimen was in a glass tube that had an internal diameter of 2.34 mm. The label 'c' indicates the cartilage tissue, which has strong orientational appearance in the magnetic field; the label 'b' indicates the bone. The bright oval feature at the bottom of the tube is the fatty component of the bone; (b) $T_2$ profiles of one specimen in the first quadrant of the 360° angle rotation. The $T_2$ anisotropy cross-sections in two zones; (c) superficial zone (SZ); (d) radial zone (RZ). These cross-sections enable the angular dependency of the tissue to be examined at a particular depth. The curve of $|(3\cos^2\theta - 1)|$ was also plotted in each graph, indicating a close correlation between the local minima in $|(3\cos^2\theta - 1)|$ and the local maxima in $T_2$. (Adapted from Ref. [21].)

is associated with groups of mutually perpendicular fibrils [21]. The results of these studies show that although μMRI does not have the resolution to identify individual collagen fibrils, the use of a $T_2$ anisotropy map and its $T_2$ anisotropy cross-section profiles can reveal the angular-dependent variations in the tissue by eliminating other biochemical and instrumental factors that contribute to the intensity variation in MR images. Moreover, there is also the potential to provide critical insights regarding the ultrastructure of macromolecules in cartilage and other biological tissues containing organized fibrils.

## 19.3
## Imaging the Proteoglycan Concentration by $T_1$ Contrast

Having understood the role of the collagen matrix in MRI of cartilage, attention is now turned to the other major macromolecule in cartilage, GAGs, the side chains of proteoglycans. As the heavily sulfated GAG molecules generate an osmotic pressure that contributes to the stiffness of articular cartilage as a load-bearing material, the reduction of GAGs in tissue will result in a biochemically and biomechanically weakened cartilage. Although proteoglycans possess no motional or orientational anisotropy in a simple environment, they may become anisotropic in the tissue. This is because the proteoglycans are compressed to only a fraction of their natural aqueous volume and entrapped to fill the extracollagen space in cartilage [23]; the highly confined proteoglycans could, therefore, be forced to align with the collagen fibrils due to their close molecular interactions.

Several approaches are available to obtain a depth-dependent profile of GAGs in articular cartilage; only one is totally nondestructive, and this is known in clinical MRI as the dGEMRIC procedure (delayed gadolinium-enhanced magnetic resonance imaging of cartilage). Based on the assumption that charged mobile ions will distribute in cartilage in an inverse relation to the concentration of the negatively charged GAG molecules, the GAG image can be constructed based on the two $T_1$ images acquired before and after the patient is injected with a charged MRI contrast agent, $Gd(DTPA)^{2-}$ [24–29]. Since $Gd^{3+}$ is a paramagnetic ion that can shorten the $T_1$ relaxation significantly, and since a small quantity of $Gd(DTPA)^{2-}$ is usually harmless to humans, the dGEMRIC protocol has become an important clinical procedure in the detection and management of joint diseases.

Our recent studies [30] in GAG imaging concerns the quantitative determination of the GAG profile by the *in vitro* version of the dGEMRIC protocol, in which an intact block of cartilage tissue is soaked directly in the $Gd(DTPA)^{2-}$. Figure 19.3 shows a set of quantitative images from one representative cartilage–bone specimen, where the Gd and GAG images in cartilage tissue were constructed two-dimensionally (Figure 19.3c and d). Figure 19.3e and f show the *averaged* profiles from three independent specimens. Several features can be observed. First, the error bars for the $T_{1before}$ and $T_{1after}$ measurement are small, which indicates the consistency of our quantitative $T_1$ measurement. Second, the concentration of $Gd(DTPA)^{2-}$ in the tissue decreases monotonically as the function of tissue depth.

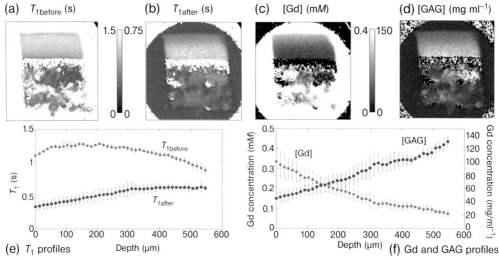

**Figure 19.3** $T_1$ images of cartilage before (a) and after (b) the specimen was immersed in 1 mM Gd(DTPA)$^{2-}$ solution; (c) The image of gadolinium concentration in cartilage; (d) The image of the GAG concentration in cartilage; (e) $T_1$ profiles before and after Gd(DTPA)$^{2-}$ immersion; (f) Concentration profiles of gadolinium and GAG in articular cartilage (averaged for three independent samples) [30].

Finally, the concentration of GAGs in cartilage is approximately a linear function, increasing from the superficial zone to the radial zone. By correlating the MRI results with GAG content in cartilage (using a histochemical method; data not shown) [30], it can be shown that the *in vitro* version of the dGEMRIC procedure can provide an accurate determination of the GAG concentration profiles in cartilage, in nondestructive manner.

## 19.4
## Imaging the Consequence of Mechanical Loading

Since the collagen fibrils can significantly influence the image intensity in MRI, might it be possible to use MRI to detect the consequence of a modified collagen matrix in cartilage? This modification to the collagen matrix could come from either of the following two origins. First, articular cartilage is a load-bearing tissue – it is bearing loads constantly, at any time a human or animal is in an upright position. Second, one of the typical events in the tissue degradation is damage to the fine structure of the collagen matrix, which reduces its resistance to loading-induced deformation.

Figure 19.4 shows a set of images, in which a healthy cartilage was imaged under static loading [31, 32]. Contrary to what has been observed in normal (unloaded) cartilage at the magic angle (Figure 19.2a), the loaded tissue at the magic angle

**(a)** µMRI maps at the magic angle

control    10% strain    20%strain

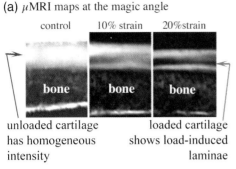

unloaded cartilage
has homogeneous
intensity

loaded cartilage
shows load-induced
laminae

**(b)** Re-orientation of fibrils

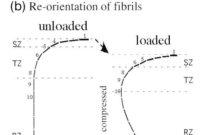

**Figure 19.4** (a) $T_2$-weighted proton images (at ~55°) of control cartilage, at ~10% and ~20% strain. The angle is defined as the angle between the normal to the articular surface of cartilage and the direction of the magnetic field ($B_0$); (b) A schematic model for the orientational adaptation of collagen fibrils across the cartilage depth as a result of mechanical compression based on $T_2$ anisotropy data. Each solid line represents the overall orientation of the collagen fibrils at this particular depth in cartilage; these lines are numbered to track the changes in zone boundaries upon compression. The left-hand figure (unloaded) shows the three classical zones in uncompressed articular cartilage (SZ, superficial zone; TZ, transitional zone; RZ, radial zone). The right-hand figure (loaded) shows the orientational changes at different depths due to external loading in cartilage. (Adapted from Ref. [31], with permission from the publisher.)

no longer exhibits a homogeneous appearance. Instead, the loaded cartilage exhibits a distinct laminar appearance *at the magic angle*. In addition, this load-induced laminar appearance becomes more profound at high strain levels. The $T_2$ anisotropy results exhibited a modified profile, which now become both depth- and load-dependent (data not shown). Based on these distinct alterations to µMRI $T_2$ anisotropy, the organizational modification of the collagen matrix due to external loading was modeled successfully (Figure 19.4b) [31, 32].

These results imply that static compression can become a controllable mechanism in MRI to induce new image contrast. This approach, using static loading as a tool to force the tissue to adapt to a new environment during microscopic imaging is, in every sense, a functional study of the tissue's structures and properties. Imaging a cartilage block while it is being loaded in µMRI is analogous to imaging a joint in a human while a force/weight is applied to the person in a clinical MRI scanner – both methods activate the primary role of articular cartilage in human, as a load-bearing media. The use of $T_2$ anisotropy maps and profiles enable the examination of fibril orientation and reorientation under loading. Moreover, a detailed knowledge of the load-modified $T_2$ anisotropy in cartilage could provide a better understanding of the structural modification to the functional adaptability of cartilage, which can facilitate the monitoring of tissue degradation more accurately and provide critical information towards an understanding and, ultimately, the prevention of arthritic diseases [33].

## 19.5
## The Power of Multidisciplinary Microscopic Imaging

During the past 15 years we have studied articular cartilage, initially using μMRI based not only on its ability to map the physical properties of viable cartilage in a near-native environment but also its role as a critical bridge between clinical MRI (which has insufficient resolution) and light microscopy (which can only be used to image thin sections). During the late 1990s, we began to incorporate PLM (the gold standard in histology) into our investigations as it can be used to image the collagen organization that modulates the μMRI signal. As these projects progressed, it became necessary to image the molecular concentrations in the early lesion cartilage as directly as possible, and at high resolution. Hence, in 2005 we began to use FTIRI, which is sensitive to the vibration of dipole moments of chemical bonds in tissue [34, 35]. In addition to these high-resolution imaging tools, we also used a variety of biomechanical and biochemical methods, each of which measured a unique aspect of the tissue's bulk properties and could be correlated with the spatially resolved changes in imaging. This multidisciplinary research approach is the mere recognition that many of today's biomedical problems are best addressed in this way, with each method having its own scientific merit.

The current, state-of-the-art understanding of articular cartilage is summarized in Figure 19.5, which highlights the many correlations among these multidisciplinary studies. Because cartilage at different depths may possess a unique combination of structures and properties (e.g. bioelectrochemical composition, solid/fluid interaction, morphological architecture, mechanical property), many aspects of these depth-dependent tissue properties can be interpreted beyond their usual meanings as measured. It is clear that, despite the fact that it is common in histology to divide noncalcified cartilage into three discrete zones, the physical properties, chemical composition and morphological features of articular cartilage vary *continuously* across its thickness (Figure 19.5a). The definition of discrete cartilage zones in histology, therefore, merely represents a conceptual 'discretization' of these continuous functions.

## 19.6
## Summary

In view of the molecular and ultrastructural changes due to early diseases, and the interdependent structure–function property relationships in tissues, the combined application of multidisciplinary techniques allows one to discriminate among the various factors/changes and how they influence the functional integrity of cartilage as a load-bearing material. Although neither μMRI, PLM nor FTIRI has sufficient resolution to identify individual collagen fibrils or other molecules, the multidisciplinary microimaging approach used in these studies can identify those subtle changes in the morphological structure and molecular concentration of cartilage

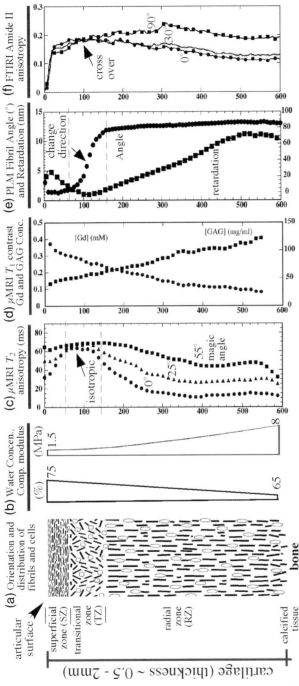

**Figure 19.5** The state-of-the-art understanding of articular cartilage based on microscopic imaging. (a) A schematic diagram of cartilage showing the orientation of the collagen fibrils (the short lines) in different histological zones and the orientation of the chondrocytes (circles and ovals) (not drawn to scale); (b) Approximate water concentration and the depth-dependent compressive modulus in cartilage [36]; (c) $T_2$ anisotropy profiles from μMRI at 14 μm resolution [20]; (d) Concentration profiles of gadolinium and GAG in articular cartilage by μMRI at 13 μm resolution [30]; (e) Angle and retardance profiles from PLM at 2.72 μm resolution [22]; (f) Amide II absorption anisotropy profiles from FTIRI at 6.25 μm resolution [34].

due to natural lesions and mechanical loading. Currently, these imaging tools are being used to probe different aspects of the structure and properties of cartilage at microscopic resolution, in order to predict and monitor the early changes in the tissue that lead to cartilage degradation as a disease entity.

## Acknowledgments

Y. Xia gratefully acknowledges support from the R01 grants from the National Institutes of Health (NIH AR045172) and thanks the students and staff in his laboratory who performed the original cartilage imaging studies. The authors are grateful to Miss Carol Searight (Department of Physics, Oakland University) and Miss Aimee Xia for editorial comments on the text.

## References

1 The Centers for Disease Control and Prevention, Press Release (2002) (http://www.cdc.gov/od/oc/media/pressrel/r021024.htm) (accessed 24 October 2002).

2 Adams, M.E., Matyas, J.R., Huang, D. and Dourado, G.S. (1995) Expression of proteoglycans and collagen in the hypertrophic phase of experimental osteoarthritis. *The Journal of Rheumatology. Supplement*, **43**, 94–7.

3 Maroudas, A. (1975) Biophysical chemistry of cartilaginous tissues with special reference to solute and fluid transport. *Biorheology*, **12**, 233–48.

4 Venn, M. and Maroudas, A. (1977) Chemical composition and swelling of normal and osteoarthritic femoral head cartilage. *Annals of Rheumatic Diseases*, **36** (2), 121–9.

5 Maroudas, A., Bayliss, M.T. and Venn, M. (1980) Further studies on the composition of human femoral head cartilage. *Annals of Rheumatic Diseases*, **39** (5), 514–34.

6 Miosge, N., Flachsbart, K., Goetz, W., Schultz, W., Kresse, H. and Herken, R. (1994) Light and electron microscopical immunohistochemical localization of the small proteoglycan core proteins decorin and biglycan in human knee joint cartilage. *The Histochemical Journal*, **26** (12), 939–45.

7 Hagg, R., Bruckner, P. and Hedbom, E. (1998) Cartilage fibrils of mammals are biochemically heterogeneous: differential distribution of decorin and collagen IX. *The Journal of Cell Biology*, **142** (1), 285–94.

8 Iozzo, R.V. (1999) The biology of the small leucine-rich proteoglycans. Functional network of interactive proteins. *Journal of Biological Chemistry*, **274** (27), 18843–6.

9 Clarke, I.C. (1971) Articular cartilage: a review and scanning electron microscope study. *The Journal of Bone and Joint Surgery*, **53B** (4), 732–50.

10 Bayliss, M., Venn, M., Maroudas, A. and Ali, S.Y. (1983) Structure of proteoglycans from different layers of human articular cartilage. *Biochemical Journal*, **209**, 387–400.

11 Maroudas, A., Wachtel, E.J., Grushko, G., Katz, E.P. and Weinberg, P. (1991) The effect of osmotic and mechanical pressures on water partitioning in articular cartilage. *Biochimica et Biophysica Acta*, **1073**, 285–94.

12 Xia, Y. (2000) Magic angle effect in MRI of articular cartilage – A review. *Investigative Radiology*, **35**(10), 602–21.

13 Mow, V.C. and Guo, X.E. (2002) Mechano-electrochemical properties of articular cartilage: their inhomogeneities and anisotropies. *Annual Review of Biomedical Engineering*, **4**, 175–209.

**14** Berendsen, H.J. (1962) Nuclear magnetic resonance study of collagen hydration. *Journal of Chemical Physics*, **36** (12), 3297–305.

**15** Fullerton, G.D., Cameron, I.L. and Ord, V.A. (1985) Orientation of tendons in the magnetic field and its effect on $T_2$ relaxation times. *Radiology*, **155**, 433–5.

**16** Peto, S., Gillis, P. and Henri, V.P. (1990) Structure and dynamics of water in tendon from NMR relaxation measurements. *Biophysical Journal*, **57** (1), 71–84.

**17** Lehner, K.B., Rechl, H.P., Gmeinwieser, J.K., Heuck, A.F., Lukas, H.P. and Kohl, H.P. (1989) Structure, function, and degeneration of bovine hyaline cartilage: assessment with MR imaging *in vitro*. *Radiology*, **170**, 495–9.

**18** Hayes, C.W. and Parellada, J.A. (1996) The magic angle effect in musculoskeletal MR imaging. *Topics in Magnetic Resonance Imaging*, **8** (1), 51–6.

**19** Modl, J.M., Sether, L.A., Haughton, V.M. and Kneeland, J.B. (1991) Articular cartilage correlation of histologic zones with signal intensity at MR imaging. *Radiology*, **181** (3), 853–5.

**20** Xia, Y. (1998) Relaxation anisotropy in cartilage by NMR microscopy (μMRI) at 14 μm resolution. *Magnetic Resonance in Medicine*, **39** (6), 941–9.

**21** Xia, Y., Moody, J. and Alhadlaq, H. (2002) Orientational dependence of $T_2$ relaxation in articular cartilage: a microscopic MRI (μMRI) study. *Magnetic Resonance in Medicine*, **48** (3), 460–9.

**22** Xia, Y., Moody, J., Burton-Wurster, N. and Lust, G. (2001) Quantitative in situ correlation between microscopic MRI and polarized light microscopy studies of articular cartilage. *Osteoarthritis and Cartilage*, **9** (5), 393–406.

**23** Muir, H. (1983) Proteoglycans as organizers of the intercellular matrix. *Biochemical Society Transactions*, **11**, 613–22.

**24** Bashir, A., Gray, M.L. and Burstein, D. (1996) Gd-DTPA$^{2-}$ as a measure of cartilage degradation. *Magnetic Resonance in Medicine*, **36** (5), 665–73.

**25** Trattnig, S., Mlynarik, V., Breitenseher, M. *et al.* (1999) MRI visualization of proteoglycan depletion in articular cartilage via intravenous administration of Gd-DTPA. *Magnetic Resonance Imaging*, **17** (4), 577–83.

**26** Nieminen, M.T., Rieppo, J., Silvennoinen, J. *et al.* (2002) Spatial assessment of articular cartilage proteoglycans with Gd-DTPA-enhanced $T_1$ imaging. *Magnetic Resonance in Medicine*, **48** (4), 640–8.

**27** Nieminen, M.T., Toyras, J., Laasanen, M.S., Silvennoinen, J., Helminen, H.J. and Jurvelin, J.S. (2004) Prediction of biomechanical properties of articular cartilage with quantitative magnetic resonance imaging. *Journal of Biomechanics*, **37** (3), 321–8.

**28** Samosky, J.T., Burstein, D., Eric Grimson, W., Howe, R., Martin, S. and Gray, M.L. (2005) Spatially-localized correlation of dGEMRIC-measured GAG distribution and mechanical stiffness in the human tibial plateau. *Journal of Orthopaedic Research*, **23** (1), 93–101.

**29** Wedig, M., Bae, W., Temple, M., Sah, R. and Gray, M. (2005) [GAG] profiles in "normal" human articular cartilage. Proceedings, 2005 Orthopedic Research Society Meeting, Washington, DC, p. 0358.

**30** Xia, Y., Zheng, S.K. and Bidthanapally, A. (2008) Depth-dependent Profiles of Glycosaminoglycans in Articular Cartilage by μMRI and Histochemistry. *Journal of Magnetic Resonance Imaging*, **28** (1), 151–7.

**31** Alhadlaq, H. and Xia, Y. (2004) The structural adaptations in compressed articular cartilage by Microscopic MRI (μMRI) $T_2$ anisotropy. *Osteoarthritis and Cartilage*, **12** (11), 887–94.

**32** Alhadlaq, H.A. and Xia, Y. (2005) Modifications of orientational dependence of microscopic magnetic resonance imaging T(2) anisotropy in compressed articular cartilage. *Journal of Magnetic Resonance Imaging*, **22** (5), 665–73.

**33** Alhadlaq, H., Xia, Y. Moody, J.B., and Matyas, J. (2004) Detecting structural changes in early experimental osteoarthritis of tibial cartilage by Microscopic MRI and polarized light microscopy. *Annals of Rheumatic Diseases*, **63** (6), 709–17.

**34** Xia, Y., Ramakrishnan, N. and Bidthanapally, A. (2007) The depth-dependent anisotropy of articular cartilage by Fourier-transform infrared imaging

(FTIRI). *Osteoarthritis and Cartilage,* **15** (7), 780–8.

**35** Ramakrishnan, N., Xia, Y. and Bidthanapally, A. (2007) Polarized IR microscopic imaging of articular cartilage. *Physics in Medicine and Biology,* **52** (15), 4601–14.

**36** Chen, S.S., Falcovitz, Y.H., Schneiderman, R., Maroudas, A. and Sah, R.L. (2001) Depth-dependent compressive properties of normal aged human femoral head articular cartilage: relationship to fixed charge density. *Osteoarthritis and Cartilage,* **9** (6), 561–9.

# 20
# MRI of Water Transport in the Soil–Plant–Atmosphere Continuum

*Henk Van As, Natalia Homan, Frank J. Vergeldt and Carel W. Windt*

## 20.1
## Introduction

The availability of water is one of the major factors that affects plant production, yield and reproductive success. Water controls, at least in part, the distribution of plants over the Earth's surface. For growth, plants need to take up $CO_2$ for photosynthesis, and evaporative water loss is an inevitable consequence of this uptake. Transpiration in higher plants accounts for about one-eighth of all of the water that evaporates to the atmosphere over the entire globe, and for about three-quarters of all water that evaporates from the land. Long-distance transport in plants thus directly affects the global water cycle, and with that, global climate [1].

Changes in global climate are expected, which include increased $CO_2$, global warming and periods of drought and flooding. These changes will affect the (composition of) ecosystems and agricultural production in many regions of the planet, and an understanding of the short- and long-term responses of plants to climate change is therefore crucial. A key parameter in this understanding is the plant hydraulic conductance [2, 3].

### 20.1.1
### Transport in the Soil–Plant–Atmosphere Continuum (SPAC)

Water and soluble compounds are passively transported inside plant xylem conduits (vessels and tracheids) in the continuum between soil and atmosphere along a water potential gradient. When this gradient becomes too steep it causes damage either by dehydration of living cells or by cavitation due to tensions (negative pressures) in the water columns of the xylem being too high [2, 3]. Cavitations result in air embolisms that inhibit water transport in the plant and affect the plants' ability to grow and take up $CO_2$. Therefore, mechanisms are needed to maintain this gradient within a nondamaging range. The most important mechanism is regulation of the stomatal aperture in the leaves, by increasing the resistance for water vapor leaving the leaves into the atmosphere with a lower water content. The

*Magnetic Resonance Microscopy.* Edited by Sarah L. Codd and Joseph D. Seymour
Copyright © 2009 WILEY-VCH Verlag GmbH & Co. KGaA, Weinheim
ISBN: 978-3-527-32008-0

hydraulic conductivity of the root and stem, together with the plants' stomatal regulation, define the water potential gradients that exist between leaf and root.

In its path from the soil to the atmosphere, water must be transported through the soil, pass living root tissue (radial transport) to reach the root xylem, move inside the xylem conduits up to the leaves, pass living tissue in the leaves, evaporate at the cell wall surfaces in the leaves and then diffuse through the intercellular air spaces and stomata to the atmosphere. In analyzing the components of the soil–atmosphere continuum, it is useful to distinguish the hydraulic conductivity ($K$, conductance per unit length) of the different components. $K$ is defined as

$$K = -\frac{Q}{(d\Psi/dx)} \tag{20.1}$$

where $Q$ is the volume flow rate, $\Psi$ the water potential driving the flow (pressure in soil and xylem, pressure and osmotic components in tissue like roots and leaves) and $x$ is the distance along the flow path. Equation 20.1 is very comparable to Darcy's law. For ease of comparison, $K$ is usually expressed relative to an area transverse the flow path. In doing so, $K$-values for soil ($K_s$), root ($K_r$), xylem ($K_x$) and leaves ($K_l$) can be compared [2].

$K_s$ and the different plant $K$ values are not constant, but depend on a number of factors, including the driving force itself. Plant $K$ values may be subject to short-term changes. $K_r$ and $K_l$ values for instance depend on the presence of aquaporins (water channel-forming membrane proteins), which can be opened or closed depending on internal (e.g. pH) and external factors (e.g. mechanical stress, temperature) [4, 5].

There are important similarities between the flow of water in soils and in the xylem [2]. Bulk flow in both media occurs through pores and is driven by pressure differences (including matrix potential for soil and tension in xylem). Depending on the plant species, the conduits in the xylem consist of relatively wide xylem vessels and/or narrower tracheids, which are interconnected by much narrower channels of the connecting pits. The permeability of the pits largely determines the total xylem conductance and depends on the $K^+$-concentration of the xylem sap.

A second path of water transport is in the phloem; this is responsible for the transport of photosynthates such as sucrose from the leaves to the rest of the plant.

## 20.1.2
### The Importance of Models

There is a strong need for eco-biophysical plant models to predict the dynamics in plant evaporation in relation to photosynthesis, growth and production, under variable climate and soil conditions, including stress conditions.

The cohesion–tension theory [6] and the resistance flow model [7] describe a plant or tree as an integrated hydraulic system. However, reality is more complex. Several types of eco-biophysical plant model have been drawn up (e.g. [1, 8–10]).

Some models already are used in horticultural practice, for example to estimate crop yield or to control greenhouse climate, while others are in use as parts of models used to predict the global water cycle and global climate. One of the most complete tree models currently available includes xylem hydraulics, leaf microclimatic factors with feedback signals from tree and soil water content [10]. However, the important interaction between xylem, phloem and sugar content/transport [8] and the dynamic behavior of phloem and xylem characteristics (e.g. $K_x$) has not yet been included. One reason for this is that little is known about the dynamic behavior of phloem and xylem transport and hydraulic conductance in the living plant, in relation to soil and plant water status (storage pools), photosynthetic activity and sugar content. Although the basic principles that underlie xylem and phloem transport have been known for about a century, their translation into accurate models has remained exceedingly difficult. The problem is that a number of key parameters cannot be measured by means of the techniques currently available, most of which are invasive, whereas both xylem and phloem are extremely sensitive to invasive experimentation [11, 12]. The pressure gradients that drive translocation in both systems are easily disturbed by cutting or puncturing.

## 20.1.3
### Existing Methods for Measuring Transport and Related Parameters in the SPAC

Among others, leaf water potential, stem water potential and trunk diameter fluctuations are in use as indicators of plant water status [13]. Most of these methods are invasive, lack automation, or are difficult to interpret. Recently, time domain reflectometry and a diversity of different capacitive techniques, by measuring dielectric constant, have been used to estimate soil and stem water content (e.g. [14]). Thermal leaf or canopy imaging as an indicator of stomatal opening is currently under evaluation to monitor water stress in plants [15].

On the cell level, the cell pressure probe has been proven to be very valuable to measure water (and solute) membrane permeability, either diffusional ($Pd$) or under hydrostatic or osmotic pressure gradients ($Pf$) [16]. On the tissue level, the pressure bomb and root pressure probe allow water potential and tissue hydraulics to be measured [17]. However, whilst these techniques can be applied to excised roots, leaves and other pieces of a plant, and are clearly very informative, they are destructive. Hydraulic conductivity has been studied by the use of the high-pressure flow meter [18], which can also only be applied to excised plant parts. Xylem pressure probes have been used to measure xylem tension in intact plants, but the practical applicability of the technique is limited [11].

For several decades heat tracer methods have been used to measure mass flow in the xylem. Here, the placement of the sensor itself is a source of error, particularly for the heat-pulse method [19]. The calculation of mass flow rates from sap velocities obtained by heat pulse techniques requires a reliable estimate of the actual sap-conducting surface area. However, an accurate estimate of the actual flow conducting area is extremely difficult to obtain, as changes in flow-conducting area occur under changing environmental conditions [3, 20], even without the occurrence of embolisms. These factors can result in substantial errors of actual sap flow measurement based on heat pulse techniques.

Thus, a truly noninvasive measurement of sap flow and active flow-conducting area in intact plants is required that will allow one to study the dynamics (day/night, stress responses) therein in relation to water content in the surrounding system. In the following we summarize several magnetic resonance imaging (MRI) methods that are currently available for this task. In addition, the water content in soils, stem tissue and leaves can be monitored; however, the measurement of transport in intact root and leaves remains a challenge.

## 20.2
### Mapping Transport and Related Parameters in the SPAC by MRI Approaches

The application of *in vivo* nuclear magnetic resonance (NMR) and MRI to plants has brought significant contributions across a wide range of topics. For some recent reviews, the reader is referred to Refs [21–25].

### 20.2.1
### The Soil–Root System

#### 20.2.1.1 Soil Water Content and Root Anatomy
Root water uptake mechanisms belong to the most challenging topics in soil science. Transport in soils is quite slow and can best be approached by temporally resolved water content imaging [26–28]. For a thorough understanding, 3-D monitoring techniques for determining water content changes must be combined with 3-D root anatomy. The determination of water content in soils is complicated by the presence of paramagnetic/ferromagnetic impurities and small and partially filled pores, both of which may cause significant increases in $1/T_2$ and $1/T_1$ [29, 30]. The MRI approaches are very comparable to those used to map moisture migration in food materials, as well as the problems of calibration in terms of water content (see Van Duynhoven *et al.*, in Chapter 21). At low water content the $T_2$ values are rather short and single-point imaging (SPI) -type experiments are needed [31]. At higher water contents (and depending on the type of soil), multiple spin-echo (MSE), turbo-SE (TSE) or RARE (Rapid Acquisition with Relaxation Enhancement) methods can be used. Root anatomy can be imaged using 3-D RARE with sufficient resolution; some combined results of multislice, multiecho imaging for soil water content and 3-D RARE for root anatomy are presented in Figure 20.1.

#### 20.2.1.2 Transport in Roots
Water transport in roots is complex, and has been described in terms of a composite root transport model [11]. This model considers three pathways for water transport: the extracellular or apoplastic pathway; the symplastic pathway; and the transcellular pathway. The latter two pathways involve cell-to-cell transport

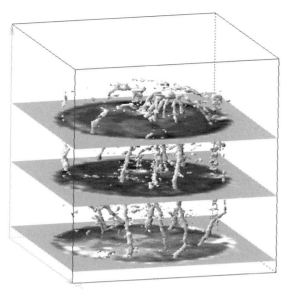

**Figure 20.1** Water content changes in a sandy soil around the root system of a *Ricinus* plant, using 3 T vertical-bore MRI. The root anatomy was obtained by 3-D RARE, and soil water content by multislice 2-D multiecho imaging ($T_E = 6.76$ ms; $N_E = 128$ with an isotropic resolution of 3.1 mm). The (extrapolated) signal intensity ($S_0$) was correlated to the water content. Field of view: $10 \times 10$ cm$^2$. Illustration courtesy of A. Pohlmeier, F.J. Vergeldt, E. Gerkema, J. Vanderborght, M. Javaux, H. Vereecken, H. Van As, M.I. Menzel, D. van Dusschoten.

and cannot be discriminated by present techniques. Apoplastic barriers occur in the endodermis and exodermis, and therefore cell-to-cell transport – including transport over membranes and via plasmodesmata – is very important. It is for this reason that membrane water permeability plays such an important role. Because of a difference in the reflection coefficient for nutrients in the apoplast and the membranes, the cotransport of water and nutrients is different in the apoplast and the cell-to-cell path, and this results in different driving forces being employed – pressure (tension) in the apoplast, and osmotic potentials in the cell-to-cell path.

Membrane permeability in tissue containing vacuolated plant cells can be obtained from the observed $T_2$ value of water in these cells. Vacuolar $T_{2,\text{obs}}$ can be described as a function of the bulk $T_2$ ($T_{2,\text{bulk}}$) of the water and the probability that water molecules reach the surrounding membrane and lose magnetization at the membrane by passing the membrane and entering a compartment with a (much) shorter relaxation time [32]. No evidence has been found that membranes themselves act as a relaxation sink [33, 34]. The probability to reach the membrane is defined by the diffusion time and is thus directly related to the compartment radii. As a result, the observed relaxation time depends, in addition to $T_{2,\text{bulk}}$, on the surface-to-volume ratio ($S/V$) and the net loss of magnetization at the compartment boundary, the so-called magnetization sink strength ($H$; units $m\,s^{-1}$) [32]:

$$\frac{1}{T_{2,\text{obs}}} = H\left(\frac{S}{V}\right) + \frac{1}{T_{2,\text{bulk}}}$$

(20.2)

where $H$ is linearly related to the actual membrane permeability [35]. Equation 20.2 is valid only if the diffusion time to traverse the cell is shorter than the bulk relaxation time in that compartment: $(R^2/2D) < T_{2,\text{bulk}}$. Care must be taken into account by using $T_{2,\text{obs}}$ from images, as the observed $T_2$ value in images with respect to its value in nonimaging NMR depends on a number of contributions that relate to details of the imaging experiment, as well as to the characteristics of the plant tissues under observation. These include diffusive attenuation in the position-encoding gradient and in local field gradients originating from suscepti-bility artifacts, in combination with longer TE values [36].

For a proper interpretation of $T_{2,\text{obs}}$ in terms of membrane permeability it is essential that the value of $S/V$ is known (cf. Equation 20.2). This information can be obtained by measuring the apparent diffusion coefficient ($D_{\text{app}}$) as a function of the diffusion labeling time, $\Delta$. For diffusion in a confined compartment, free diffusion is observed at short diffusion times. However, at increasing $\Delta$ the diffu-sion becomes restricted, but the averaging of local properties over a large enough distance does not yet occur. In that regime $D_{\text{app}}$ depends linearly on the square root of $\Delta$, and the slope is determined by $S/V$ of the compartment, irrespective of whether this compartment is connected or disconnected [37]. This $S/V$ value can be used directly to obtain $H$ from $T_{2,\text{obs}}$ (Equation 20.2). At long diffusion times, hindered diffusion is observed (cf. Figure 20.4), which is defined by the permeabil-ity, and for plants reflects cell-to-cell transport. The effective permeability $P$ can now be estimated. A number of approaches for different systems has been pre-sented [37], but all improperly ignore the effect of differences in relaxation time in the different compartments [35].

For diffusion through a geometry consisting of a series of semipermeable mem-branes (thin walls) separated by a distance $d$, Crick [38] obtained:

$$P = \frac{(D_{\text{inf}} D_0)}{(D_0 - D_{\text{inf}})d}$$

(20.3)

where $D_0$ and $D_{\text{inf}}$ are the $D$-values in the limit of $\Delta$ to zero or to infinity, respec-tively. If we know $D_0$, $D_{\text{inf}}$ and $d$, we can obtain $P$. In practice the experimental range of $\Delta$ values is limited by the relaxation times $T_2$ and $T_1$, $D(\Delta)$ is obtained over a smaller range of $\Delta$ values, and $D_{\text{inf}}$ is difficult to obtain. A fitting procedure to extract out $d$, $D_0$ and $D_{\text{inf}}$ based on a limited range of $D(\Delta)$ values has been pre-sented [39, 40]. Alternatively, $D_0$ can be estimated from the initial slope at short $\Delta$ values, if experimentally available. $P$ represents cell-to-cell transport and includes the permeability of the tonoplast, plasmalemma, walls and plasmodesmata. There-fore, $P$ is not identical to $H$ as obtained from $T_2$ measurements (Equation 20.2).

In order to discriminate between the different transport pathways/compartments, we must further unravel the NMR signals. Water in different cell compartments can best be discriminated on basis of the differences in relaxation behavior ($T_2$) and (restricted) diffusion behavior. By combined relaxation and diffusion measurements, together with analysis methods such as inverse Laplace transform (ILT) [41, 42] or alternatives [43], correlated $D$ and $T_2$ values can be generated, which greatly enhance the discrimination of different water pools in subcellular compartments. In this way an unambiguous correlation between relaxation time and compartment size can be obtained, resulting in a general approach to quantify water in the different cell compartments. This approach is very promising in nonspatially resolved measurements [44, 45], and has been shown to allow sub-pixel information in imaging mode also to be obtained [46, 47].

Some preliminary results of this approach are now available. By combining the results of $T_2$ and $D$ measurements on water in apple parenchyma (Granny Smith), $H$ was found to be approximately $1 \times 10^{-5}\,\mathrm{m\,s^{-1}}$ ($T_{2,\mathrm{obs}} = 1.25\,\mathrm{s}$; $R = 86\,\mu\mathrm{m}$). Under the assumption of parallel planes, $P = 2.9 \times 10^{-6}\,\mathrm{m\,s^{-1}}$ [45]. For Cox apple parenchyma cells a tonoplast water membrane permeability $P_\mathrm{d} = 2.44 \times 10^{-5}\,\mathrm{m\,s^{-1}}$ was reported [48] based on the Conlon–Outhred method. In maize roots, a higher value of $P$ was found of approximately $5 \times 10^{-5}\,\mathrm{m\,s^{-1}}$ [39]. Recently, values of $P$ were obtained in excised roots of normal and osmotically stressed maize and pearl millet plants: here, $P$ was approximately $3 \times 10^{-5}\,\mathrm{m\,s^{-1}}$ for both normal and stressed maize, but was about $9 \times 10^{-5}\,\mathrm{m\,s^{-1}}$ for normal pearl millet plants and $3 \times 10^{-5}\,\mathrm{m\,s^{-1}}$ for stressed plants (T.A. Sibgatullin and H. Van As, unpublished results). Ionenko *et al.* [49] reported $P$-values for water in roots of maize seedlings of $3 \times 10^{-5}\,\mathrm{m\,s^{-1}}$, but these decreased by a factor of 1.7 due to water stress or $HgCl_2$ treatment. The latter clearly demonstrates the contribution of aquaporin function towards $P$.

## 20.2.2
## The Stem

### 20.2.2.1 Water Content

The stem normally has a relatively high water content that can easily be measured quantitatively using MSE, which results in an amplitude map and a $T_2$ map. The amplitude in each pixel represents the amount of water in the pixel volume (equal to water content × tissue density). Nonquantitative SE methods have been used to study the refilling of embolized vessels by monitoring changes in signal intensity [50–53]. However, this approach proved to be insufficient because refilling does not automatically result in a restoration of flow; and because vessels that do not exhibit cavitation may stop conducting flow [54].

In addition to the water content, $T_2$ maps are obtained by MSE. As discussed in Section 20.2.1.2, $T_{2,\mathrm{obs}}$ in combination with information on the vacuole $S/V$ ratio provides access to information on membrane water permeability (cf. Equation 20.2). In intact pearl millet plants $H$ was shown to change during osmotic stress experiments; however, in maize plants no changes were observed [32, 34], which suggested that the response to stress was most likely related to changes in membrane permeability due to the presence and functioning of aquaporins (cf. Section 20.2.1.2).

In general, a monoexponential fit is used to construct the amplitude and $T_2$ maps from MSE experiments. More detailed information on water balance in the different cell compartments is available, and can be extracted by summing the signal decay of tissue types, selected on basis of the $T_2$ maps, to improve the signal-to-noise ratio (SNR) [55].

### 20.2.2.2 Axial Transport in Xylem and Phloem

Both, nonimaging and imaging NMR methods have been applied to study water transport in the xylem and phloem of intact plants (for recent overviews, see Refs [23, 25]). In these investigations, the research groups have used a variety of techniques, most of which were based on (modified) pulsed-field gradient (PFG) methods, either by using a limited number of PFG steps or by (difference) propagator approaches (see below). Flow measurements based on the uptake and transport of (paramagnetic) tracers have also been used [51].

The first (nonimaging) method to measure xylem water transport in plants, presented some 20 years ago [56, 57], is based on a series of equidistant identical radiofrequency (RF) pulses (in the range of 30°–180°) applied in the presence of a static magnetic field gradient in the direction of the flow. This method allows the averaged flow velocity and volume flow to be obtained, and also the ratio of the effective flow-conducting area. However, in order to interpret the data acquired in terms of averaged flow velocity a calibration is required. The results depend on the actual flow profile which, within the total cross-section of a stem, is not known à priori. Moreover, whilst neither xylem nor phloem flow can be discriminated using this method; rather, the sum of both is observed.

Flow profile and direction are best obtained by using PFG $q$-space or propagator techniques, whereby discrimination can be made simultaneously between non-flowing water molecules (stationary water in cells and nonactive vessels) and flow. The propagator $P(R,\Delta)$ represents the probability that a spin at any initial position is displaced by a distance $R$ in time $\Delta$ (cf. Figure 20.2a). This type of measurement has been applied to measure flow in many porous systems (e.g. [58, 59]).

One crucial step when quantifying the flow and flow-conducting area is to separate the contributions of nonflowing and flowing water, both of which are present in a single pixel in the plant situation. The propagator for free, unhindered, diffusing water has a Gaussian shape, centered around $R = 0$ (Figure 20.2a). The root mean square displacement of diffusing protons, as observed by NMR, is proportional to $\sqrt{(\Delta \times D)}$, and is directly related to the width of the Gaussian distribution (Figure 20.2a). In contrast, the mean displacement $R$ of flowing protons is linearly proportional to $(\Delta \times v_{av})$, where $v_{av}$ is the average flow velocity of the flowing protons. The fact that the propagator for stationary water is symmetric around displacement zero is used to separate the stationary from the flowing water. The signal in the nonflow direction is mirrored around zero and subtracted from that in the flow direction, to produce the displacement distribution of flowing and stationary water (Figure 20.2a). The probability (amplitude) of the propagator of

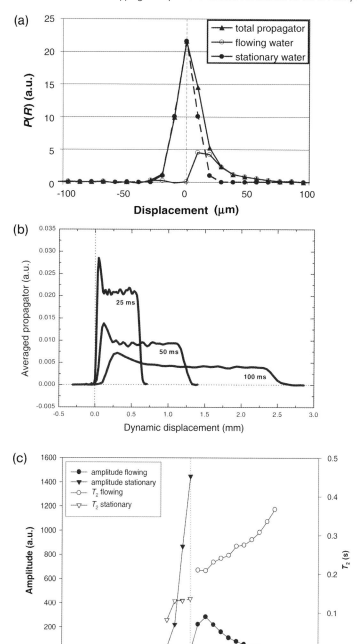

**Figure 20.2** (a) Total propagator and the propagators of
flowing and nonflowing water (deduced) in the xylem region
in the stem of a plant; (b) Propagator of laminar flow in a
capillary of 100 μm radius as a function of Δ; (c) Propagator
of flowing and nonflowing water in the stem of a tomato
plant and the correlated $T_2$ values per displacement step.

flowing water at $R = 0$ then becomes zero. For laminar flow in small capillaries ($r < 100\,\mu m$), at actual values of $\Delta$ this is a correct assumption [60] (Figure 20.2b). At increasing $\Delta$-values the diffusional averaging results in a shift of the lower velocities (or stationary water at the wall) towards higher velocities. A further argument for this assumption can be found in combined PFG propagator-$T_2$ measurements [47]. If the $T_2$ decay for each step in the propagator is analyzed, then stationary water has a different $T_2$ from flowing water, which reflects the different cell and vessel environments for water. At zero displacement, $T_2$ relates to that of stationary water only (Figure 20.2c). It is clear, however, that water in the vessels cannot be considered as totally isolated from water in the surrounding tissues, as exchange or radial transport has been shown to take place (see Section 20.2.2.3). At present, it is not clear whether this may complicate the discrimination between stationary and flowing water (see below).

Because the signal amplitude is proportional to the density of the mobile protons, the integral of the propagator provides a measure of the amount of water. The average velocity of the flowing water can be calculated by taking the amplitude-weighted average of the velocity distribution. The volume flow rate or flux is calculated by taking the integral of (amplitude × displacement of the velocity distribution). The value of $P(R,0)$ has no effect on calculating the volume flow rate, but will affect the calculated average velocity and the flow-conducting area [61]. A propagator flow-imaging method based on RARE was developed that allowed the flow profile of every pixel in an image to be recorded quantitatively, with a relatively high spatial resolution, while keeping measurement times down to 15–30 min [55, 62].

The accuracy of this approach for quantifying the sap flow and effective flow-conducting area is deduced from a comparison of the flow characteristics of water obtained by MRI, with results obtained for root water uptake and optical microscopy of the vessel dimensions. When the correct MRI settings are used the flow results agree closely with results obtained by monitoring root water uptake [54, 55]; however, care must be taken to cover the full dynamic range of the sap velocity [20, 54]. The successful discrimination of stationary and flowing water in single pixels containing a single vessel has been demonstrated by the agreement between the calculated cross-sectional area of some large vessels based on the MRI results and on microscopic inspection [54]. Parameters obtained by MRI range within 10% of those from microscopy. To the best of our knowledge, no other method is available at present to obtain this type of data in intact plants.

In MRI the relaxation behavior of water in porous materials is known to be influenced by pore diameter. If the $T_2$ relaxation behavior of xylem sap is affected by the xylem conduit diameter, and whether this must be corrected for in quantitative MR flow imaging has been investigated [63]. $T_2$-resolved flow imaging [47] has been performed on stem pieces with different conduit sizes through which water was pumped. At 0.7 T the average weighted $T_2$ of the flowing water decreased with conduit size, which confirmed that, in the xylem, a relationship between $T_2$ and conduit diameter does indeed exist. When the $T_2$ effects were not corrected for, the largest quantification error was ~30% (volume flow) in the sample with the shortest $T_2$. It was concluded that a $T_2$ correction only becomes critical when the flow labeling time approaches the $T_2$ of the flowing water. Surprisingly, in all cases the flow-conducting area (MR flow imaging) was significantly smaller than the total conduit lumen cross-sectional area (by microscopy). In samples with the widest conduits the lowest percentage of xylem lumen cross-sectional area was found to conduct water (down to 31% of the total). In samples with narrower conduits, the highest percentage of the total xylem area was found to be active, conducting water (up to 86%).

Windt *et al.* [30] further optimized the propagator fast-imaging method when quantitatively measuring, for the first time, the detailed flow profiles of phloem flow in large and fully developed plants. This approach allowed a straightforward assessment not only of the dynamics of phloem and xylem flow but also of the flow-conducting area. The observed day–night differences in flow-conducting area, which relate directly to xylem and phloem hydraulics, were among the most striking observations [20, 54], and demonstrated the potential of this method for studying hydraulics in intact plants under both normal and stress conditions.

### 20.2.2.3 Radial Transport and Exchange

As stated above (see Section 20.1.1), xylem and phloem hydraulics and plant water content/plant water potential are important parameters that determine evaporation, growth and stress responses. Water in the xylem and in the surrounding tissues are coupled by a resistance that is determined by the bordered pits in the xylem and the resistance of the (radial) cell-to-cell pathway in the stem. Recently, two MRI approaches have been used in an attempt to characterize these resistances.

In the first approach, the amount of flowing (and stationary) water is measured in a propagator as a function of $\Delta$, in analogy to the approach for studying mass transfer between water in porous beads and the flowing water in chromatography columns [64].

An example of propagators in the xylem region of the stem of a poplar tree, as a function of $\Delta$, is shown in Figure 20.3a. Due to exchange with the nonflowing water, the amplitude of the propagator of flowing water increases and the maximum velocity decreases. Diffusional averaging within the vessels then results in a decrease in the width of velocity distribution, while the amplitude of the nonflowing water decreases at increasing $\Delta$. In Figure 20.3b the amplitude of the stationary and flowing water is plotted as a function of $\Delta$, with the results showing a clear exchange between the two water pools, mainly by the passage of water over the bordered pits in the vessel walls.

In order to quantify this exchange the population of the stationary water pool is plotted as a function of $\Delta$ [64]. However, the amplitudes of stationary and flowing water are differently affected by $T_1$ relaxation. The correction of such effect by using the observed $T_1$, as suggested elsewhere [65], does not offer a solution as the observed $T_1$ values of the water pools are also affected by exchange! Hence, the need for computer simulations to better understand this problem and to find a solution, is clear.

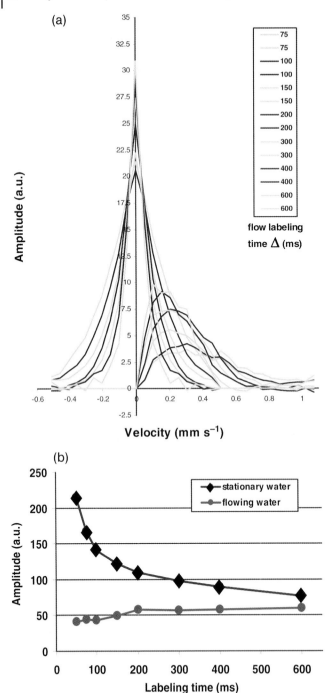

**Figure 20.3** (a) Propagators within the selected xylem region in the stem of a poplar tree as a function of Δ. The total propagator has been split into a propagator for nonflowing and flowing water; (b) Integrated amplitude of the propagators of flowing and nonflowing water as a function of Δ.

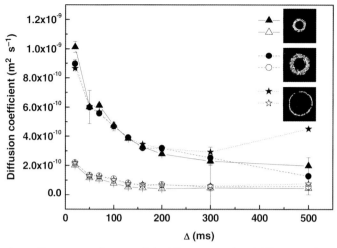

**Figure 20.4** $D_{app}$ as a function of the diffusion time $\Delta$ in three different xylem regions (see masks) in the stem of a *Viburnum* tree. The $D_{app}$ values have been resolved on the basis of $T_2$ values by combined $D$-$T_2$ measurements. Hindered diffusion is observed at long diffusion times, and can be used to define the cell-to-cell or vessel-to-vessel transport and tissue permeability.

In the second approach, $D_{app}(\Delta)$ is measured in the radial direction perpendicular to the flow in order to determine tissue permeability (cf. Section 20.2.1.2). Combined $D$-$T_2$ or flow-$T_2$ measurements have also been performed to resolve the water in vessels and in tracheids/parenchyma cells. The highest $T_2$ values observed relate to the highest $D$ value (Figure 20.4) and correspond to the $T_2$ of flowing water in xylem vessels. The lower $T_2$ values represent stationary water, most probably in tracheids and (ray) parenchyma cells. The behavior of the $D$ values was comparable to that of water in confined geometries (see Section 20.2.1.2), and an effective $P$-value could be calculated using Equation 20.3.

## 20.2.3
## Leaves

Leaf hydraulic conductance represents a crucial factor in our understanding of water transport and transpiration. As almost all water flux to and within the leaf is lost by transpiration, measurements of such flux should allow leaf transpiration to be mapped at either the plant or leaf level. However, to the best of our knowledge neither NMR nor MRI flow measurements in leaves have yet been reported.

An alternative approach would be to map the water content. The proton spin density – that is, the amount of water per unit volume – can be used in its own right as a marker of water content in leaf tissues [66]. Although MRI has been used to monitor (changes in) the water content of leaves, the disadvantage of this approach is that water transport is not measured directly; rather, it is the change in water content that is observed.

In leaves, as with all other tissues, multiexponential $T_2$ analyses may provide valuable information with regards to leaf water status and water compartments (see Section 20.2.1.2). Previously, NMR has shown an ability to measure changes in chloroplast water content, in combination with measurements of photosynthetic activity [33]. Chloroplast volume regulation represents a process by which chloroplasts import or export osmolytes to maintain a constant volume within a certain range of leaf water potential. Hence, as photosynthetic activity rates are directly coupled to changes in chloroplast volumes, these studies are of special interest for monitoring plant performance under stress conditions.

After determining combined relaxation and diffusion measurements, 2-D correlation plots between $T_1$–$T_2$ or $D$–$T_2$ or $T_2$–$T_2$-exchange [44, 65] can be generated, and this greatly enhances the ability to discriminate different pools of water in subcellular compartments; it also reveals the time scale of exchange of water between the different compartments. This approach shows great promise for nonspatially resolved measurements [44], and may even be carried out using portable unilateral NMR devices (see Section 20.3).

## 20.3
### Hardware for *In Situ* Plant Studies

Hardware solutions for laboratory-bound, dedicated intact plant NMR and MRI have been discussed elsewhere [25, 67]. In most cases, only parts of the plant (e.g. stem, leaf, petioles, seed pods, fruit stalk) will be selected for study, but not the plant as a whole. In this case, an optimal SNR would be obtained by optimizing the radius of the RF coil with respect to that part of the plant to be measured; the smaller the coil radius, the higher the SNR. In fact, the best approach would be to construct RF detector coils that closely fitted that part of the plant or tree which was to be imaged [20, 54, 55].

In many cases dedicated hardware is required in order to image intact (woody) plants. For example, Van As *et al.* [68] applied a purposely built low-field (trans)portable NMR instrument to monitor water content and water flux in intact plants in climate rooms and in greenhouses. It is to be expected that relatively cheap imaging set-ups based on permanent magnet systems will soon become available [69, 70], while for quantitative, nonspatially resolved or imaging methods specifically designed magnets and gradient coils are currently under development [71, 72]. Recently, a burst of new, small-sized magnets was unveiled at the Ninth International Conference on Magnetic Resonance Microscopy and the Seventh Colloquium on Mobile NMR, September 2007, in Aachen. These included a hinged magnet (NMR-cuff), a battery-operated pocket-sized MRI, a mobile soil NMR sensor (NMR endoscope), and a number of (homogeneous) unilateral (single-sided) NMR magnets. The latter type of magnet may be of particular interest for leaf studies, and might even be combined with fluorescence-based methods for monitoring photosynthetic activity in relation to leaf water content/chloroplast volume, perhaps even in imaging mode [73].

# References

1 Sellers, P.J., Dickinson, R.E., Randall, D.A., Betts, A.K., Hall, F.G., Berry, J.A., Collatz, G.J., Denning, A.S., Mooney, H.A., Nobre, C.A., Sato, N., Field, C.B. and Henderson-Sellers, A. (1997) *Science*, **275**, 502–9.

2 Sperry, J.S., Hacke, U.G., Oren, R. and Comstock, J.P. (2002) *Plant, Cell & Environment*, **25**, 251–63.

3 Mencuccini, M. (2003) *Plant, Cell & Environment*, **26**, 163–82.

4 Javot, H. and Maurel, C. (2002) *Journal of Experimental Botany*, **54**, 2035–43.

5 Lee, S.H., Chung, G.C. and Steudle, E. (2005) *Plant, Cell & Environment*, **28**, 1191–202.

6 Dixon, H.H., Joly, J. (1895) *Philosophical Transactions of the Royal Society of London B*, **186**, 563–76.

7 van den Honert, T.H. (1948) *Faraday Discussions*, **3**, 146–53.

8 Daudet, F.A., Lacointe, A., Gaudillère, J.P. and Cruiziat, P. (2002) *Journal of Theoretical Biology*, **214**, 481–98.

9 Tardieu, F. (2003) *Trends in Plant Science*, **8**, 9–14.

10 Zweifel, R., Steppe, K. and Sterck, F.J. (2007) *Journal of Experimental Botany*, **58**, 2113–31.

11 Steudle, E. (2001) *Annual Review of Plant Physiology*, **52**, 847–75.

12 Koch, G.W., Sillet, S.C., Jennings, G.M. and Davis, S.D. (2004) *Nature*, **428**, 851–4.

13 Turner, N.C. (1981) *Plant and Soil*, **58**, 339–66.

14 al Hagray, S.A. (2007) *Journal of Experimental Botany*, **58**, 839–54.

15 Grant, O.M., Tronina, L., Jones, H.G. and Chaves, M.M. (2007) *Journal of Experimental Botany*, **58**, 815–25.

16 Tomos, A.D. and Leigh, R.A. (1999) *Annual Review of Plant Physiology*, **50**, 447–72.

17 Henzler, T., Waterhouse, R.N., Smyth, A.J., Carvajal, M., Cooke, D.T., Schäffner, A.R., Steudle, E. and Clarkson, D.T. (1999) *Planta*, **210**, 50–60.

18 Tyree, M.T., Patino, S., Bennink, J. and Alexander, J. (1995) *Journal of Experimental Botany*, **46**, 83–94.

19 Clearwater, M.J., Meinzer, F.C., Andrade, J.L., Goldstein, G. and Holbrook, M. (1999) *Tree Physiology*, **19**, 681–7.

20 Windt, C.W., Vergeldt, F.J., de Jager, P.A. and Van As, H. (2006) *Plant, Cell & Environment*, **29**, 1715–29.

21 Shachar-Hill, Y. and Pfeffer, P.E. (1996) *Nuclear Magnetic Resonance in Plant Biology*, American Society of Plant Physiologists, Rockville.

22 Chudek, J.A. and Hunter, G. (1997) *Progress in Nuclear Magnetic Resonance Spectroscopy*, **31**, 43–62.

23 Köckenberger, W. (2001) *Trends in Plant Science*, **6**, 286–92.

24 Ratcliffe, R.G., Roscher, A. and Shachar-Hill, Y. (2001) *Progress in Nuclear Magnetic Resonance Spectroscopy*, **39**, 267–300.

25 Van As, H. (2007) *Journal of Experimental Botany*, **58**, 743–56.

26 Bottemley, P.A., Rogers, H.H. and Foster, T.H. (1986) *Proceedings of the National Academy of Sciences of the United States of America*, **83**, 87–9.

27 Brown, J.M., Kramer, P.J., Cofer, G.P. and Johnson, G.A. (1990) *Theoretical and Applied Climatology*, **42**, 229–36.

28 Van As, H. and van Dusschoten, D. (1997) *Geoderma*, **80**, 405–16.

29 Amin, M.H.G., Richards, K.S., Chorley, R.J., Gibbs, S.J., Carpenter, T.A. and Hall, L.D. (1996) *Magnetic Resonance Imaging*, **14**, 879–82.

30 Votrubova, J., Sanda, M., Cislerova, M., Amin, M.H.G. and Hall, L.D. (2000) *Geoderma*, **95**, 267–82.

31 Pohlmeier, A., Oros-Peusquens, A.M., Javaux, M., Menzel, M.I. Vereecken, V. and Shah, N.J. (2007) *Magnetic Resonance Imaging*, **25**, 579–80.

32 van der Weerd, L., Claessens, M.M.A.E., Ruttink, T., Vergeldt, F.J., Schaafsma, T.J. and Van As, H. (2001) *Journal of Experimental Botany*, **52**, 2333–43.

33 McCain, D. (1995) *Biophysical Journal*, **69**, 1111–16.

34 van der Weerd, L., Claessens, M.M.A.E., Efdé, C. and Van As, H. (2002) *Plant, Cell & Environment*, **25**, 1538–49.

35 van der Weerd, L., Melnikov, S.M., Vergeldt, F.J., Novikov, E.G. and Van As,

H. (2002) *Journal of Magnetic Resonance*, **156**, 213–21.

**36** Edzes, H.T., vanDusschoten, D. and Van As, H. (1998) *Magnetic Resonance Imaging*, **16**, 185–96.

**37** Sen, P.N. (2004) *Concepts in Magnetic Resonance. A*, **23**, 1–21.

**38** Crick, F. (1970) *Nature*, **225**, 420–2.

**39** Anisimov, A.V., Sorokina, N.Y. and Dautova, N.R. (1998) *Magnetic Resonance Imaging*, **16**, 565–8.

**40** Valiullin, R. and Skirda, V. (2001) *Journal of Chemical Physics*, **114**, 452–8.

**41** Venkataramanan, L., Song, Y.Q. and Hurlimann, M.D. (2002) *Transactions on Signal Processing*, **50**, 1017–26.

**42** Hürlimann, M.D., Venkataramanan, L. and Flaum, C. (2002) *Journal of Chemical Physics*, **117**, 10223–32.

**43** van Dusschoten, D., de Jager, P.A. and Van As, H. (1995) *Journal of Magnetic Resonance A*, **116**, 22–8.

**44** Qiao, Y., Galvosas, P. and Callaghan, P.T. (2005) *Biophysical Journal*, **89**, 2899–905.

**45** Sibgatullin, T.A., de Jager, P.A., Vergeldt, F.J., Gerkema, E., Anisimov, A.V. and Van As, H. (2007) *Biophysics*, **52**, 196–203.

**46** van Dusschoten, D., Moonen, C.T., de Jager, P.A. and Van As, H. (1996) *Magnetic Resonance in Medicine*, **36**, 907–13.

**47** Windt, C.W., Vergeldt, F.J. and Van As, H. (2007) *Journal of Magnetic Resonance*, **185**, 230–9.

**48** Snaar, J.E.M. and Van As, H. (1992) *Biophysical Journal*, **63**, 1654–8.

**49** Ionenko, I.F., Anisimov, A.V. and Karimova, F.G. (2006) *Biologia Plantarum*, **50**, 74–80.

**50** Holbrook, N.M., Ahrens, E.T., Burns, M.J. and Zwieniecki, M.A. (2001) *Plant Physiology*, **126**, 27–31.

**51** Clearwater, M.J. and Clark, C.J. (2003) *Plant, Cell & Environment*, **26**, 1205–14.

**52** Kuroda, K., Kambara, Y., Inoue, T. and Agawa, A. (2006) *International Association of Wood Anatomists*, **27**, 3–17.

**53** Fukuda, K., Utsuzawa, S. and Sakaue, D. (2007) *Tree Physiology*, **27**, 969–76.

**54** Scheenen, T.W.J., Vergeldt, F.J., Heemskerk, A.M. and Van As, H. (2007) *Plant Physiology*, **144**, 1157–65.

**55** Scheenen, T.W.J., Heemskerk, A.M., de Jager, P.A., Vergeldt, F.J. and Van As, H. (2002) *Biophysical Journal*, **82**, 481–92.

**56** Van As, H. and Schaafsma, T.J. (1984) *Biophysical Journal*, **45**, 496–72.

**57** Reinders, J.E.A., Van As, H., Schaafsma, T.J., de Jager, P.A. and Sheriff, D.W. (1988) *Journal of Experimental Botany*, **39**, 1199–210 and 1211–20.

**58** Mantle, M.D. and Sederman, A.J. (2003) *Progress in Nuclear Magnetic Resonance Spectroscopy*, **43**, 3–60.

**59** Stapf, S. and Han, S-I. (2005) *NMR Imaging in Chemical Engineering*, Wiley-VCH Verlag GmbH, Weinheim.

**60** Tallarek, U., Rapp, E., Scheenen, T., Bayer, E. and Van As, H. (2000) *Analytical Chemistry*, **72**, 2292–301.

**61** Scheenen, T.W.J., van Dusschoten, D., de Jager, P.A. and Van As, H. (2000) *Journal of Experimental Botany*, **51**, 1751–9.

**62** Scheenen, T.W.J., Vergeldt, F.J., Windt, C.W., de Jager, P.A. and Van As, H. (2001) *Journal of Magnetic Resonance*, **151**, 94–100.

**63** Windt, C.W. (2007) Nuclear magnetic resonance imaging of sap flow in plants, PhD Thesis, Wageningen University.

**64** Tallarek, U., Vergeldt, F.J. and Van As, H. (1999) *The Journal of Physical Chemistry B*, **103**, 7654–64.

**65** Washburn, K.E. and Callaghan, P.T. (2006) *Physical Review Letters*, **97**, 17550.

**66** Veres, J.S., Cofer, G.P. and Johnson, G.A. (1993) *The New Phytologist*, **123**, 769–74.

**67** Homan, N., Windt, C.W., Vergeldt, F.J., Gerkema, E. and Van As, H. (2007) *Applied Magnetic Resonance*, **32**, 157–70.

**68** Van As, H., Reinders, J.E.A., de Jager, P.A., Schaafsma, P.A.C.M. and van der Sanden, T.J. (1994) *Journal of Experimental Botany*, **45**, 61–7.

**69** Rokitta, M., Rommel, E., Zimmermann, U. and Haase, A. (2000) *Review of Scientific Instruments*, **71**, 4257–62.

**70** Haishi, T., Uematsu, T., Matsuda, Y. and Kose, K. (2001) *Magnetic Resonance Imaging*, **19**, 875–80.

**71** Raich, H. and Blümler, P. (2004) *Progress in Nuclear Magnetic Resonance Spectroscopy B*, **23**, 16–25.

**72** Blümich, B., Anferov, V., Anferova, S., Klein, M., Fechete, R., Adams, M. and Casanova, F. (2002) *Concepts in Magnetic Resonance. B15*, 255–61.

**73** Perlo, J., Casanova, F. and Blümich, B. (2004) *Journal of Magnetic Resonance*, **166**, 228–35.

# 21
# Noninvasive Assessment of Moisture Migration in Food Products by MRI

*John P.M. van Duynhoven, Gert-Jan W. Goudappel, Wladyslaw P. Weglarz, Carel W. Windt, Pedro Ramos Cabrer, Ales Mohoric and Henk Van As*

## 21.1
## Moisture Migration in Foods

### 21.1.1
### Introduction

A main challenge for the foods industry over the next decades is to address the growing concerns of society with respect to public health. Obesity, cardiovascular disease, arthrosis and diabetes are on the rise, and the link with eating patterns is scientifically well established [1, 2]. The industry is now addressing the growing demand of consumers and legislators for products that promote sustainable well-being and health. In one route, the industry is responding by reducing high levels of fat, sugar and salt, as these are clearly linked to adverse health effects. As these ingredients critically determine the taste and texture of the current generation of food products, the industry is now posing the challenge of redesigning the microstructures of these materials [3].

Water is a constituent of virtually every food product, and by its interaction with other components it exerts a critical role in determining texture [4, 5]. This applies not only to moist systems such as gels, but also to systems where water is less abundant, as in most cereal materials. Water is also involved in most transformation processes that can occur in foods, both during processing as well as during storage. During cooking, for example, proteins and carbohydrates undergo water-mediated rearrangements of their (supra-)molecular and mesoscopic structure. When processed food products are stored, structural rearrangements may also take place, mostly induced by the redistribution of moisture. The molecular events that take place during the migration of moisture are known, and can mostly be described at the thermodynamic level [6]. Thus, one can establish whether a food system is in thermodynamic equilibrium and whether migration will occur. At present, the kinetics of moisture migration is described by diffusivity parameters that are obtained from macroscopic measurements [7]. Only in few cases has it been possible to establish the underlying mesostructural and microstructural

*Magnetic Resonance Microscopy.* Edited by Sarah L. Codd and Joseph D. Seymour
Copyright © 2009 WILEY-VCH Verlag GmbH & Co. KGaA, Weinheim
ISBN: 978-3-527-32008-0

events, mainly due to a lack of appropriate measurement techniques [8]. Due to its noninvasive/perturbing nature, magnetic resonance imaging (MRI) is in a seemingly unique position, but its application to foods is often compromised with respect to resolution in time, space and molecular mobility. In this chapter we will review the issues with respect to monitoring moisture migration during storage and processing, and outline some of the recent successful attempts to address these challenges.

### 21.1.2
### Processing

Food processing is one of the oldest technologies known to man, and serves to preserve foods and to make them suitable for consumption. Many food processes have a strong impact on the distribution and status of water. Drying and freezing are commonly deployed methods to enhance the physical and microbial stability of foods. In order to obtain a pleasant and palatable texture, many food products are heat-treated by means of baking, frying and cooking. During these processes the presence of water mediates critical phase transitions. For example, during drying and freezing, water and many food ingredients undergo crystallization and glass transitions; the reverse effects take place during baking, frying and cooking, but in these cases other water-mediated phase transitions also occur. Most of the proteins present in meat, dairy, vegetables and cereals will denature and form gels with well-appreciated textures. Carbohydrates, as are abundantly present in tubers and cereals, will undergo similar processes. In cereals for example, crystalline starch is organized in granular structures, which will melt and form gels (gelatinization). Although the molecular basis of the aforementioned phase transitions is well known, food engineers continue to struggle with describing and predicting the kinetics of these processes in terms of foods mesostructure and microstructure [8–10]. One major complication here is the joint occurrence of mass/heat transfer with endothermic and/or exothermic phase transitions and the structural heterogeneity on a large span of length scales.

### 21.1.3
### Storage

The shelf-life of many food products is limited, primarily due to microbial and physical instability. In both cases water plays a critical role in technological routes to prolong shelf-life. Technologies to enhance microbial stability generally involve methods to reduce the availability and mobility of water molecules; often this is achieved by drying the food or adding salt and sugar. The physical instability of foods is typically observed when compounds are inhomogeneously distributed over a product and are thermodynamically not in equilibrium [11]. This is often the case for products where moist and dry components are brought together. The thermodynamic (non-)equilibrium of such systems can adequately be described, but again, the kinetics of moisture migration can only be described in macroscopic terms, mostly without understanding of the role of the involved mesostructures and microstructures [11].

## 21.1.4
### Water Activity, Glass–Rubber Transitions and Mobility

In all of the aforementioned processes, water plays a pivotal role. In order to describe the wide range of chemical, microbial and physical events, food researchers have taken recourse to two general concepts. For predicting chemical and microbial stability [6], one commonly deploys the concept of water activity ($a_w$), which is considered a good measure for the chemical potential of water [5]. The $a_w$ of a food product or ingredient has been defined as its water vapor pressure, divided by the vapor pressure of pure water at the same temperature. For predicting physical phenomena, such as stickiness, sogginess and crispness, the glass–rubber transition temperature ($T_g$) has been proven useful. As water is the most potent plasticizer of food polymers, $a_w$ and $T_g$ are closely related. $T_g$ considers water-induced mobility on a structural and macromolecular level [12], whereas $a_w$ is related to the molecular mobility of water itself [13]. For understanding food–water relationships at different conditions [14] it is recommended to use both $a_w$ and $T_g$ [15–17]. Several attempts have been made to relate $a_w$ and $T_g$ with molecular mobility as measured by nuclear magnetic resonance (NMR) relaxometry [5], although this has been successful only in specific cases [18–21]. Whilst no implied fundamental physical relationship exists between $a_w$ and NMR mobility, it has been stated that changing the state of water affects both water activity and NMR relaxation [5, 20]. A summary of typical $T_2$, $a_w$ and $T_g$ ranges for a variety of food materials is provided in Figure 21.1a. Here, it should be noted that, in food materials with low to intermediate moisture levels, transversal relaxation becomes very efficient – that is, a major part of the protons have a submillisecond $T_2$-values. This is illustrated in Figure 21.1b, which shows transversal relaxation decays for a cereal material equilibrated at different $a_w$ levels. In this and other cases, the plasticizing effect of water on the cereal material is clearly observable, and can also be described within a theoretical framework [23]. In general terms, however, no clear and simple relationship exists between NMR relaxation behavior and $a_w$ [5].

## 21.1.5
### Mapping of Moisture Redistribution in Food Materials by MRI

Many food products are appreciated because of their structural heterogeneity, which implies that consumers perceive contrast in taste and texture during consumption. Such multicomponent food products typically consist of crisp and soft textures, thus also involving contrast in water activity [11]. This implies that in these products a thermodynamic imbalance between the different components exists [8]. Without adequate precautions, thermodynamic equilibrium will be

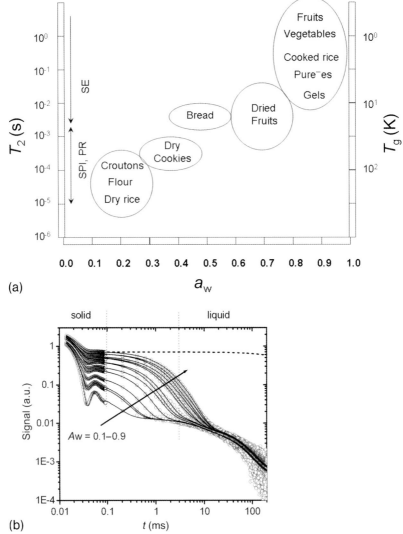

**Figure 21.1** (a) Schematic overview of typical $T_2$, $T_g$ and $a_w$ ranges for common food materials. The operational range of SPI, PR and SE MRI methods has been indicated near the vertical axis; (b) Transversal NMR signal ($T_2$) decays obtained at 0.4 T from cereal materials hydrated in $a_w = 0.1–0.9$ region. The dashed line represents the signal from bulk liquid. Reprinted with permission from Ref. [22]. Copyright (2007) Royal Society of Chemistry.

restored by migration of moisture between crisp (dry) and soft (moist) textures [11]. Elucidation of the mechanisms which govern moisture migration requires tools that are able to assess both the amount and mobility of water, at different locations, in a time-resolved manner. MRI seems ideally suited for this purpose,

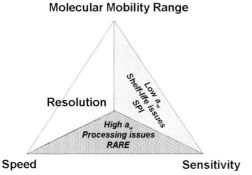

**Figure 21.2** Schematic overview of conflicting requirements when using MRI for mapping water in foods. The planes depict the situation when observing moisture migration in systems with small and large contrast in $a_w$.

due to its noninvasive and nonperturbing nature, as well as its ability to assess molecular mobilities by probing NMR relaxation rates [24–27]. As is illustrated in Figure 21.2, the use of MRI to assess moisture migration in a dynamic manner is compromised by conflicting experimental requirements. During the processing and shelf-life of typical food systems, one typically encounters two extremes: (i) rapid migration due to large $a_w$ differences; and (ii) slow migration due to small $a_w$ contrast. This places demands on the MRI measurements that can currently not be reconciled in one technique.

## 21.2
## Mapping of Moisture in the Low-*a*$_w$ Regime

### 21.2.1
### Optimization of Contrast and Sensitivity

In the low- to intermediate $a_w$ regime, conventional spin-echo techniques only work in favorable cases [28], and mostly it is necessary to use MRI methods capable of detecting signals with $T_2$ in the range of 1 ms or less. A number of solutions have been proposed to obtain images, grossly based on using strong gradients [29–32] and line-narrowing [33] techniques. The family of single point imaging (SPI) methods is most suitable for food systems [34]. Unlike most spin-echo (or gradient echo) MRI methods, which are limited to measure the liquid signal only, SPI uses the free induction decay (FID) signal [35–39]. SPI methods are based on frequently repeated measurements of a single point on the FID signal in the presence of gradients (Figure 21.3a). Figure 21.3b depicts the SPRITE (single point ramped imaging with $T_1$ enhancement) variant of SPI, where shorter measurement times can be achieved by ramping the gradients. A further gain in sensitivity can be achieved by centric samples schemes [40, 41]. Both, SPI and SPRITE are

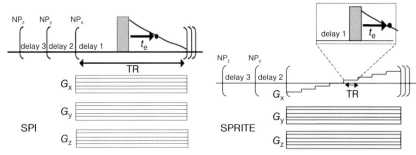

**Figure 21.3** Basic SPI pulse sequence (left) and its variant SPRITE (right) to acquire images of a matrix with $NP_x * NP_y * NP_z$ points, using an encoding time $t_e$. The repetition time TR (and therefore the $T_1$ weighting of the images) can be adjusted using a time delay 1. This delay can be much shorter in the SPRITE sequence in order to save scanning time. The other two delays are required to avoid duty cycle overloading and thermal and mechanical damage to the gradients.

attractive for imaging of materials with short $T_2$, since a short encoding time ($t_e \ll 1\,ms$) can be used. Thus, these methods can be used to image the early stages of hydration, where the NMR signal from water has a $T_2$ in order of $1\,ms$ or less (cf. Figure 21.1b). However, due to long experimental times, especially for 3-D imaging at high resolution, it has limited application for dynamic (real-time) experiments in the second–minute scale.

Even more so than in conventional spin-echo MRI, the outcome and usefulness of SPI data is strongly determined by a rational choice of experimental parameters. As indicated in Figure 21.4, before embarking on SPI experiments on low- to medium-moisture food materials one should first make an estimate of their $T_1$ and $T_2$ relaxation properties. From the material of interest, its $T_2$ will determine in which range the dephasing time needs to be chosen. The maximum gradient amplitude G needed to cover a field of view (FOV) in $n$ steps is given by:

$$G = \frac{n\pi}{\gamma t_e FOV} \tag{21.1}$$

For the short encoding times $t_e$ that are needed to observe immobile proton populations, large gradients G must be applied. This requirement becomes even more demanding when a high spatial resolution (FOV/$n$) is needed. Furthermore, in order to excite all frequencies that are introduced by the gradient, a sufficiently short excitation pulse ($t_p$) and correspondingly broad filter width (FW) are required [42]:

$$t_p \leq \frac{2t_e}{n}$$

$$FW \geq \frac{n}{4t_e} \tag{21.2}$$

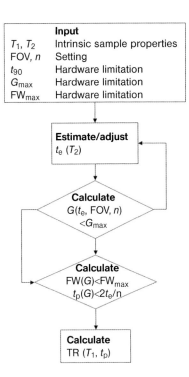

| Input | |
|---|---|
| $T_1$, $T_2$ | Intrinsic sample properties |
| FOV, n | Setting |
| $t_{90}$ | Hardware limitation |
| $G_{max}$ | Hardware limitation |
| $FW_{max}$ | Hardware limitation |

Estimate/adjust
$t_e$ ($T_2$)

Calculate
G($t_e$, FOV, n)
<$G_{max}$

Calculate
FW(G)<$FW_{max}$
$t_p$(G)<2$t_e$/n

Calculate
TR ($T_1$, $t_p$)

**Figure 21.4** Schematic depiction of parameter setting in SPI experiments on low/medium-moisture food materials.

When these requirements cannot be met, it is necessary to adjust either $t_e$, G, n or FOV (Equation 21.1). Due to the requirement for short excitation pulses, there is a need to apply small tilt angles ($\alpha = t_p/t_{90}$). Signal intensities in SPI experiments are given by [42]:

$$S = S_0 \exp\left(-\frac{t_e}{T_2^*}\right) \cdot G\left(\frac{TR}{T_1}; \alpha\right)$$

$$G\left(\left(\frac{TR}{T_1}; \alpha\right)\right) = \frac{(1-E)\cdot\sin(\alpha)}{1-\cos(\alpha)\cdot E} \tag{21.3}$$

$$E = \exp\left(-\frac{TR}{T_1}\right) \xrightarrow{ErnstAngle} \cos(\alpha) = \cos\left(\frac{t_p}{t_{90}}\right)$$

In order to achieve optimal sensitivity when a short excitation pulse ($t_p$) is deployed, a short repetition time (TR) and thus high duty cycle is required. This is illustrated in Figure 21.5a, which represents Equation 21.3 for a given $t_e$ and $T_2$. Solid-state-type radiofrequency (RF) coils (high $B_1$, short $t_{90}$ pulses) and efficient gradient coils are necessary for optimal results. An example of the dependence of SPI contrast/sensitivity when applied to a multicomponent food system is presented in Figure 21.5b, where cross-sections from 3-D SPI images of carbohydrate shells with a moist (fruit purée) filling are shown. These systems were used as models for

(a)

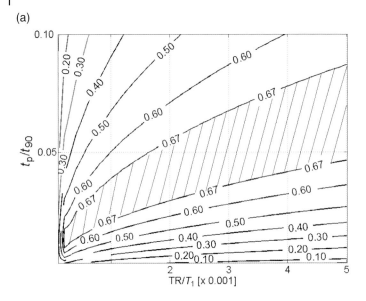

(b)

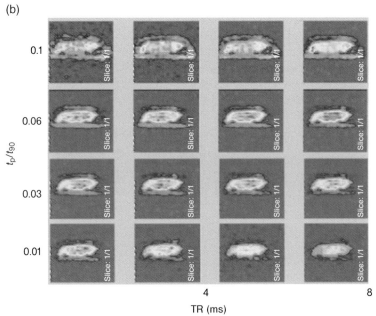

**Figure 21.5** SPI sensitivity ($\eta$) dependence on repetition time (TR) and tilt angle ($\alpha = t_p/t_{90}$). (a) Theoretical calculation; (b) Experimental dependence of SPI images of cereal shells filled with a moist fruit purée. Data were obtained at 7 T with a 3 cm coil; $t_e = 100\,\mu s$; $t_{90} = 85\,\mu s$.

multicomponent food systems where moist and dry components are combined. During the time course of the parameter optimization experiments, no migration of moisture had occurred. The approximate borders of optimal sensitivity for the carbohydrate shell are marked with dashed lines, and these correspond well with the theoretical prediction. It should also be noted here that the use of long excitation pulses, when Equation 21.2 is violated, has resulted in a diminished resolution.

## 21.2.2
### Quantification

For a good understanding of water transport in the low-$a_w$ regime, it would be helpful to develop predictive and validated physical models for moisture migration. In order to provide experimental data to allow for the verification of such models, SPI would be a good candidate. However, in most SPI applications to foods described in the literature, the resultant images were treated in either a qualitative or semiquantitative manner [43]. The dependency of SPI signal intensity on experimental and material properties is known (Equation 21.3), and in principle this should allow quantification in terms of local moisture content and hence also local $T_g$ and/or $a_w$. However, due to the low sensitivity of the SPI experiment a compromise must be made with respect to resolution of space, time and/or molecular mobility (see Figure 21.2). Such compromises have been worked out in measurement and data processing protocols for several representative moisture migration cases [44]. In the most favorable case – where moisture migration is occurring on a time scale of days or weeks – there is least compromise, and by recording SPI experiments with a range of dephasing times it is possible to obtain quantitative information on both spin density and molecular mobility. Figure 21.6 shows the results of a storage experiment where slow moisture migration in a multicomponent snack system (Figure 21.6a) was monitored by a combination of SPRITE and spin-echo (SE) experiments. In order to assess the full range of $T_2$-values a set of three SPRITE and two SE experiments was required. The individual measurement times of the different experiments ranged from 15 min to 5 h. The longest measurement time was required for the SPRITE experiment that probed $T_2$-values in the millisecond region, where the chemical shift difference of fat and water leads to oscillatory behavior in the decay, which can be deconvoluted, albeit at the expense of recording many datum points. In Figure 21.6b it can be seen how the rubbery component of bread ($T_2 = 0.1$–1 ms) increases at the expense of a mobile component ($T_2 \geq 1$ ms) in cheese.

Although this measurement protocol successfully measured the full range of $T_2$ populations [44], a strong measurement time penalty had to be paid, and a cumbersome data processing procedure was needed. In many cases, moisture migration occurs on a time scale of hours to days and, as a consequence, one either needs to compromise between resolving molecular mobility or to obtain absolute spin densities. When assumptions can be made on the relationship between signal intensity and relaxation times, it is still possible to produce semiquantitative moisture profiles [38, 44]. An example of such an approach is shown in Figure 21.7,

(a)

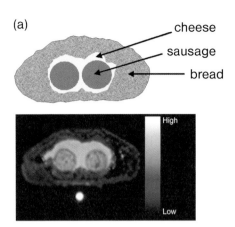

cheese

sausage

bread

(b)

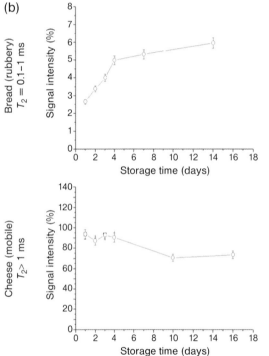

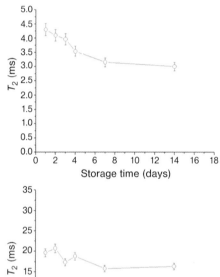

**Figure 21.6** (a) SPRITE image and schematic depiction of a snack model with small $a_w$ contrast between bread, sausage and cheese; (b) Results of quantification (spin density $M_0$ and $T_2$) of redistribution of moisture between rubbery bread ($100\,\mu s < T_2 > 1\,ms$) and mobile cheese ($T_2 > 1\,ms$) components. Data were collected at 4.7 T with a 12 cm coil, using a set of three SPRITE (SET1–3) and two SE (SET4–5) experiments. SET1: $t_e = 40$–$75\,\mu s$ in eight steps, TR = 0.75 ms; 50 ms ramp delay, FOV = 7 × 7 × 8 cm; 40 × 40 × 32 points, resolution 1.75 × 1.75 × 2.5 mm, acq. time 15 min; SET2: $t_e = 100$–$450\,\mu s$ in eight steps, TR = 1.5 ms, ramp delay 50 ms, FOV 7 × 7 × 8 cm, 64 × 64 × 32 points, resolution 1.1 × 1.1 × 2.5 mm, acq. time 40 min; SET3: $t_e = 0.5$–$3.9\,ms$ in 35 steps, TR = 4.5 ms, no ramp delay, FOV 7 × 7.8 cm; 64 × 64 × 32 points, resolution 1.1 × 1.1 × 2.5 mm; acq. time 5 h 45 min; SET4: TE = 5.6, 7.7, 11, 15, 21, 29, 40 and 50 ms, TR = 3 s, FOV = 6.4 × 6.4 cm, 64 × 64 points, resolution 1 × 1 mm, two slices of 2 mm, two averages, acq. time 50 min; SET5: TE = 50, 58, 68, 80, 94, 110, 130 and 150 ms, TR = 3 s, FOV = 6.4 × 6.4 cm, 64 × 64, resolution 1 × 1 mm, two slices of 2 mm, two averages, acq. time 50 min. Total scanning time 8 h 20 min. Reprinted with permission from Ref. [44]. Copyright (2006) American Chemical Society.

(a)

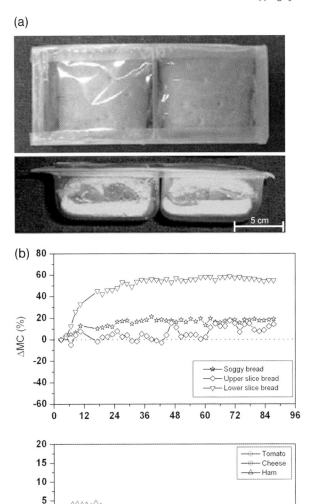

Figure 21.7 Moisture migration in sandwiches with large $a_w$ contrast between the components (bread, ham, lettuce, cheese). Treated and nontreated sandwiches (a) were studied in pairs and relative moisture content differences are plotted in (b) for different sandwich components. Images were recorded with the SPRITE sequence using $t_e = 50\,\mu s$, FOV = $9.6 \times 9.6 \times 12\,cm^3$, $128 \times 128 \times 60$ points, spatial resolution $0.75 \times 0.75 \times 2\,mm$, TR = 3 ms; 300 ms delay between SPRITE ramps, scanning time of 1 h 30 min. A doped water capillary was used to calibrate the images for absolute moisture content (per volume element). Prior to the timed series of experiments, an image of a bottle filled with the same doped solution, but occupying the complete field-of-view, was acquired using the same parameters to correct the images for spatial RF inhomogeneities. Reprinted with permission from Ref. [44]. Copyright (2006) American Chemical Society.

where moisture migration was monitored in sandwiches consisting of bread (low $a_w$), ham, lettuce and cheese (high $a_w$). In this experiment the efficacy of moisture migration control technologies (barriers, enzymatic treatment) was assessed by studying treated and nontreated sandwiches in pairs (Figure 21.7a). The moisture contents (MCs) of the different sandwich components were estimated by calibration versus a doped water capillary. The differences in MC that develop during the shelf-life (Figure 21.7b) provide a good diagnostic for the effectiveness of an attempt to control moisture migration.

As an alternative route, recourse can be taken to an external calibration, using the experimentally established relationships between moisture content and SPI signal intensity. This procedure involves the recording of SPI signal intensities of the material of interest as it has been equilibrated to different moisture contents. In general, linear relationships are obtained between MC and SPI signal intensity, as illustrated in Figure 21.8a. Such calibration curves can be used to establish local MCs in kinetic experiments. Figure 21.8b and c are examples of moisture ingress profiles into crackers with different porosity, corresponding to transport via the matrix and gas phase, respectively. The profiles represent true MCs, thus allowing the validation of models describing moisture transport in cellular cereal materials. It is important here to emphasize the unique performance of MRI in the noninvasive detection of rather subtle structural (hydration) changes in relative dry materials in a quantitative manner.

## 21.3
## Mapping of Moisture in the High-$a_w$ Regime

### 21.3.1
### Optimization of Contrast and Sensitivity

In order to make useful quantitative observations with SE MRI, investigators either emphasized spatial or time-resolution, sacrificing one feature in favor of the other (see Figure 21.2). For example, by acquiring structural information in only one [45] or two [46] spatial dimensions it was possible to monitor moisture redistribution in cereal kernels in real-time manner. These approaches are only applicable if it can be assumed that the object under study is structurally homogeneous – which is in general not the case. For real-life food systems, there is often a need to monitor the 3-D internal structure during dehydration or rehydration. Conventional SE approaches inevitably involve acquisition times which conflict with the speed of the process under study, and for this reason they are often not suitable for the real-time observation of water transport in complex food microstructures. Alternatively, real-time 3-D MRI information can be obtained by using the RARE (Rapid Acquisition with Relaxation Enhancement) (or turbo-spin-echo; TSE) sequence [47]. Within RARE, a single image is produced from signals collected from multiple (Carr–Purcell–Meiboom–Gill; CPMG)

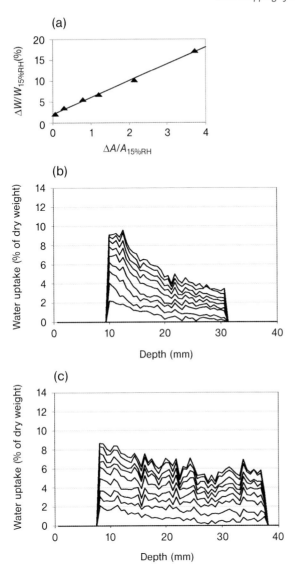

**Figure 21.8** Quantitative assessment of moisture migration into porous crackers. In (a) the calibration curve is shown that was used to construct the quantitative ingress profiles; (b,c) Ingress profiles of moisture into crackers with a dense and open porous structure, respectively. The ingress profiles were calculated on the basis of a series of 2-D SPI images, measured with an interval of 10 h. The individual images were recorded with a coil of 4 cm using the following experimental settings: $B_0 = 3$ T, $t_e = 125$ μs, FOV = 4 × 4 cm, 64 × 64 points, spatial resolution 0.63 × 0.63 mm, TR = 10 ms, 32 averages; scanning time per image 21 min.

echoes, measured at different echo times. The possibility to adjust the number of echoes (the RARE factor, $R$) and the phase-encoding scheme allows the enhancement or suppression of signals with different $T_2$. If the sample contains 'liquid' regions with different $T_2$, such as bulk water or a water-saturated porous matrix, this will result in a complicated dependence of image contrast on experimental parameters. However, if signal from a water-saturated matrix ($T_2$ in order of tens of milliseconds) needs to be enhanced over the signal from bulk water, then the temporal or/and spatial resolution must be compromised due to a limiting $R$. According to Equation 21.4, minimizing the echo time (TE) also enhances the signal from water that has ingressed into the sample. Additional suppression of the signal from bulk water can be achieved by applying a short repetition time (TR) as compared to the $T_1$ of bulk water, which typically is in order of 1–2 s.

This is illustrated in Figure 21.9, which shows RARE images of water ingress into a crunchy soup crouton [48]. In order to visualize water ingress into the crouton, short TR and TE were used, and selectively small $R$. A single scan measurement (i.e. no signal accumulation) was used, which allows for a temporal resolution of 3.5 min. The use of a higher $R$, which would allow for faster measurements, was not convenient as it suppressed the signal from the water present in the porous cereal matrix, while enhancing the signal from the bulk water which surrounded it. A short repetition time was also beneficial for a relative contrast enhancement of the water saturating the matrix.

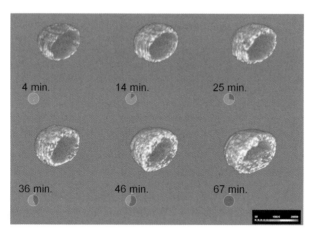

**Figure 21.9** Observation of water ingress in a crouton immersed in (20 °C) water. RARE images were recorded at 7 T with a 3 cm resonator, with a time resolution of 3.5 min. Only the signal of water in the porous matrix can be observed; the signal of surrounding bulk water has been suppressed by proper selection of acquisition parameters (TR = 200 ms, TE$_{eff}$ = 1.8 ms, $R$ = 4, FOV 13.5 × 10.8 × 13.5 mm³, resolution 0.3 × 0.3 × 0.3 mm³). Reproduced with permission from Ref. [48]. Copyright (2007) Elsevier Ltd.

21.3.2
**Quantification**

In typical SE experiments it is possible to map proton densities and relaxation times by recording images in which the SE time is varied [49]. Subsequently, proton densities and relaxation times can be obtained at pixel resolution by a simple exponential fitting of pixel intensities against SE. The effect of relaxation on the pixel intensity $S$ in RARE imaging is more complex [47], but is roughly given by:

$$\frac{S}{S_0} = \left[1 - e^{\frac{TR}{T_1}}\right] * e^{\frac{TE_{eff}}{T_2}}$$

(21.4)

where $TR$ is the repetition time and $TE_{eff}$ is the effective echo time of the RARE sequence (the time of the echo that samples the $k = 0$ in the imaging plane [50]). Within the high-$a_w$ regime it can often be tacitly assumed that $T_1$ is uniform over the sample, but for $T_2$ strong dependencies on moisture content and temperature can typically be observed. The rapid events that would preferably be monitored (on a seconds to minute scale) with RARE imaging preclude those experiments where $TE_{eff}$ is incremented. Rather, RARE images can be treated in a quantitative manner by considering the known relationships between signal intensity, MC and $T_2$. This was exploited in a recent study where the 3-D ingress of water in differently pretreated heterogeneous rice kernels during cooking was monitored in real-time manner in three dimensions by RARE imaging [51]. Here, approximately linear relationships between signal intensity $S/S_0$, $T_2$ and the moisture content $m$ (wet basis) could be assumed (Figure 21.10a). Thus, the NMR signal is also a

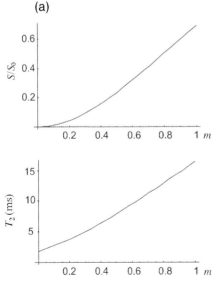

(a)

**Figure 21.10** (a) Relationship between signal intensity ($S/S_0$) and $T_2$ versus moisture content (m) for rice starch; (b, c) Water uptake of respectively a native and processed rice kernel as a function of cooking time (min), as observed at 0.7 T with a 10 mm detection coil. The color scale corresponds to relative moisture content. TR = 500 ms, R = 16, 15 × 5 × 5 mm³, 128 (read-out) × 32 × 16. The first echo was observed at 6.6 ms after excitation, subsequent echoes were 4.2 ms apart. The scanning time per 3-D image was 64 s. Reproduced with permission from Ref. [51]. Copyright (2004) Elsevier Inc.

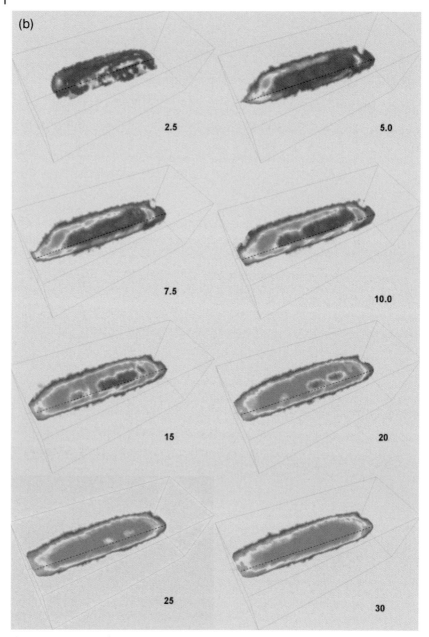

**Figure 21.10** *Continued*

(c)

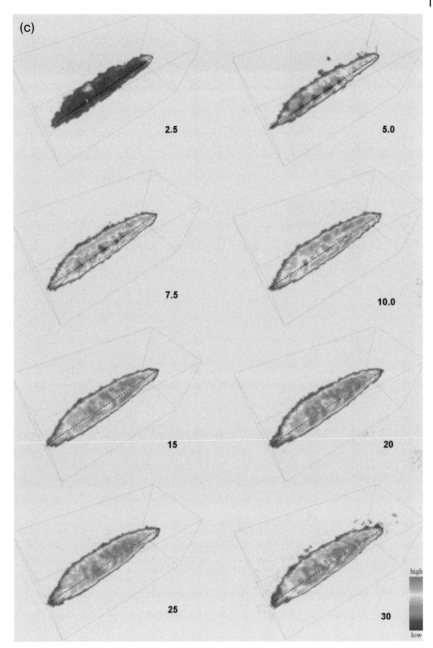

**Figure 21.10** *Continued*

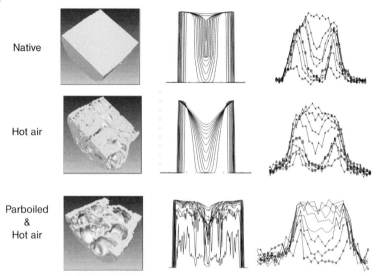

**Figure 21.11** Differently pretreated rice-kernels are represented in the separate rows. Left column: A 3-D reconstruction of the interior of rice kernels imaged by X-ray tomography (XRT); Center column: Simulations of 1-D profiles of water uptake; Right column: 1-D plots across the center of the central slice showing the (relative) moisture profile of the grain during cooking. For experimental details, see Figure 21.10.

monotonically increasing function of the MC, and the relative MC of the rice kernel can be monitored during cooking.

Representative examples are shown in Figure 21.10b and c, where it is clearly seen how the 3-D kernel microstructure is reflected in the inhomogeneous ingress of water in native and processed rice kernels. Thus, quantitative information may be gathered that can be used to verify a physical model for the ingress of water in porous cereal matrices, with varying levels of starch gelatinization. This is illustrated in Figure 21.11, where a 1-D profile across the center of the central slice taken from the 3-D RARE data set is represented for different times during cooking. The simulated ingress profiles took into account any differences in the porous properties of the kernels (middle column), as well as differences in starch crystallinity. A good match between experimental and simulated ingress profiles can be observed, indicating that the model is a good representation of the meso-structural and microstructural reality of the rice kernels.

## 21.4
## Conclusions

Today, MRI has become established as a powerful and unique tool to monitor moisture migration in food products, in noninvasive manner. However, adequate MRI methods must be chosen, and for different applications care must be taken to trade-off resolution in terms of space, time and molecular mobility. In particular, when there is a need to obtain quantitative information on mobility and abundance of water, it is necessary either to strike a compromise with time resolution or to take recourse to assumptions and/or external calibration procedures.

## Acknowledgments

Parts of these studies were supported by the Dutch BTS and IS programs (Dutch Ministry of Economical Affairs, projects BTS00103 and IS042042). W.W. acknowledges a Marie Curie Fellowship provided by the European Commission (MEIF-CT2005-009475).

## References

1 WHO/FAO (2002) Diet, Nutrition and the Prevention of Chronic Diseases: Report of a Joint WHO/FAO Expert Consultation, WHO/FAO, 2002, Geneva, Switzerland.

2 Vischer, T.L.S. and Seidel, J.C. (2001) The public health impact of obesity. *Annual Review of Public Health*, **22**, 355–75.

3 Norton, I., Fryer, P. and Moore, S. (2006) Product/process integration in food manufacture: engineering sustained health. *American Institute of Chemical Engineers*, **52**, 1632–40.

4 Lewicki, P.P. (2004) Water as the determinant of food engineering properties. A review. *Journal of Food Engineering*, **61**, 483–95.

5 Schmidt, S.J. (2004) Water and solids mobility in foods. *Advances in Food and Nutrition Research*, **48**, 1–101.

6 Roos, Y.H., Leslie, R.B. and Lillford, P.J. (2007) *Water Management in the Design and Distribution of Quality Food*, Technomic Publishing, Lancaster, USA.

7 Saravacos, G.D. and Maroulis, Z.B. (2001) *Transport Properties of Foods*, Marcel Dekker, New York.

8 Guillard, V., Broyart, B., Bonazzi, C., Guilbert, S. and Gontard, N. (2003) Evolution of moisture distribution during storage in a composite food modelling and simulation. *Journal of Food Science*, **68**, 958–66.

9 Guillard, V., Broyart, B., Bonazzi, C., Guilbert, S. and Gontard, N. (2003) Moisture diffusivity in sponge cake as related to porous structure evaluation and moisture content. *Journal of Food Science*, **68**, 555–62.

10 Roca, E., Broyart, B., Guillard, V., Guilbert, S. and Gontard, N. (2007) Controlling moisture transport in a cereal porous product by modification of structural or formulation parameters. *Food Research International*, **40**, 461–9.

11 Labuza, T.P. and Hyman, C.R. (1998) Moisture migration and control in multi-domain foods. *Trends in Food Science & Technology*, **9**, 47–55.

12 Rahman, M.S. (2006) State diagram of foods: its potential use in food processing and product stability. *Trends in Food Science & Technology*, **17**, 129–41.

13 Vittadini, E. and Chinachoti, P. (2003) Effect of physico-chemical and molecular mobility parameters on *Staphylococcus aureus* growth. *International Journal of Food Science and Technology*, **38**, 841–7.

14 Chirife, J. and Buera, M.P. (1995) A critical-review of some nonequilibrium situations and glass transitions on water activity values of foods in the microbiological-growth range. *Journal of Food Engineering*, **25**, 531–52.

15 Roos, Y.H. (1995) Characterization of food polymers using state diagrams. *Journal of Food Engineering*, **24**, 339–60.

16 Roos, Y.H. (1995) Glass transition-related physicochemical changes in foods. *Food Technology*, **49**, 97–102.

17 Sherwin, C.P. and Labuza, T.P. (2006) Beyond water activity and glass transition: A broad perspective on the manner by which water can influence reaction rates in foods, in: *Water Properties of Food, Pharmaceutical, and Biological Materials*, CRC Press, Boca Raton, FL, USA, pp. 343–62.

18 Hills, B.P., Manning, C.E. and Ridge, Y. (1996) New theory of water activity in heterogeneous systems. *Journal of the Chemical Society – Faraday Transactions*, **92**, 979–83.

19 Hills, B.P., Manning, C.E., Ridge, Y. and Brocklehurst, T. (1996) NMR water relaxation, water activity and bacterial survival in porous media. *Journal of the Science of Food and Agriculture*, **71**, 185–94.

20 Hills, B.P., Manning, C.E. and Godward, J. (1999) *A Multi State Theory of Water Relations in Biopolymer Systems*, Royal Society of Chemistry, Cambridge, UK, pp. 45–62.

21 Yoshioka, S., Miyazaki, T., Aso, Y. and Kawanishi, T. (2007) Significance of local mobility in aggregation of beta-galactosidase lyophilized with trehalose, sucrose or stachyose. *Pharmaceutical Research*, **24**, 1660–7.

22 Weglarz, W.P., Goudappel, G.J.W., Blonk, G. van Dalen, H. and J.P.M. van Duynhoven (2007) *Dynamic Visualisation of Structural Changes in Cereal Materials under High-Moisture Conditions Using 3D MRI and XRT*, The Royal Society of Chemistry, Cambridge, UK, pp. 134–40.

23 Weglarz, W.P., Inoue, C., Witek, M., Van As, H. and van Duynhoven, J.P.M. (2008) Molecular mobility interpretation of moisture sorption isotherms of food materials by means of gravimetric NMR, in: *Water Properties in Food, Health, Pharmaceutical and Biological Systems: ISOPOW 10*. Wiley, New York (in press).

24 Eads, T.M. (1998) *Principles of Nuclear Magnetic Resonance Analysis of Intact Food Materials*, Marcel Dekker, New York, pp. 1–88.

25 Hills, B.P. (2004) *Nuclear Magnetic Resonance Imaging*, Woodhead Publishing, Cambridge, UK, pp. 154–71.

26 Hills, B.P. (1998) *Magnetic Resonance Imaging in Food Science*, John Wiley & Sons, Ltd, New York.

27 Watson, A.T., Hollenshead, J.T. and Chang, C.T.P. (2001) Developing nuclear magnetic resonance imaging for engineering applications. *Inverse Problems in Engineering*, **9**, 487–505.

28 Ruan, R., Almaer, S., Huang, V.T., Perkins, P., Chen, P. and Fulcher, R.G. (1996) Relationship between firming and water mobility in starch-based food systems during storage. *Cereal Chemistry*, **73**, 328–32.

29 Demco, D.E. and Blumich, B. (2000) Solid-state NMR imaging methods. Part I: strong field gradients. *Concepts in Magnetic Resonance*, **12**, 188–206.

30 Tyler, D.J., Robson, M.D., Henkelman, R.M., Young, I.R. and Bydder, G.M. (2007) Magnetic resonance imaging with Ultrashort TE (UTE) PULSE sequences: technical considerations. *Journal of Magnetic Resonance Imaging*, **25**, 279–89.

31 Robson, M.D., Gatehouse, P.D., Bydder, M. and Bydder, G.M. (2003) Magnetic resonance: an introduction to Ultrashort TE (UTE) imaging. *Journal of Computer Assisted Tomography*, **27**, 825–46.

32 Idiyatullin, D., Corum, C., Park, J.Y. and Garwood, M. (2006) Fast and quiet MRI using a swept radiofrequency. *Journal of Magnetic Resonance*, **181**, 342–9.

33 Demco, D.E. and Blumich, B. (2000) Solid-state NMR imaging methods. Part II: line narrowing. *Concepts in Magnetic Resonance*, **12**, 269–88.

34 Cornillon, P. and Salim, L.C. (2000) Characterization of water mobility and distribution in low- and intermediate-moisture food systems. *Magnetic Resonance Imaging*, **18**, 335–41.

35 Emid, S. and Creyghton, J.H.N. (1985) High-resolution Nmr imaging in solids. *Physica B & C*, **128**, 81–3.

36 Balcom, B.J., MacGregor, R.P., Beyea, S.D., Green, D.P., Armstrong, R.L. and Bremner, T.W. (1996) Single-point ramped imaging with T-1 enhancement (SPRITE). *Journal of Magnetic Resonance Series A*, **123**, 131–4.

37  Beyea, S.D., Balcom, B.J., Prado, P.J., Cross, A.R., Kennedy, C.B., Armstrong, R.L. and Bremner, T.W. (1998) Relaxation time mapping of short T-2* nuclei with Single-Point Imaging (SPI) methods. *Journal of Magnetic Resonance*, **135**, 156–64.

38  Ziegler, G.R., MacMillan, B. and Balcom, B.J. (2003) Moisture migration in starch molding operations as observed by magnetic resonance imaging. *Food Research International*, **36**, 331–40.

39  Deka, K., MacMillan, B., Ziegler, G.R., Marangoni, A.G., Newling, B. and Balcom, B.J. (2006) Spatial mapping of solid and liquid lipid in confectionery products using a 1D centric SPRITE MRI technique. *Food Research International*, **39**, 365–71.

40  Halse, M., Rioux, J., Romanzetti, S., Kaffanke, J., MacMillan, B., Mastikhin, I., Shah, N.J., Aubanel, E. and Balcom, B.J. (2004) Centric scan SPRITE magnetic resonance imaging: optimization of SNR, resolution, and relaxation time mapping. *Journal of Magnetic Resonance*, **169**, 102–17.

41  Mastikhin, I.V., Mullally, H., MacMillan, B. and Balcom, B.J. (2002) Water content profiles with a 1D centric SPRITE Acquisition. *Journal of Magnetic Resonance*, **156**, 122–30.

42  Gravina, S. and Cory, D.G. (1994) Sensitivity and resolution of constant-time imaging. *Journal of Magnetic Resonance Series B*, **104**, 53–61.

43  Troutman, M.Y., Mastikhin, I.V., Balcom, B.J., Eads, T.M. and Ziegler, G.R. (2001) Moisture migration in soft-panned confections during engrossing and aging as observed by magnetic resonance imaging. *Journal of Food Engineering*, **48**, 257–67.

44  Ramos-Cabrer, P., van Duynhoven, J.P.M., Timmer, H. and Nicolay, K. (2006) Monitoring of moisture redistribution in multicomponent food systems by use of magnetic resonance imaging. *Journal of Agricultural and Food Chemistry*, **54**, 672–7.

45  Takeuchi, S., Maeda, M., Gomi, Y., Fukuoka, M. and Watanabe, H. (1997) The change of moisture distribution in a rice grain during boiling as observed by NMR imaging. *Journal of Food Engineering*, **33**, 281–97.

46  Stapley, A.G.F., Hyde, T.M., Gladden, L.F. and Fryer, P.J. (1997) NMR imaging of the wheat grain cooking process. *International Journal of Food Science and Technology*, **32**, 355–75.

47  Hennig, J., Nauerth, A. and Friedburg, H. (1986) Rare imaging – a fast imaging method for clinical MR. *Magnetic Resonance in Medicine*, **3**, 823–33.

48  Weglarz, W.P., Hemelaar, W.P.M., van der Linden, K., Fransiosi, N., van Dalen, G., Windt, C., Blonk, J.C.G., van Duynhoven, J.P.M. and Van As, H. (2008) Real-time mapping of moisture migration in cereal based food systems with aw contrast by means of MRI. *Food Chemistry*, **106**, 1366–74.

49  Horigane, A.K., Naito, S., Kurimoto, M., Irie, K., Yamada, M., Motoi, H. and Yoshida, M. (2006) Moisture distribution and diffusion in cooked spaghetti studied by NMR imaging and diffusion model. *Cereal Chemistry*, **83**, 235–42.

50  Callaghan, P.T. (1993) *Principles of Nuclear Magnetic Resonance Microscopy*, Clarendon Press, Oxford, UK.

51  Mohoric, A., Vergeldt, F., Gerkema, E., de Jager, A., van Duynhoven, J., van Dalen, G. and Van As, H. (2004) Magnetic resonance imaging of single rice kernels during cooking. *Journal of Magnetic Resonance*, **171**, 157–62.

# 22
# Dynamic Metabolism Studies of Live Bacterial Films

*Paul D. Majors and Jeffrey S. McLean*

## 22.1
## Introduction

Biofilms [1] are microbial assemblies that constitute the large majority of bacteria in Nature. Unlike planktonic bacteria, biofilm phenotypes attach to one another and to surfaces in a matrix of secreted extracellular polymeric substances (EPS). Biofilms constitute a heterogeneous biological tissue with a passive transport system, the overall metabolic function of which depends strongly upon its bacterial composition and spatial distribution, mass-transport properties as well as bulk growth environment. Inside the structures, they contain spatially varying metabolic growth environments (concentrations of byproducts, substrate, electron donor/acceptor and quorum sensing molecules) that are difficult to characterize due to biofilm fragility. Correspondingly, biofilm bacteria are physiologically and functionally distinct from free-floating bacteria and may be more highly resistant to antibiotics [2]. Biofilm heterogeneity and functional complexity both impedes control of detrimental forms and offers the opportunity for their exploitation in environmental and industrial settings.

The aim of these studies was to develope a technology that is useful for the development and validation of advanced models of biofilm processes. The approach taken was to: (i) culture macroscopically homogeneous biofilms that are amenable to stratified-biofilm modeling [3]; (ii) place them in a growth environment with known and controlled boundary conditions; and (iii) measure and correlate their anatomic, metabolic and transport properties using nuclear magnetic resonance (NMR) methods with high-depth resolution. While the systems studied to date have all been single-species biofilms, this technology can be extended to natural (mixed-species) biofilms of medical, environmental and industrial relevance.

*Magnetic Resonance Microscopy.* Edited by Sarah L. Codd and Joseph D. Seymour
Copyright © 2009 WILEY-VCH Verlag GmbH & Co. KGaA, Weinheim
ISBN: 978-3-527-32008-0

## 22.2
## Background

Several innovative techniques have been developed to characterize biofilms [4–6]. Invasive biofilm methods damage and/or destroy the sample and include dry weight and ash content determinations. Sample extraction methods are most often employed for high-performance liquid chromatography (HPLC), mass spectrometry (MS) and high-resolution NMR spectroscopy. Fluorescence in-situ hybridization and microautoradiography (FISH-MAR) [7] provides substrate-uptake information and species identification at the single-cell level for one labeled substrate, but is destructive. Intermediately invasive methods involve direct sample contact with microelectrodes [8] or cantilevers [9], but cause the sample to be perforated, thereby modifying its permeability and other properties. Minimally invasive, noncontacting techniques such as confocal laser scanning microscopy (CLSM) have provided considerable insight about biofilms. These include detailed structure [10], volume [11], species interactions [12, 13], viability [14], transport [15] and detachment [16]. However, optical penetration is limited by opacity and scattering phenomena to thin biofilms. Further, CLSM typically requires endogenous (genetically inserted) or exogenous fluorescent reporters that could affect metabolic activity or, in the case of green fluorescent protein (GFP), utility is limited by oxygen availability.

NMR methods applied to the study of biofilms provide noninvasive, nondestructive subatomic detection of molecular chemical, physical and transport processes. Magnetic resonance imaging (MRI) studies have included biofilm detection and visualization [17, 18] and metal ion uptake [19, 20]. NMR flow and diffusion studies have been used to study transport [21–25] and flow-dependent structure and detachment [26] and biofouling [27] properties of microbial biofilms. $^1H$ and $^{13}C$ magnetic resonance spectroscopy (MRS) studies of biofilm supernatants and extracts [28] and EPS [29] have provided details of metabolic processes. One disadvantage of NMR is its inherently low sensitivity compared to that of other higher-transition-energy methods; hence, with NMR careful optimization is required in order to reduce measurement times and lower concentration detection thresholds.

## 22.3
## Methods

The measurement capability of these studies is built upon a unique combined optical (CLSM) and NMR microscope that was developed for live cell studies. First demonstrated by Glover [30], this synergistic combination of two complementary methods allows for the correlation of high-resolution optical with moderate- resolution NMR information. The original flow tube configuration was developed to study large single cells and three-dimensional (3-D) cell agglomerates up to 800 μm in diameter [31, 32]. A planar sample chamber configuration was subsequently developed for monolayer mammalian cell cultures [33]. Earlier studies involved the characterization of water properties employing MRI and NMR diffusion measurements [34]. The planar configuration was later modified to improve its NMR spectroscopy (MRS) capabilities, and employed to study live *in situ* microbial films growing within a recirculating flow loop [35]. Although this yielded prolific biofilm growth the results were difficult to interpret due to poor environmental control.

## 22.3.1
### Biofilm Development and Environment Control

Subsequently, we developed a two-step (biofilm cultivation followed by *in vitro* measurement) procedure, as illustrated in Figure 22.1. Culturing involves growing

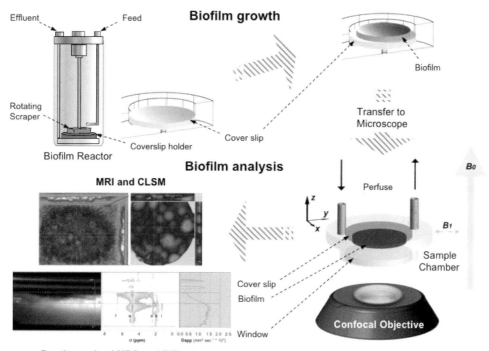

**Figure 22.1** Apparatus and procedures for biofilm preparation (top) and analysis (bottom). Biofilms are grown on coverslips by introducing the bacterial culture into a sterile biofilm reactor. A rotating scraper provides a regular shearing force to the growth surface. After a prescribed time (days), a firmly attached, monolithic biofilm is formed with a thickness set by the depth of the coverslip in its holder. The biofilm is transferred using forceps into the sample chamber (biofilm facing downward) for *in-vivo* analysis under controlled growth conditions. Measurements include microscopic MRI, CLSM and depth-resolved MRS and diffusion. Image directions are given by the *x*, *y* and *z* coordinates, and $B_0$ and $B_1$ are the NMR static and RF magnetic field directions, respectively.

the biofilm in a constant-depth (bio)film fermenter (CDFF) (Figure 22.1, top) [36] under a slowly rotating blade on a substratum (a 5 mm-diameter glass coverslip) placed within a recessed area set at a predetermined height. The uniform shearing force and careful growth media optimization results in a monolithic, firmly attached biofilm-on-glass sample that is robust to handling and transportation to the NMR sample chamber.

The mature biofilm is transferred (using forceps) to a perfused NMR sample chamber for measurement (Figure 22.1, bottom). The sample chamber is designed to support one biofilm sample attached to a 5 mm circular coverslip [37]. The biofilm faces downward into a rectangular, 4 mm-wide × 1.2 mm-deep flow channel. A transparent, bottom glass window admits CLSM laser light. Perfusion lines inject and remove growth media at opposite sides of and in a direction normal to the coverslip. This forms part of a one-pass flow system [35] (no recirculation), and flow breaks are installed to avoid any upstream microbial colonization, which would change the composition of the influent media. The flow system is operated with controlled flow rates using a pulseless dual-syringe pump (Pharmacia P-500, Uppsala, Sweden). Experimental measurements and modeling show that flow at the biofilm is laminar for all flow rates employed (1–12 ml h$^{-1}$, corresponding to dilution rates of 0.66–7.9 min$^{-1}$). The sample temperature is controlled to within 1 °C by purging the volume outside of the sample chamber with a temperature-controlled air or nitrogen gas flow integrated into the gradient insert [32]. The purging gas composition can also be adjusted to help maintain sample anaerobicity; however, the $O_2$ concentrations are currently neither controlled nor measured.

### 22.3.2
### Measurements

The NMR/optical microscope (not shown) consists of: an actively-shielded 11.7 T magnet with a 89 mm-diameter vertical bore; a home-built top-loading imaging-gradient insert; a home-built concentric top-loading RF insert containing the perfusable sample chamber, and a home-built bottom-loading CLSM system. The magnet components are integrated with a Bruker Avance spectrometer operating at a $^1$H NMR resonance frequency of 500.44 MHz. This magnet has sufficiently low fringe field to install the optical system within an optical cabinet suspended from the bottom of the magnet. This configuration simplifies optical alignment and provides vibration damping for the optics via the magnet's antivibration legs. The confocal optical microscope has two (green and red fluorescence) emission detection channels. The objective lens (Figure 22.1, bottom right) is centered in the magnet bore below the sample chamber, and is controlled by a stepper-motor-driven translation stage. The objective has a numerical aperture of 0.5. The measured resolution is approximately 0.5 µm (in plane; diffraction limited) and 6 µm (depth or Z-direction) full width at half-maximum intensity.

The NMR/optical microscope is built around a Bruker Instruments (Billerica, MA) Avance digital NMR spectrometer running Paravision version 4.0 imaging software. After sample insertion into the NMR microscope, rapid multidirectional NMR imaging is used to assess sample placement and to verify the absence of gas bubbles.

### 22.3.2.1 Biomass Volume and Distribution

Figure 22.2 (top) shows a microscopic 3-D spin-echo MRI of sample biomass. A moderate experiment repetition rate (TR/TE = 1000/5 ms) yields a bright positive

**Figure 22.2** Correlated 3-D MRI (top) and confocal optical (bottom) images for a green fluorescent protein-labeled *Shewanella oneidensis* biofilm. The symbols '1' and '2' label two matching colonies in each image for ease of comparison. (Spatial registration is approximate).

biomass image via $T_1$ contrast [18]. Sampling 128 points in each of three directions with an asymmetric field of view (2.56 mm × 7.68 mm × 7.68 mm) yields 20 μm resolution in the *z* dimension (normal to the coverslip) and 60 μm for each in-plane direction, with an acquisition time of 4.55 h. These long acquisitions are normally performed at night while the microscope is unattended. 'Rapid' (1.14–2.28 h) images are acquired during other times by reducing the resolution in one or both in-plane directions to 120 μm. These images can be processed via 3-D visualization software to determine the biomass volume and distribution, providing a semiquantitative measure of biofilm growth.

Figure 22.2 (bottom) shows the greater anatomic detail obtained by implementing the optical component of the microscope [38]. This requires a sample with a fluorescent label, and yields image contrast that depends upon the nature of that (endogenous or exogenous) label. Further, it yields a reduced field of view of 1.25 mm on each side. The in-magnet confocal system [34] has the approximate resolving power of a benchtop microscope with a 10× objective, which is good enough to resolve bacterial colonies but not individual 1 μm microbial cells. At present we are developing a system that is compatible with commercial benchtop confocal systems but capable of much higher magnification.

### 22.3.2.2 Biofilm Transport

MRI is readily combined with NMR of diffusion [39] to obtain hybrid spatially resolved water-transport rates related to biofilm density [40]. Figure 22.3 (right panel) shows the depth-resolved apparent water-diffusion rate ($D_{app}$) profile for an environmental biofilm. ($D_{app}$ is the measured local diffusion rate which includes the bounding effects of the biofilm matrix.) This 11-minute measurement procedure is tracerless and quantitative, may be performed upon the (flowing or quies-

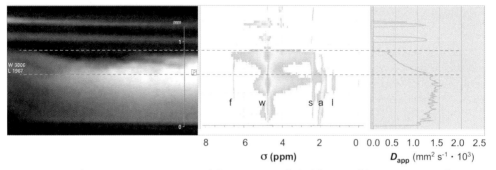

|  |  |  |  |  |  |  |  |  |  |
|---|---|---|---|---|---|---|---|---|---|
| 8 | 6 | 4 | 2 | 0 | 0.0 0.5 | 1.0 | 1.5 | 2.0 | 2.5 |

$\sigma$ (ppm)         $D_{app}$ (mm$^2$ s$^{-1}$ · 10$^3$)

**Figure 22.3** Microscopic MRI (left), MRSI (center) and diffusion-rate (right) profiles for an environmental *Shewanella oneidensis* biofilm. The biomass is attached to the bottom of two stacked coverslips (top), which appear dark in the water-selective MRI. The dashed horizontal lines intersecting all images correspond with the top and approximate bottom of the 300 μm-thick biofilm. The spectral lines for fumarate, water, succinate, acetate and lactate are each labeled by their initial letter.

cent) sealed sample, and the resulting depth profile is in general agreement with those obtained using labor-intensive microelectrode methods [41].

### 22.3.2.3 Biofilm Metabolism

NMR spectroscopy (MRS) is unique in its ability to measure, both noninvasively and quantitatively, multiple protonated metabolites within biofilms in near-real time. In our protocol, the measurement is localized to a 6–9 µl rectangular volume centered on the 5 mm-diameter coverslip and including the entire chamber height. This avoids the edges of the circular glass coverslip where local magnetic field gradients are strong and most detrimental to spectral resolution [35]. This involves a traditional single-spin-echo acquisition, employing $x$ and $y$ slice gradients: the final ($y$) slice direction coincides with the imposed flow direction in order to minimize flow-dependent signal loss and volume deformation. A prepended WET water suppression sequence [42] typically yields a factor of $10^3$–$10^4$ reduction in water signal intensity, allowing the receiver gain to be increased accordingly for improved metabolite detection and quantification. The sensitivity of the current surface-coil configuration is such that approximately 64 scans with adequate relaxation interval (e.g. 4s) must be averaged to adequately measure an average spin concentration of 3 m$M$.

As these measurements are noncontacting, nondestructive and nonsample-consuming, they can be repeated to obtain time-resolved metabolite profiles under controlled growth conditions, including sealed anaerobic environments. Bulk, localized MRS measurements have been employed for anaerobic respiration studies in *Shewanella oneidensis* strain MR-1 environmental biofilms [35] and glucose fermentation by *Streptococcus mutans* oral biofilms [38]. In a recent study [43] we demonstrated the ability for *S. oneidensis* undergoing anaerobic respiration to rapidly switch between several different electron acceptors (fumarate, dimethyl sulfoxide or nitrate), indicating that no transcriptional response was required. This system enabled multiple measurements to be acquired using the same biofilm by rapidly switching feed solutions. In comparison, traditional techniques would require the destruction of multiple parallel biofilms, introducing sample variance concerns.

Spectroscopic MRI (MRSI) methods [44] are used to measure depth-resolved metabolite concentrations [37]. The localized MRS sequence is employed while applying an MRI phase-encoding gradient normal to the coverslip. These hybrid spatial–spectral measurements provide depth-resolved metabolite profiles, currently with 22 µm resolution. Fine depth ($z$) resolution is obtained at the expense of in-plane resolution [38]. The resulting voxel volumes are 120–180 nl with corresponding isotropic dimensions of 493–565 µm, and (for a spin concentration of 3 m$M$) contain 2.1–3.3 × $10^{14}$ spins per voxel. This is consistent with our highest attainable MRI resolution, where an isotropic 15 µm MRI voxel contains 2.2 × $10^{14}$ water protons.

Figure 22.3 shows correlated MRI, depth-resolved metabolism and depth-resolved water diffusion measurements for an environmental *S. oneidensis* biofilm obtained under dynamic conditions [38]. MRI (Figure 22.3, left panel) was used to quantify and spatially locate the biomass in the chamber. Localized MRSI (Figure 22.3, center panel) shows the depth-resolved metabolite content with 31.2 µm depth resolution (vertical axis). One-dimension diffusion-weighted MRI (Figure 22.3, right panel) shows the corresponding depth-resolved apparent diffusion rate for water. The spectral lines for lactate (electron donor), acetate (lactate oxidation product generated in the biomass), fumarate (electron acceptor) and succinate (fumarate reduction product generated in the biomass) are resolved. Lactate is supplied continuously and reaches a steady-state concentration profile that is depleted at a particular depth. This depth corresponds closely with the biofilm–bulk fluid boundary and is also located in the region where the apparent diffusion rate decreases (Figure 22.3). Lactate is therefore rapidly oxidized by the cells and becomes limited in the biomass. The pronounced residual water signal in Figure 22.3 (center) is typical for the current surface–coil hardware configuration due to its strong RF magnetic field ($B_1$) gradient resulting in nonuniform water suppression with sample depth. A volume–coil configuration is being developed to address this and other performance issues.

In addition to restricting the measurement volume, the growth medium composition often requires careful consideration to avoid undue spectral broadening. Paramagnetic ions such as Mn(II) or Fe(III) can lead to extensive resolution losses at higher concentrations. For example, the linewidths decreased significantly by reducing Mn(II) from 100 µM to 1 µM in the oral biofilm growth medium. As these are necessary elements for various biofilm processes, careful media development is necessary. The two-step procedure described here can help by decoupling the media requirements during culturing and observation. Further, many standard growth media contain protein digests and gelatins which yield a viscous solution and cause a decrease in the NMR spin–spin ($T_2$) relaxation times. Our initial oral biofilms experiments [45] employed a brain–heart infusion (BHI) medium that displayed a higher apparent viscosity (even when diluted to one-tenth standard concentration) that masked the spectrum of added low-molecular-weight metabolites [45]. Most of these studies have employed chemically defined growth media containing few or no viscous agents.

## 22.4
## The Future

The current RF surface coil configuration [34] is somewhat insensitive in part because of its unilateral implementation. Further, it has a substantial $B_1$ magnetic field gradient, resulting in depth-varying sensitivity and yielding nonuniform water suppression over the 1.2 mm sample depth. At present, we are developing a new sample perfusion chamber configuration with the goals of: (i) better NMR sensitivity and resolution; (ii) better environmental control; and (iii) the ability to use thicker, natural biofilm support surfaces. A more efficient Helmholtz [46] NMR volume coil will provide improved NMR sensitivity, resolution and water suppression. The system will retain its optical measurement capability but will require the sample chamber to be removed from the NMR to a conventional benchtop optical system.

The two-step cultivation and measurement procedure (see Figure 22.1) yields experimental data favorable for modeling a biofilm with a planar geometry in an approximately 1-D growth environment with well-defined boundary conditions (a free surface with well-defined media composition and an impermeable glass slip at depth). The approach provides depth-resolved, correlated metabolism and mass-transport information amenable for the detailed modeling of biofilm processes [3].

Further, we are developing methods for the functional analysis of intact natural biofilms. Stable isotope probing (SIP) methods [47] serve to identify the active roles of key microbial-community members by the uptake and incorporation of isotope labels from substrate into their nucleic acids (RNA or DNA). Thus, NMR combined with SIP is useful for the detailed study of metabolic function in complex microbial communities [48]. Finally, the CLSM capability provides crucial species distribution and interaction information [13] for mixed-microbial modeling. These capabilities may prove useful for characterization of microbial–mammalian interactions [49, 50].

To complement the adherent-cell (biofilm) technologies discussed in this chapter, we have developed and demonstrated a live, dynamic metabolism capability for microbial suspensions under controlled-batch and continuous-culture conditions [51]. A NMR-compatible, hydrodynamically stirred bioreactor allows real-time metabolic information without the attendant chemical gradients experienced in biofilms. Thus, it might be used to distinguish biofilm phenotype and cell proximity effects upon metabolism by replicating for a live suspension the chemical environment at a particular biofilm depth.

## Acknowledgments

The authors gratefully acknowledge Prof. J. William Costerton (USC, Los Angeles CA), Dr Johannes Scholten (Merck Co. West Point PA), Prof. Wenyuan Shi (UCLA, Los Angeles CA) and Dr Robert Wind (PNNL, ret.) for their valuable contributions, and Prof. Haluk Beyenal (Washington State University, Pullman WA) for helpful discussions. We also thank Mr Mark Townsend and coworkers in the EMSL Machine Shop for advice and construction of the NMR hardware. This research was supported by NIH (NIDCR) R21 DE017232, and also by DOE's Laboratory Directed Research and Development Program at the Pacific Northwest National Laboratory, a multiprogram national laboratory operated by Battelle for the US Department of Energy under Contract DE-AC05-76RL01830. The research was performed in the Environmental Molecular Sciences Laboratory (a national scientific user facility sponsored by the Department of Energy's Office of Biological and Environmental Research) located at Pacific Northwest National Laboratory and operated for DOE by Battelle.

## References

1 Costerton, J.W., Lewandowski, Z., Caldwell, D.E., Korber, D.R. and Lappin-Scott, H.M. (1995) *Annual Review of Microbiology*, **49**, 711–45.

2 Stewart, P.S. and Franklin, M.J. (2008) *Nature Reviews Microbiology*, **6**, 199–210.

3 Beyenal, H. and Lewandowski, Z. (2005) *Chemical Engineering Science*, **60**, 4337–48.

4 Wuertz, S., Okabe, S. and Hausner, M. (2004) *Water Science and Technology*, **49**, 327–36.

5 Wagner, M., Nielsen, P.H., Loy, A., Nielsen, J.L. and Daims, H. (2006) *Current Opinion in Biotechnology*, **17**, 83–91.

6 Lewandowski, Z. and Beyenal, H. (2007) *Fundamentals of Biofilm Research*, CRC Press, Boca Raton, FL.

7 Lee, N., Nielsen, P.H., Andreasen, K.H., Juretschko, S., Nielsen, J.L., Schleifer, K.H. and Wagner, M. (1999) *Applied and Environmental Microbiology*, **65**, 1289–97.

8 Revsbech, N.P. (2005) *Methods in Enzymology*, **397**, 147–66.

9 Cross, S.E., Kreth, J., Zhu, L., Sullivan, R., Shi, W.Y., Qi, F.X. and Gimzewski, J.K. (2007) *Microbiology (UK)*, **153**, 3124–32.

10 Hall-Stoodley, L., Costerton, J.W. and Stoodley, P. (2004) *Nature Reviews Microbiology*, **2**, 95–108.

11 Staudt, C., Horn, H., Hempel, D.C. and Neu, T.R. (2004) *Biotechnology and Bioengineering*, **88**, 585–92.

12 Stoodley, P., Sauer, K., Davies, D.G. and Costerton, J.W. (2002) *Annual Review of Microbiology*, **56**, 187–209.

13 Gu, F., Lux, R., Du-Thumm, L., Stokes, I., Kreth, J., Anderson, M.H., Wong, D.T., Wolinsky, L., Sullivan, R. and Shi, W.Y. (2005) *Journal of Microbiological Methods*, **62**, 145–60.

14 Hope, C.K. and Wilson, M. (2006) *Journal of Microbiological Methods*, **66**, 390–8.

15 de Beer, D., Stoodley, P. and Lewandowski, Z. (1997) *Biotechnology and Bioengineering*, **53**, 151–8.

16 Stoodley, P., Wilson, S., Hall-Stoodley, L., Boyle, J.D., Lappin-Scott, H.M. and Costerton, J.W. (2001) *Applied and Environmental Microbiology*, **67**, 5608–13.

17 Potter, K., Kleinberg, R.L., Brockman, F.J. and McFarland, E.W. (1996) *Journal of Magnetic Resonance Series B*, **113**, 9–15.

18 Hoskins, B.C., Fevang, L., Majors, P.D., Sharma, M.M. and Georgiou, G. (1999) *Journal of Magnetic Resonance*, **139**, 67–73.

19 Nestle, N. and Kimmich, R. (1996) *Applied Biochemistry and Biotechnology*, **56**, 9–17.

20 Nott, K.P., Paterson-Beedle, M., Macaskie, L.E. and Hall, L.D. (2001) *Biotechnology Letters*, **23**, 1749–57.

21 Lewandowski, Z., Altobelli, S.A., Majors, P.D. and Fukushima, E. (1992) *Water Science and Technology*, **26**, 577–84.

22 Lewandowski, Z., Altobelli, S.A. and Fukushima, E. (1993) *Biotechnology Progress*, **9**, 40–5.

23 van As, H. and Lens, P. (2001) *Journal of Industrial Microbiology & Biotechnology*, **26**, 43–52.

24 Seymour, J.D., Codd, S.L., Gjersing, E.L. and Stewart, P.S. (2004) *Journal of Magnetic Resonance*, **167**, 322–7.

25 Nott, K.P., Heese, F.P., Hall, L.D., Macaskie, L.E. and Paterson-Beedle, M. (2005) *American Institute of Chemical Engineers*, **51**, 3072–9.

26 Manz, B., Volke, F., Goll, D. and Horn, H. (2005) *Water Science and Technology*, **52**, 1–6.

27 Seymour, J.D., Gage, J.P., Codd, S.L. and Gerlach, R. (2007) *Advances in Water Resources*, **30**, 1408–20.

28 Gjersing, E.L., Herberg, J.L., Horn, J., Schaldach, C.M. and Maxwell, R.S. (2007) *Analytical Chemistry*, **79**, 8037–45.

29 Mayer, C., Lattner, D. and Schurks, N. (2001) *Journal of Industrial Microbiology & Biotechnology*, **26**, 62–9.

30 Glover, P.M., Bowtell, R.W., Brown, G.D. and Mansfield, P. (1994) *Magnetic Resonance in Medicine*, **31**, 423–8.

31 Wind, R.A., Minard, K.R., Holtom, G.R., Majors, P.D., Ackerman, E.J., Colson, S.D., Cory, D.G., Daly, D.S., Ellis, P.D., Metting, N.F., Parkinson, C.I., Price, J.M. and Tang, X.W. (2000) *Journal of Magnetic Resonance*, **147**, 371–7.

**32** Majors, P.D., Minard, K.R., Ackerman, E.J., Holtom, G.R., Hopkins, D.F., Parkinson, C.I., Weber, T.J. and Wind, R.A. (2002) *Review of Scientific Instruments*, **73**, 4329–38.

**33** Wind, R.A., Majors, P.D., Minard, K.R., Ackerman, E.J., Daly, D.S., Holtom, G. R., Thrall, B.D. and Weber, T.J. (2002) *Applied Magnetic Resonance*, **22**, 145–58.

**34** Minard, K.R., Holtom, G.R., Kathmann, L.E., Majors, P.D., Thrall, B.D. and Wind, R.A. (2004) *Magnetic Resonance in Medicine*, **52**, 495–505.

**35** Majors, P.D., McLean, J.S., Pinchuk, G.E., Fredrickson, J.K., Gorby, Y.A., Minard, K.R. and Wind, R.A. (2005) *Journal of Microbiological Methods*, **62**, 337–44.

**36** Wimpenny, J.W. (1985) *Microbiological Sciences*, **2**, 53–60.

**37** Majors, P.D., McLean, J.S., Fredrickson, J.K. and Wind, R.A. (2005) *Water Science and Technology*, **52**, 7–12.

**38** McLean, J.S., Ona, O.N. and Majors, P.D. (2008) *International Society for Microbial Ecology Journal*, **2**, 121–31.

**39** Stejskal, E.O. and Tanner, J.E. (1965) *Journal of Chemical Physics*, **42**, 288–92.

**40** Wieland, A., de Beer, D., Damgaard, L.R. and Kuhl, M. (2001) *Limnology and Oceanography*, **46**, 248–59.

**41** Beyenal, H. and Lewandowski, Z. (2002) *Biotechnology Progress*, **18**, 55–61.

**42** Ogg, R.J., Kingsley, P.B. and Taylor, J.S. (1994) *Journal of Magnetic Resonance, Series B*, **104**, 1–10.

**43** McLean, J.S., Majors, P.D., Reardon, C.L., Bilskis, C.L., Reed, S.B., Romine, M.F. and Fredrickson, J.K. (2008) *Journal of Microbiological Methods*, **74**, 47–56.

**44** Brown, T.R., Kincaid, B.M. and Ugurbil, K. (1982) *Proceedings of the National Academy of Sciences of the United States of America*, **79**, 3523–6.

**45** Majors, P.D., McLean, J.S. and Wind, R.A. (2006) *Proceedings of International Society of Magnetic Resonance in Medicine*, **14**, 2013.

**46** Hurlston, S.E., Brey, W.W., Suddarth, S.A. and Johnson, G.A. (1999) *Magnetic Resonance in Medicine*, **41**, 1032–8.

**47** Radajewski, S., Ineson, P., Parekh, N.R. and Murrell, J.C. (2000) *Nature*, **403**, 646–9.

**48** Egert, M., de Graaf, A.A., Maathuis, A., de Waard, P., Plugge, C.M., Smidt, H., Deutz, N.E.P., Dijkema, C., de Vos, W.M. and Venema, K. (2007) *FEMS Microbiology Ecology*, **60**, 126–35.

**49** Martin, F.P.J., Dumas, M.E., Wang, Y.L., Legido-Quigley, C., Yap, I.K.S., Tang, H. R., Zirah, S., Murphy, G.M., Cloarec, O., Lindon, J.C., Sprenger, N., Fay, L.B., Kochhar, S., van Bladeren, P., Holmes, E. and Nicholson, J.K. (2007) *Molecular Systems in Biology*, **3**, 112.

**50** de Graaf, A.A. and Venema, K. (eds) (2008) *Advances in Microbial Physiology*, Vol. **53**, Elsevier Academic Press Inc, San Diego, pp. 73–314.

**51** Majors, P.D., McLean, J.S. and Scholten, J.C.M. (2008) *Journal of Magnetic Resonance*, **192**, 159–66.

# 23
# Applications of Permanent-Magnet Compact MRI Systems

*Katsumi Kose, Tomoyuki Haishi and Shinya Handa*

## 23.1
## Introduction

Today, more than 20 000 whole-body MRI (WB-MRI) systems are used routinely for clinical diagnosis throughout the world. In addition, during the past two or three decades, many nonmedical MRI applications have been proposed in academic research and industrial applications. Because these applications are usually unsuitable for the use of WB-MRI systems, our compact MRI system, which consists of a small permanent magnet utilizing Nd-Fe-B magnetic material and a portable MRI console, has been developed for these purposes [1, 2].

Applications of the compact MRI systems can be divided into two major categories, clinical and nonclinical. In this chapter, we review applications of our compact MRI systems in both categories.

## 23.2
## Clinical Applications of the Compact MRI Systems

Because of the availability of advanced WB-MRI systems, the need for compact MRI systems would seem to be limited. However, compact MRI systems have advantages over WB-MRI systems in their compactness, patient safety or comfort, and cost, as summarized in Table 23.1. Therefore, we have developed two compact MRI systems for trabecular bone microstructure measurements [3, 4] and a compact MRI system for the diagnosis of rheumatoid arthritis.

### 23.2.1
### Compact MRI System for Measurements of Trabecular Bone Microstructure in the Finger

Osteoporosis is a widespread disease characterized by age-related bone loss, and considerably increases the risk of hip fracture. Trabecular bone (TB) microstruc-

*Magnetic Resonance Microscopy.* Edited by Sarah L. Codd and Joseph D. Seymour
Copyright © 2009 WILEY-VCH Verlag GmbH & Co. KGaA, Weinheim
ISBN: 978-3-527-32008-0

**Table 23.1** Comparison of the whole-body and compact MRI systems.

| Parameter | Whole-body MRI systems | Compact MRI systems |
| --- | --- | --- |
| Design strategy | Optimization for human whole body | Optimization for a specific site |
| Size | Large | Small |
| Safety issue | Serious ($dB/dt$, SAR, acoustic noise) | Negligible for extremities |
| Cost | Expensive | Inexpensive |

SAR = specific absorption rate.

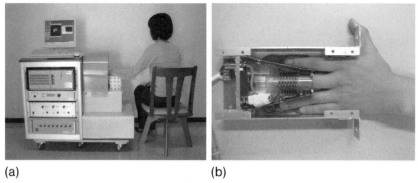

(a)                                              (b)

**Figure 23.1** (a) Overview of the compact MRI for trabecular bone microstructure measurements; (b) A flat gradient coil set and a trapezoidal RF probe box for the finger.

ture measurement is essential for the estimation of bone strength and assessments of drug therapy in osteoporosis [5, 6]. If the TB microstructure can be measured in the finger, the size and cost of microstructure measurement systems can be drastically reduced.

Figure 23.1a shows an overview of the compact MRI system for TB microstructure measurements in the finger developed by our group [3]. The system comprises a compact MRI console and a 1.0 T permanent magnet (Hitachi Metals Co., Japan). The specifications of the magnet are: field strength 1.0 T; gap width 4 cm; homogeneity 13 ppm over 13 mm dsv (diameter spherical volume); dimensions 27 (W) × 24 (H) × 18 (D) cm; and weight 85 kg. The magnet temperature is regulated at approximately 30 °C to minimize Larmor frequency drift. A flat gradient coil set and a trapezoidal radiofrequency (RF) probe box were developed for the finger, as shown in Figure 23.1b. The RF coil is a 28 mm-diameter, seven-turn solenoid.

A 3-D driven equilibrium spin-echo (3-D-DESE) sequence [TR/TE (repetition time/echo time) = 50/6 ms; image matrix = 128 × 128 × 128; voxel size 160 µm$^3$] was developed for 3-D MR microscopic imaging of the finger. The total imaging

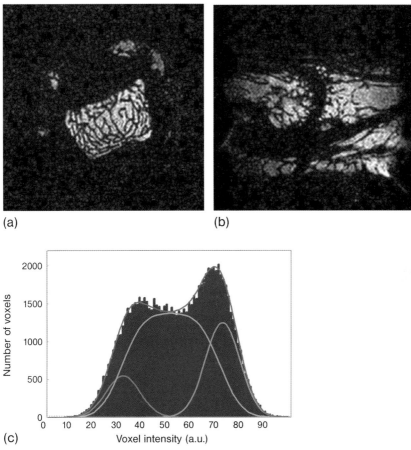

(a)  (b)

(c)

**Figure 23.2** (a) Transverse and (b) sagittal cross-sections selected from a 3-D image dataset of the middle finger; (c) Image intensity histogram in a rectangular solid region in the real-part image of the TB of the middle finger.

time is about 14 min. Figure 23.2a and b show transverse and sagittal cross-sections selected from the $256^3$-voxel 3-D image dataset obtained from a zero-filled Fourier interpolation of the $128^3$-voxel image of the middle finger of a 53-year-old man.

Figure 23.2c shows an image-intensity histogram in a rectangular solid region in the real-part image of the TB of the middle finger. The histogram of the real part image can be decomposed into two Gaussian distributions corresponding to TB bone and bone marrow, and a uniform distribution corresponding to voxels which contain both bone and bone marrow [7]. The image-intensity threshold between bone and bone marrow was calculated using this decomposition. Bone microstructure parameters are calculated using a bone structure analysis software package (TRI3D/BON, Ratoc System Engineering, Tokyo, Japan).

**Table 23.2** Mean and coefficient of variation (%) of bone microstructure parameters measured in a female subject.

| Bone structure parameter | Mean value | CV (%) |
|---|---|---|
| BV/TV | 26.1% | 7.2 |
| Tb.Th | 227 μm | 2.5 |
| Tb.N | 1.51 mm$^{-1}$ | 10.2 |
| Tb.Sp | 249 μm | 3.9 |
| SMI | 1.07 | 3.9 |

BV/TV = ratio of bone volume to tissue volume; Tb.Th = trabecular bone thickness; Tb.N = trabecular number; Tb.Sp = trabecular spacing; SMI = structure model index.

To evaluate the reproducibility of the bone microstructure measurements, the middle finger of a healthy woman (aged 44 years) was repeatedly measured 25 times after repositioning, and 15 successful measurements were used to calculate coefficient of variance. A successful scan was defined as one that gave an image-intensity histogram with two separated peaks for bone and bone marrow. The results are summarized in Table 23.2.

To evaluate the clinical efficacy of the system, a transverse study was performed using 51 healthy female volunteers (age range 19–62 years; mean ± SD age 34.1 ± 12.4 years) from the University of Tsukuba, and 230 female patients from the Nagasaki University hospital. Because the MRI measurement in the hospital was performed as one of the routine bone density measurements, the patients were not always classified as having osteoporosis.

After informed consent had been obtained from each patient, the distal middle phalanx of the middle finger of the nondominant hand was imaged using the 3-D-DESE sequence (TR/TE = 50/6 ms; acquisition matrix = 112 × 112 × 128; reconstruction matrix 256$^3$ voxels; spatial resolution 180 × 180 × 160 μm; image acquisition time 11 min). A total of 30 contiguous axial slices (a 2.4 mm-thick slab region in which the central slice was located 2.7 mm from the distal end of the middle phalanx) was analyzed using the protocol described above.

Forty-six normal subjects and 119 patients were successfully measured without motion of the finger. To compare the bone microstructure parameters with patients, 27 normal subjects aged between 20 and 44 years were selected. Correlation coefficients of the bone microstructure parameters calculated for the patients are summarized in Table 23.3. These results clearly show that all of the bone microstructure parameters have high correlations with the bone volume over total tissue volume (BV/TV), as in an example shown in Figure 23.3, that trabecular thickness (Tb.Th) and trabecular spacing (Tb.Sp) (i.e. the distance between the trabecular bones) are nearly independent, and that the structure model index (SMI; an index of trabecular bone microstructure) has a high correlation with Tb.Th.

Figure 23.4a and b show the number of patients and young normal subjects ($n = 27$) plotted against BV/TV and SMI. These histograms and Student's *t*-test for

**Table 23.3** Correlation between bone microstructure parameters.

|        | BV/TV | Tb.Th  | Tb.N   | Tb.Sp   | TBPf    | SMI     |
|--------|-------|--------|--------|---------|---------|---------|
| BV/TV  | –     | 0.6987 | 0.4473 | −0.7465 | −0.6192 | −0.5584 |
| Tb.Th  |       | –      | 0.2332 | −0.1884 | −0.7900 | −0.6907 |
| Tb.N   |       |        | –      | −0.5110 | −0.3696 | −0.4040 |
| Tb.Sp  |       |        |        | –       | 0.2291  | 0.2795  |
| TBPf   |       |        |        |         | –       | 0.9670  |
| SMI    |       |        |        |         |         | –       |

TBPf = trabecular pattern factor. For details of other abbreviations, see Table 23.2.

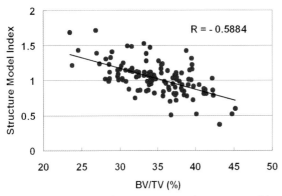

**Figure 23.3** Correlation between BV/TV and structure model index (SMI).

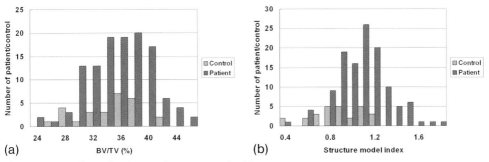

**Figure 23.4** Numbers of patients and young normal subjects ($n = 27$) plotted against (a) BV/TV and (b) structure model index (SMI).

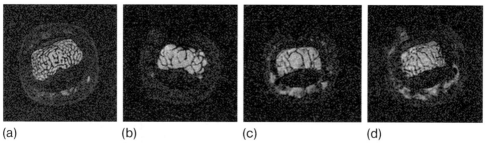

(a)  (b)  (c)  (d)

**Figure 23.5** Typical cross-sectional images with small and large SMI. (a) SMI = 0.369, normal volunteer; (b) SMI = 1.69, patient; (c) SMI = 1.72, patient; (d) SMI = 1.53, patient.

the bone microstructure parameters suggest that SMI and trabecular pattern factor (TBPf; an index of trabecular bone microstructure) are good parameters by which to characterize the pathological status of the TB, but that BV/TV, Tb.Th and Tb.Sp are not good parameters. Figure 23.5 shows typical cross-sectional images with small and large SMI values, corresponding to a normal subject (aged 22 years) and three patients. These images support the value of TB microstructure measurements in the finger as a useful clinical tool to characterize the osteoporotic status of the TB.

### 23.2.2
### Compact MRI System for Measurements of Trabecular Bone Microstructure in the Distal Radius

Although conventional bone density measurement methods such as dual energy X-ray absorptiometry (DXA) and quantitative ultrasound (QUS) for the finger are used clinically, the finger is not a well-accepted site for bone measurements. The preferred sites for bone measurements are the vertebra and proximal femur, where bone fractures caused by osteoporosis are observed frequently and, in the hip, are particularly serious. However, it is very difficult to visualize TB structure in the vertebra or proximal femur by MRI because the RF coil sensitivity for those sites is limited. For these reasons, the distal radius, distal tibia and calcaneus have been used for TB microstructure measurements by MRI, and several clinically significant results have been reported. Therefore, if we can construct a compact MRI to take measurements for TB microstructure analysis from well-accepted sites such as the distal radius, tibia and calcaneus, the system will be a useful instrument for diagnosis of osteoporosis. Hence, we developed a compact MRI system for measurements of the distal radius using a 1.0T permanent magnet [4].

Figure 23.6a shows an overview of the compact MRI system developed for TB microstructure measurements of the distal radius. The system consists of a permanent magnet (Hitachi Metals Co., Japan), a gradient coil set, RF probe and compact MRI console. The entire system can be installed in a $1 \times 2$ m space.

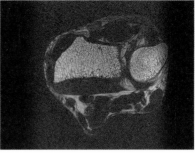

(a)                                                    (b)

**Figure 23.6** (a) Overview of the compact MRI system developed for measurements of TB microstructure in the distal radius; (b) Cross-sectional image selected from a 3-D image dataset of the distal radius acquired with the 3-D-DESE sequence.

The permanent magnet uses Nd-Fe-B material and is U-shaped. Its specifications are: field strength 1.02 T; gap width 100 mm; homogeneity 16.4 ppm over a 60 mm-diameter spherical volume; size 539 (W) × 706 (H) × 1029 (D) mm; and total weight 1340 kg. Because the magnet was designed for the *in vivo* imaging of rats and mice (see below) it was not optimized for the *in vivo* imaging of human extremities. The magnet temperature was regulated at approximately 30 °C to minimize temperature drift of the Larmor frequency.

A six-turn solenoid coil with an oval aperture (55 mm width, 75 mm height) and 40 mm length was made from Cu tape (width 5 mm, thickness 0.1 mm). The RF coil was split with 11 chip capacitors (100 pF) to obtain a sharp resonance at 43.4 MHz. The unloaded and loaded Q factors of the RF coil at 43.4 MHz were 232 and 98, respectively.

A slice-selective 3-D driven equilibrium spin-echo (3D-DESE) pulse sequence was developed with the following parameters: TR/TE = 80/10 ms; thickness of the selectively excited slab = 12 mm; image matrix = 512 × 512 × 32; voxel size = 150 × 150 × 500 μm; and total imaging time = 23 min. The acquisition dwell time was 10 μs and the pixel bandwidth 195 Hz. The excitation pulse was a 90° sinc (± 4π) pulse with an 8 kHz bandwidth and 1 ms duration.

MR images of a man (aged 25 years) were acquired after his left forearm had been placed on an arm-holding table located in the magnet gap space and his wrist inserted into the RF coil. Before 3-D image acquisition, a coronal scout view was acquired to determine the position of the distal radius.

Figure 23.6b shows a cross-sectional image selected from a 1024 × 1024 × 64 voxel 3-D image dataset of the distal radius acquired with the 3D-DESE sequence. This dataset was obtained from doubly zero-filled Fourier interpolation of the original 512 × 512 × 32 voxel image. Thus, the voxel size was 75 × 75 × 250 μm. The signal-to-noise ratio (SNR) of the bone marrow signal was about 10 (see Figure 23.6b), and close to that obtained with a 1.5 T whole-body MRI system [8–10]. This result suggests that these image data could be used for bone microstructure analysis, because the spatial resolution and SNR are sufficient for previously reported analyses [11, 12]. However, two main problems need to be solved before clinical use:

- Measurement time: To reduce the measurement time, a higher SNR is highly desirable because TR can be reduced if the SNR is sufficient. The most straightforward solution to this problem is to use a smaller-diameter RF coil which, for the distal radius, should have an open-access structure. Another solution is to improve the pulse sequence by using stronger magnetic field gradients to shorten the spin-echo time.

- Immobilization of subjects and/or correction of subject motion: Because the permanent magnet was originally designed for the *in vivo* imaging of rats and mice, the subject was required to adopt an unnatural pose for wrist imaging. The permanent magnet, gradient coil set, RF coil and positioning system should be redesigned to allow a natural positioning of the subjects for clinical use.

Nevertheless, we believe that our results have clearly demonstrated the feasibility of using a permanent magnet compact MRI system as a clinical instrument for distal radius bone microstructure measurements.

### 23.2.3
### Compact MRI System for the Diagnosis of Rheumatoid Arthritis

Rheumatoid arthritis (RA) is a chronic inflammatory disorder that causes the patient's immune system to attack their joints. RA is a disabling and painful inflammatory condition that can lead to substantial loss of mobility because of pain and joint destruction. However, recent studies have clarified that an early diagnosis of RA and the use of biological agents dramatically improve the status of the condition. Hence, an early diagnosis of RA is highly desirable.

Although, to date, the diagnosis of RA has been achieved by using a blood test and clinical data, the results of recent studies have shown that a hand MR examination is very useful for an early diagnosis [13, 14], as the earliest symptoms appear in the hands. Unfortunately, MR examinations using whole-body MRI scanners require the patient to assume a painful position because they must place their hands at the center of the magnets. One of the best solutions to this problem is to use an MRI system which has been specially designed for human extremities.

Several commercially available extremity MRI systems have been used for hand examinations in the diagnosis of RA [15]. However, for early detection a whole-hand examination is indispensable because the locations of lesions caused by RA in the hand cannot be predicted. Commercially available extremity MRI systems have a serious disadvantage in this regard, because their field of view (FOV) is too small to assess the hand and wrist in one examination. To overcome this problem, we have developed compact MRI systems using 0.2 and 0.3 T permanent magnets [16]; in the following section we describe the system using a 0.3 T magnet.

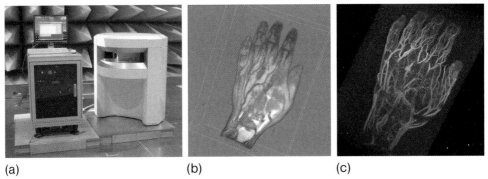

(a)                              (b)                           (c)

**Figure 23.7** (a) Overview of a compact MRI system for the diagnosis of rheumatoid arthritis; (b) Volume-rendered image made from a 3-D image dataset acquired with a gradient-echo sequence; (c) 3-D maximum intensity projection image made from a 3-D image dataset acquired with a STIR-3-DFSE sequence.

Figure 23.7a shows an overview of the system developed to diagnose RA. The specifications of the permanent magnet are: magnetic field strength 0.3 T; gap width 13 cm; homogeneity <50 ppm over a $22 \times 22 \times 8$ cm-diameter ellipsoidal volume; and weight 600 kg. The transverse gradient ($G_x$ and $G_y$) coils were designed using the target-field approach [17], and the axial gradient ($G_z$) coil was designed using a genetic algorithm [18]. The maximum field gradient strengths were 18, 18 and 28 mT m$^{-1}$ for $G_x$, $G_y$ and $G_z$, respectively, when a 10 A constant power supply was used for each channel. A 14-turn solenoid RF coil with an oval aperture (width 12.5 cm, height 6.5 cm, length 22 cm) was developed for whole-hand imaging.

Figure 23.7b shows a volume rendered image made from a 3-D image dataset from a 54-year-old man acquired with a gradient echo sequence (TR/TE/FA (flip angle) = 35 ms/8 ms/50°; image matrix = $512 \times 256 \times 32$; voxel size = $0.4 \times 0.8 \times 1.6$ mm; acquisition time = 7 min 20 s). The anatomical structures of the whole hand are clearly visualized. Figure 23.7c shows a 3-D maximum intensity projection image made from a 3-D image dataset acquired with a short $T_1$ inversion recovery 3-D fast spin-echo (STIR-3DFSE) sequence (TR/TE/TI = 1000/80/110 ms; ETL (echo train length) = 16; image matrix = $256 \times 320 \times 32$; voxel size = $0.8 \times 0.8 \times 1.6$ mm; acquisition time = 10 min 30 s). The bone marrow and subcutaneous fat signals are well suppressed, while tissues with long $T_2$ such as joint fluid and blood are clearly visualized. These images demonstrate the great potential of our system for the diagnosis of RA in patients.

## 23.3
### Nonclinical Applications of Compact MRI Systems

Compact MRI systems have found various nonclinical applications, including small animal imaging [19], food science/engineering, flow measurements and plant science. The remarkable advantages of the compact permanent-magnet MRI systems over MRI systems with superconducting magnets are their 'openness' and 'portability'. Below, we describe two typical nonclinical applications of these compact MRI systems.

### 23.3.1
### Compact MRI System for Outdoor In-Situ Measurements of Trees

The compact MRI systems can be used outdoors, and consequently many novel applications may be explored. For example, some preliminary investigations were conducted of *in-situ* MRI measurements of tree internal structures [20].

Figure 23.8a shows an overview of a portable MRI system developed for this purpose. The system consists of a permanent magnet (field strength 0.3 T; gap width 80 mm; homogeneity 50 ppm over 30 mm dsv; weight 60 kg), gradient coil set, openable RF probe, portable MRI console and a portable electric generator (EU9i; Honda, Japan). The total weight of the system is about 150 kg, and it can be moved easily to any outdoor location. The openable RF probe consists of a split acrylic bobbin, solenoid coil and tank circuit (see Figure 23.8b and c); the diameter of the solenoid is 20 mm.

Imaging experiments were performed using a water phantom and plant samples under several combinations of experimental conditions: using an AC power source or the electric generator; indoors or outdoors; and *in-vitro* or *in-vivo* tree samples. To enable an extended-time image accumulation, an internal NMR locking technique was applied between serial image acquisition sequences.

(a)  (b)  (c)

**Figure 23.8** (a) Overview of the portable MRI system developed for *in-situ* MR imaging of trees; (b) The openable RF probe; (c) Signal detection set-up.

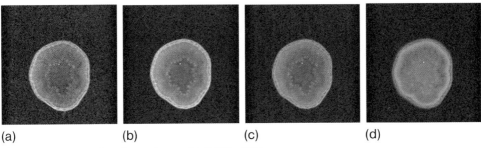

(a)          (b)          (c)          (d)

**Figure 23.9** Cross-section of a maple tree. (a) TR/TE = 100/
12 ms, NEX = 40; (b) TR/TE = 250/12 ms, NEX = 16; (c) TR/
TE = 500/12 ms, NEX = 8; (d) TR/TE = 2000/12 ms, NEX = 2.

Figure 23.9 shows cross-sections of a tree (maple, diameter 10.0–11.5 mm) acquired in an outdoor environment using an AC power source. The slice thickness, image matrix and pixel size were 3 mm, 128 × 128 and 150 × 150 μm, respectively. When the electric generator was used, external noise came into the bandwidth of the NMR signal through the part of the plant located outside the RF probe. This noise was probably due to the spark or switching noise produced by the generator. Although the problem of noise from the generator has not yet been removed, we have demonstrated that the compact MRI can be used for outdoor *in-situ* measurements of trees.

### 23.3.2
### Compact MRI System for Mice and Rats

Although both mice and rats are used widely to develop models of human disease, full-blown MRI systems cannot be used practically in such investigations. However, the use of a compact and inexpensive MRI system for such purpose would be highly desirable. Here, we describe such as system which uses a yokeless permanent magnet and can be applied to mice and rats [19].

The system consists of a yokeless permanent magnet (Hitachi Metals Co. Japan), a gradient coil assembly, an RF probe and a portable MRI console (Figure 23.10), and was installed in a 2 × 1 m space. The specifications of the permanent magnet were as follows: field strength 1.04 T at 25 °C; gap 9 cm; homogeneity 10 ppm over 30 mm dsv; magnet size 57.4 (W) × (H) × 48 (D) cm; weight 980 kg. A home-built planar gradient coil set was fixed on the faces of the pole pieces of the magnet. An RF shield box made from 0.3 mm-thick brass plates was fixed between the gradient coil planes. For mouse brain imaging, an eight-turn solenoid coil (32 mm diameter, 50 mm length) was fixed at the center of the RF shield box.

Animal studies were performed under the guidelines of the University of Tsukuba. The male ICR mice (aged 8–13 weeks) used for these MRI studies were anesthetized with isoflurane (3–5%) in air; anesthesia was maintained by reducing the isoflurane:air ratio to 0.5–1.5%. While anesthetized, the mouse was placed in a mouse-holder that was custom-designed to fit inside the RF coil, and imaged

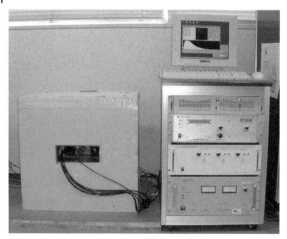

**Figure 23.10** Overview of the compact MRI system developed for imaging mice and rats.

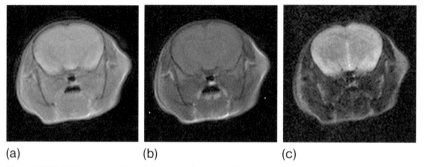

(a)                                (b)                                (c)

**Figure 23.11** 2-D cross-sectional images of a mouse brain
selected from 3-D image datasets acquired with 3-D spin-echo
sequences. (a) TR/TE = 2000/15 ms; (b) TR/TE = 500/15 ms;
(c) TR/TE = 2000/80 ms.

using 3-D spin-echo or 3-D FLASH sequences with an internal NMR lock technique. All animals recovered fully after the imaging experiments, which were of a few hours' duration.

Figure 23.11 shows the 2-D cross-sectional images of a mouse brain selected from 3-D image datasets acquired with 3-D spin-echo sequences. All images were acquired with the following parameters: FOV $25.6 \times 25.6 \times 32$ mm; image matrix $128 \times 128 \times 16$; voxel size $200 \times 200 \times 2$ mm; and number of excitations (NEX) = 1. Figure 23.11a–c show a proton-density-weighted image, and $T_1$- and $T_2$-weighted images of the same slice; the imaging times were about 70, 17 and 70 min, respectively.

Figure 23.12 shows 2-D cross-sectional images selected from 3-D image datasets of a mouse body acquired with 3-D FLASH sequences without respiratory gating. The imaging times were 14 and 3 min, respectively.

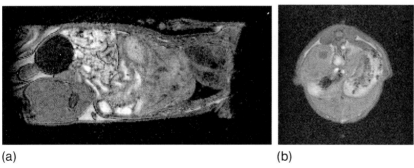

(a)                                                (b)

**Figure 23.12** 2-D cross-sectional images selected from 3-D image datasets of a mouse body acquired with 3-D FLASH sequences without respiratory gating. (a) Coronal slice. TR/TE = 30/5 ms; (b) Axial slice. TR/TE = 30/5 ms.

Although the SNR of the MR images obtained with this system may be lower than could be achieved with MRI using high-field superconducting magnets, this system has several advantages over such systems:

- Compactness; the system is installed in a $2 \times 1$ m space; by comparison, a high-field superconducting magnet system requires more than $20\,m^2$.

- Mouse accessibility; the distance between the magnet center and opening is 24 cm, and the gap is 24 cm wide and 9 cm high. Hence, mouse access is easier than with a high-field superconducting animal MRI system (where distance between magnet center and opening is usually >50 cm).

- Similarity of image contrast to clinical MRI systems; image contrast in MRI is largely affected by the resonance frequency (magnetic field strength), because $T_1$ and $T_2^*$ depend on resonance frequency. Thus, image contrasts between MR images acquired with high-field animal MRI systems or with clinical MRI systems (usually use a 1.5 T magnetic field) cannot be compared directly. In our system the magnetic field strength is also close to that of clinical MRI.

- Biological isolation; the mice used often require strict biological isolation from the outside environment. In this case the of superconducting MRI systems is difficult, as cryogen refill or refrigerator maintenance may destroy the biological isolation.

- Magnet maintenance; although a permanent magnet requires some electrical power for temperature regulation, cryogen refill or cold-head replacement of the refrigerator are not required.

Figure 23.13 shows permanent magnets developed for commercial compact MRI systems for imaging mice and rats. Although the prototype permanent magnet had only horizontal apertures, the first commercial magnet (Figure 23.13a) had both horizontal and vertical apertures as it had been designed for various MRI

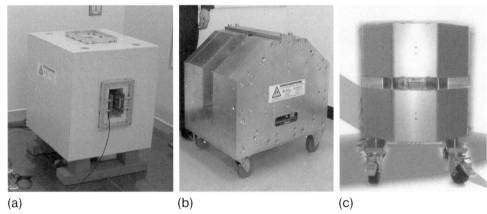

(a)                           (b)                           (c)

**Figure 23.13** Commercialized permanent magnets for imaging mice and rats. (a) 1.0 T, 100 mm gap; (b) 1.0 T, 100 mm, gap; (c) 1.5 T, 60 or 80 mm gap.

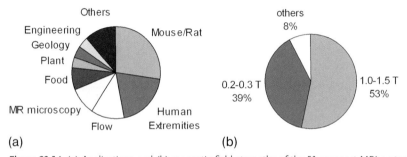

(a)                                              (b)

**Figure 23.14** (a) Applications and (b) magnetic field strengths of the 51 compact MRI systems.

applications. The second magnet (Figure 23.13b) had a U-shaped magnetic circuit with a horizontal magnetic field, while the third magnet (Figure 23.13c) had a C-shaped magnetic circuit, to improve animal accessibility. The magnetic field strength of the first and second magnets was 1.0 T, while that of the third magnet was increased to 1.5 T. Although a permanent magnet with a 2.0 T field strength has been developed for imaging mice [21], the 1.5 T permanent magnet will most likely be the standard magnet for commercial compact mouse MRI systems.

## 23.4
## Statistical Aspects of Compact MRI Systems

The first compact MRI system (our definition) was constructed in 1998 [1], since which time 13 systems have been developed for research purposes at the MR laboratory at University of Tsukuba, and 38 have been shipped from MRTechnology, Inc. The applications of the 51 systems are classified as shown in Figure 23.14a;

details of the magnetic field strengths used are shown in Figure 23.14b. As magnets with fields >1.0 T have yokeless magnetic circuits, about half of the 51 systems have magnets of this type, this being the result of developments made during the past decade – a situation which will doubtless change over the next 10 years.

## 23.5
## Conclusions

Although the absolute SNR of the NMR signal is limited by available magnetic field strength (at most, 2 T), we believe that compact MRI systems using permanent magnets will continue to be developed and utilized in novel applications of MRI.

## Acknowledgments

The authors acknowledge Prof. M. Ito (Nagasaki University) for evaluating the finger MRI system, Mr M. Aoki (NEOMAX) for permanent magnet development, Mr M. Aoki (DSTechnology Inc.) for RF electronics development, Mr M. Marutani (Jyonan Electric Laboratory) for technical support and encouragement, and E. Fukushima (NMR) for encouragement and advice. They also acknowledge Drs A. MacDowell (ABQMR) and S. Utsuzawa (NMR) for the software developments, Mr T. Shirai and Dr Y. Matsuda for the mouse system, Dr S. Tomiha for the bone microstructure system, Ms N. Iita, Mr K. Ohya and Ms M. Uruchida for the finger system, and Ms F. Okada for the *in-situ* measurements of trees.

## References

1 Haishi, T., Uematsu, T., Matsuda, Y. and Kose, K. (2001) *Magnetic Resonance Imaging*, 19, 875–80.

2 Kose, K. (2006) *NMR Imaging in Chemical Engineering* (eds S. Stapf and S.-I. Han), Wiley-VCH Verlag GmbH.

3 Iita, N., Handa, S., Tomiha, S. and Kose, K. (2007) *Magnetic Resonance in Medicine*, 57, 272–7.

4 Handa, S., Tomiha, S., Haishi, T. and Kose, K. (2007) *Magnetic Resonance in Medicine*, 58, 225–9.

5 Majumdar, S. (2002) *Topics in Magnetic Resonance Imaging*, 13, 323–34.

6 Wehrli, F.W., Saha, P.K., Gomberg, B.R., Song, H.K., Snyder, P.J., Benito, M., Wright, A. and Weening, R. (2002) *Topics in Magnetic Resonance Imaging*, 13, 335–55.

7 Iita, N., Handa, S., Tomiha, S., Kose, K. and Haishi, T. (2007) Proceedings, 15th ISMRM, p. 2630.

8 Ma, J., Wehrli, F.W. and Song, H.K. (1996) *Magnetic Resonance in Medicine*, 35, 903–10.

9 Techawiboonwong, A., Song, H.K., Magland, J.F., Saha, P.K. and Wehrli, F.W. (2005) *Journal of Magnetic Resonance Imaging*, 22, 647–55.

10 Magland, J., Vasilic, B. and Wehrli, F.W. (2006) *Magnetic Resonance in Medicine*, 55, 465–71.

11 Hwang, S.H. and Wehrli, F.W. (1999) *International Journal of Imaging Systems and Technology*, 10, 186–98.

**12** Hwang, S.H. and Wehrli, F.W. (2000) *Magnetic Resonance in Medicine*, **47**, 948–57.

**13** Sugimoto, H. Takeda, A. and Hyodoh, K. (2000) *Radiology*, **216**, 569–75.

**14** Klarlund, M., Ostergaard, M., Gideon, P., Sorensen, K., Jensen, K.E. and Lorenzen, I. (1999) *Acta Radiologica*, **40**, 400–9.

**15** American College of Rheumatology Extremity Magnetic Resonance Imaging Task Force (2006) *Arthritis Rheumatism*, **54**, 1034–47.

**16** Handa, S., Yoshioka, H., Tomiha, S., Haishi, T. and Kose, K. (2007) *Magnetic Resonance in Medical Sciences*, **6**, 113–20.

**17** Turner, R.A. (1986) *Journal of Physics D: Applied Physics*, **19**, 147–51.

**18** Goldberg, D.E. (1989) *Genetic Algorithms in Search, Optimization and Machine Learning*, 1st edn, Addison-Wesley Longman Publishing, Boston.

**19** Shirai, T., Haishi, T., Utsuzawa, S., Matsuda, Y. and Kose, K. (2005) *Magnetic Resonance in Medical Sciences*, 137–43.

**20** Okada, F., Handa, S., Tomiha, S., Ohya, K., Kose, K., Haishi, T., Utsuzawa, S. and Togashi, K. (2006) 6th Colloquium on Mobile Magnetic Resonance, Aachen, Germany.

**21** Haishi, T., Aoki, M. and Sugiyama, E. (2005) Proceedings, 13th ISMRM, p. 869.

# Part Five   Materials Science

# 24
# Magnetic Field Control of Chemical Waves

*Melanie Britton and Christiane Timmel*

## 24.1
## Introduction

Traveling chemical waves and fronts form in reactions where there is a coupling between autocatalysis and diffusion [1] and are a type of reaction–diffusion (RD) phenomenon. Traveling fronts occur when the reaction is initiated, or 'excited', in a localized region. The concentration of an autocatalytic species rapidly increases in this region, and as the autocatalyst diffuses into neighboring regions the auto-catalytic process is repeated, resulting in the propagation of a chemical front. Multiple waves occur in systems where the reacting solution returns to its initial state ready for another excitation to occur. The most famous example of this type of reaction is the Belousov–Zhabotinksy reaction [2].

Reaction–diffusion processes enable the spreading of molecules or ions to occur more rapidly than via diffusion alone. This enhancement in propagation arises because only small amounts of the autocatalyst need diffuse into a region before the feedback step takes over and the concentration of that species rapidly increases. Reaction–diffusion processes are believed to underlie the signaling mechanisms in many biological systems, which exploit this enhanced propagation velocity [3]. It is the application of these nonlinear and oscillatory chemical reactions as models for wave and oscillatory behavior in biological processes, such as chemotaxis and calcium waves, that is the driving force behind much of the research in this area.

We have investigated a system which produces a traveling front during the reaction between (ethylenediaminetetraacetato)cobalt(II) ($Co(II)EDTA^{2-}$) and hydrogen peroxide, at pH 4.2:

$$2Co(II)EDTA^{2-} + H_2O_2 \rightarrow 2Co(III)EDTA^- + 2OH^- \qquad (24.1)$$

Hydroxide ions autocatalyze this reaction and a propagating front is produced by introducing sodium hydroxide solution to a localized region of the reactive $Co(II)EDTA^{2-}/H_2O_2$ solution. During the reaction, paramagnetic Co(II) ions (high

*Magnetic Resonance Microscopy.* Edited by Sarah L. Codd and Joseph D. Seymour
Copyright © 2009 WILEY-VCH Verlag GmbH & Co. KGaA, Weinheim
ISBN: 978-3-527-32008-0

spin, $d^7$) are oxidized to diamagnetic Co(III) ions (low spin, $d^6$) and it is this transition that enables the wave to be observed using magnetic resonance imaging (MRI). This is because the nuclear magnetic resonance (NMR) relaxation times of protons in solvent molecules are significantly shorter when surrounding paramagnetic Co(II) ions than diamagnetic Co(III) ions. It is this difference in relaxation times that produces the image contrast necessary to visualize the traveling front.

As the front propagates, a gradient in both concentration, of Co(II) and Co(III) ions, and magnetic susceptibility is produced. Nagypal and coworkers first postulated [4] that it is this combination of gradients that is responsible for the magnetic field effects observed for this wave in the presence of an inhomogeneous magnetic field. They showed that when the Co(II)EDTA$^{2-}$ and hydrogen peroxide solution was studied as a shallow layer in a Petri dish, the propagating wave could be manipulated by placing a horseshoe-magnet underneath the dish [4, 5]. In these experiments the autocatalytic reaction was initiated with a drop of sodium hydroxide solution added to the center of the Petri dish. In the absence of any applied magnetic fields, the resulting wave propagated isotropically (Figure 24.1a). By placing a horseshoe-magnet underneath the solution, an inhomogeneous magnetic field is produced across the Petri dish and the wave is seen to propagate in a manner sensitive to the inhomogeneity of the field (Figure 24.1b). A qualitative explanation for this effect has been postulated by Nagypal and coworkers, based on the different behavior of the paramagnetic and diamagnetic cobalt ions in a magnetic field where there is a concentration gradient [4, 5]. They proposed that,

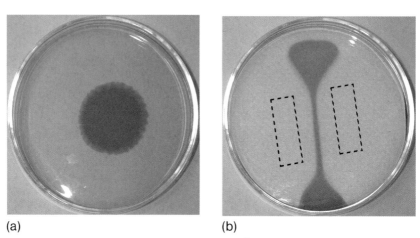

(a)     (b)

**Figure 24.1** Photographs of the reaction of Co(II)EDTA$^{2-}$ with H$_2$O$_2$ in a shallow layer in a Petri dish 26 min after mixing, in the absence (a) and presence (b) of an applied magnetic field. The light regions are areas of unreacted Co(II)EDTA$^{2-}$; the dark regions are areas of reacted Co(III)EDTA$^{-}$ solution. The orientation of the poles of the horseshoe-magnet used for image (b) is indicated by the dotted lines.

as the reaction progresses, the diamagnetic Co(III)EDTA$^-$ ions produced are pushed down the magnetic field gradient and away from the region of high magnetic field, between the poles of the magnet. Attractive forces then cause the paramagnetic Co(II)EDTA$^{2-}$ ions to move into regions of high magnetic field.

In this chapter we present details of experiments conducted to demonstrate control of the propagation of the Co(II)/Co(III) chemical front, using the gradients of the MRI spectrometer. Our work shows that it is possible to control both the velocity and direction of the front, and that the geometry of the front and its orientation with respect to the magnetic field gradient is important in determining its behavior. The degree of control possible in the Co(II)EDTA$^{2-}$/H$_2$O$_2$ system has presented the question of whether other chemical fronts might also be controllable by using magnetic fields. There are a number of chemical fronts formed in reactions where a transition metal ion changes its oxidative state with a consequent production of a gradient in magnetic susceptibility. One example is the front produced during the oxidation of Fe(II) ($d^6$ high spin) with nitric acid to Fe(III) ($d^5$ high spin). The first MR images of traveling front in this reaction are presented here, along with investigations of magnetic field effects in this system.

## 24.2
## Experimental

The reagents for the cobalt(II) EDTA/H$_2$O$_2$ reaction are sodium hydroxide, EDTA, cobalt(II) chloride and hydrogen peroxide (35% solution by volume). All materials were of A.C.S. grade, obtained from Aldrich, and used without further purification. Solutions of 0.02 $M$ Co(II)EDTA$^{2-}$ were prepared by dissolving equimolar quantities of EDTA and CoCl$_2$ in deionized water and adjusting to pH 4.2. The reacting solution was made from 0.02 $M$ Co(II)EDTA$^{2-}$ and hydrogen peroxide solutions, in a 9:1 ratio. The concentration of sodium hydroxide solution used to initiate the traveling wave was approximately 0.02 $M$. The reagents for the iron(II)/nitric acid reaction were ammonium iron(II) sulfate, nitric acid, hydrazine and sodium nitrite. A solution of 0.1 $M$ Fe(II), 1 m$M$ hydrazine and 2 $M$ nitric acid was prepared in deionized water. The hydrazine was used to remove any HNO$_2$ impurities, which would cause spontaneous autocatalysis, and was used at concentrations below an level which affects wave velocities. The front was produced, by initiating with a drop of 0.01 $M$ NaNO$_2$ solution.

MRI experiments were performed on a Bruker DMX-300 spectrometer equipped with a 7.0 T superconducting magnet, operating at a proton resonance frequency of 300 MHz. The cobalt(II) EDTA/H$_2$O$_2$ reaction was studied in 5 mm NMR tubes, using a 25 mm radiofrequency coil. The front was initiated inside the magnet using an injection device to deliver the sodium hydroxide to the top of the Co(II)EDTA$^{2-}$/H$_2$O$_2$ solution. The hydroxide solution used was less dense, thus preventing any initial density fingering, which is a possibility in tubes of the diameter used. The iron(II)/nitric acid reaction was studied in capillaries (1 mm i.d.), in a 25 mm radiofrequency coil. The front was initiated outside the magnet, from either the bottom or the top of the capillary, and then placed in the magnet.

Inversion recovery and CPMG experiments [6] were used to measure the relaxation times of both oxidative states of the iron and cobalt solutions used in the reactions. At 0.02 M, a solution containing paramagnetic Co(II) ions gave relaxation times of $340 \pm 1$ ms $(T_1)$ and $33 \pm 1$ ms $(T_2)$ whilst a solution containing diamagnetic Co(III) ions demonstrated relaxation times of $622 \pm 4$ ms $(T_1)$ and $398 \pm 3$ ms $(T_2)$. At 0.1 M, a solution containing Fe(II) ions gave relaxation times of $19.8 \pm 0.4$ ms $(T_1)$ and $7.6 \pm 0.2$ ms $(T_2)$ whilst a solution containing Fe(III) ions produced relaxation times of $0.5 \pm 0.1$ ms $(T_1)$ and $0.3 \pm 0.1$ ms $(T_2)$.

Images for the Co(II)/Co(III) front were obtained using the fast imaging, multiple spin-echo sequence (RARE) [7]. Here, a single excitation pulse is used to excite the spins in the sample and multiple echoes are acquired, which were encoded for position using magnetic field gradients. The signal was reconstructed using two-dimensional (2-D) Fourier analysis, producing a 2-D image. By collecting multiple echoes, after the initial excitation, $T_2$ relaxation contrast is possible, where regions of longer $T_2$ appear brighter (predominantly Co(III) ions) than regions of shorter $T_2$ (predominantly Co(II) ions).

Images for the Fe(II)/Fe(III) front were collected using a spin-echo experiment, where only a single echo was acquired for each excitation, but the repetition time was kept short $(T_R = 250$ ms). Image contrast was produced through $T_2$ differences, because the echo time of the experiment $(T_E = 7.6$ ms) was comparable with the $T_2$ of the solutions. Regions containing Fe(III) solution, which had a very short $T_2$, appeared dark, while the Fe(II) solution appeared brighter due to the longer $T_2$.

The orientations of images are shown in Figure 24.2. Both *zy* and *xy* images were obtained for the Co(II)/Co(III) front, and only *zy* images for the Fe(II)/Fe(III)

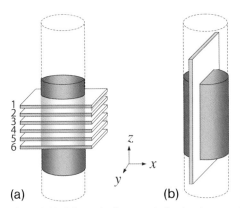

**Figure 24.2** Schematic diagrams indicating image orientation and fields-of-view for multiple horizontal *xy* slices (a) and a vertical *zy* slice (b). In both diagrams the gray area represents the field-of-view, which is the region of the tube held within the radiofrequency coil.

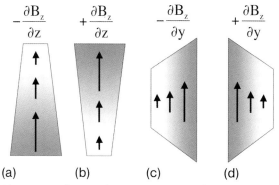

$$-\frac{\partial B_z}{\partial z} \qquad +\frac{\partial B_z}{\partial z} \qquad -\frac{\partial B_z}{\partial y} \qquad +\frac{\partial B_z}{\partial y}$$

(a)        (b)        (c)        (d)

**Figure 24.3** Schematic diagram representing the variation in magnetic field for a set of vertical (a, b) and horizontal (c, d) gradients. Both negative (a, c) and positive (b, d) gradients are shown.

front. Heating of the sample due to the imaging sequence and gradient pulses was negligible.

The $zy$ images had a slice thickness of 1 mm and were positioned in the center of the tube. For the Co(II)/Co(III) front experiments, the pixel size was $195 \times 195\,\mu m$. Multiple-slice $xy$ images were also acquired; each slice had a thickness of 1 mm and separation of 1.2 mm, with a pixel size of $78 \times 78\,\mu m$. For the Fe(II)/Fe(III) front experiments, the pixel size was 195 (vertically) $\times\,91\,\mu m$ (horizontally).

To follow the effect of magnetic field gradients on the traveling wave, trains of gradient pulses were applied between image acquisitions. Gradient trains were generated using the imaging gradients of the spectrometer and comprised pulses which were switched on for 2 ms and off for 1 ms, cycled 2000 (Co(II)/Co(III)) or 500 (Fe(II)/Fe(III)) times, with amplitude of $+0.2\,T\,m^{-1}$ or $-0.2\,T\,m^{-1}$ (see Figure 24.3). Gradients were applied either in the vertical or horizontal direction [8, 9]. Relatively long time intervals between images were chosen to minimize the influence on the wave from the magnetic field gradients involved in the *imaging* sequence.

## 24.3
## Results

### 24.3.1
### Co(II)/Co(III) Fronts

After initiation of the reaction with NaOH solution, a flat interface for the front formed (Figure 24.4a). This front propagated downwards and maintained its structure for approximately 10–20 min, after which the front became distorted and

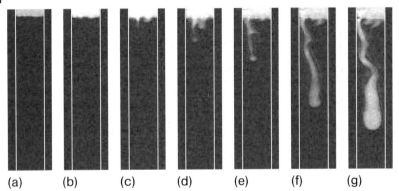

(a)      (b)      (c)      (d)      (e)      (f)      (g)

**Figure 24.4** Time series of seven MRI *zy* images of a traveling wave formed in the reaction of Co(II)EDTA$^{2-}$ with H$_2$O$_2$. The slice thickness of each image is 1 mm and a region of 27.3 mm (vertical) × 7.8 mm (horizontal) is displayed. Image (a) was taken 17 s after the wave was initiated by the addition of a drop of NaOH solution. Subsequent images were taken at intervals of 51 s following image (a). The signal intensity is high (bright) where Co(III)EDTA$^-$ ions predominate, and low (dark) where Co(II)EDTA$^{2-}$ ions predominate. The vertical white lines in each image indicate the sides of the NMR tube.

eventually a finger formed. The fingering is believed to arise through hydrodynamic instabilities caused by the reaction where a layer of more dense liquid is above a less dense liquid [8]. This behavior is seen in a number of traveling front reactions [1, 10, 11]. Images of the propagation of a typical finger are shown in Figure 24.4b–d. The velocity of the finger was considerably faster due to these hydrodynamic effects, than the flat front, which was propagating only through a RD mechanism [3].

A series of experiments were performed, whereby the effect of a magnetic field gradient applied in the vertical direction was tested. No magnetic field effects were observed when the front propagated with a flat interface, and an average front velocity of $8.2 \pm 1\,\mu m\,s^{-1}$ was measured [8]. Once the finger formed, a dependency of the front velocity on the presence or absence and orientation of the applied magnetic field gradient was observed. Front velocities after fingering were on average $135.1 \pm 3.3\,\mu m\,s^{-1}$ in the absence of a gradient. Average velocities were $173.2 \pm 17.7\,\mu m\,s^{-1}$, for fronts propagating from higher to lower magnetic field [i.e. when the field increased down the tube (Figure 24.3a) and $125.8 \pm 3.6\,\mu m\,s^{-1}$] for fronts propagating from lower to higher magnetic field (i.e. when the field decreased down the tube; Figure 24.3b [8]).

An explanation for the acceleration of the front *down* a magnetic field gradient is believed to lie with an increase in the transport of the autocatalytic hydroxide ions. The finger is more diamagnetic than the surrounding solution and so experiences a magnetic force directed down the field gradient that would bring hydroxide ions to the reacting front, thus propagating the wave more quickly. An interesting feature of this magnetic effect is that it only occurs once the finger has formed – that is, once a 'symmetry-breaking' process has occurred.

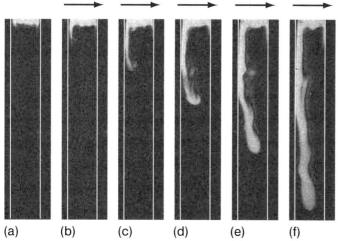

**Figure 24.5** Time series of six MRI $zy$ images of the Co(II)/Co(III) front, of 1 mm slice thickness; showing a region of 7.8 (horizontal) × 33.2 mm (vertical). Image (a) is taken 17 s after the wave was initiated by the addition of a drop of NaOH solution. Following image (a) magnetic field gradient trains were applied between imaging experiments. The applied gradients increased the magnetic field from left to right, as indicated by the arrows. Images were collected every 51 s.

To further investigate the control of these fronts, horizontal field gradients were applied, according to $B_z = B_0 + y \partial B_z / \partial y$. These gradients were applied as soon as the front formed, prior to finger formation. A typical series of images is shown in Figure 24.5. For this series of experiments gradients were applied horizontally so that the magnetic field increased from left to right (Figure 24.5b–f). Distortion of the front occurred soon after the gradients were applied, with the left side progressing quicker than the right side. As more gradients were applied this became increasingly exaggerated, until a finger was forced [9]. It was found that these horizontal gradients could reproducibly force a finger to form on the side where the magnetic field was least. By taking a series of $xy$ images following trains of $\partial B_z / \partial y$ gradients, we could show that the finger develops in a fixed position and that no other fingers were formed outside of the central $yz$ slice. Figure 24.6 shows a set of these images wherein the $\partial B_z / \partial y$ gradients produced a magnetic field that was greater on the right-hand side of the image than the left (according to Figure 24.3d).

It was subsequently found that the path of a finger could be further controlled once it was forced, by switching the direction of the gradients midway through an experiment. Figure 24.7 shows images from a typical series of experiments. The first image was taken at 20 s after the wave was initiated, prior to the application of any gradient pulse trains. The subsequent two images were taken following the application of a set of horizontal gradients, which increase the magnetic field from right to left (according to Figure 24.3c). The resultant finger

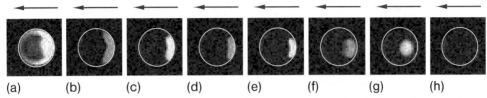

(a)    (b)    (c)    (d)    (e)    (f)    (g)    (h)

**Figure 24.6** Multiple *xy* slice images, of the Co(II)/Co(III) front, taken down the length of a chemical finger formed 650 s after wave initiation and following the application of magnetic field gradient trains. Applied gradients increased the magnetic field from right to left, as indicated by the arrows. Each image has a slice thickness of 1 mm, with a separation of 1.2 mm between images, and a field of view of 12.5 mm in both *x* and *y* directions. Slice position 1 is the highest and is positioned nearest to where the wave was initiated.

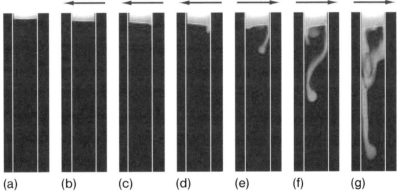

(a)        (b)        (c)        (d)        (e)        (f)        (g)

**Figure 24.7** Series of MRI *zy* images of the Co(II)/Co(III) front; a region of 7.8 mm (horizontal) × 26.3 mm (vertical) is displayed. Image (a) was taken 20 s after wave initiation. Following image (a), magnetic field gradient trains were applied between imaging experiments and subsequent images collected at 71 s (b), 123 s (c), 175 s (d), 226 s (e), 277 s (f) and 328 s (g) after wave initiation. For the images shown in (b–d) the applied gradients increased the magnetic field from right to left. Following image (d) the direction of the gradients were switched so that the magnetic field increased from left to right. The arrows indicate the direction in which the magnetic field increases during the gradient trains, which were applied prior to acquiring the images.

can be clearly seen in Figure 24.7d. Following this image the direction of the horizontal gradients was switched so that the field then increased from left to right. The position of the finger then clearly switched to the left side of the tube, in the next two images (Figure 24.7e and f), and continued to propagate down this side. It is important to note that only the tip of the wave was manipulated and that part of the finger which contained the fully reacted Co(III)EDTA⁻ solution remained fixed on the left side of the tube. Furthermore, a second finger began to form on the right-hand side of the tube from the interface at the top (Figure 24.7f).

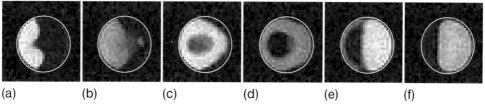

(a)       (b)       (c)       (d)       (e)       (f)

**Figure 24.8** Multiple $xy$ slice images taken through the length of a chemical finger, formed during the reaction of Co(II)EDTA$^{2-}$ with H$_2$O$_2$, following a switching of the applied horizontal gradients. The finger had initially been formed down the left side of the tube and then switched to the right side following the change in direction of the horizontal gradient, which produced a higher magnetic field on the left side than the right. Imaging parameters are the same as Figure 24.5.

The effect of switching the sign of the magnetic field gradient during an experiment is further illustrated by the six $xy$ images in Figure 24.8. The initial image (Figure 24.8a) shows how a gradient magnetic field, of the geometry shown in Figure 24.3c, has forced the formation of a finger on the left-hand side of the tube. As the tip of the finger moves down the tube in images b–f, it is progressively pushed towards the right-hand side of the tube by the application of gradients of opposite sign to those used initially (Figure 24.3d).

In our first experiments, using vertical magnetic field gradients, the velocity of wave propagation could be affected by the magnetic field only after finger formation had occurred. In contrast, the application of horizontal gradients forced a finger to form immediately on the side of lowest magnetic field. So, the question arose as to the importance of hydrodynamic effects in the magnetic field control of this front. To explore this, several experiments were performed on the traveling front in a porous medium, which would suppress convection. The choice of porous medium was difficult, as many materials – such as silica gel or glass beads – promoted the reaction to 'clock' immediately. The best medium found was packing foam, as this did not stimulate the reaction to clock immediately but allowed the front to be observed for several minutes. However, the length of time over which the traveling front could be studied was limited by the presence of additional initiation sites producing further fronts in the foam. A number of experiments were performed, studying the effect of horizontal gradients on a front propagating through the porous medium, and a typical set of images is presented in Figure 24.9. The velocity of this front, averaged over its width, was $0.8 \pm 0.2\,\mu\mathrm{m\,s}^{-1}$, which was slower than that found in the liquid phase, even prior to fingering. The origin of this reduction in wave velocity is, as yet, uncertain, although is quite likely to be due to the tortuosity of the pore space. However, in analogy to the liquid-phase experiments, the application of horizontal gradients forced the front propagation down the side of lowest magnetic field.

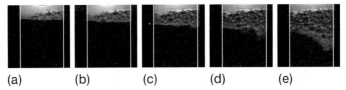

(a)          (b)          (c)          (d)          (e)

**Figure 24.9** Time series of *zy* images showing a Co(II)/Co(III) traveling front propagating through foam. The observed region is 12.5 mm². The image in (a) was taken 14 s after the front was initiated, and the subsequent images were taken at 51 s intervals. Horizontal gradients were applied following the image in (a) and increased the magnetic field from right to left.

### 24.3.2
### Fe(II)/Fe(III) Fronts

The exceptional magnetic control possible in the Co(II)/Co(III) front, opens up the question of other chemical fronts possessing similar behavior. A number of reactions produce waves and fronts involving the change in oxidative state of a transition metal ion, a typical example being the oxidation of Fe(II) in nitric acid. This reaction is autocatalytic and also produces a front. One advantage that this reaction has over the $Co(II)EDTA^{2-}/H_2O_2$ reaction is that its chemistry and mechanism are much better understood.

Previous interest in the Fe(II)-nitric acid reaction has focused on the presence of convection and its influence on both ascending and descending waves. The velocity of waves moving both upwards and downwards are greater than when studied in a medium that suppresses convection, such as silica gel [10]. This is because these wave velocities are enhanced by hydrodynamic effects arising from the reaction. Traveling fronts in this reaction have previously only been observed optically [12]. Here, we present the first study using MRI to visualize fronts in this reaction.

Figure 24.10 and Figure 24.11 show typical images for ascending and descending waves, respectively. The velocity of the front is dependent on its direction, with ascending fronts propagating at a velocity of $30 \pm 3 \, \mu m \, min^{-1}$ and descending fronts at $190 \pm 10 \, mm \, min^{-1}$. The structure of the front is dependent on the direction of propagation, as ascending and descending waves are differently influenced by convective flow [12]. Descending waves propagate faster than pure RD waves because the reacted solution has a higher density than the unreacted solution, due to compositional differences; hence, convection occurs. The descending front therefore moves like a chemical finger and, as the tube is narrow, only a single finger forms which pushes unreacted fluid up the sides. The ascending wave is also affected by convection, although it is the density differences associated with thermal differences, as the reacted solution is warmer and less dense, which are important here. The hydrodynamic flow associated with this configuration leads

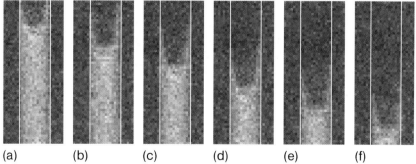

**Figure 24.10** Time series of images for a descending front formed during the oxidation of iron(II) with nitric acid. A slice thickness of 1 mm was used and a region of 2.4 (horizontal) × 11.7 mm (vertical) is displayed. Signal intensity is high (bright) where Fe(II) ions predominate and low (dark) where Fe(III) ions predominate. The vertical white lines in each image indicate the sides of the capillary tube.

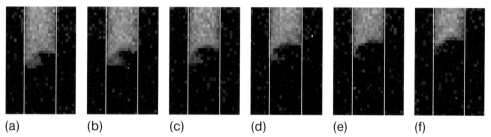

**Figure 24.11** Time series of images for an ascending Fe(II)/ Fe(III) front. A slice thickness of 1 mm was used and a region of 2.8 (horizontal) × 7.0 mm (vertical) is displayed. Images are shown at 25 s intervals. The interface between reacted (dark) and unreacted (bright) solution fluctuates, indicating the presence of convection.

to greater mixing of the autocatalyst [12] and an enhancement of the wave propagation compared to a pure RD wave. This increased mixing is indicated in the images in Figure 24.11, where the wave front is shown to fluctuate as it propagates upwards, through the formation of a tendril. Figure 24.12a shows the interface after a further 11 min, when the tendril can be seen to have switched sides.

Magnetic field gradients were applied in the horizontal direction, to see if the finger associated with the ascending wave could be affected. Although a number of experiments were performed, the position of the finger was seen to be unaffected by the gradients, as was the velocity of the wave. Hence, further experiments must be conducted in order to fully rule out any magnetic field effects in

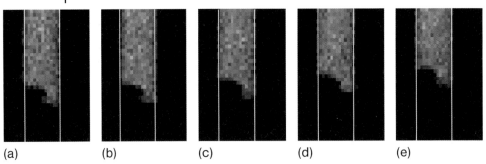

(a)  (b)  (c)  (d)  (e)

**Figure 24.12** Time series of images for an ascending Fe(II)/
Fe(III) front, continuing on from the images shown in Figure
24.10. Imaging parameters are the same as Figure 24.10,
except that images are shown at 16.7 s intervals. Image (a) is
taken 676 s after the image in Figure 24.10f. It can be seen
from these images that the fluctuations at the interface
continue, although the tendril has now switched sides.

this reaction and to determine why this front may not be as field-sensitive as the
Co(II)/Co(III) front.

## 24.4
## Discussion

We have shown that the Co(II)EDTA$^{2-}$/H$_2$O$_2$ autocatalytic reaction can be affected
dramatically by the application of magnetic field gradients. Here, we propose that
the magnetic field effects observed in this system are due to the action of the
Maxwell force according to

$$\mathbf{F}_M = k\,(\mathbf{B}\cdot\nabla)\mathbf{B} = \begin{pmatrix} F_x \\ F_y \\ F_z \end{pmatrix} = k \begin{pmatrix} B_x\partial B_x/\partial x + B_y\partial B_x/\partial y + B_z\partial B_x/\partial z \\ B_x\partial B_y/\partial x + B_y\partial B_y/\partial y + B_z\partial B_y/\partial z \\ B_x\partial B_z/\partial x + B_y\partial B_z/\partial y + B_z\partial B_z/\partial z \end{pmatrix} \qquad (24.2)$$

where $k$ is a collection of constants $(V/\mu_0)\Delta\chi_V$. The action of any magnetic force
in the system therefore relies on the existence of a magnetic susceptibility gradient
in the sample, $\Delta\chi_V$ and requires at least one term of the form $B_p\partial B_q/\partial p$ to be
nonzero (hence, producing a force $F_q$ on the front in a given direction $q$). The
former condition is clearly fulfilled as the transition from paramagnetic Co(II) to
diamagnetic Co(III) produces a magnetic susceptibility gradient across the wave
front (the difference in volume magnetic susceptibilities for the concentration of
pure Co(II)EDTA$^{2-}$ and Co(III)EDTA$^-$ solution as used in these experiments is
$0.184 \times 10^{-5}\,\mathrm{m}^3\,\mathrm{kg}^{-1}$). However, the effect of different terms of the form $B_p\partial B_q/\partial p$
is more complex and shall be dealt with in turn for the different experiments dis-
cussed above.

## 24.4.1
## Application of Longitudinal Gradients, $B_z = B_0 + zG_z$

In these experiments, the sample was subjected to a longitudinal magnetic field gradient superimposed on the homogeneous magnetic field $B_0$ produced by the vertical superconducting magnet of the MRI spectrometer oriented along the axis (z) of the magnet bore, that is $B_z = B_0 + zG_z$, where $B_0 = 7.0\,T$ and $G_z = \partial B_z / \partial z = \pm 0.2\,T\,m^{-1}$ (see Figure 24.3a and b), respectively. According to Equation 24.2, a force in z-direction results[1]

$$F_z \propto B_z \partial B_z / \partial z = (B_0 + zG_z)G_z \approx B_0 G_z \qquad (24.3)$$

and using $B_0 \gg zG_z$. It is hence in agreement with these considerations that this longitudinal geometry allows manipulation of the finger in the z-direction accelerating or retarding the velocity of the spreading front depending on the sign of the gradient employed.

## 24.4.2
## Application of Transverse Gradients, $B_x = B_y = 0$, $B_z = B_0 + yG_y$

Transverse gradients are applied to create a field $B_z = B_0 + yG_y$, where $G_y = \partial B_z / \partial y = \pm 0.2\,T\,m^{-1}$ and $-2\,mm < y < 2$ mm. Thus, the magnetic field in the z-direction varies in a linear fashion *across* the cylindrical sample tube rather than along its length (see Figure 24.3c and d).

Although Figures 24.5–24.9 clearly prove that transverse gradients (Figure 24.3c and d) have a significant effect on the propagation of the wave front, a quick glance at Equation 24.2 seems to indicate that the application of such transverse gradients should have no effect on the velocity of the front propagation: the only obvious nonzero gradient, $\partial B_z / \partial y$, contributes only via the $B_y \partial B_z / \partial y$ term in Equation 24.2 and the spectrometer field $B_x = B_y = 0$ renders the $F_z$ term zero also.

Maxwell's equation, $\nabla \times \mathbf{B} = 0$, however, means that the following relationship must hold in the absence of electric currents in the sample or time-dependent electric fields:

$$\frac{\partial B_p}{\partial q} = \frac{\partial B_q}{\partial p}, \quad p, q = x, y, z \qquad (24.4)$$

Therefore, when a magnetic field with transient gradient such as $B_z = B_0 + yG_y$ is applied to the sample, there *must* also be a 'concomitant' field gradient in the y-direction according to $B_y = zG_y$ and

1) As $\nabla \cdot \mathbf{B} = 0$ has to be true, it is clear that there are distance-dependent forces also acting in the x- and y-directions. However, these are negligible compared to the force acting in the z-direction as the distances across the sample tubes (~0.0025 m) are small, as are $B_{x,0}$ and $B_{y,0}$.

$$F_y \propto B_z \partial B_y / \partial z = (B_0 + \gamma G_y)G_y \approx B_0 G_y \qquad (24.5)$$

(with $G_y = 0.2\,\mathrm{T\,m^{-1}}$ and $|y| < 2\,\mathrm{mm}$, it follows that $\gamma G_y \ll B_0$ and $F_z = k B_y G_y = kz G_y^2 \ll F_y$). It is known that these so-called concomitant magnetic fields create a number of artifacts in MRI at low fields, particularly in experiments sensitive to phase evolution [13]. In the study presented here the concomitant gradients do not generate any detectable artifacts of this type, but instead are responsible for the ability to manipulate the front using magnetic forces that are transverse to the direction of propagation.

It follows from Equation 24.5 that $F_y$ is the dominant force component for this system, changing direction when the applied gradient is reversed. As depicted in Figures 24.5– 24.9, the sign of the transverse gradient, $G_y$, determines the initial point of formation for the chemical finger as well as its consequent motion within the tube. As expected, the finger (of diamagnetic reaction product) is formed exclusively in the region of lowest field. The consequent application of magnetic field gradients allows the finger to be 'pushed' around within the tube as the (diamagnetic) finger consistently moves towards the lowest field region of the tube.

However, only the *leading edge* of the finger is affected by the gradient trains. As demonstrated in Figure 24.7, only the most recently reacted part of the finger is affected by the applied field when the gradient direction is switched during an experiment, whereas those regions of the finger which formed prior to switching the gradients remain unaffected. Possible explanations for this phenomenon might be based on the increased significance of hydrodynamic instabilities at the leading edge of the finger, or the curvature of the wave front.

At present it is not clear why there is no magnetic field effect observed on the Fe(II)-nitric acid reaction. At first glance, the reaction lends itself perfectly for such study given the nonlinear nature of the kinetics as well as the change in oxidation state and hence susceptibility of the ions involved. However, whilst in the cobalt reaction there is a large change in susceptibility (as a paramagnetic ion containing three unpaired electrons is oxidized to a diamagnetic ion), the iron reaction shows a far smaller change in susceptibility as high-spin $d^6$ ion is oxidized to high-spin $d^5$ species (neglecting any orbital contributions in this discussion). Although the effect is expected to be smaller, its complete absence is somewhat surprising. It might well be that other factors such as the characteristics of the hydrodynamic flow in the systems cause the difference between the two systems. At this point it is important to note that the Co(II)/Co(III) finger travels at a velocity of about $135.1 \pm 3.3\,\mu\mathrm{m\,s^{-1}}$ whilst the Fe(II)/Fe(III) front is faster at $500\,\mu\mathrm{m\,s^{-1}}$ when ascending and a factor of 20 faster (at $3000\,\mu\mathrm{m\,s^{-1}}$). Indeed, the magnetic field effect in the cobalt travelling front is also significantly reduced when working in foam. It is not surprising therefore that it is most certainly a delicate interplay of kinetics, convection and diffusion which must be satisfied in order for a magnetic field to have its most pronounced effect. Simulations and further experiments exploring these factors are presently under way in our laboratories.

## 24.5
## Conclusions

It has been shown that the direction and velocity of a descending front formed during the autocatalytic oxidation of $Co(II)EDTA^{2-}$ can be controlled by applying magnetic field gradients of appropriate orientation. The observed phenomena are interpreted qualitatively, based on the action of the Maxwell stress. It was postulated that the experimental methods used to control the Co(II)/Co(III) front might be suitable also to manipulate other traveling chemical waves, as long as the reactants and products of the underlying chemical reaction are characterized by different magnetic susceptibilities. Surprisingly, the application of magnetic field gradients to the traveling front produced during the oxidation of Fe(II) in nitric acid did not allow any spatial control of this chemical wave. Further studies into the nature of these effects are therefore under way.

## Acknowledgments

The authors thank Robert Evans. M.M.B. thanks EPSRC for an Advanced Research Fellowship and Prof. Lynn Gladden and the Magnetic Resonance Research Center, Cambridge for support. C.R.T. thanks the Royal Society for a University Research Fellowship and the EPSRC for financial support.

## References

**1** Epstein, I.R. and Pojman, J.A. (1998) An introduction to nonlinear chemical dynamics, in *Topics, in Physical Chemistry* (ed. D.G. Truhlar), Oxford University Press, Oxford.

**2** Zaikin, A.N. and Zhabotinsky, A.M. (1970) *Nature*, **225**, 535–7.

**3** Scott, S.K. (1994) *Oscillations, Waves, and Chaos in Chemical Kinetics. Oxford Chemistry Primers*, Oxford University Press, Oxford.

**4** Boga, E., Kádár, S., Peintler, G. and Nagypál, I. (1990) *Nature*, **347**, 749–51.

**5** He, X., Kustin, K., Nagypál, I. and Peintler, G. (1994) *Inorganic Chemistry*, **33**, 2077–8.

**6** Fukushima, E. and Roeder, S.B.W. (1981) *Experimental Pulse NMR: A Nuts and Bolts Approach*, Addison-Wesley, Reading, MA.

**7** Hennig, J., Naureth, A. and Friedburg, H. (1986) *Magnetic Resonance in Medicine*, **3**, 823–33.

**8** Evans, R., Timmel, C.R., Hore, P.J. and Britton, M.M. (2004) *Chemical Physics Letters*, **397**, 67–72.

**9** Evans, R., Timmel, C.R., Hore, P.J. and Britton, M.M. (2006) *Journal of* the *American Chemical Society*, **128**, 7309–14.

**10** Bazsa, G. and Epstein, I.R. (1985) *Journal of Physical Chemistry*, **89**, 3050–3.

**11** Horvath, D., Bansagi, T. and Toth, A. (2002) *Journal of Chemical Physics*, **117**, 4399–402.

**12** Pojman, J.A., Nagy, I.P. and Epstein, I.R. (1991) *Journal of Physical Chemistry*, **95**, 1306–11.

**13** Norris, D.G. and Hutchison, J.M.S. (1990) *Magnetic Resonance Imaging*, **8**, 33.

# 25
# Fluid Distribution and Movement in Engineered Fibrous Substrates by Magnetic Resonance Microscopy

*Johannes Leisen and Haskell W. Beckham*

## 25.1
## Introduction

Nuclear magnetic resonance (NMR) has been used widely as a tool for the characterization of porous materials [1–3]. While much research effort has been directed towards porous rocks, sandstones, sands or foams, studies of engineered fibrous substrates via methods of magnetic resonance microscopy (MRM) have received much less attention.

Engineered fibrous substrates are defined as materials whose structure is composed of, and whose functionality is achieved through, fibers. Scattered MRM studies of these materials range from textile fabrics [4, 5] used in apparel and technical applications (e.g. filtration materials [6–8]) to pulp/paper [9–14], carpets [15–18] and diapers [19]. Common to all of these studies is the relevance and importance of the interaction of the fibrous substrates with fluids [20]. On the one hand, the production of fibrous substrates can involve fluid-based processes such as dyeing and laundering. On the other hand, fibrous substrates might be specifically designed to provide a desired end-use interaction with fluids. Examples here range from fabrics with special wicking behavior for use in sports apparel to technical fabrics used for filtration purposes. Thus, as knowledge of fluid location and movement within fibrous substrates is important for optimizing manufacturing processes and designing improved products, MRM holds great potential for studies of these soft porous materials.

From a purely theoretical point of view, fibrous substrates are a very interesting group of porous materials for the following reasons:

(i) Fibrous substrates are *soft* porous materials, the pore structure of which can change upon fluid incorporation or movement through the material, resulting in swelling and deformation of samples under investigation.

(ii) As fibrous substrates are man-made materials, a diverse variety of samples with well-defined pore structures and fluid–solid interactions can easily be fabricated.

*Magnetic Resonance Microscopy.* Edited by Sarah L. Codd and Joseph D. Seymour
Copyright © 2009 WILEY-VCH Verlag GmbH & Co. KGaA, Weinheim
ISBN: 978-3-527-32008-0

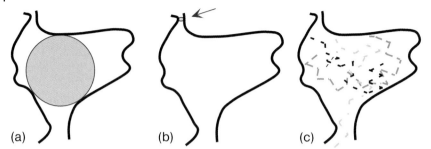

**Figure 25.1** Schematic representation of an irregular-shaped pore formed by two fibers, and how its measured size depends on the technique used to probe it. (a) Image analysis often relies on optimizing the position and size of defined objects (e.g. spheres, ellipsoids, etc.) into the pore space, which are then used to compute the pore size distribution [21, 22]; (b) Capillary flow porometry measures the pore throat sizes [23]; (c) Diffusion NMR or *q*-space imaging probes the size and connectivity of the pore space through Brownian motion [24].

(iii) As with some other porous materials, it is difficult to identify unambiguously clear boundaries for the pores in a fibrous substrate. Often, the pores are defined only by the analytical methods used to measure them (Figure 25.1) [23].

Magnetic resonance microscopy measurements of time-dependent phenomena such as fluid uptake, removal and transport in fibrous substrates provide valuable information that is often difficult or impossible to gain through any other method. In many cases, this information is simply measured as the fluid concentration as a function of position and time. Knowledge of the fluid concentration as a function of position can be used to extract information on the underlying pore structure, which is of relevance for a large variety of engineered fibrous substrates. For example, as one component of quality control, the pore size distribution is routinely measured in the nonwovens industry. The conventionally used porometric techniques are based on the intrusion or extrusion of fluids into the substrate [25], which can lead to erroneous – if not irreproducible – results due to the characteristics of the substrates listed above. In particular, the *soft* nature of the substrates can result in changes in sample dimensions (and underlying pore structure) due to the removal or uptake of fluids during the porometric procedures. MRM techniques can be used to measure *static* fluid distributions and content for well-defined fluid concentrations, from which information on the underlying pore structure can be extracted. Thus, as the substrates (and underlying pore structure) do not change during such static MRM measurements, the drawbacks of the porometric techniques are avoided.

This chapter provides examples in which the use of MRM is demonstrated for the characterization of fluids in fibrous substrates. One of the key aspects of this work (as for fluids in many other materials) is the quantitative measurement of fluid concentrations in complex porous media. This is associated with a number of challenges, which can be tackled using different strategies depending on the details of the sample and the fluid. In order to provide a practical guide for conducting MRM experiments on fluids in fibrous substrates, we have organized the applications reviewed in this chapter according to whether qualitative, semiquantitative or quantitative determinations of fluid concentrations are required.

## 25.2
## Signal Contrast and Fluid Concentration

For this discussion we concentrate on a simple spin-echo sequence [24, 26] defined by a repetition delay, TR, and an echo time, TE. As this sequence provides the basis for many more advanced MRM experiments, the points raised below will also be valid for most of these sequences. For the following discussion it is not relevant whether echoes are produced using radiofrequency (RF) pulses or gradients; therefore the following arguments hold equally for the entire class of gradient echo techniques.

As measured in an MRM experiment, the intensity of any voxel, $I$, at a position, $\mathbf{r}$, is given by a series of factors ($F$) that depend on the experimental set-up and the sample characteristics, as defined by the $T_1$ and $T_2$ relaxation times and the self-diffusion constant, $D$ [16]:

$$I(\mathbf{r}) \propto \left( \sum_i^n w_i(\mathbf{r}) S_i(\mathbf{r}) F_i^{T1}(\mathbf{r}) F_i^{T2}(\mathbf{r}) F_i^{D}(\mathbf{r}) \right) F^{hardware}(\mathbf{r}) \tag{25.1}$$

Each voxel contains $n$ different components distinguished by their relaxation and/or diffusion characteristics. The separate concentrations of these $n$ components are accounted for by the weighting factors $w_i$, and $S_i$ is the spin density of each respective component $i$. The weighted addition of these component spin densities according to

$$S(\mathbf{r}) = \sum_i^n w_i(\mathbf{r}) S_i(\mathbf{r}) \tag{25.2}$$

provides a measure of the total number of spins in any voxel – that is, the sum of fluid and fiber. However, in most instances, the $T_2$ values for solid fibers are much less than 1 ms, such that for technically feasible values of TE (>1 ms), the factor $F_{fiber}^{T2}$ approaches 0. Therefore, Equation 25.1 represents what we now refer to as the fluid components $i$ contained within a fibrous substrate. Ideally, the experiment is conducted with all remaining factors, $F_i^{T1,T2,D} = 1$, so that the measured signal intensity for each voxel is quantitatively proportional to the amount of fluid contained at that position in the substrate.

For each component $i$ the longitudinal relaxation factor, $F_i^{T1}$, at a given position $\mathbf{r}$ is defined by

$$F_i^{T1}(\mathbf{r}) = 1 - \exp\{-TR/T_{1,i}(\mathbf{r})\} \tag{25.3}$$

The contribution of this factor to the signal attenuation can be effectively eliminated by simply setting the TR duration sufficiently longer than the longest $T_{1,i}$ of the sample. If there is an experimental need for fast imaging, the $T_1$ relaxation times may be shortened by addition of relaxation or contrast agents. Strictly speaking, Equation 25.3 only holds for complete excitation of the spin system (i.e. perfect $\pi/2$ pulses). The use of very short excitation pulses is also an efficient means for conducting rapid MRM experiments; as only a portion of the overall magnetization is excited (and therefore available for detection), most of the equilibrium magnetization is available for subsequent excitation so that short repetition delays can be utilized.

The contribution to the transverse $T_2$ relaxation is given by

$$F_i^{T2}(\mathbf{r}) = \exp\left\{-\frac{TE}{T_{2,i}(\mathbf{r})}\right\} \tag{25.4}$$

The influence of this factor may be minimized by setting values for TE as short as possible. Note, however, especially in cases where fluid-substrate binding takes place, $T_{2,i}$, will assume values in the range or even below the minimum settable values of TE using current MRM equipment. Hence, there are cases where $F^{T2}$ may lead to unavoidable and substantial deviations from the correct quantitative proportionality between $S$ and the detected signal intensity, $I$.

Similar considerations exist for the contribution of molecular self diffusion to the image contrast. The attenuation of the signal will depend on the overall durations and magnitudes of the gradient pulses. For a simple spin-echo sequence in which the read gradient $G$ is permanently on, $F_i^D$ is given by [26]

$$F_i^D(\mathbf{r}) = \exp\left\{-\gamma^2 G^2 D_i(\mathbf{r})\frac{TE^3}{12}\right\} \tag{25.5}$$

The effect of the diffusion factor will be reduced by using a short TE duration and a low gradient strength $G$. However, the choice of $G$ is determined by the required spatial resolution, and independently modifying it is not a useful option. For high spatial resolutions, which require high gradient strengths, $F^D$ may very well have a larger influence on signal attenuation than $F^{T2}$.

Lastly, one must consider the overall imperfection of the hardware set-up, $F^{hardware}$, in accounting for image contrast. Spatial inhomogeneities in excitation fields or detection can be problematic depending on the RF coil design. While such inhomogeneities can never be fully avoided, they may be minimized through an appropriate choice of RF coil, sample size, and placement within the coil. For quantitative studies it is important to know exactly the characteristics of a specific coil so that the sample can be placed into those areas where the RF excitation is as homogeneous as possible.

All things considered, the most serious obstacle to the measurement of quantitative fluid concentrations in fibrous substrates is the inability to record images for TE $\rightarrow$ 0 and thereby to eliminate the influence of the factors $F^{T2}$ and $F^D$ on the signal contrast. The creation of data by extrapolation to TE = 0 from a series of images measured with different TE values is often ambiguous. This is especially true when components $i$ with very different relaxation times contribute to detectable intensity within a particular voxel – that is, a multiexponential decay must be extrapolated over a substantial time of several milliseconds.

In summary, attributing measured voxel intensities, $I$, to concentrations, $S$, may not be straightforward. The true spin density is attenuated by a number of factors that depend on the sample characteristics, hardware, and imaging sequence durations. It is important to be aware of these factors so that experimental strategies can be judiciously designed for quantitative measurements.

## 25.3
## Qualitative Investigations of Fluid Distributions

For some applications, qualitative information about the distribution of fluids in fibrous substrates is sufficient. Indeed, it is common to extract important and relevant information solely from determination of the fluid location within a sample. Examples include the study of: (i) fluid ingress into and ultimate distribution within surface-treated carpets [15]; (ii) drying patterns of water contained in paper sheets to determine underlying preferred fiber orientations [10]; and (iii) average inter-pore spacings in fabrics by calculation of fluid-density autocorrelation functions from MRM images [4]. In all of these examples, information on the solid structure of the fibrous substrate is contained in the MRM data. For ordered woven or knit textile fabrics, information on the relative positioning and spacing of the two principal yarns – referred to as fabric geometry – is provided. Such knowledge is important for building computational models for studying the mechanical and flow-resistance properties (e.g. for filtration applications) of fibrous materials. This is particularly relevant for textiles that are still developed for technical applications by trial-and-error processes. The design and optimization of textile products and production processes via 'virtual textiles' is currently just emerging [27–30].

Conventionally, solid fabric structures are determined by techniques based on light microscopy in which fabric samples are first embedded in a resin and then mechanically sliced using a microtome. Micrographs of the slices are then digitized and used for the reconstruction of a 3-D model of the fabric. Whilst this approach provides images with high resolution it is, needless to say, very labor-intensive. This disadvantage has been somewhat overcome by a commercially available system referred to as digital volumetric imaging (DVI), in which the slicing and imaging steps are automated [31]. Even so, to be generally applicable, DVI requires the use of a suitable dye that must stain, and a curable resin that must wet, any given sample.

As long as the required resolution of the digital model is above the lower limits of MRM, the latter can be readily used to assess the solid structure of fibrous substrates. The sample is immersed in a protonated fluid and an MR image recorded to depict the fluid surrounding the fibers; the negative of this image

(a)                         (c)                         (d)

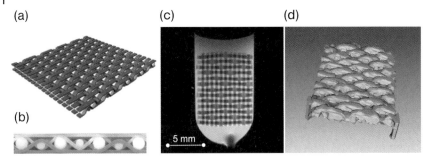

(b)

**Figure 25.2** Translation of a 3-D solid fabric structure into stereolithographic data for computer-aided-design environments. The fabric, a technical textile used as paper machine clothing, is shown in (a) as a graphic representation and in (b) as an optical micrograph of the cross-section; (c) MR micrograph (2-D spin-echo, TE = 12 ms, TR = 500 ms) of the fabric immersed in an aqueous solution; (d) 3-D rendered image of the fabric constructed from a 3-D gradient-echo data set using the Mimics software package (Materialise, Leuven, Belgium). Panels (a) and (b) reprinted with permission from AstenJohnson, Kanata, Ontario.

provides the structure of the fibrous substrate. Typically, 3-D imaging techniques that record isotropic voxels are preferred over multislice methods. Using this approach, we were able to measure an MR image of a woven fabric with sufficiently high precision to convert into a stereolithography format for use in computational modeling studies (Figure 25.2) [32]. This technical fabric is used in paper manufacturing, and its 3-D digital model was used for computational fluid dynamics (CFD) studies of through-fabric flow that were compared with our MR flow measurements [33]. For generating such 3-D digital models of soft fibrous substrates, MRM is competitive with optical techniques if the time required for sample preparation and data acquisition is taken into account.

## 25.4
## Semiquantitative Studies of Fluid Distribution

### 25.4.1
### Feasibility

An intuitive analysis of many simple spin-echo images of fluids in textiles suggests that the measured voxel intensities correspond well to fluid concentrations. As explained in Section 25.2, this is only true if all fluid molecules exhibit similar relaxation and diffusion behavior (i.e. $n = 1$ in Equation 25.1) and $F_1^{T1,T2,D}$ is independent of $\mathbf{r}$. More commonly, fluid molecules in fibrous substrates exist in sufficiently diverse environments that they exhibit a broad range of diffusion and $T_2$ relaxation behavior ($i = 1\ldots n$). This leads to several different attenuation factors, $F_i^{T2}$ and $F_i^{D}$, with significant influence on the sum in Equation 25.1. This effect was investigated for a series of fabrics made of fibers with varying degrees of

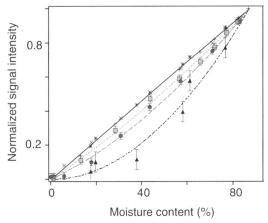

**Figure 25.3** Normalized NMR signal intensity versus
gravimetric moisture content for cotton (*, ▲), nylon (x, •) and
polypropylene (+, ■) fabrics measured with Bloch decays
(crossed symbols) and spin-echoes (solid symbols; TR = 10 s,
TE = 6.5 ms). Reprinted with permission from Ref. [16];
copyright SAGE Publications.

hydrophobicity (Figure 25.3) [16]. The total intensity of spin-echo profiles was
determined for different fluid contents in some preconditioned textile fabrics. For
fibers that experience no or only a minor interaction with the fluid (e.g. polypropylene and water) a reasonable proportionality is found between the water content
and the detected MRM signal. This is not the case for cotton, which is known to
bind water into its complex structure [34]. According to Figure 25.3, semiquantitative information on the concentration of fluids in fibrous substrates can be obtained
using spin-echo sequences when there is no significant binding interaction
between the fiber and the fluid. It should be noted that fiber hydrophilicity/hydro-
phobicity is often modified by surface treatments. For instance, hydrophilic
wicking finishes are applied to polypropylene fabrics used in sports clothing so
that they transport perspiration. When studying commercial fibrous products, care
must be taken in drawing conclusions based on the chemical structure of the fibers
alone. Even in cases where no binding interaction between fibers and fluid is
expected, it is important to perform an experimental validation of the proportionality between MR signal intensity and fluid content.

## 25.4.2
## Wicking

An example of a semiquantitative study of fluid concentrations in fibrous substrates is described here for wicking – that is, the spontaneous imbibition of fluids
into textiles. This phenomenon is of general interest for understanding the ingress
of curable liquid polymers into fibrous substrates for the production of high-

performance fiber–polymer composites [35]. Wicking is also important for a variety of consumer products, including paper towels, diapers and so-called 'wicking fabrics' used in comfort and athletic wear. Although many products are sold with claims of superior wicking abilities, to date there is no accepted industry standard to adequately measure wicking [36]. Published wicking tests are based on measuring the apparent height of the front of an upward-moving fluid through a porous material for a defined time period. This simple test does not yield an intensive or bulk property of the fibrous material, and often leads to contradicting results when different arbitrarily defined time periods are used [37]. Fluid transport by wicking is governed by the permeability ($k$), fluid saturation ($S$) and capillary pressure ($p_c$) of the fibrous material [38]. Small pores lead to large capillary pressures but low permeabilities. For a typical porous substrate with a distribution of pore sizes, the capillary network accessed by the fluid is governed by the fluid saturation. Hence, the $k$-$S$-$p_c$ relationships are needed to adequately describe the wicking ability of a porous material. An experimental approach to measure these properties on textile fabrics has been developed and is based on the gravimetric monitoring of horizontal and then downward wicking of fluids after a variable upward wicking segment in which the height is adjusted to set the saturation [39].

Upward wicking of fluid into fabrics, the initial stages of which occur on timescales of seconds to minutes, was investigated with spin-echo imaging. An aqueous CuSO$_4$ solution was used as the wicking fluid; the short $T_1$ relaxation time of this solution enables the recording of fully relaxed images ($F^{T1} \to 1$) for repetition durations of less than 100 ms. The overall time requirement to record a 2-D image consisting of $128 \times 128$ pixels was 5 s; hence, 128 consecutive images (frames) were recorded to produce a data set from which a movie was constructed to show the wicking process.

In order to most accurately capture the initial stages of wicking, the fluid should be added after the sample and probe have been loaded into the magnet. It is also important for the fluid reservoir to be kept at a constant level during the experiment so that the hydrostatic head remains steady and the fluid front can be determined from a fixed reference height. If the end of a fabric strip were to be immersed in a small, fluid-filled container, wicking would lead to removal of the fluid accompanied by a drop in the reservoir level. These problems were avoided by siphoning fluid from a large external reservoir to a glass tube extending into the RF coil (Figure 25.4). The fabric strip was mounted inside this glass tube; in this way fluid contact to the bottom end of the fabric strip, and the reservoir level at the line of contact, could be easily controlled by raising or lowering the height of the external reservoir.

Mounting the fabric strip inside the glass tube was more challenging than anticipated, as the intake of fluid into these soft, deformable porous materials often leads to swelling, curling or twisting. Fastening the ends of the sample strips between two clamps turned out to be problematic as the sample deformation caused it to move during the experiment. Hence, it was necessary to mount the fabric on a stiff, solid support. Initial experiments were conducted using a flat Teflon sheet, but this proved to be unfeasible as water became trapped

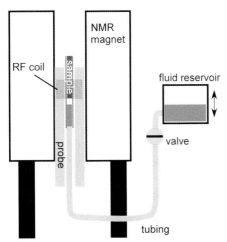

**Figure 25.4** The set-up for MRM wicking experiments. The sample is mounted in the RF coil and the fluid level is controlled by the height of the large external reservoir to about 2 cm below the MRM field-of-view. Before conducting the experiment, all fluid is contained in the fluid reservoir and the valve is closed. Opening the valve allows the fluid to move into the magnet and contact the bottom end of a sample strip in a controlled manner and at a defined time.

between the fabric strip and the Teflon surface. The best results were achieved by placing the fabrics on a stiff mesh made from hydrophobic fibers and containing large holes. The imaging slice thickness was set to several times the thickness of the fabric; in this way the measured overall fluid concentration within the fabric plane was insensitive to minor deformations of the fabric samples.

At the start of each wicking experiment no fluid was contained within the RF coil; this prevented the recording of scout images to properly align the sagittal imaging plane with the plane of the fabric. A small strip of silicone rubber, characterized by long $T_1$-values, was glued to the rear side of the solid support mesh. Then, by using long repetition delays, images of the rubber strip could be measured and used to properly align the imaging plane with the sample. As short repetition delays were used for the wicking experiments, no signal for the rubber scout appeared in the fluid images.

The last remaining experimental challenge is to determine the reservoir level with respect to the flow front detected during the wicking experiment. This was resolved by raising the siphon by 2 cm after completion of the experiment; this resulted in a movement of the reservoir meniscus into the MR field-of-view, and allowed one additional image to be collected. The position of the reservoir meniscus in the measured MR image was then used as a reference line below which (by 2 cm) the reservoir level during the wicking experiment was determined. An additional important piece of information from this last image was the signal intensity for the fabric sample immediately above the meniscus, reasonably assumed to represent 100% saturation ($S = 1$). The voxel intensities for all images were then normalized with respect to those for which $S = 1$, thereby allowing semiquantitative reporting of saturations on a scale of 0 to 1.

Figure 25.5 presents a series of MR images of fluid wicking into two knit fabrics marketed for use in sportswear. Although the fibers – nylon/Spandex or polyester – are different for the two fabrics, the use of hydrophilic surface finishes can easily

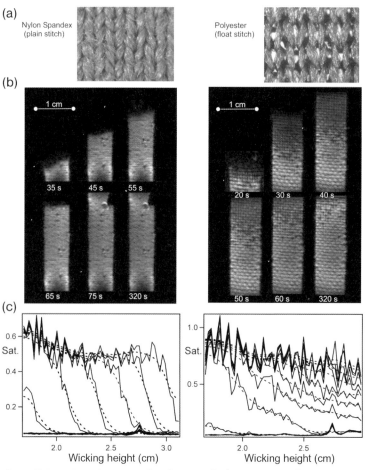

**Figure 25.5** (a) Optical micrographs of two knit fabrics used for MRM wicking studies. Note the larger pores in the float-stitch fabric; (b) A series of MR micrographs recorded while upward wicking of aqueous $CuSO_4$ solution into both fabrics. In the images of the float-stitch fabric, brightness continues to increase after the fluid front has passed through the field-of-view. From these images, fluid saturation profiles were extracted and are shown in (c). Characteristic transient and equilibrium saturation gradients exist for both fabrics. The spikes in the data are not noise but reflections of the underlying structure of the fabric. Details as to the calculation of these profiles (dotted lines) are given in the text.

mask any wicking disparities due to the fibers. The major distinction between the two fabrics is in the knit construction. As seen in the optical micrographs (Figure 25.5a), the float-stitch polyester fabric contains larger pores than the plain-stitch nylon/Spandex fabric. Hence, the float-stitch fabric is likely characterized by a broader pore size distribution.

No significant changes in the images for the plain-stitch fabric were observed after the fluid front passed through the fabric section located in the MR field-of-view. For the float-stitch fabric, however, bright spots indicative of a high fluid content appeared and grew in the MR images long after the front passed through (cf. the upper section of the images recorded at 60 and 320 s displayed in Figure 25.5b). These spots are attributed to the filling of larger pores with the wicking fluid. The fluid-concentration profiles calculated for both fabrics are shown in Figure 25.5c. For both fabrics, sharp fluid fronts (which are commonly assumed) were not seen to exist. Transient concentration gradients exist at the front of wicking fluids in textile fabrics. When compared to the plain-stitch fabric, the gradient was larger for the float-stitch fabric and likely related to its broader pore size distribution. It should be noted that the seemingly large (but regular) scatter of signal intensities in the profiles was not due to poor signal-to-noise ratios (SNRss) but to the fluid distributions as directed by the underlying fabric structures.

The ultimate goal of these wicking studies was the extraction of $k$-$S$-$p_c$ relationships from transient wicking data. In Ref. [40] a method is described for determining permeabilities as a function of saturation from transient upward wicking data collected gravimetrically from cut sections of separate fabric strips through which fluid was allowed to wick for various time intervals. The approach requires the solution of a time-dependent form of Darcy's equation:

$$\frac{\partial S}{\partial t} + \frac{\partial}{\partial z}\left[\frac{K}{\Phi}\left(\frac{\partial h}{\partial S}\frac{\partial S}{\partial z} - 1\right)\right] = \frac{\partial S}{\partial t} + \frac{\partial}{\partial z}\left[\frac{k}{\Phi\mu}\left(\frac{\partial p_c}{\partial S}\frac{\partial S}{\partial z} - \rho g\right)\right] = 0 \qquad (25.6)$$

where $\Phi$ is the porosity of the material formed by the fibers, $K$ is the hydraulic conductivity, $\mu$ is the fluid viscosity, $\rho$ is the fluid density, $z$ is the wicking height, $h$ is the equilibrium wicked height and $\partial p_c/\partial S$ is the change in capillary pressure with saturation determined at equilibrium. The saturation is needed as a function of wicking height, $z$, and time, $t$. The experimental approach of Ref. [40] is time-consuming and subject to inaccuracies due to evaporative weight losses during data collection. MRM does not suffer these disadvantages, as the entire $S(z,t)$ data set may be measured quickly on a single fabric sample. Challenges remain, nevertheless, in using standard MRM for rapidly determining these characteristic $k$-$S$-$p_c$ relationships in textile fabrics. Foremost among these is probably the need for equilibrium $\partial p_c/\partial S$, or equivalently, $\partial h/\partial S$ data.[1] For

1) At equilibrium, the cumulative fluid-filled pore volume at a given height, $h$, is determined by the balance between the capillary pressure, $p_c$, and the hydrostatic pressure, $p = \rho g\,h$, where $g$ is the gravitational constant and $\rho$ is the fluid density. Thus, $\partial pc/\partial S = \rho g\partial h/\partial S$.

most fabrics, the length over which fluid saturation data is needed is much longer than the homogeneous field-of-view of standard RF coils used in traditional vertical magnets. One solution for obtaining the full $\partial h/\partial S$ data set[2] is to record multiple images at different heights above the fluid reservoir; these images can then be used to construct a composite image representing the saturation distribution over the required fabric length. An alternative solution is to reconstruct the $\partial p_c/\partial S$ data from the pore structure of the fabric, which can be determined separately from porometric techniques [23, 25], diffusion NMR/$q$-space imaging [24], or even the analysis of MR images. As these solutions involve a substantial experimental effort and require some model assumption, they are not particularly attractive, and therefore alternative approaches are currently being sought.

## 25.5
## Quantitative Studies

### 25.5.1
### Using Calibration Curves

As discussed above, MR images of fluids in fibrous substrates may not provide even semiquantitative information if the fluid strongly binds the fibers. For example, as very strong binding interactions exist between water and cellulosic fibers, an especially challenging problem is to determine the distribution of moisture in the densely packed cellulosic fibers of paper. As wetting and subsequent moisture movement into cellulosic fibrous substrates affects their mechanical properties, this is especially relevant for paper-based packaging materials used in high-humidity climates.

The adsorption of moisture from air and its ingress into thick paper samples (~1.2 mm) was investigated using MRM [11]. For this, paper specimens were exposed to humid air from either one side or both sides, and the through-plane moisture distribution (i.e. presumptive) was measured with 2-D spin-echo imaging as a function of exposure time (Figure 25.6). During the early stages of the study, the low moisture contents in the paper samples meant that images of sufficient quality required 15–30 min to be recorded. For the best resolution along the path of moisture transport, gradient strengths were higher along this axis than in the perpendicular in-plane direction. No slice selection was used. Although it may have been sufficient simply to record 1-D profiles along the

**2)** An equilibrium $S(h)$ data set is also a convenient source from which pore size distributions may be estimated. Using $pc = 2\gamma\cos\theta/R$ ($R$ = average effective pore radius, $\gamma$ = fluid surface tension, $\theta$ = contact angle), and since $p_c = p = \rho g\, h$, then $h$ may be directly converted into an average effective pore size. As $S$ represents the normalized volume, $V$, of the pore space with that effective pore size, then $S(h) \rightarrow V(R)$, the cumulative normalized pore size distribution.

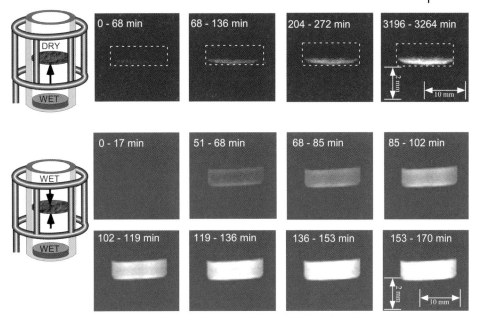

**Figure 25.6** Spin-echo images (TE = 1.4 ms, TR = 2 s) representing moisture ingress into a thick paper sample from a high-humidity (relative humidity ~100%) environment at either one surface (top) or both surfaces (bottom). For the sample exposed on one surface (top), the physical dimensions of the paper are outlined with a dashed rectangle. Images were recorded using an anisotropic field-of-view and no slice selection. Reprinted with permission from Ref. [11]; copyright American Chemical Society, 2002.

direction of moisture transport, mounting and maintaining the paper samples exactly perpendicular to the direction of moisture transport was very difficult as the samples tended to swell and curl as they adsorbed moisture. To ensure that pure through-plane profiles could be extracted, complete 2-D images were recorded; these images were then used to calculate the moisture profiles shown in Figure 25.7c. In an attempt to convert the signal intensities to actual moisture contents, a calibration curve was established between spin-echo signal intensity and moisture content measured gravimetrically (Figure 25.8). Unfortunately, below moisture contents of 4–6%, water-cellulose binding leads to such short $T_2$ values that there is virtually no detectable signal ($F^{T2} \rightarrow 0$). Since it is impossible to calculate moisture contents below the detection threshold of the MR experiment, reconstruction of the 'true' moisture-content profiles from MR data is an ill-posed problem. Alternatively, the data were analyzed by comparison with simulations of moisture content as a function of time and position. This is shown in Figure 25.7 for a very simple model; here, moisture transport through the pore space of the paper occurs only in the vapor phase, while moisture sorption by the paper occurs as a secondary, slower process. Applying the experimental calibration curve (Figure 25.8) to the simulated moisture

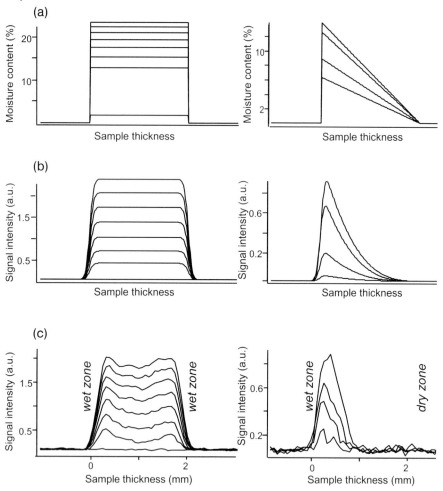

**Figure 25.7** Moisture ingress into a paper sample from a high-humidity environment at either one surface (right) or both surfaces (left). (a) Theoretical moisture profiles as predicted by a simple model; (b) Theoretical MR signal intensity profiles calculated from the theoretical moisture profiles using an experimental calibration function relating gravimetric moisture content to MR signal; (c) Experimental moisture profiles calculated from the images shown in Figure 25.6. Reprinted with permission from Ref. [11]; copyright American Chemical Society, 2002.

profiles (Figure 25.7a) leads to a set of theoretical MR signal intensity profiles (Figure 25.7b), which can be readily compared to the experimental MR profiles (Figure 25.7c).

In determining calibration curves such as that shown in Figure 25.8, evaporative weight losses may occur while transferring samples between the spectrome-

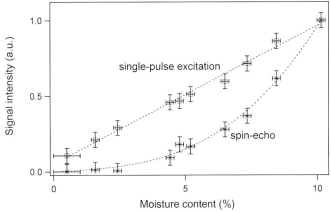

**Figure 25.8** Normalized NMR signal intensities versus gravimetric moisture content for a paper sample measured with Bloch decays and spin echoes (TE = 1.4 s, TR = 3 s). Reprinted with permission from Ref. [11]; copyright American Chemical Society, 2002.

ter, the balance, and the conditioning chamber.[3] It is therefore advisable to record simple Bloch decays along with the signal using the selected MRM sequence. The total spin density as determined by Bloch decays should be linearly proportional to the gravimetric moisture content, and is therefore a convenient internal check for such evaporative weight losses. This may not be the case, however, if the RF coil does not fully excite all nuclei in a sample composed of mobile ($\Delta v \leq 1\,\text{kHz}$) and rigid ($\Delta v \geq 35\,\text{kHz}$) components across the entire relative concentration of each, or if the added moisture plasticizes some fraction of the rigid components and makes them detectable as mobile components at higher moisture contents [41].

## 25.5.2
### Using NMR Techniques that Minimize Signal Contrast Factors

Establishing calibration curves can be very time-consuming and wrought with uncertainty, as a series of samples must be prepared with different fluid contents distributed homogeneously. The ideal situation would be to have an NMR imaging experiment that allows elimination of the contrast factors of Equation 25.1 (i.e. $F^{T1}$, $F^{T2}$, $F^{D} \to 1$) so that the signal intensities, $I$, are directly proportional to the spin densities, $S$. For routine spin-echo and gradient-echo techniques, $F^{T1}$ can generally be removed by setting a sufficiently long repetition delay (TR), but $F^{T2}$

3) Note that the spectrometer probe head could simultaneously serve as the conditioning chamber [41].

and $F^D$ cannot be removed as it is often not possible to set TE sufficiently short. This is not the case for the class of single point imaging (SPI) techniques [42, 43], in which a single complex data point is sampled at a time $t_p$ after an excitation pulse. As $t_p$ may be in the range of microseconds, signal attenuation due to $T_2$ relaxation and diffusion can be minimized or neglected. For the SPI techniques, spatial encoding is achieved through phase-encoding gradients applied during acquisition of the single data point; each dimension must be sampled point by point. One complete dimension with $N$ datum points, which can be sampled with a single scan using either gradient-echo or spin-echo sequences, requires a set of $N$ scans using SPI. In principle, therefore, images take longer to record with SPI than with spin-echo or gradient-echo sequences. These long time requirements are offset by the use of short RF pulses, making rapid scanning possible (i.e. short TR). Further developed SPI schemes such as Spiral-SPRITE [44] switch read gradients in an effective pattern to allow the sampling of 2-D or 3-D space in a relatively short time frame. Due to the use of excitation pulses shorter than $\pi/2$, the SNRs for SPI images are generally inferior to comparable gradient-echo/spin-echo images measured during the same time period. In fact, it was not possible to use any of the SPI methods in the study of moisture ingress into paper (cf. Section 25.5.1) due to the low fluid contents of the samples and the low SNR sensitivities of the SPI sequences. Our experience has taught us that SPI techniques are most appropriate for the quantitative characterization of fluid distributions, as long as the fluid concentrations are not too low and the time-resolution requirements are moderate.

These criteria are met – and SPI is indeed appropriate – for studies of moisture removal from wet-processed fibrous substrates. For example, the industrial production of carpet includes an aqueous-based coloration step that is followed by moisture removal. Even after mechanically squeezing the carpet and passing it over vacuum slots, moisture contents of 70–120%[4] remain and must be removed in an energy-consuming drying step. Figure 25.9 shows a series of MR profiles of the moisture distribution in a carpet sample recorded while drying [17, 18]. The data shown in Figure 25.9 were collected to facilitate a modeling effort. The influence of a variety of process parameters (e.g. temperature, air flow velocity and direction) on local moisture concentrations and overall drying time may be examined with MRM. The optimization of such industrial drying processes is important for product quality, production cost-savings and energy conservation.

Drying studies using MRM can be problematic if significant temperature variations exist across the sample, or if localized temperatures change during the drying process. Strictly speaking, Equation 25.1 only holds for a sample at a given temperature, since the $T_1$, $T_2$ and $D$ are all temperature-dependent. Signal intensities also depend on the temperature through the Boltzmann distribution of spin states; detectable increases in signal intensity can occur when the sample temperatures are lowered. If significant temperature variations exist,

---

**4)** Moisture content = (moisture weight/dry sample weight) × 100.

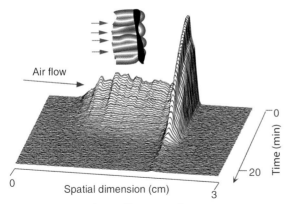

**Figure 25.9** Spin-echo profile images of moisture content in a carpet sample measured as a function of drying time. The moisture distribution in this fibrous substrate directly reflects the underlying anisotropic structure of this porous material. This study was part of an effort to optimize an industrial drying process for time and energy conservation [17, 18].

then Equation 25.1 should be modified to account for these with an additional contrast factor, $F^{Boltzmann}(T,\mathbf{r})$. During drying studies, evaporative cooling can certainly cause decreases in sample temperature. In our carpet-drying study, forcing air through the sample at room temperature caused its temperature to drop by about 15 °C. Such a drop would be detectable in the signal intensity and may be used for the temperature mapping of samples, depending on the SNR and relative changes in other contrast parameters. In our carpet-drying study, time-resolved spin-echo profiles were measured and initially exhibited increases in signal intensity, which were found to be more related to shortened $T_1$-values at the lower temperatures as opposed to changes in the Boltzmann distribution.

Drying curves are shown in Figure 25.10 for carpet samples with known initial moisture contents. Bloch decays were collected in tandem with either spin-echo (Figure 25.10a) or SPI (Figure 25.10b) profiles as a function of drying time. For both experiments, Fourier transformation of the Bloch decays gave peaks with linewidths <1 kHz; based on such calibration curves as those shown in Figures 25.3 and 25.8, the integrated intensity of this peak was assumed to be proportional to the actual moisture content in the sample. By using the known initial moisture contents it was then possible to calculate total actual moisture content versus drying time. The same approach was followed when determining the drying curves from the spin-echo and SPI profiles. Whilst the spin-echo sequence underestimates the true moisture content over most of the drying period, the SPI sequence provides the same moisture contents as the Bloch decays, thereby confirming its use for quantitative studies of moisture distributions in fibrous substrates [17, 18].

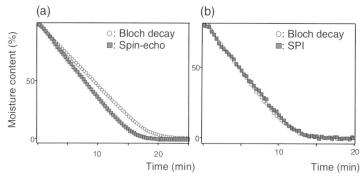

**Figure 25.10** Drying curves for wet carpet measured with a tandem sequence consisting of Bloch decays (open symbols) and a 1-D imaging sequence. (a) Spin echo (TE = 6.5 ms, TR = 20 sec); (b) Single point imaging (SPI) ($t_p$ = 37 µs, TR = 50 ms, flip angle = 3.5°). Data measured with Bloch decays and SPI quantitatively reflect the actual moisture content. Reprinted with permission from Ref. [16]; copyright SAGE Publications.

## 25.6
## Summary

Conventional studies of fluids in fibrous substrates rely on light microscopy, bulk gravimetry or cutting and weighing wet samples as a function of fluid content. These methods are difficult to automate, not easily employed for temporal studies, and provide information with limited spatial resolution. MRM is a convenient, nondestructive, time- and labor-efficient tool for examining fluid concentrations and movement in porous fibrous substrates. As fibrous substrates are *soft* porous media in which a wide variety of fluid–solid interactions are possible, MRM applications may take advantage of the more highly developed applications to biological and geological materials. Basic studies using MRM provide information on the underlying pore structure and fundamental fluid behavior – information that can greatly aid material design and development. Applied studies using MRM have provided information on manufacturing and end-use properties, thereby facilitating optimization as well as time and energy savings. Specific examples were presented in this chapter to show how MRM can provide qualitative, semiquantitative and quantitative information on fluid-containing fibrous substrates. Extracting semiquantitative or quantitative fluid concentrations from MRM data requires a knowledge of how the interplay between sample characteristics (e.g. $T_1$, $T_2$, $D$) and pulse sequence durations (e.g. TR, TE) affect the signal intensity through a set of signal contrast factors. By using standard spin-echo and gradient-echo imaging sequences, the signal loss due to $T_2$ relaxation and diffusion is generally more difficult to overcome than that due to $T_1$ relaxation. By utilizing SPI sequences, signal loss due to $T_2$ and $D$ can also be managed so that quantitative fluid concentrations can be measured. The emphasis in this chapter was on measuring fluid concentrations; fluid movement was simply inferred from time-resolved series of fluid concentrations. In other MRM applications to fluid-containing fibrous materials, fluid flow is characterized directly and the detection of concentrations is of minor relevance [32, 33, 45].

## Acknowledgments

In applying magnetic resonance techniques to fibrous substrates, we are grateful to the many collaborators with whom we have worked and published. Funding for the studies described in this chapter has been provided by the National Science Foundation, the National Textile Center, the Georgia Consortium for Competitiveness of the Apparel, Carpet, and Textile Industries, and the Institute of Paper Science and Technology at Georgia Tech.

## References

1 Fantazzini, P. (Ed.) (2003) Proceedings of the Sixth International Conference on MR Applications to Porous Media. *Magnetic Resonance Imaging*, **21**, 159–450.

2 Fantazzini, P., Gore, J. and Korb, J.-P. (Eds) (2005) Proceedings of the Seventh International Conference on Recent Advances in MR Applications to Porous Media. *Magnetic Resonance Imaging*, **23**, 121–444.

3 Fantazzini, P. and Gore, J. (Eds) (2007) Proceedings of the Eighth International Bologna Conference on Magnetic Resonance in Porous Media. *Magnetic Resonance Imaging*, **25**, 439–592.

4 Leisen, J. and Beckham, H.W. (2008) Void structure in textiles by nuclear magnetic resonance I. Imaging of imbibed fluids and image analysis by calculation of fluid density autocorrelation functions. *Journal of the Textile Institute*, **99** (3), 243–51.

5 Tutunjian, P.N., Borchardt, J.K., Prieto, N.E., Raney, K.H. and Ferris, J.A. (1994) Laundering and deinking applications of $^1$H NMR imaging. *Journal of Magnetic Resonance, Series A*, **106**, 209–13.

6 Hoferer, J., Lehmann, M.J., Hardy, E.H., Meyer, J. and Kasper, G. (2006) Highly resolved determination of structure and particle deposition in fibrous filters by MRI. *Chemical Engineering & Technology*, **29**, 816–19.

7 Lehmann, M.J., Hardy, E.H., Meyer, J. and Kasper, G. (2005) MRI as a key tool for understanding and modelling the filtration kinetics of fibrous media. *Magnetic Resonance Imaging*, **23**, 341–2.

8 Dirckx, C.J., Clark, S.A., Hall, L.D., Antalek, B., Tooma, J., Hewitt, J.M. and Kawaoka, K. (2000) Magnetic resonance imaging of the filtration process. *AIChE Journal*, **46**, 6–14.

9 Heikkinen, S., Alvila, L., Pakkanen, T.T., Saari, T. and Pakarinen, P. (2006) NMR imaging and differential scanning calorimetry study on drying of pine, birch, and reed pulps and their mixtures. *Journal of Applied Polymer Science*, **100**, 937–45.

10 Leisen, J., Hojjatie, B., Coffin, D.W. and Beckham, H.W. (2001) In-plane moisture transport in paper detected by magnetic resonance imaging. *Drying Technology*, **19**, 199–206.

11 Leisen, J., Hojjatie, B., Coffin, D.W., Lavrykov, S.A., Ramarao, B.V. and Beckham, H.W. (2002) Through-plane diffusion of moisture in paper detected by magnetic resonance imaging. *Industrial & Engineering Chemistry Research*, **41**, 6555–65.

12 Harding, S.G., Wessman, D., Stenström, S. and Kenne, L. (2001) Water transport during the drying of cardboard studied by NMR imaging and diffusion techniques. *Chemical Engineering Science*, **56**, 5269–81.

13 Bernada, P., Stenström, S. and Månsson, S. (1998) Experimental study of the moisture distribution inside a pulp sheet using MRI. Part II: drying experiments. *Journal of Pulp and Paper Science*, **24**, 380–7.

14 Bernada, P., Stenström, S. and Månsson, S. (1998) Experimental study of the moisture distribution inside a pulp sheet using MRI. Part I: principles of the MRI technique. *Journal of Pulp and Paper Science*, **24**, 373–9.

15 Leisen, J., Beckham, H.W., Good, J., Warner, S.B. and Carr, W.W. (1999) Magnetic resonance imaging of water ingress and distribution in fluorochemical-finished polyester cut-pile carpet. *Textile Chemist and Colorist*, **31**, 21–6.

16 Leisen, J. and Beckham, H.W. (2001) Quantitative magnetic resonance imaging of fluid distribution and movement in textiles. *Textile Research Journal*, **71**, 1033–45.

17 Lee, H.S., Carr, W.W., Leisen, J. and Beckham, H.W. (2001) Through-air drying of unbacked tufted carpets. *Textile Research Journal*, **71**, 613–20.

18 Lee, H.S., Leisen, J., Carr, W.W. and Beckham, H.W. (2002) A model of through-air drying of tufted textile materials. *International Journal of Heat and Mass Transfer*, **45**, 357–66.

19 Beckham, H.W., Schauss, G., Stanley, C. and Leisen, J. (2007) Magnetic resonance imaging applications in textile and fiber engineering: fabrics and diapers, *AATCC Review*, **8**(5), 32–6.

20 N. Pan and P. Gibson (2006) *Thermal and Moisture Transport in Fibrous Materials*, Woodhead Publishing, Cambridge.

21 Timonen, J., Kippo, K., Glantz, R. and Pakkanen, T. (2005) Combination of 3D MRI and connectivity analysis in structural evaluation of cancellous bone in rat proximal femur. *Journal of Materials Science: Materials in Medicine*, **12**, 319–25.

22 Leisen, J., Leonas, K. and Beckham, H.W. (1999) Fluid distributions and pore sizes in nonwoven fabrics detected by magnetic resonance. Proceedings, 9th

Annual TANDEC Nonwovens Conference, November 10–12, Knoxville, TN, vol. 9, 5.1-1.

23 Jena, A. and Gupta, K. (2002) Characterization of pore structure of filtration media. *Fluid Particle Separation Journal*, **4**, 227–41.

24 Callaghan, P. (1994) *Principles of Nuclear Magnetic Resonance Microscopy*, Oxford University Press, Oxford, UK.

25 Venkataraman, C. and Gupta, K. (2002) Characterization of pore structures in woven fabrics. *Advances in Filtration and Separation Technology*, **15**, 54–61.

26 Blümich, B. (2000) *NMR Imaging of Materials*, Oxford University Press, Oxford, UK.

27 Vahedi Tafreshi, H. (Ed.) (2007) A series of papers on modeling of nonwovens. International Nonwovens Technical Conference, September 24, Atlanta, GA. Conference proceedings, distributed on CD.

28 Maze, B., Vahedi Tafreshi, H. and Pourdeyhimi, B. (2007) Geometrical modeling of fibrous materials under compression. *Journal of Applied Physics*, **102**, 073533/073531–073533/073539.

29 Wang, Q., Maze, B., Vahedi Tafreshi, H. and Pourdeyhimi, B. (2007) On the pressure drop modeling of monofilament-woven fabrics. *Chemical Engineering Science*, **62**, 4817–21.

30 Lu, W.-M., Tung, K.-L. and Hwang, K.-J. (1996) Fluid flow through basic weaves of monofilament filter cloth. *Textile Research Journal*, **66**, 311–23.

31 Cuy, J.L., Irvin, C.A., Sundh, M.T. and Hauch, K.D. (2005) Digital volumetric imaging: automated serial sectioning and block face imaging for three dimensional reconstruction and visualization of the biomaterials interface. *Microscopy and Microanalysis*, **11**, 1262–3.

32 Leisen, J. and Farber, P. (2005) A synergistic approach by MRI and CFD for the study of micro-flow in textiles. Proceedings, Industrial Simulation Conference, June 9–11, Berlin.

33 Leisen, J. and Farber, P. (2006) Micro-flow in textiles. Proceedings, Techtextil North America, March 28–30, Atlanta, GA; conference proceedings distributed on CD.

**34** Topgaard, D. and Söderman, O. (2001) Diffusion of water absorbed in cellulose fibers studied with proton NMR. *Langmuir*, **2001**, 2694–702.

**35** Neascu, V., Leisen, J., Beckham, H.W. and Advani, S. (2007) Use of magnetic resonance imaging to visualize impregnation across aligned cylinders due to capillary forces. *Experiments in Fluids*, **42**, 425–40.

**36** Beckham, H.W. (2006) Should high-tech fabrics require high-tech labeling? Proceedings, AATCC International Conference & Exhibition, October 31–November 2, Atlanta, GA, pp. 252–4.

**37** Miller, B. (2000) Critical evaluation of upward wicking tests. *International Nonwovens Journal*, **2000**, 35–40.

**38** Bear, J. (1972) *Dynamics of Fluids in Porous Media*, Dover, New York.

**39** Simile, C. (2004) Critical evaluation of wicking in performance fabrics, M.S. Thesis, Georgia Institute of Technology, Atlanta.

**40** Ghali, K., Jones, B. and Tracy, J. (1994) Experimental techniques for measuring parameters describing wetting and wicking in fabrics. *Textile Research Journal*, **64**, 106–11.

**41** Leisen, J., Beckham, H.W. and Benham, M. (2002) Sorption isotherm measurements by NMR. *Solid State Nuclear Magnetic Resonance*, **22**, 409–22.

**42** Emid, S. and Creyghton, J.H.N. (1985) High resolution NMR Imaging in solids. *Physica B*, **128**, 81–3.

**43** Gravina, S. and Cory, D.G. (1994) Sensitivity and resolution of constant-time imaging. *Journal of Magnetic Resonance, Series B*, **104**, 53–61.

**44** Chen, Q., Halse, M. and Balcom, B.J. (2005) Centric scan SPRITE for spin density imaging of short relaxation time porous materials. *Magnetic Resonance Imaging*, **23**, 263–6.

**45** Bijeljic, B., Mantle, M.D., Sederman, A.J., Gladden, L.F. and Papathanasiou, T.D. (2004) Slow flow across macroscopically semi-circular fibre lattices and a free-flow region of variable width-visualisation by magnetic resonance imaging. *Chemical Engineering Science*, **59**, 2089–103.

# 26
# Imaging of Water in Polymer Electrolyte Membrane in Fuel Cells

*Shohji Tsushima and Shuichiro Hirai*

## 26.1
## Introduction

Polymer electrolyte membrane fuel cells (PEMFCs), often called proton-exchange membrane fuel cells, have attracted much attention due to their strong potential for automobile applications and on-site power generation. Of particular note is their high energy efficiency and rapid start-up/shut-down characteristics during low-temperature operation (generally <100 °C). In PEMFCs, protons (H$^+$) generated in the anode catalyst layer are transported to the cathode across a polymer electrolyte membrane (PEM) and react with oxygen, resulting in water in the cathode catalyst layer (see Figure 26.1). In order to reduce electrical resistance loss originating from the polymer electrolyte membrane in PEMFCs, it is of great importance to keep the membrane hydrated as much as possible, because proton conductivity in the membrane decreases with membrane dehydration [1]. However, it is also necessary to exhaust generated water from the cell in order to avoid liquid water accumulation in the cell, which gives rise to plugged pore channels in the gas diffusion layers and catalyst layers. When the gas diffusion layers and/or catalyst layers have become flooded with liquid water, it is difficult to supply enough reactant gas to the reaction sites in the catalyst layers, and this results in a deterioration of fuel cell performance. Therefore, delicate water management for membrane hydration without flooding of the gas diffusion layers and the catalyst layers is needed for stable fuel cell operation.

During fuel cell operation, water distribution in the membrane is dominated by several water transport processes, including electro-osmosis, diffusion, vaporization and condensation. From this point of view, a fundamental understanding of water transport in the membrane under fuel cell operating conditions is essential for establishing a method for water management of the fuel cell over a wide range of operating conditions. However, due to difficulties of optically viewing the membrane in the cell, few measurements of water distribution in the membrane under fuel cell operating conditions have been reported previously [2–4].

*Magnetic Resonance Microscopy.* Edited by Sarah L. Codd and Joseph D. Seymour
Copyright © 2009 WILEY-VCH Verlag GmbH & Co. KGaA, Weinheim
ISBN: 978-3-527-32008-0

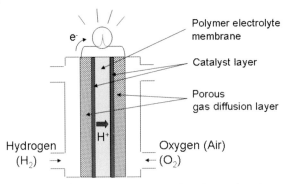

**Figure 26.1** Schematic representation of a polymer electrolyte membrane fuel cell (PEMFC).

Recently, we introduced magnetic resonance microscopy (MRM) as a useful diagnostic tool to probe the behavior of water in the operating fuel cell [5–19]. Previously, it had been reported that a PEM that was not within the fuel cell assembly could be visualized using a MRM diffusion-weighted imaging sequence [20]. More recently, other MRM studies of PEM visualization have been reported. For example, Seymour *et al.* demonstrated material heterogeneity in PEMs via $T_2$ and diffusion mapping by MRM [21], while Feindel *et al.* conducted MRM measurements in an operating fuel cell which they had constructed and reported the importance and advantages of MRM for studying PEMFCs [22, 23]. Balcom and coworkers also performed MRM experiments to measure the water content across a PEM in an operational PEMFC [24]. These authors introduced double half k-space (DHK) spin-echo (SE) single-point imaging (SPI) for high spatial resolution imaging across the membrane, and also developed a radiofrequency (RF) resonator assembled in the fuel cell. All MRM applications reported recently have demonstrated the strong potential of MRM as a diagnostic tool for fuel cell research and development as it covers not only material development but also the system design of fuel cells.

In this chapter we present details of MRM applications to visualize the transversal and lateral water content distribution in PEMs under fuel cell operating conditions. We also describe our experimental set-up to perform MRM experiments of PEMs in operational fuel cells that were specially designed. Finally, we describe an advanced MRM diagnostic method to visualize proton and water transport in PEMs by using a nuclear labeling method, and discuss upcoming demands on the further development of MRM techniques for PEMFC research and development.

## 26.2
### Experimental Apparatus

In order to perform MRM visualization of water content distribution in PEMs in operational PEMFCs, it was necessary to build fuel cells from nonmagnetic materials because the MRM system (INOVA 300 SWB, 7.05[T]; Varian Associates Inc.) generates strong magnetic fields that can be distorted by magnetic materials. The first of our custom-built fuel cells for the MRM experiments [6] is shown in Figure 26.2, together with a general view of the cell. A PEM was sandwiched between carbon gas diffusion layers (GDLs) on which platinum particles were dispersed as a catalyst. A 340 μm-thick membrane made from perfluorinated sulfonic acid (PFSA) was used in these experiments. The PFSA membrane has been widely used as an electrolyte in PEMFCs. It is also noteworthy that the water content distribution in a hydrocarbon membrane (which has attracted much attention recently owing to its low cost and advantageous handling and synthesis of a wide variety of chemical structures) can also be visualized using MRM [17]. The reactive area of the membrane was approximately 2.0 cm$^2$. The current collectors in the cell were made from copper that was coated with gold to prevent the copper from dissolving into the membrane.

In order to obtain magnetic resonance images of the water content in the membrane under fuel cell operation, the fuel cell was placed into the cylindrical test section of the MRM system, which was approximately 56 mm in diameter and 80 mm in axial length. Pure hydrogen and oxygen were then supplied to the fuel cell. During the experiments, the cell current was controlled by using an electric load connected to the cell. The cell was operated at room temperature and no humidification was applied, although the typical operational temperature of PEMFCs was approximately 80 °C in industrial applications. It should be noted that recently we had developed a temperature-controlled PEMFC for MRM experiments in which the supply gases were able to be fully humidified up to 80 °C; details of these cell and the gas supply unit can be found elsewhere [18].

For MRM visualization, we used a conventional spin-echo sequence with a repetition time (TR) of 0.5 s and an echo time (TE) of about 10 ms. The nucleus observed in these MRI experiment was the hydrogen atom, $^1$H, and an excellent correlation between measured MRI signal intensity and water content in the membrane was confirmed in preliminary experiments. Because the $T_1$ relaxation time of water in the membrane was about 0.1s in the experimental set-up, measured spin-echo images were used with a short echo time and a long repetition time to minimize magnetic relaxation and diffusion weighting impacts on the water content distributions ($^1$H density images) of the membranes in the fuel cell. A spatial plane was imaged perpendicular to the static magnetic ($B_0$) field, as shown in Figure 26.2a. The slice, phase and read encode gradient orientations with respect to the fuel cell correspond to the $Z$, $Y$ and $X$-axes, as indicated in the figure.

The spin-echo signal was read out in 256 real and imaginary points, while the phase-encoding gradients were incremented in 128 steps. Consequently, a 128 × 128 point matrix was formed to image the field of view (FOV) of 5.12 × 0.32 cm with a slice thickness of 2.5 mm. Thus, the pixel size was 400 × 25 micro;m. In order to increase the signal-to-noise ratio (SNR) under steady-state operation, a total of 16 MR images was averaged, with a mean period 18 min being required for image acquisition.

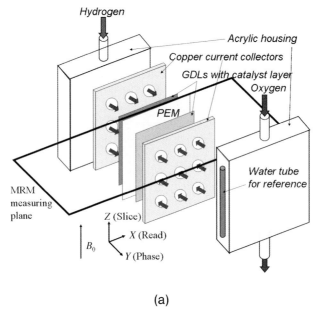

(a)

(b)

**Figure 26.2** A polymer electrolyte membrane (PEM) fuel cell for MRM measurement. A PEM was sandwiched with gas diffusion layers (GDLs), in which fine platinum particles are dispersed as a catalyst facing the membrane. MRM visualization was conducted in a plane perpendicular to the static magnetic field in the MRM system with a conventional spin-echo imaging sequence. All gradient orientations corresponding to the slice, read and phase encoding are also illustrated. (a) A schematic diagram of the fuel cell; (b) Direct photograph of the fuel cell.

## 26.3
## Results and Discussion

### 26.3.1
### Visualization of Transversal Water Content Distributions in PEMs

Figure 26.3 shows the measured water content distributions in the membrane under the operational PEMFC. The output current density of cell was increased from left to right (see Figure 26.3), the water content in the membrane being defined as the ratio of the number of water molecules in the membrane to the number of charged ($SO_3^-$) sites; therefore, the unit of water content was expressed by $H_2O/SO_3^-$. The right-hand side in each image is the cathode, where water is generated by the electrochemical reaction:

$$2H^+ + \frac{1}{2}O_2 + 2e^- \rightarrow H_2O.$$

The anode side of the membrane was found to become more depleted of water as the output current density was increased. In the images, the anode side of the membrane is more dehydrated compared to the cathode side, especially at the higher output current density in Figure 26.3d. This can be explained using the electro-osmotic effect and the production of water at the cathode as follows. With an increase in output current density, the number of water molecules accompanied by a proton due to electro-osmotic drive also increases, and this causes dehydration of the membrane on the anode side. In addition, water production

(a)　　(b)　　(c)　　(d)

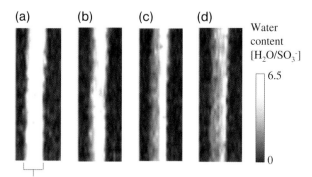

Water content [$H_2O/SO_3^-$]

6.5

0

Polymer electrolyte membrane (340 μm)

**Figure 26.3** Water content in the membrane under fuel cell operation with four values of output currents. The left side of each image corresponds to the anode. The spatial resolution is 25 micron along the direction from the anode to the cathode. The output current increases from left to right: (a) 0 mA cm$^{-2}$; (b) 89 mA cm$^{-2}$; (c) 178 mA cm$^{-2}$; (d) 267 mA cm$^{-2}$. (Reproduced by permission of The Electrochemical Society.)

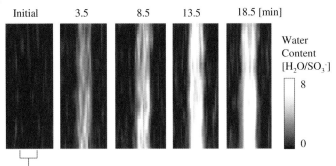

Polymer electrolyte membrane (340μm)

**Figure 26.4** Time-resolved MRM images of transversal water content distribution in the membrane after the flow of humid gas starts. The dew point of the supplied gas after humidification was kept at 60 °C.

due to electrochemical reactions occurs in the cathode, resulting in a water concentration gradient from the anode to the cathode, as observed in Figure 26.3.

MRM experiments were also performed to determine the water transfer coefficient through the membrane and GDL. The hydration process of the membrane in the fuel cell was examined using time-resolved MRM visualization. In this experiment, the membrane in the fuel cell was initially kept dry, after which humidified nitrogen gas was gradually supplied to the fuel cell. The gas flow rate was 200 ml min$^{-1}$ and the pressure 0.1 MPa, while the dew point (the temperature at which dew forms) of the gas was regulated to stay within 40–60 °C. In this experiment, the spatial resolution of the MRM was reduced and the water distribution of the PEM measured in the fuel cell at 5 min intervals. The time variation of transverse water content distributions in the membrane assembled in the fuel cell when the flow of humid gas had started is shown in Figure 26.4. Here, the membrane hydration process can be clearly observed, and the membrane reached equilibrium within about 15 min.

The water content in the membrane at equilibrium was also shown to vary with the dew point of the humid gas (Figure 26.5). A higher dew point corresponds to a higher relative humidity, which in turn results in an enhancement of membrane hydration. Based on these MRM observations, numerical simulations of the membrane hydration process were conducted to obtain the water transfer coefficient through the membrane and GDL. Details of the numerical analysis can be found elsewhere [10]. As shown in Figure 26.5, the best fits of the numerical model to the MRM data gave a water transfer coefficient, $h$, of $10^{-4}$ cm s$^{-1}$. The value of $h$ depends heavily on the porosity and thickness of the GDL and catalyst layer, and also affects the overall water transport process in PEMFCs. Notably, when using MRM visualization the water content increase in the PEM can be measured directly (unlike other methods), and thus $h$ can be monitored as described above.

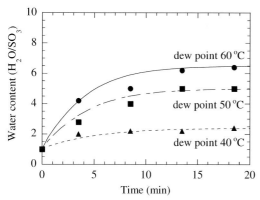

**Figure 26.5** Average water content in the membrane after the flow of humid gas starts. The points are MRM data for the three dew points and the curves are simulation results that use $h = 10^{-4}\,cm\,s^{-1}$ as the water transport constant. (Reproduced by permission of The Electrochemical Society.)

## 26.3.2
### Visualization of Lateral Water Content Distributions in PEMs

When establishing a reliable method for water management in PEMFC, it is especially important to have a basic understanding of both lateral and transverse water distribution in the membrane. Hence, a series of three-dimensional magnetic resonance imaging (3-D-MRI) studies was conducted to measure lateral water distribution in the membrane under fuel cell operation. Subsequently, investigations were made on the effect of gas flow channel configuration on lateral water distribution with both typical parallel flow and serpentine flow [15].

The reconstructed spatial distribution of water in the operating fuel cell measured by 3-D-MRI is shown in Figure 26.6. Both, the water in the membrane and condensed water in the gas flow channels in the cathode can be separately visualized. However, as the operating fuel cell contains a large quantity of water in both the gas flow channel and the membrane, it is difficult to measure lateral water distribution in the membrane by using a single-slice spin-echo sequence. This problem occurs because the slice thickness should be adjusted exactly equal to that of the membrane thickness. Therefore, 3-D-MRI was applied to monitor lateral water distribution in the membrane.

The lateral water content distribution of the membrane in an operational PEMFC with a parallel flow field is shown in Figure 26.7a. For comparison, a serpentine flow configuration is also depicted in Figure 26.7b. The cell current was $0.1\,A\,cm^{-2}$ in both cases and, in the parallel flow, the inlet side of the membrane was dehydrated under fuel cell operation. During operation, water droplets were observed to condense in the outlet region of the cell's gas flow channels.

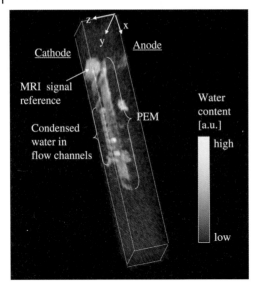

**Figure 26.6** Visualization of spatially resolved water distribution in the fuel cell by 3-D-MRI. (Reproduced by permission of The Electrochemical Society.).

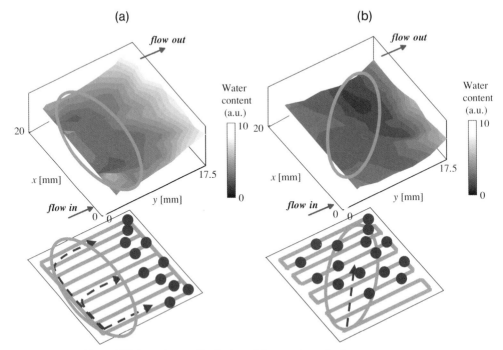

**Figure 26.7** Lateral water distribution of the membrane in the operating fuel cell with parallel flow with different flow configurations. (a) Parallel flow; (b) Serpentine flow. (Reproduced by permission of The Electrochemical Society.).

In the MRM experiment the fuel cell was vertically aligned in the system, while both reactive gases were supplied from the top of the cell and passed through nine straight channels out to the bottom of the cell. In this way the condensed water droplets were trapped at the bottom of the cell (the outlet side of the gas flow channels), and this resulted in membrane hydration on the outlet side of the cell. In contrast, fresh hydrogen and oxygen were supplied on the inlet side of the membrane, and this may have enhanced the local electrochemical reaction. As shown in the transverse water content distribution in Figure 26.3, an increase in cell current gives rise to a partial dehydration of the membrane at the anode due to an electro-osmotic drag effect. Therefore, in the parallel flow configuration the inlet side of the fuel cell showed less hydration due to an electro-osmotic drag effect caused by the electrochemical reaction, and this was clearly seen in the MRM experiment.

In the case of serpentine flow, a dehydration area across the membrane was seen, from the flow inlet to the outlet, as indicated by an oval in Figure 26.7b. This could not be expected if were assumed that the supply gas flows along the serpentine flow channels, in zigzag manner, from the inlet to the outlet. Under this type of assumed zigzag flow pattern, dehydration of the membrane from the inlet to the outlet along the serpentine flow channel should be observed, and not across the membrane. However, MRM visualization of the water content distribution in the membrane showed dehydration from the inlet to the outlet across the membrane, and suggested an intensive electrochemical reaction in this region of the membrane. This could be explained in relation to the behavior of the liquid water droplets in the gas flow channels. When water has condensed in the serpentine flow channel and the water droplets plug the channel, then the flow patterns in the fuel cell are greatly influenced in the case of a serpentine flow configuration. If the channel is plugged with liquid water, the supply gas may pass through the GDL and reach the adjacent channel. Thus, a flow from the inlet to the outlet over the membrane may have occurred, which might be consistently established by the pressure gradient induced in the cell.

These behaviors – whether the supplied gases pass through the GDL in the fuel cell, or not – are clearly closely related to local pressure gradients and the wettability of the surface of materials used in the fuel cell. MRM visualization clearly shows a variation of lateral water content distribution in the membrane under different flow channel configurations, thereby providing a fundamental understanding of the effects of flow pattern on membrane hydration under fuel cell operation.

## 26.3.3
### Nuclei-Labeling MRI for Visualization of Proton and Water Transport

Proton conductivity of the PEM not only has a direct influence on the performance of a PEMFC, but also provides motivation for understanding the mechanism of proton transport in water-immersed PFSA membranes. As PEM proton conductivity depends on membrane water content, this implies a central role for the aqueous phase in proton transport. Protons travel in aqueous solutions at an unexpectedly

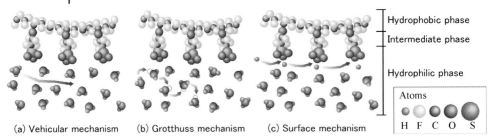

(a) Vehicular mechanism      (b) Grotthuss mechanism      (c) Surface mechanism

**Figure 26.8** Proton-conducting mechanisms in perfluorinated sulfonic acid membranes. (a) Vehicular mechanism; (b) Grotthuss mechanism, (c) Surface mechanism. (Reproduced by permission of The Electrochemical Society.).

high rate compared to ions similar in size to hydronium ions, which makes it unlikely that their transport relies solely on the normal vehicular mechanism (Figure 26.8a) of ion transport. A relay system, known as the Grotthuss mechanism (Figure 26.8b), has been proposed as a proton-specific alternative by which protons 'hop' from one water molecule to another by the formation and breakage of hydrogen bonds – a process which is much faster than the diffusion of hydronium ions in water. In PFSA membranes, an additional surface mechanism in which protons are conducted along the array of sulfonic acid groups, hopping from one anion moiety to another (Figure 26.8c), is also thought to contribute to proton transport.

In order to examine the proton-conduction process in PEMs under fuel cell operation, we applied the technique of nuclei-labeling magnetic resonance imaging (NL-MRI) [16]. For this, we supplied an isotopically labeled (deuterium; $^2$D) material to the PEMFC and monitored $^2$D permeation (which is directly related to the proton conduction mechanism in the membrane) by using MRI, which detected hydrogen ($^1$H). There was no variation in fuel cell performance when either hydrogen or deuterium was supplied, which indicated that water transport process was not greatly affected by a deuterium supply, despite the self-diffusion coefficient of deuterium being less than that of hydrogen.

Figure 26.9 shows the time variation of the transverse profiles of proton ($^1$H) content per sulfonic acid group ($^1$H/SO$_3^-$) in the membrane after changing the supply gas at the anode from hydrogen (H$_2$) to deuterium (D$_2$). Initially, approximately 10-fold more protons were associated with water molecules than there were ionized protons. The substitution of water protons was seen to occur at the anode in preference to the cathode. Depending on the suggested proton-conducting mechanisms (Figure 26.8), the $^2$D permeation behavior and resultant macroscopic $^1$H distribution in the membrane (visualized using NL-MRI) should differ. Assuming that either the surface mechanism or/and vehicular mechanism accounted for proton conduction in the membrane, then the MRI signal intensity should decrease at the cathode. Otherwise, a proportion of the hydrogen atoms, which play a role in proton conduction, would be replaced imme-

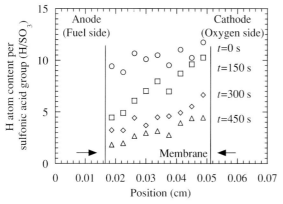

**Figure 26.9** Proton content distribution and time variation in the membrane. The time from the start of $^2$D supply is shown. The proton content is the sum of ionized and water protons. (Reproduced by permission of The Electrochemical Society.).

diately with $^2$D, due to proton movement from anode to cathode. However, experimental observations using NL-MRI (see Figure 26.9) have shown the MRI signal intensity gradually to decrease from the anode (Figure 26.9, left side) with $^2$D permeation, such that all of the $^1$H was replaced by $^2$D in approximately 450 s. Taken together, these results confirm that the Grotthuss mechanism, acting through hydrogen bonds, is the dominant proton-conducting mechanism in the PEM.

More recently, we have demonstrated the existence of membrane hydration paths by using NL-MRI [18]. These findings should prove especially valuable for our understanding of membrane hydration processes as they allow the examination of membrane hydration pathways, thereby distinguishing the effects of the three water sources: water vapor supplied with humidified anode gas; water vapor supplied with humidified cathode gas; and water generated as a result of the electrochemical reaction at the cathode. These three sources of water were labeled on the molecular level with $D_2$ and $D_2O$ and, by using various combinations of $D_2$ fuel and $D_2O$ for inlet humidification, NL-MRI was employed successfully to identify each water source and transport mechanism.

## 26.4
## Summary

During recent years, MRM has shown great promise and unique potential for the visualization of water content in PEMs in operational fuel cells, and so has attracted much attention from the fuel cell community [25]. MRM visualization can provide fundamental insights into both water and proton transport during fuel cell operation, and a wide diversity of MRM applications in this field, notably associated with further developments of hardware and software, can be expected in the near future.

Among recent developments of PEMFCs for use in automobiles, PEMs of <50 μm thickness have been developed not only to reduce ohmic voltage losses associated with proton transport across the membrane, but also to enhance membrane hydration by back-diffusion from cathode to anode. In order to observe the water content in thinner membranes when using MRM, further developments to increase not only MR signal intensity but also spatial resolution will be required. An additional point to be explored for water management in PEMFCs is the static and dynamic behaviors of liquid water in both porous catalyst layers and GDLs. The accumulation of liquid water in both such layers causes cell performance to deteriorate due to the blockage of gas transport in pore channels. The operation of PEMFCs at subzero (freezing) temperatures is also related to the liquid water inside both porous matrices; a major challenge here when investigating water behavior in catalyst layers and GDLs is the use of MRM with short acquisition times.

## References

1 Zawodzinski, T.A., Derouin, C., Radzinski, S., Sherman, R.J., Smith, V.T., Springer, T.E. and Gottesfeld, S. (1993) *Journal of The Electrochemical Society*, **140**, 1041–7.

2 Watanabe, M., Igarashi, H., Uchida, H. and Hirasawa, F. (1995) *Journal of Electroanalytical Chemistry*, **399**, 239–41.

3 Satija, R., Jacobson, D.L., Arif, M. and Werner, S.A. (2004) *Journal of Power Sources*, **129**, 238–45.

4 Mukundan, R., Davey, J.R., Rockward, T., Spendelow, J.S., Pivovar, B.S., Hussey, D.S., Jacobson, D.L., Arif, M. and Borup, R. (2007) *ECS Transactions*, **11**(1), 411–22.

5 Tsushima, S., Teranishi, K. and Hirai, S. (2005) *Energy*, **30**, 235–45.

6 Tsushima, S., Teranishi, K. and Hirai, S. (2004) *Electrochemical and Solid-State Letters*, **7** (9), A269–72.

7 Teranishi, K., Tsushima, S. and Hirai, S. (2002) *Thermal Science and Engineering*, **10** (4), 59–60.

8 Teranishi, K., Tsushima, S. and Hirai, S. (2005) *Electrochemical and Solid-State Letters*, **8** (6), A281–4.

9 Teranishi, K., Tsushima, S. and Hirai, S. (2003) *Thermal Science and Engineering*, **11** (5), 35–6.

10 Teranishi, K., Tsushima, S. and Hirai, S. (2006) *Journal of The Electrochemical Society*, **153** (4), A664–8.

11 Teranishi, K., Tsushima, S. and Hirai, S. (2004) *Thermal Science and Engineering*, **12** (4), 91–2.

12 Tsushima, S., Teranishi, K., Nishida, K. and Hirai, S. (2005) *Magnetic Resonance Imaging*, **23**, 255–8.

13 Tsushima, S., Teranishi, K. and Hirai, S. (2003) *Thermal Science and Engineering*, **11** (4), 31–2.

14 Mibae, T., Tsushima, S. and Hirai, S. (2004) *Thermal Science and Engineering*, **12** (4), 89–90.

15 Tsushima, S., Nanjo, T., Nishida, K. and Hirai, S. (2005) *Electrochemical Society Transactions*, **1** (6), 199–205.

16 Tsushima, S., Teranishi, K. and Hirai, S. (2006) *Electrochemical Society Transactions*, **3** (1), 91–6.

17 Tsushima, S., Hirai, S., Kitamura, K., Yamashita, M. and Takase, S. (2007) *Applied Magnetic Resonance*, **32** (1), 233–41.

18 Kotaka, T., Tsushima, S. and Hirai, S. (2007) *Electrochemical Society Transactions*, **11** (1), 445–50.

19 Tsushima, S., Nanjo, T. and Hirai, S. (2007) *Electrochemical Society Transactions*, **11** (1), 435–44.

**20** Zawodzinski, T.A., Springer, T.E., Uribe, F. and Gottesfeld, S. (1993) *Solid State Ionics*, **60**, 199–211.

**21** Howe, D.T., Seymour, J.D., Codd, S.L., Busse, S.C., Peterson, E.S., Were, E.H. and Taylor, B.F. (2005) Proceedings, 8th International Conference on Magnetic Resonance Microscopy, O-34.

**22** Feindel, K.W., Bergens, S.H. and Wasylishen, R.E. (2006) *ChemPhysChem*, **7**, 67–75.

**23** Feindel, K.W., LaRocque, L.P.A., Starke, D., Bergens, S.H. and Wasylishen, R.E. (2004) *Journal of the American Chemical Society*, **126**, 11436–7.

**24** Zhang, Z., Martin, J., Wang, H., Promislow, K. and Balcom, B.J. (2007) Proceedings, 9th International Conference on Magnetic Resonance Microscopy, 43.

**25** Wills, J. (2005) *The Fuel Cell Review*, **2** (5), 27–9.

# 27
# NMR of Liquid Crystals Confined in Nano-Scaled Pores

*Farida Grinberg*

## 27.1
## Introduction

During the past two decades, molecular self-assembly on mesoscopic and macro-scopic length scales has been successfully exploited in molecular engineering and in the production of new materials with tailored properties [1–4]. Examples include new methods of synthesis of novel mesoporous materials based on liquid crystal (LC) templates or by the manufacture of stimuli-responsive ('intelligent') nano-sized containers for drug delivery. Hierarchic molecular self-assembly is, in turn, accompanied by multiple molecular (individual or collective) dynamic processes that tend to range over many time decades. Although the structural properties of these materials are rapidly becoming increasingly accessible due to advances in optical and light-scattering techniques, one of the main remaining challenges in the physics of soft matter systems relates to understanding their complex dynamic features at the molecular level, as well as to addressing structure–property relation-ships with predictable power.

The specific functional properties of such new materials often exploit the order-ing molecules in contact with solid interfaces. In particular, this refers to LCs that are confined to nanostructured materials used in electro-optical devices and in sensors. Today, it is widely acknowledged that LCs constrained within microscale or nanoscale environments tend to exhibit partial orientational (paranematic) order [5–7] induced by surface interactions. This type of ordering can extend over mesoscopic length scales, and is observed for temperatures well above the bulk isotropization temperature, $T_{NI}$.

Nuclear magnetic resonance (NMR) has proven to be a powerful tool for study-ing confined LC [5], and especially informative in this context are those methods that permit one to access molecular dynamics within the low-frequency limit, down to a few kHz. In soft materials, the low-frequency range is characteristic of the unique mechanisms of molecular motions that are not observed in liquids of low viscosity. Such mechanisms are represented, for example, by the orientational director fluctuations (ODFs) in nematics, or by the long back-bone reptation-like

*Magnetic Resonance Microscopy.* Edited by Sarah L. Codd and Joseph D. Seymour
Copyright © 2009 WILEY-VCH Verlag GmbH & Co. KGaA, Weinheim
ISBN: 978-3-527-32008-0

motions of macromolecular chains in polymer melts. Another example is given by the so-called reorientations mediated by translational displacements (RMTD) mechanism, which applies to polar liquids [8, 9] and LCs [10] confined to mesoscopic pores. In the case of polar liquids, under strong adsorption conditions diffusive displacements along the surface are described with non-Gaussian propagators and obey an anomalous time dependence of the mean-square displacement values. In the case of LCs, molecular ordering occurring at the interfaces gives rise to the ultra-longlived correlations in the molecular reorientational dynamics and produces strong relaxation mechanisms in the frequency range below 1 MHz.

The longest time scale accessible to NMR is addressed by application of the three 90°-pulse sequence, and is limited by the rate of longitudinal relaxation. Therefore, in systems in which longitudinal relaxation is much slower than the transverse counterpart, the longest observation time allowed by this sequence exceeds that of the other Hahn spin-echo sequences limited by the rate of transverse relaxation. For this reason, the three 90°-pulse sequence (also known as the 'stimulated echo' sequence) is often used in combination with the strong pulsed magnetic field gradients [11, 12] in order to enable the measurement of long-timescale translational molecular dynamics. It is worth noting, however, that the surface-layer molecules in confined systems are not easily accessible in pulsed-field gradient NMR diffusion studies because their relative fraction in the whole sample often tends to be insufficient for a direct detection. The influence of surface ordering on the molecular diffusion of confined LCs is masked by dominating nonspecific effects of geometric confinements and porosity [10].

One powerful tool for monitoring ultra-slow molecular reorientational dynamics is provided by the technique known as the dipolar correlation effect on the stimulated echo [13, 14]. In respect to LCs, this technique permits experimental proof of the existence of collective molecular motions as predicted by theory [15], and also an estimation of the cross-over between the 'bulk-dominated' and 'surface-dominated' regimes of the ODFs under confinement. Another application of the stimulated echo pulse sequence was used to evaluate cross-relaxation rates and also to determine experimentally the nematic correlation length in the confined LC [16].

Among various low-frequency NMR techniques, a special emphasis should be placed on field cycling (FC) NMR relaxometry [8] which is suited to probing molecular dynamics in the frequency range covering several orders of magnitude. This technique is extremely sensitive, even to a relatively small fraction of molecules oriented by the surface, and permits the probing of mechanisms such as RMTD [9, 17] over a broad time scale. Several groups have reported thorough examinations of the relaxation mechanisms in confined LCs with the help of FC relaxometry [14, 18–20]. However, any systematic view which includes pore-size effects on these mechanisms remains rather poor. The aim of this study was to shed more light on the major properties governing low-frequency relaxation mechanisms in confined nematic LCs. We report here the results obtained with FC relaxometry and Monte-Carlo computer simulations for the mesoporous confinements over a broad range of pore sizes. The main features exhibited by the experimental relaxation dispersions are analyzed in the context of three pore-size ranges, referring to small, large and intermediate pores.

## 27.2
## Materials and Methods

The samples studied were controlled porous glass (CPG), Bioran and Vycor meso-porous glasses filled with nematic liquid crystal 4'-n-pentyl-4-cyanobiphenyl (5CB). The $T_{NI}$ of 5CB is 308.5 K. The mean pore radii of the CPG samples denoted as CPG-1.5 and CPG-4 were 1.5 nm and 4 nm, respectively. The mean pore radius of the Vycor sample denoted as Vycor-2 was 2 nm. The mean pore radii of the Bioran samples denoted as Bioran-5, Bioran-15, Bioran-35 and Bioran-100 were 5, 15, 35 and 100 nm, respectively. The frequency-dependences of the spin-lattice relaxation rates $T_1^{-1}$ were measured using the home-built FC relaxometer at the University of Ulm.

## 27.3
## Monte-Carlo Simulations

In our simulations, dipolar correlation functions are defined as

$$G_k^{red}(t) \equiv \frac{\langle Y_{2,k}(0) Y_{2,k}^*(t) \rangle}{\langle Y_{2,k}^2(0) \rangle},$$  (27.1)

where $Y_{2,k}$ ($k = 0,1,2$) are second-order spherical harmonics describing the instantaneous orientation of the internuclear vector relative to the external magnetic field. Random-walk Monte-Carlo simulations were performed for a spherical cavity of a given radius $R$. In each cycle time $\Delta t \ll t$ a random step of a fixed length $\Delta l \ll R$ was generated and $G_k^{red}(t)$ was evaluated as described elsewhere [16, 20]. All relevant time and length scales are expressed in terms of the predefined values of $\Delta l$ and the diffusion coefficient $D$. The (predefined) value of the diffusion coefficient was equal to $10^{-10}\,\mathrm{m^2\,s^{-1}}$.

The simulated correlation functions were used to evaluate the laboratory frame spin-lattice relaxation rates [21]:

$$\frac{1}{T_1(\omega_0)} = K[I_1(\omega_0) + 4I_2(2\omega_0)]$$  (27.2)

Here, $\omega_0$ is the Larmor frequency, $K = \frac{3}{20}\gamma^4\hbar^2 r^{-6}\left(\frac{\mu_0}{4\pi}\right)^2$, where $\gamma$ is the gyromagnetic ratio of the nuclei, $\hbar$ is the Planck constant divided by $2\pi$, $r$ is the internuclear distance, and

$$I_k(\omega) = \int G_k(t)\exp(i\omega t)\,dt. \tag{27.3}$$

## 27.4
## RMTD and the Exchange Model

The following approach was used in order to account for surface-ordering effects on the nuclear magnetic relaxation of liquid molecules adsorbed into pores. Inside the pore, the liquid was assumed isotropic (bulk-like), and any molecular reorientations were assumed to be fast and not contributing to the observed relaxation. In the simulation procedure this was approached by replacing the spherical harmonics by zero at any time when a spatial position of the molecule was in the bulk-like fraction. In contrast, for molecules directly at the surface or within the surface-ordered layer of thickness, $\delta r$, the value of $Y_{2k}$ was determined by the preferential orientation (in this case, perpendicular) relative to the surface at the instantaneous position of the random walker. Besides the 'orienting' property, the surface was also ascribed an 'adsorbing' property. The latter term means that the probability for the molecules at the direct surface position to leave the surface within the next random step was governed by the parameter $W_{s-b} \leq 1$. The situations with $W_{s-b} < 1$ correspond to 'adsorption' by the surface and may be relevant, for instance, for polar liquids adsorbed by active surface sites [9].

Generally, the total correlation function $G_k^{red}(t)$ can be decomposed into four partial correlation functions [20] related to the two fractions of molecules initially and finally in the same phase (surface layer or bulk) and the two fractions of molecules initially and finally in different phases. Under these conditions, $G_k^{red}(t)$ reduces to [20]:

$$G_k^{red}(t) = f_{s,s}\langle Y_{2,k}(0)Y_{2,-k}(t)\rangle_{RMTD} \tag{27.4}$$

where $f_{s,s}$ is the fraction of molecules that are initially and finally in the surface layer. Here, RMTD represents the mechanism [8] describing molecular reorientations due to diffusion between the surface sites with different preferred molecular orientations (different directors of the surface order). Between times 0 and $t$, the correlation to the initial molecular orientation in the surface layer may temporarily be lost, as molecules perform extended excursions to the bulk-like area, although it tends to restore at much longer times, as molecules repeatedly return to the ordered layer and adopt the preferential orientations again. The correlation function for the RMTD process thus depends on factors such as the surface topology, geometry of the pore space, molecular diffusivity and interactions with the surface.

The fraction $f_{s,s}$ is generally a function of time. At very short times, for which the probability of the exchange between the ordered and the bulk-like populations tends to zero, $f_{s,s} \approx f_s$, where $f_s$ is the (constant) population of the surface layer. At long times, for which the initial and final probabilities to be in the surface layer become independent of each other, $f_{s,s}(t) \approx f_s^2$. The function $f_{s,s}(t)$ in Equation 27.4 thus decays from the initial value $f_s$ to the value $f_s^2$ and resembles an exchange-loss process; that is, the loss of molecules populating the surface phase at time 0 but being rather in the bulk-like phase at time $t$. Thus, for the time scale exceeding characteristic exchange time, $G_k^{red}(t)$ as the function of $t$ is entirely determined by the RMTD process

$$G_k^{red}(t) = f_s^2 \langle Y_{2,k}(0) Y_{2,-k}(t) \rangle_{RMTD}. \tag{27.5}$$

A typical shape of the simulated correlation function $G_0^{red}(t)$ is demonstrated in Figure 27.1 for a sphere of $R = 50\,nm$ and $\delta r = 3\,nm$. The correlation function clearly exhibits two components with different characteristic decay rates. The initial rapid loss of correlations $(t < 10^{-6}\,s)$ is due to the exchange losses represented by the factor $f_{s,s}(t)$. The long-time tail of the correlation functions is unaffected by the exchange losses and is governed entirely by the RMTD process. The latter determines the correlation loss between the orientations adopted by the random walker in the surface-ordered layer in course of the random walk process. If the molecular orientation at any position of the surface layer was the same (in other words, the RMTD mechanism was inactivated), then Equation 27.4 would reduce merely to the function $f_{s,s}(t)$. This situation allows one to distinguish the factors $f_{s,s}(t)$ and $\langle Y_{2,k}(0) Y_{2,-k}(t) \rangle_{RMTD}$ from each other.

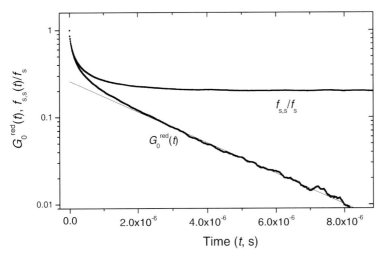

**Figure 27.1** The correlation function, $G_0^{red}$, and the (normalized) exchange-loss factor, $f_{s,s}/f_s$, as a function of time (Equation 27.4), simulated for the spherical surface with $R = 50\,nm$ and $\delta r = 3\,nm$. The straight line is the fit of the function $C \exp(-t/\tau_{RMTD})$ to the long tail of the correlation function.

The simulation result for $f_{s,s}(t)/f_s$ is demonstrated in Figure 27.1. The crossover between the regime of a rapid decrease of $f_{s,s}(t)/f_s$ in the short time limit and the regime of the time-independent behavior in the long time limit is clearly distinguishable. This provides an estimate of the mean exchange time between two phases of about $1 \mu s$ (for conditions applied in this simulation). The long time tails $\langle Y_{2,k}(0) Y_{2,-k}(t) \rangle_{RMTD}$ of the correlation functions were fitted by an exponential function $C\exp(-t/\tau_{RMTD})$, as shown by the straight line. The values of the prefactor $C$ depend on the thickness of the oriented layer (or on the population of the ordered molecules) and $W_{s-b}$. The values of $\tau_{RMTD}$ were evaluated for a broad range of $R$, $\delta r$ and $W_{s-b}$.

Figure 27.2 shows $\tau_{RMTD}$ as a function of $R$ for $W_{s-b} = 1$ ($\delta r$ was set to 0.1 $R$) and for $W_{s-b} = 0.5$ and 0.01 (with $\delta r$ set to 0, that is, only the molecules directly at the surface were oriented). In the case of $W_{s-b} = 1$ (no specific adsorption at the surface), the dependence $\tau_{RMTD} = f(R)$ is clearly quadratic, as shown by the straight line, $\tau_{RMTD} \propto R^2$. For $W_{s-b} < 1$, the behavior of the function $\tau_{RMTD} = f(R)$ deviates from the quadratic law. Remarkably, in the range of $R < R^*$ it follows the linear dependence (shown by the dashed lines in Figure 27.2) but tends to become quadratic again for larger pores, $R > R^*$. Here, $R^*$ denotes the crossover between the two regimes. For $R < R^*$, the values of $\tau_{RMTD}$ are larger for smaller $W_{s-b}$, as would be expected since adsorption tends to slow down the effective diffusion along the surface and thus the process of reorientation and the correlation loss.

The power law approximations for $\tau_{RMTD}$ as a function of $R$, that is, $\tau_{RMTD} \propto R$ and $\tau_{RMTD} \propto R^2$, should be compared to the analytical results (e.g. see Ref. [9]) approaching the RMTD process in terms of the surface structure factor and the mode correlation times for surface diffusion. Two cases are considered in which

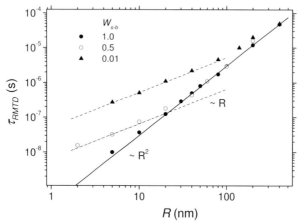

**Figure 27.2** Values of $\tau_{RMTD}$ as a function of the sphere radius evaluated from the Monte-Carlo simulations. The curve parameter is $W_{s-b}$. The values of $\delta r$ were set to 0.1 $R$ for $W_{s-b} = 1$ and to zero for $W_{s-b} < 1$.

the surface propagators are described either by a Gaussian displacement probability density

$$P(s,t) = \frac{1}{4\pi} \frac{\exp[-s^2/(4Dt)]}{Dt} \tag{27.6}$$

or by a Cauchy displacement probability density:

$$P(s,t) = \frac{1}{2\pi} \frac{ct}{\left[(ct)^2 + s^2\right]^{3/2}} \tag{27.7}$$

where $s$ has a dimensionality of a distance and constant $c$ has a dimensionality of a velocity. The Gaussian distribution corresponds to a (ordinary) random walk process, whereas the Cauchy distribution is an appropriate function for describing Lévi walk statistics [22] in the strong adsorption limit. Within the $k$-space presentations of the propagators, Equation 27.6 and Equation 27.7, the mode correlation times are given as

$$\tau_k = \frac{1}{Dk^2} \text{ for Gaussian propagator} \tag{27.8}$$

and

$$\tau_k = \frac{1}{ck} \text{ for Cauchy propagator,} \tag{27.9}$$

where $k$ is the wave number.

The values of $\tau_{RMTD}$ evaluated in our simulations for a simple spherical geometry can be interpreted as an analogue of the mode correlation time for the smallest value of $k_{min} \propto R^{-1}$. The quadratic dependence of $\tau_{RMTD}$ on $R$ is thus related to the Gaussian propagator characteristic of the conventional random walks, whereas the linear dependence of $\tau_{RMTD}$ on R indicates the regime associated with the Lévi-walk statistics. The Monte-Carlo simulations in Figure 27.2 allow one to distinguish the conditions at which one or another regime (Gauss or Cauchy) dominates.

The values of $\tau_{RMTD}$ determine the cut-off frequency of the dispersions (the crossover between the plateau and the power-law dependence) of the relaxation rates. Figure 27.3 shows the simulated frequency dependences of $T_1^{-1}$ normalized by the low-frequency values (curves 1–7). Simulation data are shown for various thicknesses $\delta r$ within a cavity of the same size ($R = 30$ nm) (curves 3–6), and for the cavities of different sizes ($R = 60$ nm, 50 nm, 20 nm) (curves 1, 2 and 7). The data in Figure 27.3 show that, given the same cavity size, the width of the surface-ordered layer strongly affects the dispersion slopes (cf. curves 1–4), but not the cut-off frequency of the dispersion. The slopes 'span' the power laws between $v^{-2}$ and $v^{-0.5}$ represented graphically by the straight lines. Thinner

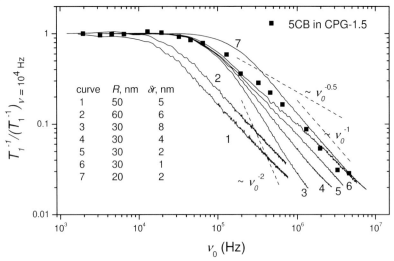

**Figure 27.3** Normalized frequency dependencies of the spin-lattice relaxation rates, curves 1–7, evaluated from the simulated correlation functions. The curve parameter refers to different values of $\delta r$ and $R$. The experimental data points refer to 5CB in CPG-1.5 at 323 K.

layers are characteristic of flatter slopes as a result of the increasing fraction of the bulk-like phase. Increasing the pore radius shifts the cut-off frequency towards lower values, but it does not significantly affect the slope, provided that the thickness of the ordered layer remains in the same proportion to the pore size.

## 27.5
## Low-Frequency Relaxation in Confined 5CB

The relaxation behavior of confined 5CB in the low-frequency range is rather complex, and is related predominantly to the following features: (i) an existence of the weak orientational order above $T_{NI}$; and (ii) a depression of $T_{NI}$ in small pores followed by a gradual evolution of the order parameter [18, 19]. Figure 27.4 shows a series of low-frequency relaxation dispersions observed in the range between 2 kHz and 7 MHz for bulk and confined 5CB at temperatures above $T_{NI}$. A strong impact of confinements on the low-frequency relaxation rates is clearly observed.

In contrast to bulk 5CB, where the relaxation rates are practically independent of the frequency (as typically expected for ordinary liquids below the mHz range), the confined samples exhibit a pronounced dispersion depending on pore size. At the lowest frequency limit, the values of $T_1^{-1}$ were highest for the sample with

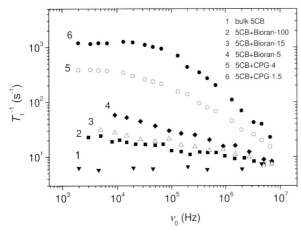

**Figure 27.4** Experimental frequency dependencies of the spin-lattice relaxation rates of bulk 5CB at 323 K and confined 5CB at temperatures above $T_{NI}$: 323 K for CPG-1.5 and CPG-4 and 316 K for Bioran-5, Bioran-15 and Bioran 100.

the smallest pore size (1.5 nm) and exceeded that of bulk 5CB by more than two orders of magnitude. A strong enhancement of the relaxation rates of confined samples in the kHz range is usually attributed to the orientational ordering at the surface and the RMTD mechanism [18, 23, 24]. The orientational anisotropy stabilized by interactions at solid interfaces prevents a complete averaging of dipolar interactions by molecular motions. Diffusion between the regions with different surface orientations and different levels of orientational anisotropy (local order parameter) then produces a strong relaxation mechanism in the kHz frequency range. In samples with larger pores – for example, in the Bioran glasses – the interior middle area is becoming less ordered so that the relative fraction of the more 'bulk-like' area increases for larger pores. Consequently, the decrease in relaxation rate with increasing the pore size is due to the averaging effect of translational molecular diffusion between the regions with different local order parameters.

The pore size influences strongly not only the value of the relaxation rates but also their temperature-dependence. Three qualitatively different scenarios of temperature behavior of the dispersions, $T_1^{-1} = f(v, T)$, were observed when comparing them at temperatures above and below $T_{NI}$. In the following, these scenarios will be analyzed for the cases of small, large and intermediate pores.

## 27.5.1
## Small Pores

As 'small', we denote pores with $R < \xi$, where $\xi$ is the nematic correlation length introduced in the frame of the Landau–de Gennes theory [15]. This theory predicts that, in the pore, the local order parameter $S$ decreases exponentially with increasing distance $r$ from the wall

$$S(r) \propto \exp\left(-\frac{r}{\xi}\right),\tag{27.10}$$

where $\xi$ is the so-called nematic correlation length. In the investigated materials this quantity was estimated as approximately 3 nm at 313 K [16], which was close to its theoretical value of about 5 nm.

The regime of small pores among these data is represented by samples CPG-1.5 and Vycor-2, with pore radii of 1.5 nm and 2 nm, respectively. Since $R < \xi$, the dominating fraction of confined molecules should exhibit a relatively strong ordering. The low-frequency dispersions of 5CB in these samples are shown in Figure 27.5a, where no essential changes were observed for the dispersion curves of the confined samples while lowering the temperature from above $T_{NI}$ (323 K) to below $T_{NI}$ (303 K). This is in contrast to bulk 5CB, where a low-frequency plateau (typical for isotropic liquids) transforms below $T_{NI}$ to a well-known inverse square root law owing to the ODFs of the nematic LCs. Thus, the relaxation mechanisms governing the low-frequency dispersion in the confined samples are the same below and above $T_{NI}$, indicating that the nematic–isotropic transition in such small pores is damped. This finding agrees with results obtained on the basis of the transverse relaxation and cross-relaxation studies [16, 23]. The fact that $T_1^{-1}$ in CPG-1.5 depends only slightly on temperature supports the idea [7] that the order parameter profile in the molecular layer nearest to the wall is constant and independent of temperature; the exponential decrease of the order parameter is thought to begin not directly at the wall but at distances exceeding roughly one molecular size.

The normalized dispersion curve of the CPG-1.5 sample at 323 K is shown again in Figure 27.3, together with the simulated data. Here, the low-frequency value of $T_1^{-1}$ of the bulk sample was subtracted from the measured values of relaxation rates of the confined samples in order to account for the contribution of the fast bulk-like motions (these are not taken into account in the simulations). Clearly, the observed dispersion cannot be described as a Lorentzian-type mechanism (slope −2) but rather exhibits a power law frequency dependence with the slope between −1 and −0.5. This correlates with the results reported in Ref. [18] for 5CB confined in much larger (compared to the present samples) pores of 72 nm. As shown in Ref. [18], the RMTD mechanism may give rise to the inverse square root frequency dispersion for surface structures imposing equal weights to diffusion modes with different wave numbers.

The cut-off frequency observed for the CPG sample corresponds to a much larger radius used in the simulated data than to its actual pore size. However, this is not surprising as diffusion in these samples is not restricted to a single cavity of a spherical shape. During the typical observation time, the molecules probe much larger distances (exceeding 100 nm) of the random pore space compared to the pore radii. The cut-off frequency, however, permits an estimate to be obtained of the maximal curvilinear displacement along the surface between the two

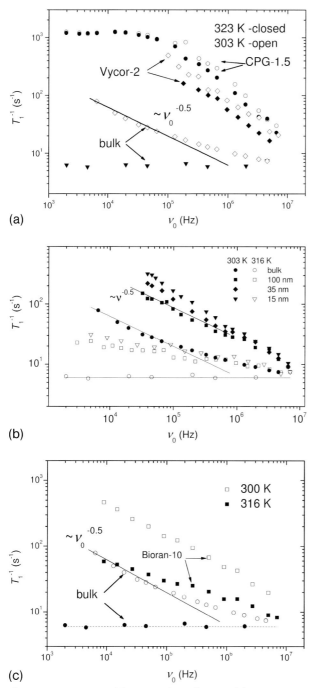

**Figure 27.5** Experimental frequency dependencies of the spin-lattice relaxation rates of bulk 5CB and 5CB confined in: (a) CPG-1 and Vycor-2 at 303 K and 323 K; (b) Bioran-15, Bioran-35 and Bioran-100 at 316 K and 303 K; (c) Bioran-10 at 300 K and 316 K. The bulk values refer to 303 K and 323 K.

correlated orientations (corresponding to the longest mode correlation time), which tends to be of the order of a few tenths of a nanometer [18, 19, 24].

### 27.5.2
### Large Pores

The opposite limit to that of the small pores was investigated for the Bioran samples with radii between 15 nm and 100 nm; in this case, $R \gg \xi$. Figure 27.5b shows frequency dependences of the spin-lattice relaxation rates of 5CB in bulk and confined in Bioran glasses at 303 K and 316 K. Above $T_{NI}$ (316 K), only slight dispersions are observed in the confined samples, and their relaxation rates in the kHz range exceed the bulk values by a factor of three to four. The enhancement of relaxation rates in large pores is thus much weaker than in small pores (by two orders of magnitude), mainly because the fraction of molecules ordered by the surface is smaller relative to the bulk-like fraction in the interior part of the pore. In turn, the contribution of ordered molecules to the overall relaxation rates averaged by molecular diffusion between the surface and interior regions becomes smaller for bigger pores.

Below $T_{NI}$, a square-root dependence characteristic of ODFs dominates the spin-lattice relaxation of bulk 5CB at the low-frequency end of the dispersion. Both, in bulk and in confined samples, the dispersion curves exhibit a sharp jump-like change (indicating phase transition) as the temperature falls below $T_{NI}$. In the confined samples, the slope of the dispersion remains rather close to the inverse square root law observed for the bulk sample. The absolute values of relaxation rates in confined samples, however, exceed the bulk values by a factor of between four and five. This means that both mechanisms – the ODFs and the RMTD – are essential for relaxation below $T_{NI}$ in 5CB when confined to large pores. In the range below 1 MHz, the values of $T_1^{-1}$ become somewhat larger for smaller pores, which probably indicates that the relative contribution of the RMTD mechanism becomes larger with increasing relative surface fraction in smaller pores.

### 27.5.3
### Intermediate Pores

The intermediate or crossover pore regime is represented by the sample of 5CB confined in Bioran-5. Here, the pore size is about the length of the nematic correlation length and, with decreasing temperature this sample exhibits a gradual change of the dispersion slope, as shown in Figure 27.5c. This is in contrast to both regimes described above; that is: (i) only a slight change of the dispersion curves with changing temperature observed for the small pores; and (ii) the jump-like change of the dispersion slopes observed for the large pores when temperature falls below $T_{NI}$. The temperature dependence of the relaxation rate, measured at a frequency of 32.6 kHz, is shown in Figure 27.6. Here a dramatic enhancement of $T_1^{-1}$ by more than one order of magnitude is observed with decreasing temperature in the range between 5 K above and 15 K below $T_{NI}$. The inset shows the difference

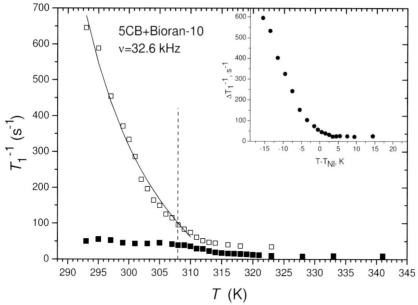

**Figure 27.6** Spin-lattice relaxation rate of bulk 5CB (closed symbols) and 5CB confined in Bioran-10 (open symbols) as a function of temperature. The solid curve represents the fit of Equation 27.11 to the data points below 310 K. The inset shows the temperature dependence of the difference of the relaxation rates of the confined and bulk materials.

in relaxation rates in the confined sample and in bulk. The observed enhancement should be attributed to a gradual increase in the number of molecules in the oriented state. Indeed, according to Equations 27.3 and 27.5, $T_1^{-1} \propto f_s^2$. Thus, the temperature dependence of the relaxation rate should depend on the evolution of $f_s$ as a function of temperature.

Earlier, a strong increase in transverse relaxation rate was reported for confined 5CB in the pretransitional temperature range [10]. This observed enhancement was interpreted [10] by assuming that $f_s = \dfrac{2\xi}{R} \propto \sqrt{\dfrac{T^*}{T-T^*}}$ which should be valid for $R \gg \xi$ and the temperatures above $T^* \approx T_{NI} - 1\,\text{K}$. Under these assumptions, the relaxation rate as a function of temperature behaves as

$$T_1^{-1} \approx A + B f_s^2 \approx A + B\frac{T^*}{T-T^*} \tag{27.11}$$

This model is of course too simple to be applicable to the present experimental data as it does not take into account the depression of $T_{NI}$. Neither is the condition $R \gg \xi$ fulfilled in our case, and therefore – as expected – we could not obtain a reasonable fit of Equation 27.11 to the experimental data. However, the solid curve in Figure 27.6 shows that the shape of the observed temperature dependence of relaxation rate in the range below 310 K can well be described by Equation 27.11. The solid curve represents Equation 27.11, where $T^*$ was replaced by the arbitrary parameter $T^{depr}$ which was fitted to the experimental points. The best fit was obtained for $T^{depr} = 278$ K (about 30 K below $T_{NI}$). This value might provide a rough idea of the depression range of the transition in the investigated sample.

## 27.6
## Summary

Ordering effects and low-frequency molecular dynamics in the nematic LCs confined in mesoscopic pores were studied using FC NMR relaxometry and Monte-Carlo simulations. Proton relaxation rates were measured above and below $T_{NI}$ in the broad frequency range between a few kHz and 7 MHz. Several mesoporous glasses with average pore radii between 1.5 nm and 100 nm were used as confining matrices. The relaxation dispersion curves in the confined materials exhibited strong deviations from the behavior in bulk. In the few kHz range, a dramatic enhancement in relaxation rate was observed which depended on the pore size. The enhancement factor compared to the bulk sample below the MHz regime was between two and four in relatively large pores, and up to two orders of magnitude in small pores. The temperature behavior of the dispersion curves was analyzed for three ranges of pore size. With pores smaller than the nematic correlation length, the observed very strong frequency dispersion was only slightly affected by lowering the temperature below $T_{NI}$. In contrast, with pores much larger than $\xi$ a jump-like change from a relatively weak to a much stronger dispersion was observed when reducing the temperature below $T_{NI}$. Relatively high relaxation rate values prompted the suggestion that both the ODFs and the RMTD process were responsible for relaxation mechanisms in the considered range of pores. The intermediate, crossover, case was represented by a Bioran sample with pore radii of 5 nm (comparable to $\xi$). In contrast to the limits of small and large pores, dispersion curves in this sample showed a strong, gradual change with decreasing temperature such that a gradual evolution of the oriented molecular fraction with decreasing temperature was considered. These experimental findings were interpreted in terms of the surface-induced orientational order and diffusion between sites with different orientations of local directors. The analysis was supported by Monte-Carlo simulations of the RMTD mechanism in spherical cavities.

## Acknowledgments

The author thanks Prof. Dr. R. Kimmich for the most fruitful cooperation during many years; the experimental parts of these studies were performed in his laboratory at the University of Ulm. Thanks are also expressed to Dr. T. Link and Prof. Dr. S. Stapf for cooperation and help with experiments, and to Prof. Dr. J. Kärger for providing excellent research conditions at the University of Leipzig.

# References

1 Woltman, S.J., Jay, G.D. and Crawford, G.P. (2007) Liquid-crystal materials find a new order in biomedical applications. *Nature Materials*, **6**, 929–38.

2 Mann, S. (1993) Molecular tectonics in biomineralization and biomimetic materials chemistry. *Nature*, **365**, 499–505.

3 Li, M., Schnablegger, H. and Mann, S. (1999) Coupled synthesis and self-assembly of nanoparticles to give structures with controlled organization. *Nature*, **402**, 393–5.

4 Yang, K.-L., Cadwell, K. and Abbott, N.L. (2005) Use of self-assembled monolayers, metal ions and smectic liquid crystals to detect organophosphonates. *Sensors and Actuators*, **B104**, 50–6.

5 Vilfan, M., Vrbančič-Kopač, N., Ziherl, P. and Crawford, G.P. (1999) Deuteron NMR relaxometry applied to confined liquid crystals. *Applied Magnetic Resonance*, **17**, 329–44, and references therein.

6 Žumer, S., Ziherl, P. and Vilfan, M. (1997) Dynamics of microconfined nematic liquid crystals and related NMR studies. *Molecular Crystals and Liquid Crystals*, **292**, 39–59.

7 Crawford, G.P. and Žumer, S. (eds) (1996) *Liquid Crystals in Complex Geometries*, Taylor & Francis, London, p. 584.

8 Kimmich, R. (1997) *NMR: Tomography, Diffusometry, Relaxometry*, Springer, Heidelberg, p. 526.

9 Zavada, T. and Kimmich, R. (1998) The anomalous adsorbate dynamics at surfaces in porous media studied by nuclear magnetic resonance methods. The orientational structure factor and Lévi walks. *The Journal of Chemical Physics*, **109**, 6929–39.

10 Vilfan, M., Apih, T., Gregorovic, A., Zalar, B., Lahajnar, G., Zumer, S., Hinze, G., Boehmer, R. and Althoff, G. (2001) *Magnetic Resonance Imaging*, **19**, 433–8.

11 Kärger, J., Grinberg, F. and Heitjans, P. (eds) (2005) *Diffusion Fundamentals*, Leipziger Universitätsverlag, Leipzig, p. 615.

12 Heitjans, P. and Kärger, J. (eds) (2005) *Diffusion in Condensed Matter*, Springer, Berlin., p. 965.

13 Grinberg, F. and Kimmich, R. (1996) Pore size dependence of the dipolar-correlation effect on the stimulated echo in liquid crystals confined in porous glass. *The Journal of Chemical Physics*, **105**, 3301–6.

14 Grinberg, F., Vilfan, M. and Anoardo, E. (2003) The low-frequency NMR relaxometry of spatially constrained oriented fluids, in *NMR of Orientationally Ordered Liquids* (eds E. Burnell and C. de Lange), Kluwer Academic Publishers, Dordrecht, p. 488.

15 de Gennes, P.G. (1974) *The Physics of Liquid Crystals*, Clarendon Press, Oxford, p. 346.

16 Grinberg, F. (2002) Monitoring ultraslow motions in organized liquids. NMR experiments and computer simulations, in *Magnetic Resonance in Colloid and Interface Science* (eds J. Fraissard and O. Lapina), Kluwer Academic Publishers, Dordrecht, p. 672.

17 Stapf, S., Kimmich, R. and Seitter, R.-O. (1995) Deuteron field-cycling NMR relaxometry of liquids in porous glasses: evidence for Lévy-walk statistics. *Physical Review Letters*, **75**, 2855–8.

18 Sebastião, P.J., Sousa, D., Ribeiro, A.C., Vilfan, M., Lahajnar, G., Seliger, J. and Žumer, S. (2005) Field-cycling NMR relaxometry of a liquid crystal above $T_{NI}$ in mesoscopic confinement. *Physical Review E*, **72**, 061702-1–11.

19 Vilfan, M., Apih, T., Sebastião, P.J., Lahajnar, G. and Žumer, S. (2007) Liquid crystal 8CB in random porous glass: NMR relaxometry study of molecular diffusion and director fluctuations. *Physical Review E*, **76**, 051708-1–15.

20 Anoardo, E., Grinberg, F., Vilfan, M. and Kimmich, R. (2004) Proton spin-lattice relaxation in a liquid crystal–Aerosil complex above the bulk isotropization temperature. *Chemical Physics*, **297**, 99–110.

**21** Abragam, A. (1961) *Principles of Nuclear Magnetism*, Clarendon Press, Oxford, p. 618.

**22** Klafter, J., Shlesinger, M.F. and Zumofen, G. (1996) *Physics Today*, **49**, 33–9.

**23** Grinberg, F. and Kimmich, R. (2001) Surface effects and dipolar correlations of confined and constrained liquids investigated by NMR relaxation experiments and computer simulations. *Magnetic Resonance Imaging*, **19**, 401–4.

**24** Grinberg, F. (2007) Surface effects in liquid crystals constrained in nano-scaled pores investigated by field-cycling NMR relaxometry and Monte-Carlo simulations. *Magnetic Resonance Imaging*, **25**, 485–8.

# 28
# NMR Imaging of Moisture and Ion Transport in Building Materials

*Leo Pel and Henk Huinink*

## 28.1
## Introduction

Our world's cultural heritage is under constant threat from its environment. Salts are widely recognized as a major cause of the loss of many historical objects, such as statues, buildings and other artworks. While a porous material is drying, salt crystallization may occur at the surface (efflorescence) resulting in visual damage or just below the surface (subfluorescence), where it may cause structural damages, for example, delamination, surface chipping or disintegration, with consequent loss of detail. Many contemporary buildings and civil constructions also suffer from salt-induced damage processes. Salt weathering can therefore be considered as a common hazard with significant cultural and economic implications. Hence, a detailed knowledge of moisture and salt transport is essential for understanding the durability of these materials.

In order to measure the combined moisture and ion transport, destructive methods are often used that usually involve drilling or grinding of the sample, gravimetrically determining its moisture content, and chemically determining its salt concentration. These destructive methods not only complicate any time-dependent measurements but also usually lack sufficient resolution (of the order of 10 mm). Hence, for such studies it is important to measure the dynamic moisture and ion profiles quantitatively. A variety of methods exist for this purpose, including those based on the dielectric properties of water such as capacitance measurement or time-domain reflectometry (TDR), gamma-attenuation and nuclear magnetic resonance (NMR) [1] to measure the moisture profiles in a nondestructive manner. Among these techniques, however, only NMR offers the possibility of measuring moisture and ion transport simultaneously.

Unfortunately, serious complications may occur if the materials under investigation contain large amounts of paramagnetic ions, as is the case for many common building materials such as fired-clay brick (e.g. up to 4% Fe) and mortar. The red color of fired-clay brick and the gray color of mortar is due to the presence of Fe.

*Magnetic Resonance Microscopy.* Edited by Sarah L. Codd and Joseph D. Seymour
Copyright © 2009 WILEY-VCH Verlag GmbH & Co. KGaA, Weinheim
ISBN: 978-3-527-32008-0

The short transverse relaxation time and broad resonance linewidth of the hydrogen nuclei in these materials preclude the use of standard NMR imaging techniques. In addition, many nuclei of interest in the study of ion transport, as for example Na, have a quadrupole moment. This can result in quadrupolar splitting and a complex relaxation behavior. In this chapter we will first discuss the problems related to imaging moisture and ion profiles with NMR, after which some typical examples will be given of moisture and ion transport in porous building materials.

## 28.2
## NMR and Porous Building Materials

### 28.2.1
### Moisture Measurement

In NMR experiments, the magnetic moment of the nuclei are manipulated by suitably chosen radiofrequency (RF) fields, resulting in a so-called spin-echo experiment. The resonance frequency of a certain type of nucleus, called the Larmor frequency, is determined by the magnitude of the applied magnetic field $B$:

$$f_i = \gamma_i B \tag{28.1}$$

where the index $i$ refers to the type of nucleus (e.g. $^1H$ or $^{23}Na$), $f_i$ (in Hz) is the Larmor frequency, $\gamma_i$ is the gyromagnetic ratio of the nucleus ($\gamma_H/2\pi = 42.58\,MHz\,T^{-1}$; $\gamma_{Na}/2\pi = 11.27\,MHz\,T^{-1}$) and $B$ (in Tesla) is the strength of the magnetic field.

Assuming that both transverse and longitudinal relaxation mechanisms give rise to a simple exponential relaxation, and that spin–lattice relaxation is much slower than the spin–spin relaxation, the magnitude of the NMR spin-echo signal is given by (see e.g. [2, 3]):

$$S \sim G\rho[1 - \exp(-TR/T_1)]\exp(-TE/T_2) \tag{28.2}$$

In this expression, $G$ is the relative sensitivity of the nucleus in comparison to hydrogen (for $^1H$, $G = 1$; for $^{23}Na$, $G = 0.093$), $\rho$ is the density of the nuclei, $T_1$ the spin-lattice or longitudinal relaxation time, $TR$ the repetition time of the spin-echo experiments, $T_2$ the spin–spin or transverse relaxation time, and $TE$ the so-called spin-echo time. Obviously, small $T_2$ values lead to a decrease in the spin-echo signal whereas, on the other hand, small $T_1$ values are preferred as this parameter limits the repetition time (usually $TR \sim 4T_1$) and hence the rate at which the moisture and ion profiles can be scanned.

For the spin-echo decay in the presence of a uniform magnetic-field gradient, three length scales are important (see e.g. [4–9]):

- The diffusion length, $l_D = \sqrt{6Dt}$
- The structural length, $l = V/S$, which is equal to $R/3$ for a spherical pore with volume $V$, surface area $S$, and radius $R$;
- The dephasing length, $l_g = \sqrt[3]{\dfrac{D}{\gamma g}}$ where $g$ is the gradient strength.

It should be noted that the diffusion length is a function of time – the longer the molecules can diffuse, the larger the distance they can travel. The structural length determines how far the molecules can travel because of the restricted geometry. The dephasing length indicates how far a particle has to travel to dephase by $2\pi$. Only particles with less dephasing contribute significantly to the spin-echo decay.

The shortest of the length scales determines the dominant mechanism of the spin-echo decay. Free diffusion will occur when the diffusion length is the shortest length scale, which will occur always for very small times. Motional averaging will occur when the structural length is the shortest length scale, which means that the particles can probe the complete pore space, without significant dephasing. In this case one can apply the well-known Brownstein–Tarr model, that is, [10]:

$$T_{1,2} \approx \frac{1}{\rho_{1,2}} \frac{V}{S} \qquad\qquad (28.3)$$

where $\rho_{1,2}$ is the surface relaxivity for the $T_1$ or for the $T_2$ relaxation, respectively. In this model the source for the relaxation is the presence of magnetic impurities along the pore wall, which is often the case in building materials. Hence, the relaxation is proportional to the $V/S$, which is the structural length of a pore. In this way it is possible to use the relaxation distribution as an indication for the pore water distribution so that, by measuring the relaxation one can study the pore size distribution during the hydration of cements [11, 12], or pore water distribution during drying [13].

Finally, the localization regime will occur when the dephasing length is the shortest length scale. This is the case for many for many building materials which contain large amounts of paramagnetic ions and significantly large pores. As a result of this there will be a large susceptibility contrast between the pore fluid and the porous matrix, resulting in local field gradients. Hence, the particles will dephase significantly before they reach the pore wall.

In Table 28.1 an overview is provided of measured effective relaxation times, magnetic susceptibilities and porosities. For fired-clay brick a high $\chi$ is measured which can be attributed to the presence of inclusions (mass fraction $10^{-4}$) of metallic Fe, which are magnetically ordered at room temperature. As a result the dephasing length is much less than the diffusion length, and $T_2$ no longer contains any pore size information. As a result, for those materials which have large pores (i.e. on the order of $1\,\mu m$) the repetition time which reflects the pores size must be taken in the order of 1.5 s. On the other hand, in order to have sufficient signal the spin-echo time should be on the order of $T_2$ (i.e. $300\,\mu s$). Therefore, when using standard NMR imaging techniques, which usually employ a $TE$ of a few

**Table 28.1** The effective relaxation times of hydrogen nuclei in various types of common porous building materials determined from NMR measurements at $B = 0.78$ T, assuming a single exponential relaxation, together with the porosities of these materials, determined from vacuum saturation and the differential magnetic susceptibility $\chi = dM/dHg^{-1}$ at $B = 0.78$ T and 293 K as measured by SQUID magnetometer.

| Material | $T_1$ (ms) | $T_2$ (ms) | $\theta$ (m$^3$ m$^{-3}$) | $\chi$ ($10^{-6}$ emu Gs$^{-1}$ g$^{-1}$) |
|---|---|---|---|---|
| Fired-clay brick A | 290 | 1.9 | 0.19 | 2.7 |
| Fired-clay brick B | 180 | 2.3 | 0.27 | 2.5 |
| Fired-clay brick C | 300 | 8.1 | 0.2 | 4.2 |
| Fired-clay brick D/ fired-clay brick E | 73 | 1.7 | 0.27 | 3.6 |
| Fired-clay brick F | 310 | 1.6 | 0.28 | 3.4 |
| Calcium silicate brick | 30 | 7.5 | 0.27 | 0.5 |
| Mortar | 35 | 14 | 0.18 | 0.13 |
| Gypsum | 100 | 22 | 0.28 | −0.26 |

milliseconds or more, almost no signal will be found for these types of material. In contrast a material such as gypsum, which has almost no impurities and a comparable pore size to that of fired-clay brick, the $T_2$ is very long.

In general, for a material under investigation the content of magnetic impurities is not known. Hence, great care must be taken when interpreting the $T_2$ relaxation distribution as reflecting the pores size distribution. Often, additional measurements must be performed, such as cryoporometry [14] or mercury intrusion porosimetry to obtain additional information on the pore size distribution.

Due to the magnetic impurities in building materials such as fired-clay brick, the natural linewidth described by $T_2^*$ is found to be in the order of 20–30 µs, corresponding to a linewidth in the order of 8 kHz. Hence, in order to achieve a spatial resolution of approximately 1 mm, magnetic field gradients of about 0.2 T m$^{-1}$ (20 G cm$^{-1}$) are needed. As a consequence, in order to measure moisture profiles with NMR it is not possible to use standard medical magnetic resonance imaging (MRI) equipment, and usually homebuilt or specially adapted NMR equipment is employed. Also, because of the presence of magnetic impurities, diffusion measurements can only be made by applying high static field gradients and using a stimulated echo sequence. It has been shown that such experiments can both probe the diffusion coefficient and amplitude of magnetic fields induced by paramagnetic impurities [15].

28.2.2
**Ion Measurement**

In the study of salt-weathering deterioration there are many relevant salts such as, NaCl, $Na_2SO_4$ and $Na_2CO_3$. As the relative sensitivity of Na is high, it is often chosen as the nucleus to image in order to study combined moisture and ion transport. The relaxation behavior of sodium is more complex, because it is a quantum spin number $I = 3/2$ nuclei and has a quadrupole moment. At low field (i.e. in the order of 1 T), the additional quadrupole interaction will be small compared to the Zeeman interaction, and in that case the quadrupole interaction can be considered as a perturbation on the Zeeman interaction. The resulting energy levels are shown schematically in Figure 28.1. Here, three transition frequencies are possible which satisfy the selection rule $\Delta m \pm 1$. If all nuclei were to experience the same electric field gradient (EFG), the NMR spectrum would therefore consist of three different transition frequencies, namely the Larmor frequency and two sidebands [16].

During the NMR experiments, however, the ions in the solution are not static but move randomly because of Brownian motion; indeed, this movement of ions can average out the EFG experienced by the ion. In general, two distinct regimes of quadrupolar ions in solution can be distinguished [16]:

- In the slow modulation regime the ions experience a net EFG which results in an energy level splitting as described above, and the magnetization decay can no longer be described by a single $T_{1,2}$ value. An example of this effect has been studied by Delville and coworkers [17] in laponite clay, where the negative charge of the clay results in a large electric field gradient, which is not averaged out by the movement of the ions.
- In the fast modulation regime a quite different behavior is observed, whereby the movement and tumbling of the ions occurs rapidly compared to the timescale

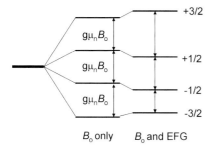

$B_o$ only     $B_o$ and EFG

**Figure 28.1** The energy levels of a nucleus with a quantum spin number $I = 3/2$ which has a Zeeman and a quadrupole interaction. The NMR frequency is determined by the distance between the energy levels. It can be seen that if no electric field gradient (EFG) is the case then one single frequency exists, whereas three frequencies exist if an EFG is present.

of the NMR experiment, and therefore the ions do not experience a net EFG. As a result no splitting occurs and a monoexponential relaxation behavior is observed. In the fast modulation regime the relaxation can again be interpreted using the Brownstein–Tarr model; in other words, the relaxation can provide information on pore ion distribution [18].

In practice, it is impossible to tell clearly in which regime the Na measurement will be present for a material under investigation, and therefore a calibration must always be performed – that is, the signal must be obtained as a function of a known Na-content. Although the relative sensitivity of $^{23}$Na is much lower (i.e. 0.093) in comparison to $^{1}$H, the signal can be acquired much faster due to the much lower $T_1$ (usually on the order of 20–40 ms). Consequently, in experiments with building materials, almost comparable signal-to-noise ratios (SNRs) can be obtained for the $^{1}$H and $^{23}$Na signals, for the same measurement time [19].

## 28.2.3
### NMR Techniques

Although MRI is a well-established, three-dimensional (3-D) medical diagnostic imaging technique, traditional MRI methods may fail when applied to many porous building materials due to their short relaxation times. As this also precludes the use of spatial encoding, it would be necessary to switch gradients to within 1–2 μs in order to use these methods. Conventional MRI methods therefore will usually fail, or at best produce a poor image. It is also necessary to be aware that, whereas in conventional MRI the interest is not often targeted at quantitative information, for moisture and ion transport there is a clear need for quantitative data. For example in the case of water absorption there will be an abrupt change in moisture content as the wetting front is passing, and this will be reflected in a large variation in the dielectric permittivity of the sample. As a result, the tuned inductive/capacitive (LC) circuit of the NMR can be detuned, and this will result in an apparent loss of signal. Pel *et al.* [20] have solved this problem by adding a Faraday shield to the LC circuit. The difficulty can also be overcome by lowering the quality factor $Q$ of the circuit, or by lowering the filling factor of the NMR coil. One must also be aware that, especially in salt solutions, eddy currents can be generated that result in RF power losses. Generally, it is always advisable to include a reference sample that can be used to correct for possible detuning.

Among the first to show that NMR could be used for imaging moisture transport during absorption in porous building materials were Gummerson *et al.* [21]. This group used a constant magnetic field and measured the free induction decay (FID), with spatial resolution being determined by the length of the RF coil. In order to obtain a higher spatial resolution, Pel and coworkers [20] used a constant gradient field of 0.3 T m$^{-1}$ to give 1-D resolution on the order of 1 mm, whilst to measure the moisture content they used a Hahn spin-echo time of approximately 200 μs. For measuring the moisture and Na-content, the same group [19] used a specially designed RF circuit with which the tuned LC circuit of the NMR set-up could be toggled between the resonance frequency for $^1$H and $^{23}$Na, thus allowing quasi-simultaneous measurement. In this way, for every position the moisture content was first measured, after which the LC circuit was toggled and the Na content measured. In order to overcome problems with short relaxation times, Fagan et al. [22, 23] used a continuous-wave technique which allowed solid-state imaging. By using a back-projection reconstruction method (also known as radon transform), the group obtained a 2-D distribution of the moisture and ions. In order to image ions it necessary to change the resonance frequency of the RF circuit manually. In another study, Balcom et al. [24, 25] used single-point imaging (SPI), a pure phase-encoding MRI technique where one point is taken during each FID, thus allowing short effective echo-times. By changing the resonance frequency of the RF circuit in a single point method the ions can be measured separately.

## 28.3
## Examples of Moisture and Ion Transport

In this section we will provide some typical results of some experiments relevant for salt weathering studies, that is, drying and crystallization. A schematic representation of the drying process for a sample saturated with a salt solution is shown in Figure 28.2; during the process, moisture will be transported to the drying surface. If gravity is neglected, the moisture transport for the 1-D problem considered here can be described by a nonlinear diffusion equation (see, e.g. Refs [26–29]):

$$\frac{\partial \theta}{\partial t} = \frac{\partial}{\partial x}\left(D(\theta)\frac{\partial \theta}{\partial x}\right)$$

(28.4)

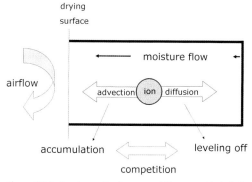

Figure 28.2 A schematic diagram of the one-sided drying of a sample saturated with a salt solution.

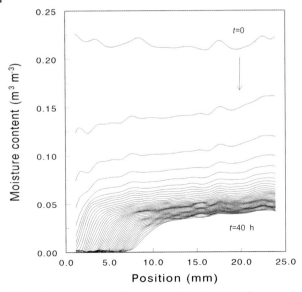

**Figure 28.3** Moisture profiles measured during drying for fired-clay brick. The time between subsequent profiles is 1 h, and the profiles are given for a period of 40 h. The sample is dried from the top ($x = 0$), while the bottom ($x = 25$) is sealed off.

In this equation $\theta$ ($m^3\,m^{-3}$) is the volumetric liquid moisture content, $D(\theta)$ ($m^2\,s^{-1}$) is the so-called isothermal moisture diffusivity. In this 'lumped' model all mechanisms for moisture transport – that is, liquid flow and vapor diffusion – are combined into a single moisture diffusivity which depends on the actual moisture content. Here, the transport of water due to a salt gradient has been neglected. However, the salts will have a direct influence on capillary action, viscosity and permeability and, therefore, on the moisture diffusivity.

In Figure 28.3 an example is given of the moisture profiles as measured during the drying of a sample saturated with pure water. The variations which repeat from profile to profile reflect the inhomogeneities, such as small stones or holes, in the material. Clearly, two drying phases can be distinguished. Initially, the moisture transport is externally limited and the moisture profiles in the brick remain almost flat, indicating that liquid transport in the material is much faster than evaporation at the surface. After some time – that is, on the order of 5 h – a receding drying front is observed, indicating the second phase where the moisture transport is internally limited. In this phase the moisture must evaporate at the drying front and be transported through the porous material by the much slower vapor transport process. By using these moisture profiles it is possible to determine the moisture diffusivity [29].

During the drying of a salt-saturated sample the ions will be transported by advection with the moisture flow and diffusion within the liquid regions. If the adsorption of ions to the walls is neglected (which is justified for fired-clay brick because of the low specific surface absorptivity), the ion transport can be described by (see e.g. Ref [26]):

$$\frac{\partial(\theta c)}{\partial t} = -\frac{\partial}{\partial x}\left(\theta\left(cv_l - D_c\frac{\partial c}{\partial x}\right)\right) - R \tag{28.5}$$

where $c$ (mol l$^{-1}$) is the ion concentration in the water, $D_c$ (m$^2$ s$^{-1}$) is the diffusion coefficient of the ions in the moisture in the porous medium, and $R$ (mol l$^{-1}$ s$^{-1}$) is a term reflecting the crystallization rate. As can be seen from Equation 28.5, during a drying experiment there will be a competition between advection, which transports ions to the top of the sample and thereby causes accumulation, and diffusion, which levels off any accumulation (see also Figure 28.2).

For the competition between advection and diffusion in a porous material during drying a Peclet number (Pe) can be defined as (see Refs [26, 30]):

$$Pe \equiv \frac{hL}{\theta D_c} \tag{28.6}$$

where $h$ (m$^3$ m$^{-2}$ s$^{-1}$) is the drying rate, $L$ (m) is the length of the sample, and $\theta$ (m$^3$ m$^{-3}$) the maximum fluid content. It should be noted that the Peclet number is a macroscopic parameter–that is, it is valid at continuum level–and is not defined at the pore scale, that is for moisture and ion transport within in the pore. For Pe < 1 diffusion dominates and the ion profiles will be uniform, whereas for Pe > 1 advection dominates and ions will be transported to the drying surface. As ions cannot leave the material they will accumulate at the drying surface. Then, as soon as the maximum solubility is reached, the ions will start to crystallize, becoming visible as efflorescence at the surface.

As an example of a drying experiment, we will discuss the results for fired-clay bricks saturated with a 3 M NaCl solution. Fired-clay bricks were used in these experiments because the amount of ions that are bound to the pore wall is low. Some representative moisture profiles are plotted in Figure 28.4. These data reveal that the moisture profiles are almost flat, indicating that moisture distribution within the sample remains homogeneous during drying; that is, for up to 14 days no receding drying front is observed. This effect is attributed to the wetting properties of the NaCl solution and the low drying rate used in these experiments. The corresponding Na ion profiles are plotted in Figure 28.5. With the NMR settings used in these experiments only the Na nuclei in the solution are measured–that is, no signal is obtained from NaCl crystals. Inspection of these data shows that within one day after the start of the drying process the Na content develops a peak just below the drying surface, but after 12–15 days the Na profiles become flat again. This effect can be better observed in the Na concentration profiles shown in Figure 28.6. Here, the concentration is obtained via point-by-point division of the

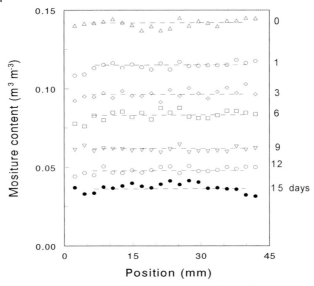

**Figure 28.4** Moisture profiles measured during drying of a fired-clay brick sample of 45 mm length after 0, 1, 3, 6, 9, 12 and 15 days. The drying surface is at 0 mm. The lines are provided as a guide to the eye.

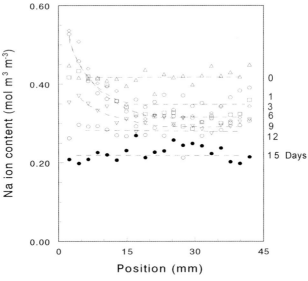

**Figure 28.5** Absolute content of dissolved Na ion profiles measured during drying of a fired-clay brick sample of 45 mm length after 0, 1, 3, 6, 9, 12 and 15 days. The drying surface is at 0 mm. The lines are provided as a guide to the eye.

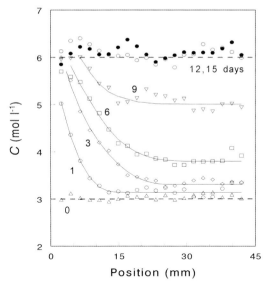

**Figure 28.6** NaCl concentration profiles measured during drying of a fired-clay brick sample of 45 mm length after 0, 1, 3, 6, 9, 12 and 15 days. The drying surface is at 0 mm. Initially for days 0–3 the Pe was >1, whereas for days 6–15 the Pe was <1. The lines are provided as a guide to the eye.

corresponding Na and H profiles. It is clear that, during the initial drying, Na ions are advected to the surface (position 0 mm) and the NaCl concentration increases to 6 *M*, which is the saturation value for a NaCl solution. At this point any additional advection will result in crystallization at the top of the sample, which is observed as efflorescence. From this point on, the NaCl concentration profile in the sample starts to level off until, after 15 days, the total sample is at 6 *M*. This leveling off is related to a drop in the evaporation rate, due to which advection loses its dominance over diffusion and a homogeneous distribution of ions is promoted.

As salts begin to crystallize in a porous material (as is the case at the surface of the previous experiment), they are confined and as a consequence mechanical pressure can build up. Such pressure causes the solubility of the salt inside the pore to differ from the bulk solubility. Thermodynamically, this pressure can be related to the solubility of the salt, and hence [31]:

$$P_c = \frac{nRT}{v_c} \ln\left(\frac{C}{C_o}\right) \qquad (28.7)$$

where $P_c$ is the crystallization pressure, $C$ is the saturation concentration in the pore, $C_o$ the concentration in the bulk solution, $v_c$ the molar volume of the salt and $n$ is the number of the ions in the salt. Hence, as a crystallization pressure is built up the concentration in the pores will increase – a situation which is often referred to as supersaturation. Such supersaturation can be measured nondestructively in porous materials by using NMR.

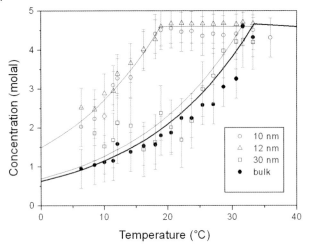

**Figure 28.7** The measured solubility by NMR of $Na_2CO_3$ in bulk (•) and inside a series of model porous material Nucleosil with pores sizes 10 nm (o), 12 nm (Δ) and 30 nm (□).

There are two ways to induce crystallization: (i) by drying, as in the previous experiment; and (ii) by changing the temperature. However, the latter approach requires that there is a significant variation in solubility with temperature.

In order to obtain an unambiguous value of concentration in the pores, a model porous material (Nucleosil) was used which has pores of well-determined sizes (10, 12 and 30 nm) rather than a broad pore size distribution, as do most building materials. In order to create supersaturation by temperature changes in the range of 0 to 40 °C, sodium carbonate ($Na_2CO_3$) was used (NaCl is not used as within this temperature range its solubility varies only slightly). A saturated sodium carbonate (prepared at 40 °C) was used to saturate the samples, which were then cooled to 2 °C. When nucleation had occurred the samples were heated slowly, such that equilibrium was always guaranteed. The results (see Figure 28.7) showed that, for bulk $Na_2CO_3$, the previously reported solubility curve was reproduced, but for the $Na_2CO_3$ solution in pores of 10 and 12 nm a significant increase in solubility was found, indicating a build-up of pore pressure. Based on these measurements it can be estimated that a crystallization pressure build-up of 4 MPa occurred in the 10 nm pores.

## 28.4
## Conclusions

The use of standard NMR techniques for studying transport phenomena in porous building materials is limited due to the fact that many common building materials contain large amounts of paramagnetic ions. Hence, specially adapted NMR set-ups must be used. Moreover, the magnetic impurities in these materials complicate the relaxation behavior such that the latter is often rather a reflection of the magnetic impurities and not of pore water distribution. Nonetheless, NMR represents a powerful tool for studying combined moisture and ion transport in porous materials as it can be used to measure both moisture and ion contents quasi-simultaneously.

## Acknowledgments

Part of this research was supported by the Dutch Technology Foundation (STW), TNO Built Environment and Geosciences and the EU Directorate General Research.

## References

1 Roels, S., Carmeliet, J., Hens, H., Adan, O., Brocken, H., Cerny, R., Pavlik, Z., Ellis, A.T., Hall, C., Kumaran, K., Pel, L. and Plagge, R. (2004) *Journal of Thermal Envelope and Building Science*, **27**, 261–76.

2 Hahn, E.L. (1950) Spin echoes. *Physical Review*, **80**, 580–94.

3 Vlaardingerbroek, M.T. and den Boer, J.A. (1999) *Magnetic Resonance Imaging*, 2nd edn, Springer, New York.

4 Hürlimann, M.D. (1998) *Journal of Magnetic Resonance*, **131**, 232–40.

5 Sen, P.N., André, A. and Axelrod, S. (1999) *The Journal of Chemical Physics*, **111**, 6548–55.

6 Zielinski, L.J. and Sen, P.N. (2000) *Journal of Magnetic Resonance*, **147**, 95–103.

7 Sen, P.N. (2004) *Concepts in Magnetic Resonance A*, **23A**, 1–21.

8 Valckenborg, R.M.E., Huinink, H.P., v.d. Sande, J.J. and Kopinga, J.J. (2002) *Physical Reviews E*, **65**, 021306.

9 Valckenborg, R.M.E., Huinink, H.P. and Kopinga, K. (2003) *The Journal of Chemical Physics*, **118**, 3243–51.

10 Brownstein, K.R. and Tarr, C.E. (1979) *Physical Reviews A*, **19**, 2446–53.

11 Yehng, J.Y. (1995) Microstructure of wet cement pastes: a nuclear magnetic resonance study, Ph.D. Thesis, North-western University, Evanston IL, Chicago.

12 Bhattacharja, S., Moukwa, M., D'Orazio, F., Jehng, J.-Y. and Halperin, W.P. (1993) *Advanced Cement-Based Materials*, **1**, 67–76.

13 Valckenborg, R., Pel, L., Hazrati, K., Kopinga, K. and Marchand, J. (2001) *Materials and Structures*, **34**, 599–604.

14 Valckenborg, R.M.E., Pel, L. and Kopinga, K. (2002) *Journal of Physics D: Applied Physics*, **35**, 249–56.

15 Petkovic, J., Huinink, H.P., Pel, L. and Kopinga, K. (2004) *Journal of Magnetic Resonance*, **167**, 97–106.

16 Slichter, C.P. (1990) *Principles of Magnetic Resonance*, Springer, New York.

17 Porion, P., Al Mukhtar, M., Meyer, S., Faugere, A.M., van der Maarel, J.R.C. and Delville, A. (2001) *The Journal of Physical Chemistry B*, **105**, 10505–14.

18 Rijniers, L.A., Huinink, P.C.M.M., Magusin, H.P., Pel, L. and Kopinga, K. (2004) *Journal of Magnetic Resonance*, **167**, 25–30.

19 Pel, L., Kopinga, K. and Kaasschieter, E.F. (2000) *Journal of Physics D: Applied Physics*, **33**, 1380–5.

20 Kopinga, K. and Pel, L. (1994) *Review of Scientific Instruments*, **65**, 3673–81.

21 Gummerson, R.J., Hall, C., Hoff, W.D., Hawkes, R., Holland, G.N. and Moore, W.S. (1979) *Nature*, **281**, 56–7.

22 Fagan, A.J., Davies, G.R., Hutchison, J.M.S., Glasser, F.P. and Lurie, D.J. (2005) *Journal of Magnetic Resonance*, **176**, 140–50.

**23** Fagan, A.J., Davies, G.R., Hutchison, J.M.S. and Lurie, D.J. (2003) *Journal of Magnetic Resonance*, **163**, 318–24.

**24** Balcom, B.J., Barrita, J.C., Choi, C., Beyea, S.D., Goodyear, D.J. and Bremner, T.W. (2003) *Materials and Structures*, **36**, 166–82.

**25** de J. Cano, F., Bremner, T.W., McGregor, R.P. and Balcom, B.J. (2002) *Cement and Concrete Research*, **32**, 1067–70.

**26** Bear, J. and Bachmat, Y. (1990) *Introduction to Modeling of Transport Phenomena in Porous Media*, Kluwer, Dordrecht, The Netherlands.

**27** Philip, J.R. and de Vries, D.A. (1957) *Transactions of the American Geophysical Union*, **38**, 222–32.

**28** Huinink, H.P., Pel, L. and Michels, M.A.J. (2002) *Physics of Fluids*, **14**, 1389–95.

**29** Pel, L., Brocken, H. and Kopinga, K. (1996) *International Journal of Heat and Mass Transfer*, **39**, 1273–80.

**30** Pel, L., Huinink, H. and Kopinga, K. (2002) *Applied Physics Letters*, **81**, 2893–5.

**31** Rijniers, L.A., Huinink, H.P., Pel, L. and Kopinga, K. (2005) *Physical Review Letters*, **94**, 075503.

# 29

# Magnetic Resonance Studies of Drop-Freezing Processes

*Michael L. Johns, David I. Wilson and Jason P. Hindmarsh*

## 29.1
## Introduction

The use of magnetic resonance (MR) techniques for the investigation of freezing/solidification phenomena from a melt or liquid has found wide application. Predominantly, these studies focus on exploiting the very short $T_2^*$ of ice (~5 µs), which effectively eliminates it from contributing to the MR signal, enabling any unfrozen liquid to be selectively detected. Often, such unfrozen liquid results from the freeze concentration of a solute or solutes, thus suppressing the freezing point of the remaining unfrozen liquid material.

Freezing processes are common in the food industry and thus a substantial body of work exists, applying MR techniques to such systems. This includes imaging of freezing/thawing meat products [1–3], cheese [4], bread [5], corn [6] and fruit [7]. Another productive area of research has been sea ice. MR imaging and diffusion measurements have been used to elucidate the microstructure of brine-filled pores [8], and the Earth's magnetic field has been used to elucidate microstructure via restricted self-diffusion measurement for *in situ* sea ice [9, 10]. As sea ice freezes, the increasing brine concentration of the residual liquid can result in excessive dielectric losses; Aussillous *et al.* [11] used model glucose solutions to follow the formation of a solidifying mushy layer cooled from above. The formation of chimneys in these layers due to buoyancy-driven convection was observed, while MR velocimetry was used to image the velocity field in these structures. Freezing of the liquid in porous support structures has also been investigated using a range of MR techniques. Georgiadis and Ramaswamy [12] imaged freezing of the interstitial water in various packed beds, while Ozeki *et al.* [13] considered snowpack structures. SPRITE imaging techniques have been successfully applied to the freezing and thawing of concrete and cement (e.g. Ref. [14]), systems in which unfavorable signal relaxation necessitates the use of single-point methods. The effect of pore size on freezing point depression has also facilitated cryoporometry, a technique where MR is used to measure the unfrozen liquid signal as a function of temperature and hence infer a pore size distribution [15]. One of the

*Magnetic Resonance Microscopy.* Edited by Sarah L. Codd and Joseph D. Seymour
Copyright © 2009 WILEY-VCH Verlag GmbH & Co. KGaA, Weinheim
ISBN: 978-3-527-32008-0

more unusual studies was that of Rubinsky *et al.* (1994) [16], who imaged the freezing and thawing of baby turtles.

While the above description of MR application with respect to freezing/solidification processes is not exhaustive, it does convey the vast range of systems studied. Here, we consider the freezing of suspended drops; this is designed to mimic the drop-freezing process that occurs during spray-freezing operations. Spray-freezing can simply be described as the solidification of a liquid by atomization in a relatively cold atmosphere. It is a process researched and developed for the production of powders of a range of materials including foodstuffs [17], pharmaceuticals [18] and for biological cell preservation [19]. The resultant powder is often subsequently dried to leave a highly porous microstructure, which improves the rate of dissolution if required in subsequent use. The comparatively low operating temperatures of this combined operation typically result in favorable volatiles (e.g. flavors) retention.

Essential for predictive modeling of spray-freezing is a detailed description of the drop solidification kinetics. The aim of the studies summarized here was to provide such data via various rapid MR measurements. Equally important in terms of the final powder product's functionality is an ability to characterize its microstructure as formed by the solidified ice network; this was achieved via a suitable combination of MR methods.

## 29.2
## Background

Figure 29.1 shows the typical temperature transitions observed in water (Figure 29.1a) and 20% (w/v) sucrose solution (Figure 29.1b) drops, each with a diameter of 2 mm, as they freeze in −25 °C and −16 °C air flows, respectively [20]. The drops are suspended from the end of a thermocouple which thus measures the drop internal temperature. The drop-freezing process can be described by five distinct stages:

(i)   *Liquid cooling and supercooling,* during which the liquid drop is cooled from its initial state to a temperature below the equilibrium freezing point.

(ii)  *Nucleation,* where there is sufficient supercooling for spontaneous crystal nucleation to occur. Note that the nucleation temperature is stochastic and hence will occur over a relatively broad range.

(iii) *Recalescence,* during which supercooling drives rapid kinetic crystal growth from the crystal nuclei. There is an abrupt temperature rise as this growth liberates the latent heat of fusion. This stage is terminated when the supercooling is exhausted and the drop has reached an equilibrium freezing temperature. Recalescence is slightly slower for the sucrose solution drop due to reduced molecular diffusion.

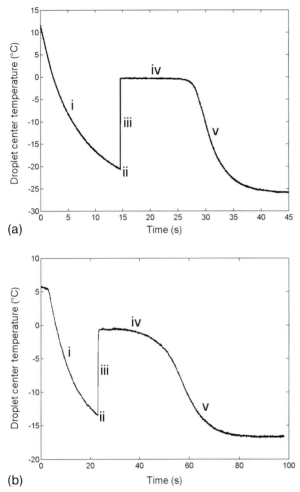

(a)

(b)

**Figure 29.1** Temperature variations as a function of time for
(a) water drop and (b) 20wt% sucrose solution drop of
diameter 2mm frozen in −25 °C and −16 °C air, respectively.
The five stages of freezing are highlighted [20].

(iv) *Heat transfer controlled freezing*, where further growth of the solid phase is
governed by the rate of heat transfer to the environment from the drop. In
the case of the water drop (Figure 29.1a) the temperature remains constant,
but in the case of the sucrose solution drop (Figure 29.1b) the temperature
decreases as solute concentration increases and hence progressively greater
freezing point depression occurs.

(v) *Solid cooling or tempering*, where the temperature of the solidified drop reduces
to a steady-state value approaching that of the ambient air.

## 29.3
## Methodology

All MR experiments were performed using a Bruker DMX 300 spectrometer featuring a 7.14 T vertical bore magnet fitted with either a 5, 10 or 20 mm inner diameter (i.d.) $^1$H (used exclusively) birdcage radiofrequency (RF) coil and a Bruker BVT 3000 temperature control unit. Temperature was controlled via a heater/cold airflow apparatus attached to the bottom of the magnet. A schematic of the rig used to deliver and suspend a drop inside the magnet is shown in Figure 29.2, along with a photograph of a single drop. The principle is that a tube is inserted into the top of the magnet bore which acts as a guide to deliver the drop suspension carriage to the top of the RF coil. The suspension carriage has a drop suspended from the tip of a fine glass filament (replacing the metal thermocouple used in generating Figure 29.1) and is prefitted to locate the drop at the center of the RF coil when the carriage rests on top of it. The droplet suspension carriage and delivery tube are made from Perspex (nonmagnetic). A bead of glue (~100 μm diameter) is attached to the tip of the glass filament (50 μm diameter), from which the drop is suspended.

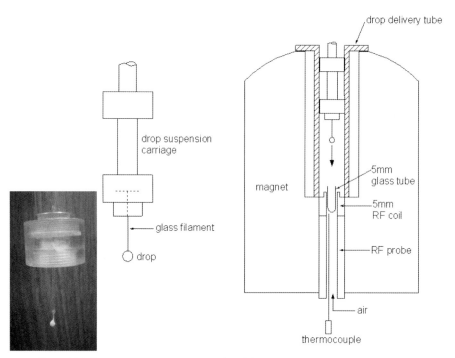

**Figure 29.2** Drop suspension carriage used to deliver a single drop into the center of the RF coil where it was frozen in cold air. A photograph of a ~2 mm-diameter suspended drop is also shown.

Drops have been formed in a range of sizes from 2 to 10 μl and were composed of water, sucrose solution, sucrose solution plus oil emulsion (model ice cream), or coffee solution. The bulk of the data presented in this chapter will, however, consider a 20 wt% sucrose solution drop, with or without oil droplets (1–10 μm diameter) forming an emulsion, and with a diameter of ~2 mm. A variety of MR protocols were employed, including rapid spectroscopy, rapid imaging, chemical shift imaging, emulsion droplet sizing based on pulsed-field gradient (PFG) measurements, restricted self-diffusion based on PFG measurements and rapid velocimetry/dispersion measurement. Further details of each method are contained in the relevant results sections below.

## 29.4
## Freezing Kinetics

In order to sequentially image the freezing drops, fast-imaging protocols are required due to their rapid solidification, as shown in Figure 29.1. All 2-D imaging of the drop-freezing process was conducted using the RARE (rapid acquisition with relaxation enhancement) pulse sequence [21]. 2-D projected images (i.e. no soft pulse/slice selection was used) were acquired in the vertical direction parallel to the axis of the drop, at an isotropic spatial resolution of typically 32 μm and a repetition time of 0.5 s during stage (iii) recalescence and of 5 s during stage (iv) heat transfer controlled freezing. By way of example, Figure 29.3a–f shows a selected series of images during the freezing of a 20 wt% sucrose solution drop at 0, 5, 35, 80, 145 and 205 s after nucleation in a −8 °C air stream. The signal is received only from the unfrozen sucrose solution, and the position of the suspending capillary and gradual dendritic ice growth are evident. It is possible to extract out the temporal variation in the total signal intensity from these images; this is presented in Figure 29.3g. The recalescence stage (iii) is distinctly evident as a rapid decrease in signal following nucleation due to rapid ice growth.

Such imaging protocols can highlight where unexpected drop-freezing occurs due to heterogeneities in either the drop structure or concentration distribution. Such a system is presented in Figure 29.4a, which shows images of a freezing 20 wt% dodecane-in-water emulsion drop where 'compartments' are observed to form and freeze separately; this effect was reproducible. Figure 29.4b shows phase stratification within a 20 wt% dodecane in sucrose (20 wt%) solution emulsion drop, and consequently different freezing profiles for the different layers. Increasing the surfactant (Tween 20) concentration from 0.1 wt% to in excess of 0.3 wt% eliminated the heterogeneity in both cases.

Whilst such rapid imaging can provide us with a qualitative picture of freezing, and certainly can indicate a heterogeneous or poorly formed drop, it is difficult to make the images quantitative due to relaxation effects and the need to use a long echo time during the required rapid image acquisition. This makes the procedure of limited use for drop-freezing model development and validation. Quantitative data for model comparison is, however, accessible using simple time-resolved $^1$H

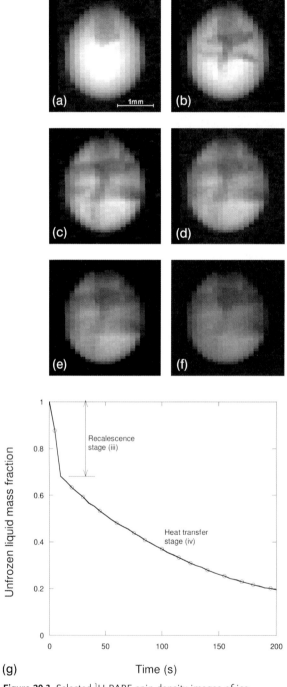

**Figure 29.3** Selected ¹H RARE spin density images of ice growth in a freezing 20wt% sucrose solution drop at the following times: (a) 0 s, (b) 5 s, (c) 35 s, (d) 80 s, (e) 145 s and (f) 205 s. The integrated total signal variation with time is shown in (g). The drop was ~2 mm in diameter and the resolution was 62 μm [22].

(a)

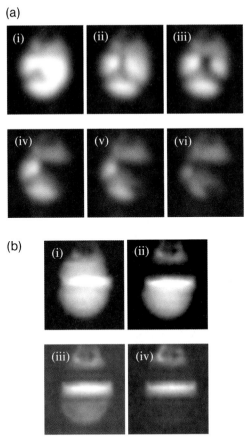

(b)

**Figure 29.4** (a) ¹H spin-density images of a 20 wt% dodecane/water drop freezing in still air at −15 °C Time of frames: (i) 0 s, (ii) 5 s, (iii) 30 s, (iv) 60 s, (v) 75 s and (vi) 100 s; (b) ¹H spin-density images of a 20 wt% dodecane/20 wt% sucrose solution drop freezing in still air at −15 °C. Time of frames: (i) 0 s, (ii) 5 s, (iii) 85 s and (iv) 200 s. The drops were ~2 mm in diameter and the resolution was 62 μm. The images have been interpolated to aid visualization.

spectroscopy. Free induction decay (FID) signal acquisition is started at 20 μs after excitation, effectively eliminating any contributions from the solid ice component, and the effect of temperature on the $T_1$ spin–lattice relaxation is accounted for, as detailed in Ref. [20]. Sample data are presented in Figure 29.5 for a 20 wt% dodecane in sucrose (20 wt%) solution emulsion droplet frozen in −40 °C air. Chemical shift differences allow separate spectral peaks due to water (4.6 ppm), sucrose (3–4 ppm) and dodecane (1–2 ppm) to be assigned. Following nucleation and freezing of the water, there is a sharp decrease in the sucrose signal at approximately 70 s. The normal glass transition temperature for saturated sucrose solutions is reported as −32 °C [24]. This sucrose signal reduction is accompanied by a reduction of the freezing rate of the water, as indicated by the arrow. At an even later time, the

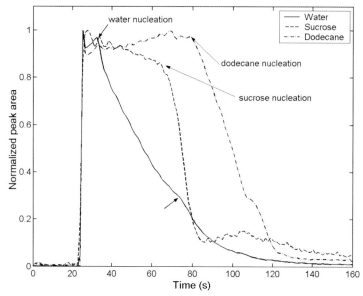

**Figure 29.5** Time evolution of the normalized NMR peak area of water, sucrose and dodecane peaks during the freezing of a drop of 20 wt% dodecane-in-20 wt% sucrose solution after being plunged into air at −40 °C [23].

dodecane droplets (1–10 μm in diameter) are observed to freeze (normal freezing point of −9.6 °C). Their smaller size presumably reduces the chance of a nucleation event occurring; the shape of the freezing curve for the dodecane will be a function of the droplet size distribution. The freezing of this 20 wt% dodecane in sucrose (20 wt%) solution emulsion was studied in this manner as a function of air temperature and the result data are collectively shown in Figure 29.6. The freezing rate and frozen fraction of all constituents clearly increases with lower air temperatures. The arrows indicate the reduction in water freezing rate, observed in Figure 29.5, at the range of temperatures considered.

## 29.5
## Modeling Drop Freezing

Desirable for any attempt to effectively model or simulate the freezing profiles of the drops is an understanding of mass transfer and motion within them. Rapid 1-D MR phase shift velocimetry across the diameter of the drop is possible; such data are presented in Figure 29.7 for a freezing water drop in −15 °C air. These data were acquired using the 13-interval APGSTE sequence [26] with the addition of a 1-D imaging gradient, producing a resolution of 391 μm. In Figure 29.7, the drop is observed initially to experience a solid-body rotation around the axis

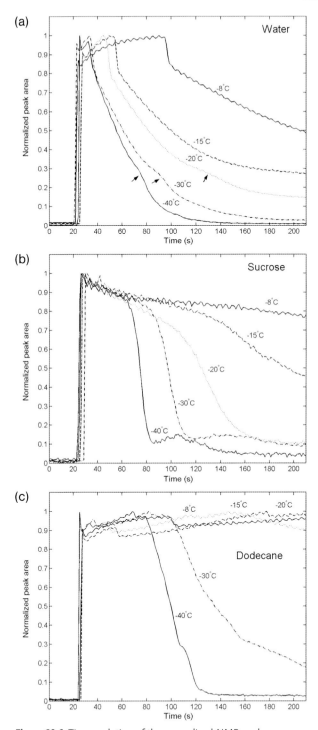

**Figure 29.6** Time evolution of the normalized NMR peak area during freezing a drop of 20 wt% dodecane-in-20 wt% sucrose solution emulsion at a range of air temperatures. (a) Water peak; (b) Sucrose peak; (c) Dodecane peak [23].

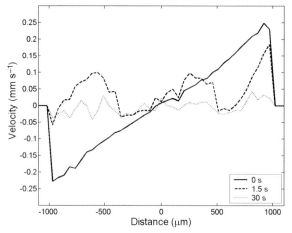

**Figure 29.7** The *x*-direction (out of the page) velocity profile in a slice across the center of a 2 mm-diameter water drop. The times refer to the elapsed time after introduction of the drop [25].

through the suspending capillary (presumably due to slightly uneven surface shear). Rapid chaotic motion is observed during nucleation/recalescence (at 1.5 s after introducing the drop; at stage ii/iii); this however rapidly reduces to below the detection threshold during heat transfer controlled freezing (stage iv). Such data are compromised however by the need to relate the 'average' phase shift in the acquired signal to a mean velocity (which is not exact for a nonsymmetric velocity distribution) as well as the effect of acceleration on the measured phase shift.

Consequently, 'dispersion' measurements were made during the freezing process using the 13-interval APGSTE pulse sequence. Signal attenuation was measured at two values of applied gradient strength ($g$ = 2 and 10 G cm$^{-1}$), a gradient duration ($\delta$) of 2 ms, and an observation time ($\Delta$) of 100 ms. A recycle time of 1.5 s gave a total measurement time of 3 s. The time-resolved dispersion coefficient, $D$, was then extracted using the conventional Stejskal–Tanner equation [27]. In the context of its application to drops, $D$ indicates the variance in molecular displacement due to both velocity variations and self-diffusion; in the absence of internal drop flow, it will tend to $D_0$, the self-diffusion coefficient of the unfrozen solution at the relevant temperature.

Figure 29.8a presents the value of $D$ measured as a function of time for a water drop frozen at −15 °C. The dashed line indicates the value of $D_0$ for water at 0 °C. After the drop had been introduced the value of $D$ decreased with time, which suggests minimal internal motion and simply that $D_0$ was reducing in response to cooling. Following nucleation and during nucleation, however, there was a marked increase in the value of $D$, indicating a significant flow of material within the drop. However, this rapidly dampened out as the value of $D_0$ was measured

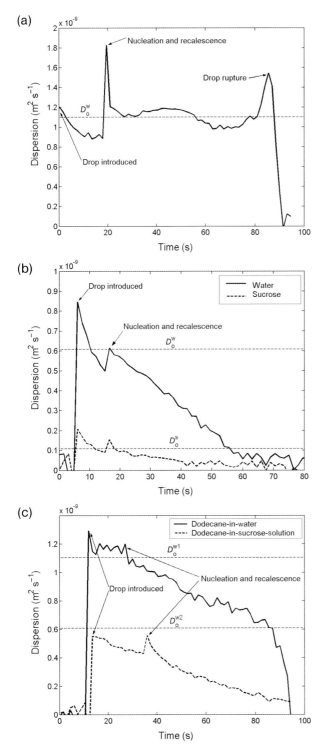

**Figure 29.8** Temporal evolution of the dispersion coefficient
(D) for drops composed of (a) water, (b) sucrose solution
and (c) dodecane emulsions. The dashed lines indicate the
value of $D_0$ for the initial concentration at the equilibrium
freezing temperature [25].

for the remainder of freezing, indicating minimal internal motion. Rupture of the drop (due to expulsion of internal air) naturally resulted in a sudden increase in $D$. The corresponding data are presented in Figure 29.8b and c for sucrose solution and dodecane emulsion drops, respectively. The effect of freeze concentration is evident by the observed reduction in $D$ over time, which is consistently lower than the value of $D_0$ relevant to the initial concentration at its equilibrium temperature (indicated by the dashed lines). Thus, no evidence was presented that suggested any significant motion within the drop following recalescence.

The amount of ice, $f_s$, formed during recalescence can be estimated as follows:

$$f_s = \frac{C_{pl}\rho_l(\Delta T_s)}{\rho_s \Delta H_{fus}}$$

(29.1)

where $C_{pl}$ and $\Delta H_{fus}$ are the liquid heat capacity and latent heat of fusion, respectively, $\Delta T_s$ is the degree of supercooling, and $\rho$ is density. Equation 29.1 consists of an energy balance between the sensible heat of supercooling and the latent heat release required by ice formation [20]. Figure 29.9 shows the suitability of this energy balance to predict $f_s$ (as measured using NMR spectroscopy) for water droplets which have nucleated (randomly) at a range of temperatures. The agreement is reasonable.

Modeling of the combined recalescence and heat transfer controlled freezing stages is shown in Figure 29.10 for (a) water and (b) 20 wt% sucrose solution drops. The model is entirely predictive (i.e. it contains no free parameters) and considers

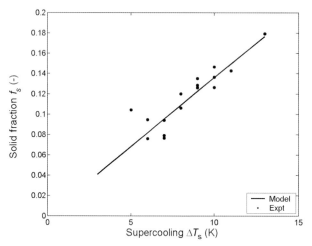

**Figure 29.9** Solid fraction, $f_s$, formed from recalescence of water drops for different degrees of supercooling. The solid line indicates the prediction of Equation 29.1. Agreement is good [28].

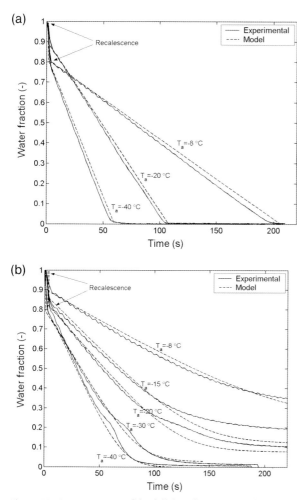

**Figure 29.10** Comparison of the full drop-freezing model with experimental data with respect to the time-resolved unfrozen fraction for drops composed of (a) water and (b) 20 wt% sucrose solution drops [20].

heat and mass transfer (evaporation) as well as radiation; further details can be accessed from Ref. [20]. Agreement with the experimental data for water in Figure 29.10a is excellent. However, there are reasonably small discrepancies with the experimental data for the sucrose solution in Figure 29.10b which we believe are a consequence of several model simplifications: (i) including the assumption that recalescence is instantaneous; (ii) not including the heat required for any sucrose phase transformation; and (iii) not considering mass transfer limitation within the drop as freeze concentration occurs.

## 29.6
## Drop Microstructure

Cryo-scanning electron microscopy (SEM) images reveal skin formation for frozen sucrose solution drops which have been fractured to reveal their internal microstructure [29]. A sample image is presented in Figure 29.11a. 1-D chemical shift profiling of the *in-situ* frozen drop was used to measure the composition of this skin layer. The resultant data are presented in Figure 29.11b and c, which correspond to before and after freezing of a 20 wt% sucrose solution drop, respectively. The uniform sucrose/water ratio before freezing is replaced by surface regions which are rich in sucrose, thus confirming the fact that the surface layer corresponds to a concentrated sucrose solution. This surface layer is believed to be due to liquid exclusion upon ice formation.

By using PFG methods it is possible to determine droplet size distributions for emulsions based on an interpretation of the restricted diffusion behavior of the liquid molecules in the discrete droplet phase. A review of this technique was presented by Johns and Hollingsworth [30]. Figure 29.12a shows the droplet size distribution for the inner water droplets within a dodecane-in-sucrose solution emulsion and a dodecane-in-water emulsion (Figure 29.12b), frozen at a range of air temperatures. What is immediately obvious is that the water emulsion results in significant droplet coalescence upon freezing, the extent of which decreases as freezing occurs more rapidly at lower temperatures. Regarding the sucrose solution emulsions, there is minimal change in the droplet size distribution, the partially frozen sample and the concentrated sucrose solution have cushioned the droplets and reduced their coalescence. It is also possible to follow the evolution in droplet size distribution as a function of freeze–thaw cycles – such data is presented in Hindmarsh *et al.* [23].

PFG methods can also be used to quantify the internal microstructure of porous materials [31]. The restricted diffusion experienced by the pore space liquid can be interpreted so as to quantify the surface-to-volume (S/V) ratio and tortuosity of the confining pore walls. Such an approach can be applied to the partially frozen sucrose solution drops, and relies on the restricted self-diffusion of the unfrozen concentrated sucrose solution content. Potential errors in such an approach include exchange between the ice and the unfrozen water, and also significant surface signal relaxation at the ice–fluid interface. To assess the influence of these factors, the data acquired were compared directly to random walk diffusion simulations in the unfrozen pore space, the required lattice being provided by 2-D cryo-SEM images of fracture planes through the drops. The resultant data are presented in Figure 29.13a (along with the binary gated fracture planes used in the simulations). The acquired experimental data consist of the apparent diffusion coefficient (determined as $\mathbf{q}$ tends to zero using the conventional Stajskal–Tanner equation) against the observation time of diffusion ($\Delta$). This is presented for 20 wt% sucrose solution drops frozen at various temperatures. The equivalent data produced by the simulations is also shown, and was extracted from the simulated displacement propagators with $D$ being determined using the following version

(a)

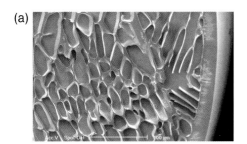

(b)

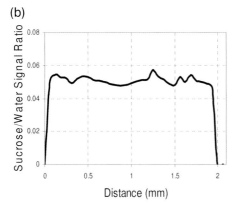

(c)

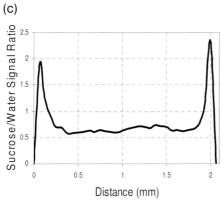

**Figure 29.11** (a) Fracture plane of a 20 wt% sucrose solution drop, as imaged using cryo-SEM; a skin layer is evident. (b,c) 1-D chemical shift imaging enables the relative concentration of sucrose to water to be determined as a function of position across the drop before (b) and after (c) freezing of the drop. An increased sucrose content of the surface layer is readily detected [22].

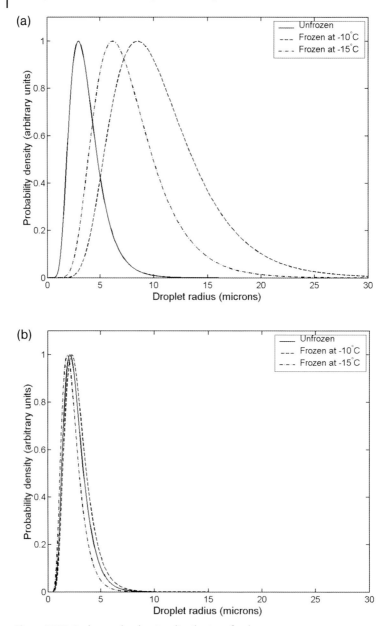

**Figure 29.12** Dodecane droplet size distributions for drops composed of (a) water and (b) sucrose solution as a function of freezing temperature [23].

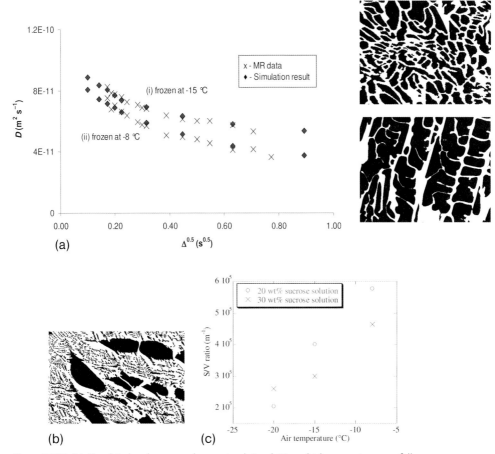

(a)

(b)

(c)

**Figure 29.13** (a) Simulated and measured values of $D$ as a function of the observation time, $\Delta$. The lattices used in the simulation are also shown; (b) An example of a simulation lattice which was not successfully simulated; (c) Variation in S/V ratio for sucrose solution drops, as measured using PFG.

of the Stajskal–Tanner equation as $\delta$ effectively equals 0 in the simulations conducted:

$$\frac{S}{S_0} = \exp\left(-(\gamma g \delta)^2 D\Delta\right) \tag{29.2}$$

The simulations are averaged over the various directions to produce an isotropic result. In Figure 29.13a, there is excellent agreement between simulation and experiment, which suggests that the 2-D lattices extracted from the fracture planes are representative of the full 3-D structure. This excellent agreement was generally replicated for a wider range of temperatures and concentrations, except at those equal to or exceeding 30 wt% sucrose. Figure 29.13b shows a fracture plane through a 30 wt% sucrose solution drop; it is plausible that the very fine ice structure observed in this plane is not adequately captured in the sample preparation and imaging or represented in the simulations conducted. Variations in the S/V ratio of the internal pore space, as measured using PFG, are presented in Figure 29.13c as a function of air temperature.

## 29.7
## Conclusions and Future Potential

By using a wide range of NMR and MRI techniques, we have been able to extract valuable and unique information with respect to freezing drops. In particular, we were able to track the kinetics of solidification of various drop constituents (hence requiring the use of rapid NMR/MRI techniques) and to describe the microstructure of the final product.

Although the studies presented here have focused on sucrose solution (plus oil) compositions, we have also applied the range of techniques to coffee solutions [32], which represents another potential product for spray-freezing. Future applications might also be in biotechnology for the preservation of various biological materials, as the droplet-sizing technique used to monitor coalescence in freezing emulsion drops could be used productively in this context to monitor cell survival rates.

Some of the NMR/MRI techniques developed, applied and demonstrated in this chapter have recently also been applied to conventional spray-drying, which is a much more widely established technology (e.g. for detergents [33]). This includes the rapid evolution in microstructure monitoring during drying, as probed using Difftrain, a pulse sequence capable of the rapid measurement of diffusion [34].

## Acknowledgments

The authors thank EPRSC for funding, and Mr Richard Thowig for the simulations of S/V ratios in fracture planes.

## References

1 Evans, D.S., Nott, K.P., Kshirsagar, A.A. and Hall, L.D. (1998) *International Journal of Food Science and Technology*, 33, 317–28.

2 Nott, K.P., Evans, S.D. and Hall, L.D. (1999) *Magnetic Resonance Imaging*, 17 (3), 445–55.

3 Renou, J.P., Foucat, L. and Bonny, J.M. (2003) *Food Chemistry*, 82, 35–9.

4 Kuo, M.-I., Anderson, M.E. and Gunasekaran, S. (2003) *Journal of Dairy Science*, 86, 2525–36.

5 Lucas, T., Grenier, A., Quellec, S., Le Bail, A. and Davenel, A. (2005) *Journal of Food Engineering*, 71, 98–108.

6 Borompichaichartkul, C., Moran, G., Srzednicki, G. and Price, W.S. (2005) *Journal of Food Engineering*, 69, 199–205.

7 Hills, B.P. and Remigereau, B. (1997) *International Journal of Food Science and Technology*, **32** (1), 51–61.

8 Menzel, M.I., Han, S.I., Stapf, S. and Blümich, B. (2000) *Journal of Magnetic Resonance*, **143**, 376–81.

9 Callaghan, P.T., Eccles, C.D., Haskell, T.G., Langhorne, P.J. and Seymour, J.D. (1998) *Journal of Magnetic Resonance*, **133** (1), 148–54.

10 Mercier, O.R., Hunter, M.W. and Callaghan, P.T. (2005) *Cold Regions Science and Technology*, **42** (2), 96–105.

11 Aussillous, P., Sederman, A.J., Gladden, L.F., Huppert, H.E. and Worster, M.G. (2006) *Journal of Fluid Mechanics*, **552**, 99–125.

12 Georgiadis, J.G. and Ramaswamy, M. (2005) *International Journal of Heat and Mass Transfer*, **48**, 1064–75.

13 Ozeki, T., Kose, K., Haishic, T., Nakadsubod, S., Nishimurad, K. and Hochikuboe, A. (2003) *Cold Regions Science and Technology*, **37**, 385–91.

14 Prado, P.J., Balcom, B.J., Beyea, S.D., Bremner, T.W., Armstrong, R.L. and Grattan-Bellew, P.E. (1998) *Cement and Concrete Research*, **28** (2), 261–70.

15 Strange, J.H. and Webber, J.B.W. (1997) *Measurement Science and Technology*, **8** (5), 555–61.

16 Rubinsky, B., Hong, J.S. and Storey, K.B. (1994) *American Journal of Physiology-Regulatory Integrative and Comparative Physiology*, **36** (4), R1078–88.

17 Windhab, E.J. (1999) *Journal of Thermal Analysis and Calorimetry*, **57**, 171–80.

18 Leuenberger, H. (2001) Business briefing pharmatech 2001: new technologies for the manufacture of nano-structured drug carriers, Report for World Markets Research Centre, (WMRC).

19 Fields, S.D., Strout, G.W. and Russell, S.D. (1993) Proceedings–51st Annual Meeting, Microscopy Society of America, Cincinnati, Ohio, pp. 134–5.

20 Hindmarsh, J.P., Wilson, D.I., Johns, M.L., Russell, A.B. and Chen, X.D. (2005) *AIChE Journal*, **51** (10), 2640–8.

21 Hennig, J., Nauerth, A. and Friedburg, H. (1986) *Magnetic Resonance in Medicine*, **3**, 823–33.

22 Hindmarsh, J.P., Buckley, C., Russell, A.B., Chen, X.D., Gladden, L.F., Wilson, D.I. and Johns, M.L. (2004) *Chemical Engineering Science*, **59** (10), 2113–22.

23 Hindmarsh, J.P., Hollingsworth, K.G., Wilson, D.I. and Johns, M.L. (2004) *Journal of Colloid and Interface Science*, **275** (1), 165–71.

24 Sahagian, M. and Goff, H.D. (1994) *Thermochimica Acta*, **246**, 271–83.

25 Hindmarsh, J.P., Sederman, A.J., Gladden, L.F., Wilson, D.I. and Johns, M.L. (2005) *Experiments in Fluids*, **38** (6), 750–8.

26 Cotts, R.M., Hoch, M.J.R., Sun, T. and Markert, J.T. (1989) *Journal of Magnetic Resonance*, **83**, 252–66.

27 Stejskal, E.O. and Tanner, J.E. (1965) *The Journal of Chemical Physics*, **42**, 288–92.

28 Hindmarsh, J.P., Wilson, D.I. and Johns, M.L. (2005) *International Journal of Heat and Mass Transfer*, **48** (5), 1017–21.

29 Hindmarsh, J.P., Russell, A.B. and Chen, X.D. (2007) *Journal of Food Engineering*, **78** (1), 136–50.

30 Johns, M.L. and Hollingsworth, K.G. (2007) *Progress in Nuclear Magnetic Resonance Spectroscopy*, **50** (2–3), 51–70.

31 Mitra, P.P., Sen, P.N. and Schwartz, L.M. (1993) *Physical Review B*, **47**, 8565–74.

32 MacLeod, C.S., McKittrick, J.A., Hindmarsh, J.P., Johns, M.L. and Wilson, D.I. (2006) *Journal of Food Engineering*, **74** (4), 451–61.

33 Griffith, J.D., Bayly, A.E. and Johns, M.L. (2007) *Journal of Colloid and Interface Science*, **315** (1), 223–9.

34 Davies, C.J., Griffith, J.D., Sederman, A.J., Gladden, L.F. and Johns, M.L. (2007) *Journal of Magnetic Resonance*, **187** (1), 170–5.

**Part Six    Hardware**

# 30
# High-Performance Shimming with Permanent Magnets

*Ernesto Danieli, Juan Perlo, Federico Casanova and Bernhard Blümich*

## 30.1
## Introduction

Portable open NMR probes built from permanent magnets offer several advantages over conventional NMR systems. For example, they can be used to study arbitrarily large samples, are small, inexpensive, and robust [1, 2]. However, the magnetic field generated by open magnets is believed to be inherently inhomogeneous, precluding the acquisition of chemical shift-resolved NMR spectra. In the absence of chemical shift resolution, sample composition can only be determined by measuring indirect parameters, such as signal amplitudes, relaxations times or molecular self-diffusion coefficients [1–3].

Although several attempts to recover high-resolution spectra in inhomogeneous magnetic fields have been reported [4–7], these unfortunately rely on intramolecular interactions (which are not necessarily present in general) and become inadequate when the main magnetic field is relatively low. A breakthrough in the development of methodologies for NMR spectroscopy in inhomogeneous magnetic fields was the discovery that nutation echoes, which are generated by combining nutation of the magnetization in the $B_1$ field with precession in the $B_0$ field, preserves the chemical shift information [8]. By taking advantage of what nowadays is called the *ex situ* methodology [8–10], it was possible to recover spectroscopic information by matching the inhomogeneous $B_0$ field generated by an open magnet with the similarly inhomogeneous field generated by a surface radiofrequency (RF) coil [11]. The important step towards achieving the high precision in the matching condition was the realization that permanent magnets, which are usually considered to be rigid units, could be equipped with small movable magnet blocks to introduce control variables that would shape the spatial dependence of the static magnetic field. Although this approach was used by one group to shim the static field to a desired spatial dependence (that of the RF field) [11], it led us to consider the possibility of using the same approach to shim the field to high homogeneity. In this chapter we present details of the magnet components, as well as the functioning of the shim unit, by which we could obtain sub-ppm

*Magnetic Resonance Microscopy.* Edited by Sarah L. Codd and Joseph D. Seymour
Copyright © 2009 WILEY-VCH Verlag GmbH & Co. KGaA, Weinheim
ISBN: 978-3-527-32008-0

homogeneity outside the magnet, such that $^1$H chemical shift-resolved spectra could be measured for the determination of chemical structures of dissolved molecules [12].

## 30.2
## Shimming Magnetic Fields with Permanent Magnets

A standard and robust approach to adjust the magnetic field for high resolution in conventional NMR magnets uses shim coils to generate fields with known spatial dependences [13]. The straightforward adaptation of this approach to shim the field inhomogeneities generated by open magnets [14–17] must be discarded simply because of excessive requirements for the shim currents. However, we found that a current loop of 1000 A could be replaced by an approximately 1 mm-thick NdFeB permanent magnet block, and single-sided shimming could be achieved by a suitable arrangement of such blocks. Those magnets have a permanent polarization, providing on the one hand the equivalent of a 1000 A direct current (DC) source without a power supply, although on the other hand the current value could be changed, as in conventional shimming. Nevertheless, adjustable shim fields can be generated by the proper movement of such magnet blocks.

At this point, before entering a more detailed description of the shim hardware, the concept will be outlined. First, the magnet is composed of two parts or units, with the main unit generating the main magnetic field and the shim unit being used to correct the inhomogeneity of the main field. Second, the shim unit is specially designed for a given geometry of the main magnet and will work only in combination with it. This means that a detailed knowledge of the main magnetic field properties is required for designing the shim unit. The approach of having a universal shim unit with $N$ movable magnets blocks to generate several shim field components will not necessarily work, even for rather large $N$. Finally, the positions of the magnet blocks forming the shim unit need to be adjustable because of unavoidable inaccuracies in the polarization, size and positioning of the magnet pieces. Small displacements of the magnet blocks around their optimum positions allow one to reach the performance experimentally that is calculated numerically.

### 30.2.1
### The Main Unit

In these studies, the main unit is a U-shaped, single-sided magnet consisting of two permanent magnet blocks with anti-parallel polarization placed on an iron yoke with a gap between them [18, 19]. The direction along the gap is called $x$, the direction along the depth $y$, and $z$ is the direction across the gap. All distances are measured from a reference system the origin of which coincides with the geometric center of the magnet in an $xz$ plane placed at the magnet surface. The region of interest is a slice parallel to the magnet surface located at a depth $y_0$, with the size of the slice and the depth $y_0$ small compared to the magnet gap. In good approximation, the magnetic field in this region can be considered to be oriented along the $z$ axis and has its dominant gradient component along the depth direction. Due to the symmetry of the magnet along $z$, the spatial dependence of the field along this direction has only even terms in its Taylor expansion. As a function of position the field is lower at the center of the magnet ($z = 0$) and increases as the permanent magnets blocks are approached. This is mainly due to the inverse dependence of the field on the distance to the magnetic source. This suggests that the field is shaped like an 'arms-up' parabola centered at $z = 0$. Moreover, also due to symmetry the field dependence along $x$ must be an even function of $x$. However, in this case the field decreases when moving away of the center, reflecting the finite size of the magnet along $x$. Therefore, the field behaves like an 'arms-down' parabola along $x$. This spatial dependence gives rise to the well-known horse-saddle shape of the magnetic field in the $xz$ plane. The field shape described above can formally be expressed by expanding the magnitude of the magnetic field $B_0(\vec{r})$, around $\vec{r}_0 = (0, y_0, 0)$ as

$$B_0(\vec{r}) = B_{00} + G_y(y_0)(y - y_0) + \alpha_z(y_0)z^2 + \alpha_x(y_0)x^2 + \dots, \qquad (30.1)$$

where $B_{00} = B_0(\vec{r}_0)$, $G_y(y_0) = \left.\dfrac{\partial B_0(\vec{r})}{\partial y}\right|_{r_0}$, and $\alpha_k(y) = \left.\dfrac{\partial^2 B_0(\vec{r})}{\partial k^2}\right|_{r_0}$, with $k = x, z$.

Some specifications of the set-up used in Ref. [12] are the following: at a depth $y_0 \sim 5\,mm$, the average field is around $B_{00} \sim 0.25\,T$ (10 MHz $^1$H Larmor frequency), and the inhomogeneities are characterized by $G_y \approx -1000\,ppm\,mm^{-1}$, $\alpha_z \approx 300\,ppm\,mm^{-2}$, and $\alpha_x \approx -30\,ppm\,mm^{-2}$. Those data reveal that the main source of inhomogeneity is the gradient along the depth direction, and is the reason why a slice in the $xz$ plane was chosen as the region of interest for shimming. The size of the volume where field homogeneity of 1 ppm is obtained (no shimmed situation) is about $180 \times 60\,\mu m^2$ along the lateral directions and $1\,\mu m$ thick along the depth direction. These dimensions provide some insight into the magnitude of the inhomogeneities associated with single-sided magnets.

## 30.2.2
### The Shim Unit

The field generated by the shim unit must reproduce the spatial dependence of the main field having, in principle, the smallest average field strength as possible. By setting the polarization of the shim unit opposite to that of the main field, the inhomogeneities of the main field are corrected while the total field strength is maintained at an acceptable magnitude. Basically, the shim unit follows the same geometry as the main magnet—in this case a U-shaped geometry—but with different aspect ratio and size (Figure 30.1a).

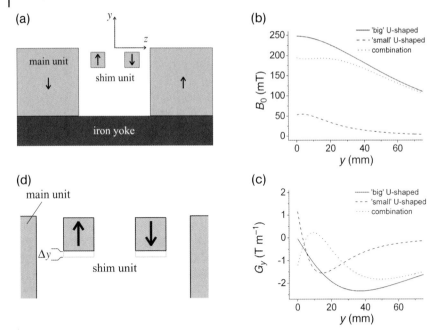

**Figure 30.1** (a) Sensor array: The main unit is a conventional U-shaped magnet [18, 19], while the shim unit consists of a pair of magnet blocks also in a U-shape configuration, but without the iron yoke. The arrows show the direction of polarization of the magnets; (b) Dependence of the magnetic field strength (magnitude) for the main and the shim units, shown by solid and dashed lines, respectively. The dotted line shows the combination of both fields; (c) Magnetic field gradient along the depth direction corresponding to the fields of (b); (d) Schematic illustration of the shim unit displacement $\Delta y$ that generates the y shim component.

### 30.2.2.1 The y Shim Component

Equation 30.1 contains three coefficients, $G_y$, $\alpha_x$ and $\alpha_z$ that characterize the field inhomogeneities. Let us start by focusing on how to cancel the gradient $G_y$ along the depth direction. Figure 30.1b shows the field strength as a function of depth for the main magnet ('big' U-shaped) and the shim unit ('small' U-shaped). For each unit the field has a maximum along the depth direction that depends on the dimensions of the permanent magnets and the gap. These variables determine the position $y_{max}$ for the maximum field of each unit. An increase in the blocks size shifts the maximum closer to the surface, while an increase in the gap does the opposite. The use of an iron yoke has mostly two effects: an increment of the field strength; and a displacement of the maximum to smaller depths. $G_y$ is positive for $y < y_{max}$, and becomes negative for $y > y_{max}$, (Figure 30.1c). The main unit has the field maximum at the surface of the magnet, which explains why $G_y^{main}$ does not assume positive values in the depth range shown in the figure. However, $G_y^{shim}$ takes positive values for $y < 3\,mm$ and becomes negative for $y > 3\,mm$. At $y_0 \sim 5\,mm$ the gradients of both fields match. When both units are brought together and their polarizations are in opposite directions, the gradient of the total field becomes zero at this particular depth. Therefore

$$G_y(y_0) = G_y^{main}(y_0) + G_y^{shim}(y_0) = 0. \tag{30.2}$$

Following this idea, one can design a sensor with no linear variation of the magnetic field along the depth direction at a particular position $y_0$ [14]. Although the combination of two nested units can be used in a numerical simulation to generate a so-called 'sweet spot',[1)] the crucial step towards matching the gradient of both units at one depth in a real magnet requires a variable to control the field dependence of at least one of the units. Such control can be acquired by considering the movement of the whole shim unit along the depth direction (Figure 30.1d). When the shim unit is displaced a distance $\Delta y$ in the positive $y$ direction, the dashed line in Figure 30.1c shifts to the right. Note that the reference system is fixed to the main magnet. Thus, at the position $y_0$ the gradient generated by the shim unit is smaller in magnitude than that of the main magnet, when the gradient of the total field becomes negative. On the other hand, if the shim unit is displaced by a distance $\Delta y$ in the negative $y$ direction, the dashed line in Figure 30.1c shifts to the left. In this case, at position $y_0$ the gradient of the shim unit dominates over that of the main unit, so that the gradient of the total field becomes positive. In this way the $y$ shim component can be generated and its amplitude is proportional to $\Delta y$, the variable of control.

### 30.2.2.2 Generation of the Shim Component $z^2$

The second largest source of inhomogeneity is related to the field variation across the gap direction, which is characterized by $\alpha_z$. The field generated by the shim unit possesses the same spatial dependence as that generated by the main magnet, and can also be characterized by a coefficient $\alpha_z^{shim}$, which generally does not match $\alpha_z^{main}$. Let us consider the shim unit introduced above, which was designed to cancel the main gradient at a depth $y_0$, and assume that at this particular depth $|\alpha_z^{shim}| > \alpha_z^{main}$. If when starting in these conditions the size of the shim unit is reduced, the gradient along the depth $|G_y^{shim}|$ becomes weaker. But then, if the gap is decreased, $|G_y^{shim}|$ increases until the value required to cancel the gradient along $z$ of the main unit is reached (Figure 30.2a). Such a modification leaves $G_y^{shim}$ unchanged but leads to a smaller $|\alpha_z^{shim}|$ with a lower curvature than the main field (Figure 30.2b).

Hence, there must be an intermediate shim unit that fulfils $|\alpha_z^{shim}| = \alpha_z^{main}$. Then, as the polarizations of the main and shim units are set in opposite directions, the quadratic coefficients along $z$ cancel each other. If the initial condition is $|\alpha_z^{shim}| < \alpha_z^{main}$ the same procedure should be followed, but in the reverse direction. Figure 30.2 also shows that it is possible to generate a shim field

$$\alpha_z(y_0) = \alpha_z^{main}(y_0) + \alpha_z^{shim}(y_0) = 0. \tag{30.3}$$

---

**1)** A 'sweet spot' is called the point in space where all first-order derivatives of the magnetic field magnitude become zero.

(a)

main unit

shim unit

(b)

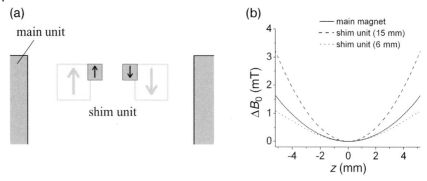

Figure 30.2 (a) Two shim unit configurations generating at $y_0 = 5\,mm$ a gradient $G_y^{shim}$ that fulfils Equation 30.2. The large unit (light gray) is $15 \times 15\,mm^2$ in size, and has a gap $d_z$ of 40 mm. The small unit is $6 \times 6\,mm^2$ in size and has a gap of 30 mm; (b) Dependence of magnetic field strength (magnitude) as a function of the lateral direction $z$ for the two shim units and the main magnet. The offset field value at $z = 0$ was subtracted in order to facilitate the comparison.

Once the optimum dimensions of the shim magnets are found in the simulations, control of the $z^2$ component in a real magnet is achieved by varying the gap of the shim unit. In this way, an increment of the gap size leads to a positive $z^2$ shim field $(|\alpha_z^{shim}| < \alpha_z^{main})$, while reducing it leads to a negative $z^2$ shim field $(|\alpha_z^{shim}| > \alpha_z^{main})$. It must be mentioned that a change in the gap value leads to a small, but undesired variation of $G_y^{shim}$, which needs to be readjusted by varying the position of the shim unit along $y$. This repositioning slightly modifies the $z^2$ term, thus requiring an iterative procedure to cancel the field variations along both $y$ and $z$ axes.

### 30.2.2.3 Generation of the Shim Component $x^2$

The last coefficient to be canceled in Equation 30.1 is $\alpha_x$. Although this coefficient is at least one order of magnitude smaller than $\alpha_z$ it must be canceled while keeping the values $G_y^{shim}$ and $\alpha_z^{shim}$ constant. Fortunately, it is possible to vary $\alpha_x^{shim}$ over quite a large range almost without modifying the other two coefficients. By changing the length of the shim unit designed to cancel $G_y^{main}$ and $\alpha_z^{main}$ along $x$, the condition $\alpha_x^{shim} = |\alpha_x^{main}|$ can be achieved. Considering once again the fact that the polarization of the main and the shim units opposes, we can write

$$\alpha_x(y_0) = \alpha_x^{main}(y_0) + \alpha_x^{shim}(y_0) = 0. \tag{30.4}$$

The natural variable of control of the shim term $x^2$ is the length of the magnet blocks. However, it cannot be taken as a real variable, unless a very large collection of blocks with different lengths would be required at the time of shimming the magnet. Another way of controlling the $x^2$ coefficient is by splitting the blocks, generating a second gap $d_x$ along the $x$ direction (Figure 30.3). For small increments of $d_x$ the effect on the field is equivalent to incrementing the length of the

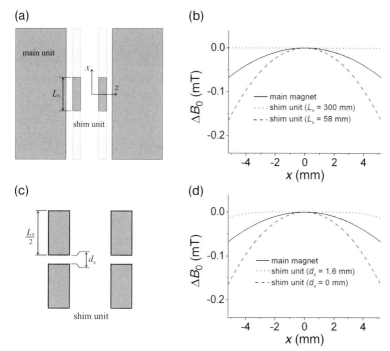

**Figure 30.3** (a) Top view of the sensor array of Figure 30.1a; (b) Dependence of magnetic field strength (magnitude) as a function of $x$ for the two different sizes $L_x$ of the shim unit, and the main magnet. The offset field value (at $x = 0$) was subtracted in each case to facilitate the comparison; (c) The shim unit is split into two identical pairs to introduce the gap $d_x$ along the $x$ direction; (d) Same as (b) but for two different values of $d_x$ with $L_x = 58$ mm. For both shim configurations: (a) with $L_x = 73$ mm, (c) with $L_x = 58$ mm and $d_x = 1$ mm, the field curvatures of the main and shim units are matched.

blocks. As $d_x$ cannot take negative values, it is important to design the shim unit to cover a sufficiently large range for $d_x$. The dependence of the shim term $x^2$ on $d_x$ is opposite to that one of $z^2$ on $d_z$. Thus, an increment of $d_x$ leads to a negative $x^2$ shim field and a reduction in the gap to a positive $x^2$ shim field.

Until now we have shown how to design a shim unit to compensate the field inhomogeneity of an ideal U-shaped magnet within a certain region around $\vec{r}_0 = (0, y_0, 0)$. Imperfections in the magnet blocks used to build either the main magnet or the shim unit are expected to lead to variations in the values of $G_y$, $\alpha_x$ and $\alpha_z$. Therefore, in a real magnet variables of control are required to adjust the shim terms $y$, $x^2$ and $z^2$, respectively. Although it is true that such imperfections unavoidably affect the mentioned field coefficients, they surely also introduce inhomogeneities not considered in the expansion of Equation 30.1. For instance, the symmetry arguments used to assume the dependence of the field to be an even function of $z$ and $x$ are no longer valid. Consider, for example, that one of the magnet blocks of the main unit has a higher polarization than the other; then, a constant gradient along $z$ will be observed. In general, any imperfection of the magnet pieces will introduce an asymmetry in the field that can be characterized (up to order two) by linear components $x$ and $z$, and by cross-terms $xy$, $xz$ and $yz$.

### 30.2.2.4 Generation of Linear Terms along x and z

The shim unit was designed to cancel the main field inhomogeneities described by Equation 30.1, leading to the requirements specified by Equations 30.2, 30.3 and 30.4. The same unit can be used to compensate linear field variations along $x$ and $z$. Let us consider the $z$ direction for the present analysis (a similar argument holds for the $x$ direction). A displacement $\Delta z$ of the whole shim unit leads to a total field given by

$$B_0(0, y_0, z) = B_0^{main}(0, y_0, z) + B_0^{shim}(0, y_0, z - \Delta z)$$

$$= B_{00}^{main} + \alpha_z^{main} z^2 + B_{00}^{shim} + \alpha_z^{main}(z - \Delta z)^2$$

$$= \left(B_{00}^{main} + B_{00}^{shim} + \alpha_z^{shim}\Delta z^2\right) + \left(\alpha_z^{main} + \alpha_z^{shim}\right)z^2 - \left(2\alpha_z^{shim}\Delta z\right)z. \quad (30.5)$$

The first term in brackets represent the magnitude of the total field independent of $z$ (spatially homogenous term). The second term in bracket vanishes if Equation 30.3 is fulfilled, and the last term is the desired shim field with a linear variation along $z$. The shim field coefficient depends on the magnitude and direction of the displacement $\Delta z$ enabling positive as well as negative $z$ corrections. In the same way, a linear term along $x$ is achieved by displacing the whole unit along the $x$ axis.

### 30.2.2.5 Generating Cross-Terms xy, xz and yz

In contrast to the generation of linear terms, cross-terms are obtained by moving the magnet block pairs in asymmetric ways. For example, while the generation of an $x$ term requires a movement of all four magnets along $x$, an $xz$ term is achieved by displacing the shim unit blocks that have opposite polarization in opposite directions along $x$ (in Figure 30.3c, the two blocks to the right are displaced towards the bottom and those to the left towards the top). The dependence of the field after an asymmetric displacement of the shim unit can be understood by considering the effect of each pair in the total field. As can be noted, while the magnet pair displaced along the positive $x$ direction (left pair) contributes with a $G_x^{left}$ that is negative and decreases (in magnitude) when moving from left to right, the pair moved along the negative $x$ axis (right pair) generates a $G_x^{right}$ that is positive and decreases when moving from right to left. By symmetry, both gradients match in amplitude at $z = 0$, canceling each other and leading to a $G_x = 0$ at this position. For $z > 0$, $|G_x^{right}| > |G_x^{left}|$ and a $G_x > 0$ is generated by the shim magnets. On the other hand, for $z < 0$, $|G_x^{right}| < |G_x^{left}|$ and the opposite behavior is obtained. This dependence of the magnetic field shows that in a first approximation the magnetic field is proportional to $xz$. The proportionality constant–or, in other words, our variable of control–is the value of the displacement along $x$. The other two cross-terms can be generated using the displacements described in Table 30.1.

The data in Table 30.1 summarize all of the movements explained above (the magnet numbers are specified in Figure 30.4). The shim unit displacements

**Table 30.1** Shim components and required magnet displacements.

| Shim component | Magnet | | | |
|---|---|---|---|---|
| | 1 | 2 | 3 | 4 |
| $x$ | $-\Delta x$ | $-\Delta x$ | $-\Delta x$ | $-\Delta x$ |
| $y$ | $-\Delta y$ | $-\Delta y$ | $-\Delta y$ | $-\Delta y$ |
| $z$ | $\Delta z$ | $\Delta z$ | $\Delta z$ | $\Delta z$ |
| $x^2$ | $\Delta x$ | $\Delta x$ | $-\Delta x$ | $-\Delta x$ |
| $z^2$ | $\Delta z$ | $-\Delta z$ | $\Delta z$ | $-\Delta z$ |
| $xy$ | $-\Delta y$ | $-\Delta y$ | $\Delta y$ | $\Delta y$ |
| $zy$ | $\Delta y$ | $-\Delta y$ | $\Delta y$ | $-\Delta y$ |
| $xz$ | $\Delta x$ | $-\Delta x$ | $\Delta x$ | $-\Delta x$ |

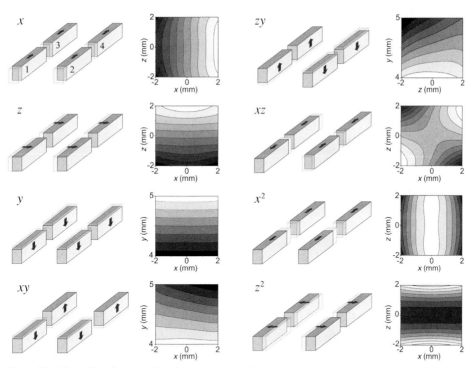

**Figure 30.4** The effect of moving the shim magnets in the directions indicated by the arrows is shown by means of 2-D maps of the total magnetic field strength. The field strength (magnitude) increases are indicated by color changes from black (weak) to white (strong). The magnet numbers are shown in the plot for the linear $x$ term.

together with 2-D maps of the total field obtained after the corresponding movement are shown in Figure 30.4. In the next section, the set-up built to implement this shim concept is described and experimental results of $^1$H high-resolution spectroscopy are provided.

## 30.3
## Experimental Set-Up

The main magnet has a classical U-shaped geometry with dimensions $280 \times 120 \times 280 \, mm^3$ and gap $d_z^{main} = 100 \, mm$ (Figure 30.5). It is built from a NdFeB alloy with a remnant flux density of 1.33 T, coercive field strength of $796 \, kA \, m^{-1}$, and a temperature coefficient $\kappa = -1200 \, ppm \, °C^{-1}$. The working volume is situated at a depth $y_0 \sim 5 \, mm$, the average field there is $B_{00} \sim 0.25 \, T$ (10 MHz $^1$H Larmor frequency), and the field inhomogeneities are characterized by $G_y \approx -1000 \, ppm \, mm^{-1}$, $\alpha_z \approx 300 \, ppm \, mm^{-2}$ and $\alpha_x \approx -30 \, ppm \, mm^{-2}$. This magnet has been used previously for imaging [19] and single-sided spectroscopy with the field-matching techniques [11].

The shim unit was designed following the procedure described above. Although, in principle, we have shown how four magnet blocks (two identical pairs) cancel the spatial variations of the main field, we found it more convenient to use two sets of four magnets (Figure 30.5). These two sets will be called the 'lower' and 'upper' shim units. The lower shim unit (the bulky one) is located at the bottom of the gap, and its main function is to cancel the strong gradient along the depth direction. The positions of the magnet blocks are fixed, and in practice this set can be considered to be a part of a new main unit with the effect of reducing the gradient along the depth direction. The upper shim unit is also placed in the gap, but directly underneath the sensor surface. These four blocks are movable along the three Cartesian directions in order to generate the desired shim components (see

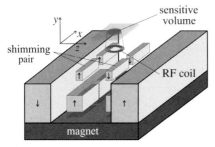

**Figure 30.5** Magnet array used to generate a volume of highly homogeneous magnetic field external to the magnet. The arrows show the direction of polarization of the magnets. The four pairs of shim magnets placed in the gap compensate for the inhomogeneity of the magnetic field of the main magnet.

Table 30.1). Due to the presence of the lower shim unit, the upper shim unit becomes smaller than it would be if it were alone. This implies that the required displacements for shimming are larger, which brings along more precision in shimming the field. The dimensions and positions of all shim magnets were obtained by computer simulations considering the field of the main magnet as the target field to be shimmed.

A four-turn surface RF coil with an outer diameter of 7 mm placed at 3 mm from the magnet surface is used for excitation and detection. The natural lateral selection of the RF coil is combined with a 90° soft pulse for excitation (slice) to achieve volume selection. The sensitive volume has lateral dimensions of about $5 \times 5$ mm$^2$ and a thickness of 0.5 mm. It is located 5 mm from the surface of the magnet where the $^1$H Larmor frequency is 8.33 MHz (see Figure 30.5). The sensor is also provided with three single-sided shim coils placed inside the main gap (not shown in Figure 30.5), that can produce pulsed gradients along the Cartesian directions. The coils producing the gradients along the $x$ and $z$ directions are similar to those described in Ref. [19] but with different aspect ratios. The coil generating the gradient along the depth is similar to the $z$-direction gradient coil but it is wound in an anti-parallel configuration. These coils are used in the final stage of the shimming procedure.

Once the shim units had been mounted and placed in their optimum positions (the calculated ones), the final shim configuration was obtained in two stages:

- In *stage 1*, the total magnetic field was scanned via measurement of the resonance frequency of a tiny water sample (~1 mm$^3$). Subsequently, the positions of the magnet pairs for shimming were adjusted for maximum compensation of the field variations. This procedure was repeated several times until deviations from the average field smaller than 10 ppm were obtained.

- In *stage 2*, to further improve the resolution, the current through the single-sided shim coils was optimized, by maximizing the signal peak in the frequency domain of a large sample.

Figure 30.6a shows the spectrum of a water sample much larger than the sensitive volume (an arbitrarily large sample) placed on top of the sensor. The linewidth is 2.2 Hz, corresponding to a spectral resolution of about 0.25 ppm. For comparison, the state-of-the-art spectrum for single-sided NMR, measured recently by nutation echoes in the presence of spatially matched static and RF fields [11], is also shown. The spectral resolution has been improved by a factor of about 30, with a concomitant fivefold extension of the excited volume. Together with a sensitivity optimized surface RF coil, the increased size of the sensitive volume leads to an appreciably higher signal-to-noise ratio (SNR).

The sub-ppm resolution achieved in these studies allowed us to resolve different molecular structures such as toluene and acetic acid (Figure 30.6b and c). The experiments were carried out in the laboratory, without the need for a temperature chamber. The insulation provided by a 1 cm-thick polystyrene foam layer, added to the large thermal inertia of the magnet, allowed us to average the signal over 1 min, without appreciable frequency drift.

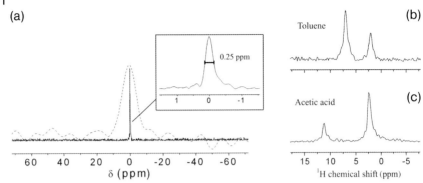

Figure 30.6 (a) Solid line: magnitude spectrum of a water sample much larger than the sensitive volume, placed on top of the RF coil. The spectrum is the Fourier transform of the Hahn echo signal from 64 scans acquired with a repetition time of 5 s to improve the SNR. The full width at medium height of the line (inset) is 0.25 ppm at a proton resonance frequency of 8.33 MHz. Dashed line: best spectrum obtained by *ex situ* spectroscopy following the nutation-echo method. The line width is 8 ppm [11]. Both data sets were obtained in the same measuring time, using sensors of the same size and working at comparable depths. Therefore, spectral resolution and sensitivity can be compared quantitatively; (b) $^1$H NMR spectra of different liquid samples obtained within a measuring time of 1 min. The chemical-shift differences and the relative peak intensities are in good agreement with the results obtained using conventional high-resolution NMR spectrometry.

The toluene $^1$H spectrum exhibits two lines at 7.0 and 2.1 ppm, with relative intensities of 5 : 3, corresponding to the aromatic and methyl protons, respectively. In the case of acetic acid, the two lines correspond to methyl and carboxylic protons, which appear at 2.3 and 11.3 ppm, with an intensity ratio of 3 : 1 [12].

## 30.4
## Conclusions

In this chapter the construction of a mobile open sensor for $^1$H NMR spectroscopy is described, with particular attention being paid to the use of movable small permanent magnet blocks to shim the stray field of a conventional U-shaped magnet. By combining the different movements of the shim magnets it is possible to generate and control several shim terms up to order two, including linear ($x$, $y$ and $z$), quadratic ($x^2$ and $z^2$) and cross-terms ($xy$, $zy$ and $xz$). The size of the sensitive volume where high homogeneity was achieved is about 10 mm$^3$, and is placed 5 mm above the magnet surface. The volume selection is achieved by means of a surface coil and a soft excitation pulse. In this way, spectra with 2.2 Hz (0.25 ppm) resolution can be obtained simply by placing an arbitrarily large sample on top of the sensor. This resolution is sufficient to determine the molecular composition of many liquids from $^1$H NMR spectra. The most important advantage of having high-field homogeneity available for *ex situ* NMR is that established techniques of multidimensional NMR spectroscopy and imaging can now be implemented in a straightforward manner for the nondestructive testing of large objects.

## References

**1** Blümich, B. (2000) *NMR Imaging of Materials*, Clarendon Press, Oxford.

**2** Blümich, B., Perlo, J. and Casanova, F. (2008) *Progress in Nuclear Magnetic Resonance Spectroscopy*, **52**, 197–269.

**3** Callaghan, P. T. (1991) *Principles of Nuclear Magnetic Resonance Microscopy*, Clarendon Press, Oxford.

**4** Weitekamp, D.P., Garbow, J.R., Murdoch, J.B. and Pines, A. (1981) *Journal of the American Chemical Society*, **103**, 3578–9.

**5** Balbach, J.J., Conradi, M.S., Cistola, D.P., Tang, C., Garbow, J.R. and Hutton, W.C. (1997) *Chemical Physics Letters*, **277**, 367–74.

**6** Hall, L.D. and Norwood, T.J. (1987) *Journal of the American Chemical Society*, **109**, 7579–81.

**7** Vathyam, S., Lee, S. and Warren, W.S. (1996) *Science*, **272**, 92–6.

**8** Meriles, C.A., Sakellariou, D., Heise, H., Moulé, A.J. and Pines, A. (2001) *Science*, **293**, 82–5.

**9** Shapira, B. and Frydman, L. (2004) *Journal of the American Chemical Society*, **126**, 7184–5.

**10** Topgaard, D., Martin, R., Sakellariou, D., Meriles, C.A. and Pines, A. (2004) *Proceedings of the National Academy of Sciences of the United States of America*, **101**, 17576–81.

**11** Perlo, J., Demas, V., Casanova, F., Meriles, C.A., Reimer, J., Pines, A. and Blümich, B. (2005) *Science*, **308**, 1279.

**12** Perlo, J., Casanova, F. and Blümich, B. (2007) *Science*, **315**, 1110–12.

**13** Chmurny, G.N. and Hoult, D.I. (1990) *Concepts in Magnetic Resonance*, **2**, 131–49.

**14** Fukushima, E. and Jackson, J.A. (2002) Unilateral magnet having a remote uniform field region for nuclear magnetic resonance. US Patent 6, 489, 872.

**15** Perlo, J., Casanova, F. and Blümich, B. (2006) *Journal of Magnetic Resonance*, **180**, 274–9.

**16** Manz, B., Coy, A., Dykstra, R., Eccles, C.D., Hunter, M.W., Parkinson, B.J. and Callaghan, P.T. (2006) *Journal of Magnetic Resonance*, **183**, 25–31.

**17** Marble, A.E., Mastikhin, I.V., Colpitts, B.G. and Balcom, B.J. (2006) *Journal of Magnetic Resonance*, **183**, 240–6.

**18** Popella, H. and Henneberger, G. (2001) *COMPEL: The International Journal for Computation and Mathematics in Electrical and Electronic Engineering*, **20**, 269–78.

**19** Perlo, J., Casanova, F. and Blümich, B. (2004) *Journal of Magnetic Resonance*, **166**, 228–35.

# 31
# Inversion of NMR Relaxation Measurements in Well Logging

*Lizhi Xiao, Guangzhi Liao, Ranhong Xie and Zhongdong Wang*

## 31.1
## Overview of NMR Logging

At present, many commercial NMR logging tools have been used in oil fields to provide a high-quality logging service worldwide. Some of the systems currently in use are discussed:

- Combinable Magnetic Resonance (CMR), which is Schlumberger's commercial NMR logging tool, has the following characteristics: (i) A magnetic field intensity of 470 G and a resonance frequency of 2000 KHz; (ii) a pad tool which is forced against the borehole; (iii) a vertical resolution of 15 cm, with a much smaller volume than the Magnetic Resonance Imaging Logging (MRIL) system (see below) and a lower SNR; (iv) there is a homogeneous magnetic field distribution with a single frequency; and (v) the system is affected by hole rugosity.

- The MRIL–which is Halliburton's commercial NMR logging tool–has the following characteristics: (i) The gradient magnetic field intensity is $17 \, G \, cm^{-1}$, ranging from 150 to 200 G, and the five center frequencies are 590, 620, 650, 680 and 760 kHz; (ii) it is run with a centralizer in the bore hole; (iii) the vertical resolution is 60 cm, with a 1 mm-wide sensitive volume, 35–43 cm from the tool center; (iv) it can perform up to multifrequency measurements at different depths; (v) there is a better SNR than with the CMR system; and (vi) it is unaffected by hole rugosity.

- The Magnetic Resonance Explorer (MREX), which is Baker Hughes's new commercial MR Explorer tool, operates at multiple frequencies, allowing faster logging and multiple NMR measurements in a single logging pass. This tool uses a side-looking antenna design and a gradient magnetic field to provide NMR data for formation evaluation and fluid analysis.

- The Magnetic Resonance Scanner (MR-Scanner), which is Schlumberger's new generation wire-line NMR logging tool, uses a side-looking antenna (containing a main antenna and two high-resolution antennae) and performs multifrequency measurements in a gradient-field design. The MR-Scanner

*Magnetic Resonance Microscopy.* Edited by Sarah L. Codd and Joseph D. Seymour
Copyright © 2009 WILEY-VCH Verlag GmbH & Co. KGaA, Weinheim
ISBN: 978-3-527-32008-0

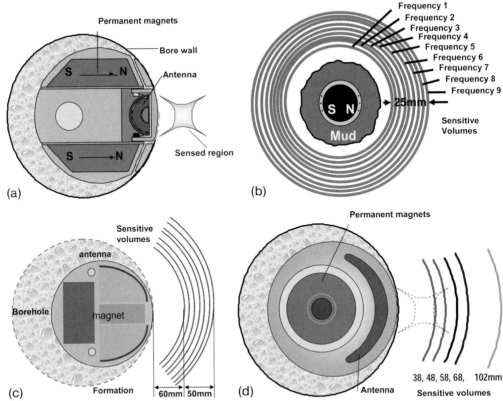

**Figure 31.1** Contrast of CMR (a), MRIL (b), MREX (c) and MR-Scanner (d). CMR is a pad NMR logging tool with single frequency. MRIL-P is a centralized NMR tool with multifrequencies in nine sensitive volume cylinders (each 1 mm thick at approximately 1 mm spacing). MREX and MR-Scanner are pad NMR logging tools with seven sensitive volume cylinders and five sensitive volume cylinders, respectively. The illustrations are adopted from the websites of Schlumberger, Halliburton and Baker Hughes.

performs investigations at multiple depths of investigation (DOI; which is temperature-dependent, and ca. 16 cm for a 15 cm magnet) in a single pass. Moreover, the measurement sequence of the MR-Scanner allows a profiled view of the reservoir fluids. Resonance volumes for each tool can be estimated exactly, and all of the tools are designed for $T_2$ relaxation measurement. Figure 31.1 shows the different magnetic designs of the CMR (a), MRIL (b), MREX (c) and MR-Scanner (d) (these figures are adopted from the websites of Schlumberger, Halliburton and Baker Hughes).

Since the year 2001, many well-logging companies have also provided their own Logging While drilling (LWD) NMR logging tools for commercial service, including Halliburton's 'Magnetic Resonance While Drilling' (MR-WD), Schlumberger's 'ProVision' and Baker Hughes's 'Mag-Track'. As an example, MR-WD–Halliburton's first LWD MRIL tool–provides direct measurement of the reservoir while drilling. The $T_1$ signature of this measurement can provide real-time information identification.

## 31.2
## Description of Inversion of NMR Relaxation in Well Logging

The NMR relaxation measurements, whether longitudinal ($T_1$) or transverse ($T_2$), in porous media can be expressed as a multiexponential function. The $T_2$ relaxation signal measured by the Carr–Purcell–Meiboom–Gill (CPMG) spin echo method can be written as [1–4]:

$$\text{For } T_1 \text{ measurement,} \quad \sum_{T1\,j=T1\,min}^{T1\,max} f(T_{1j})[1-c\exp(-t_i/T_{1j})]=M(t_i) \tag{31.1}$$

$$\text{For } T_2 \text{ measurement,} \quad \sum_{T2\,j=T2\,min}^{T2\,max} f(T_{2j})\exp(-t_i/T_{2j})=M(t_i) \tag{31.2}$$

where $T_{1min}$, $T_{2min}$, $T_{1max}$ and $T_{2max}$ are the minimum and maximum of $T_1$ and $T_2$, respectively, $M(t_i)$ is the measured magnetization at time $t_i$, $f(T)$ is the amplitude associated with the variable $T_1$ or $T_2$ to be solved, and $c$ is a constant. Equation 31.1 indicates that $T_1$ is measured by saturation recovery if $c = 1$, or inversion recovery if $c = 2$. Equations 31.1 and 31.2 are the discrete forms of a Fredholm integral of the first kind:

$$\int_{T2\,min}^{T2\,max} f(T_2)\exp(-t/T_2)dT_2 = M(t) \tag{31.3}$$

It can also be cast in the following general form:

$$\sum_j f(T_{1,2j})[c_1-c_2\exp(-t_i/T_{1,2j})]=g_i=M(t_i)+\varepsilon_i; \quad i=1,\dots,m; \quad j=1,\dots,n. \tag{31.4}$$

where $m$ is the number of echoes and $n$ is the number of relaxation times, $t_i$ is the time when the $i$-th echo is acquired [normally, $t_i$ is the integral multiple of the echo space (TE)], $g_i$ is the amplitude of the $i$-th echo, $T_{1,2j}$ is a set of n pre-selected $T_1$ or $T_2$ relaxation times equally spaced on a logarithmic scale, $f(T_{1,2i})$ is the amplitude corresponded to $T_{1j}$ or $T_{2j}$, and $\varepsilon_i$ is the Gaussian noise. For longitudinal relaxation time, $c_1 = 1$, $c_2 = 1$ in Equation 31.4, if measured by saturation recovery method; and $c_1 = 1$, $c_2 = 2$, if measured by inversion recovery method. For transverse relaxation time, $c_1 = 0$, $c_2 = -1$ in Equation 31.4. For $T_2$ measurements, Equation 31.4 can be written in the following simultaneous equations:

$$\begin{cases} f_1 \cdot e^{-t_1/T_{21}} + f_2 \cdot e^{-t_1/T_{22}} + \dots + f_n \cdot e^{-t_1/T_{2n}} = g_1 = M(t_1) + \varepsilon_1 \\ f_1 \cdot e^{-t_2/T_{21}} + f_2 \cdot e^{-t_2/T_{22}} + \dots + f_n \cdot e^{-t_2/T_{2n}} = g_2 = M(t_2) + \varepsilon_2 \\ \quad \vdots \qquad \qquad \vdots \qquad \cdots \qquad \vdots \qquad \vdots \qquad \vdots \\ f_1 \cdot e^{-t_m/T_{21}} + f_2 \cdot e^{-t_m/T_{22}} + \dots + f_n \cdot e^{-t_m/T_{2n}} = g_m = M(t_m) + \varepsilon_m \end{cases} \tag{31.5}$$

and in vector form as: $\quad Af = g = M + \varepsilon \tag{31.6}$

Since all $e^{-t_i/T_2 j}$'s are known and we are solving for $f_j$, it is a linear inversion problem. A least-squares fit is used to minimize the following sum [4]:

$$\min\left\{\phi(f) = \sum_{i=1}^{m}\frac{1}{\sigma^2}\left(\sum_{j=1}^{n}f_j e^{-t_i/T_2 j} - g_i\right)^2\right\} \tag{31.7}$$

where $\sigma^2$ is the variances of $\varepsilon_i$.

## 31.3
## Inversion Methods of NMR Relaxation in Well Logging

### 31.3.1
### Inversion Methods

Equation 31.6 is a well-known ill-posed inversion problem. One of the popular methods for solving this problem is the 'singular value decomposition' (SVD) method [5–8]. The SVD theorem in linear algebra [8] states that that any real m × n matrix **A** (m ≥ n) can be written as the product of an m × m orthonormal matrix U, an m × n diagonal matrix Λ with positive or zero diagonal elements, and the transpose of an n × n orthonormal matrix V: that is

$$A = U\Lambda V^T \tag{31.8}$$

where $U^T U = 1$, $V^T V = 1$ and $\Lambda = diag(\lambda_1, \lambda_2, \ldots, \lambda_m)$. The diagonal elements $\lambda_1 \geq \lambda_2 \geq \ \geq \lambda_m \geq 0$ are called the 'singular values' of the matrix **A**. After calculating the noise level of $\varepsilon_i$, it is necessary to determine how to cut off the singular values in the SVD inversion procedure. By using Equation 31.9 [4, 7], we can choose a suitable lower cut-off $\lambda_r$ of singular value:

$$\frac{\lambda_1}{\lambda_r} = \frac{\|g\|}{\sigma} \tag{31.9}$$

where the singular values smaller than $\lambda_r$ which are discarded in the restricted sums, are effectively replaced by 0 in the inverse matrix $\Lambda^{-1}$. Hence, the solution $f$ is given by

$$f = A^{-1}g \approx V_r \Lambda_r^{-1} U_r^T g \tag{31.10}$$

Normally, some 'regularization' methods which depend on penalty function have also been used to solve Equation 31.6 by minimizing Equation 31.11,

$$\min\left\{\phi(f) = \frac{1}{2}\sum_{i=1}^{m}\left(\sum_{j=1}^{n}f_j a_{ij} - g_i\right)^2 + \frac{\alpha}{2}\sum_{j=1}^{n}f_j^2\right\} \tag{31.11}$$

where this has been named 'norm' smoothing [4], $a_{ij} = e^{-t_i/T_j}$, and $\alpha$ – which is often referred to as a 'smoothing' or a 'regularization' parameter – is a constant optimally chosen to be commensurate with the measurement error. Bergman *et al.* [4] had suggested a method to implement the BRD method [9], which works as follows:

(1) For a fixed $\alpha$, we find an initial solution $c$ which satisfies, $(K + I\alpha)c = g$, where
$$K_{ij} = \sum_{x=1}^{n} a_{ix} a_{jx}.$$

(2) The $f_x$ is then given by $f_x = \max\left(0, \sum_{i=1}^{m} c_i a_{ix}\right)$.

(3) Then, update the matrix element $K_{ij} = \sum_{x}^{i} a_{ix} a_{jx}$; this is summed over only those $x$-values which give positive values for $\sum_{i=1}^{m} c_i a_{ix}$. The new $K_{ij}$ is used for a new $c$.

(4) Repeat the process, until $c$ stops changing. The final value for $f_x$ is given by (2).

Although these two methods usually work quite well, all have disadvantages such as being too sensitive to the SNR and inflexible to bin selection.

## 31.3.2
### New Inversion Method

During their early studies Dines and Lyttle introduced an improved algebraic-reconstruction technique (ART) algorithm [10–12]. By making use of their algorithm, a new method has been developed for NMR multiexponential inversion, the implementation steps of which are as follows:

First, an initial model $f^0$ (normally set with zero) is given to compute an error $\Delta g$ from the forecast relaxation signal $g^0$ and the measured relaxation signal $g$, as Equation 31.12:

$$\Delta g = g - g^0 \tag{31.12}$$

$$\sum_{j=1}^{n} a_{ij} \Delta f_j = \Delta g_i \tag{31.13}$$

For matrix **A**, all of the elements $a_{ij}$ are greater or equal to zero, and $\Delta g_i$ can therefore be attributed to each $\Delta f_j$ according to the magnitude of $a_{ij}$. It is essential for the new method to compute the correct magnitude $\Delta f_j$ of relaxation time from $\Delta g_i$, so to maintain the continuity of NMR relaxation time distribution and ensure the inversed results physics system.

Second, by allowing $\Delta f_j = \lambda_i a_{ij} \Delta g_i$, and taking it into Equation 31.13, we have

$$\lambda_i = \frac{1}{\displaystyle\sum_{j=1}^{n} a_{ij}^2} \tag{31.14}$$

where $\lambda_i$ is the coefficient for attributing $\Delta g_i$ to $f_j$. The correction values for all rows (i.e. the correction value for each relaxation signal) are estimated respectively. These corrections are then averaged according to the attributing coefficients and the number of nonzero elements at the $j$-th column. The correction for $f_j$ is [10],

$$\Delta f_j^{(q+1)} = \frac{1}{LA(j)} \sum_{i=1}^{n} \left[ \lambda_i a_{ij} \Delta g_i^{(q)} \right] \tag{31.15}$$

where $LA(j)$ is not a fixed value $m$ (as in the ART method) but rather the number of nonzero elements (larger than very small positive number) of the $j$-th column in matrix **A**. $\Delta g_i^{(q)}$ is the residual for the $i$-th relaxation signal computed at the $q$-th iteration. Because matrix **A** has many zero elements, $LA(j)$ is normally less than $m$; the iteration speed of the new method is faster than that of the ART method [11].

The expression for the corrected relaxation signal components is

$$f^{(q+1)} = f^{(q)} + \Delta f \tag{31.16}$$

Equations 31.15 and 31.16 are the iteration formulae for inversion. The new algorithm is known as the simultaneous iterative reconstruction technique (SIRT) because the approximate solutions to Equation 31.6 are not calculated until all of the equations are processed.

The amplitude in the NMR $T_1$ and $T_2$ spectra cannot be less than zero in a physical context. Therefore, $f_j (j = 1, \ldots, m)$ must be non-negative in the inversion process, and is assigned very small positive number, for example, 1.0E-30, when $f_j^{(q)}$ is negative. The iteration computation is then restarted until the variation of the correction for the equation solutions $(\Delta f / f_j^{(q)})$ satisfies the requirements, or the preset iteration time finishes.

### 31.3.3
### Numerical Results and Discussion

The inversion results of SVD, BRD and SIRT algorithms with different SNR have been contrasted. Figure 31.2a–c show three sets of $T_2$ spectra calculated using these three methods, respectively. The SNR of the echo data vary from $\infty$ through to 5 (where SNR = $\infty$ means no noise at all). The influences of SNR on these three methods have been found to differ; for example, with a decreasing SNR of the echo data the $T_2$ spectra calculated using the SVD method become smoother. The inversion result has come to be much smoother when the SNR is extremely low, because the relationship $\lambda_1 / \lambda_{cut} = \|g + \varepsilon\|_{\infty} / \sigma$ exists between the cut-off of the singular value and the noise level, where $\lambda_{cut}$ is the cut-off of the singular value and $\sigma$ is the noise level and $\|g + \varepsilon\|_{\infty} / \sigma$ reflects the SNR. As the SNR decreases the sharp peaks of the $T_2$ spectra calculated using the BRD method become sharper

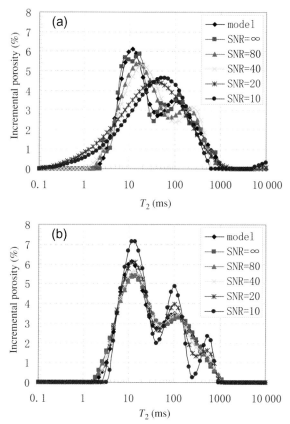

**Figure 31.2** The inversion results of SVD, BRD and SIRT methods for spin-echo strains with SNR reducing from ∞ to 5 in (a), (b) and (c); (d) This is the result of $T_1$ inversion with the SIRT method.

and wildly oscillatory (Figure 31.2b). The reason for this is that the penalty terms do not suppress the oscillatory behavior of the unknown function $f(s)$ adequately. Figure 31.2c shows that $T_2$ spectra from SIRT with different SNRs are more reproducible than using the SVD and BRD methods. $T_2$ spectra converted by SIRT retain the double-peak feature of the model, and they are also similar to one another. These test results indicate that the new method is suitable for different SNR NMR measurements, even when the SNR is very low, for example less than 5.

As is shown in Figure 31.2d, the $T_1$ spectra inverted by SIRT retain the bimodal feature, but the correct peaks of the $T_1$ spectra gradually rise as the noise level increases, which indicates that when using this method a long $T_1$ is sensitive to noise.

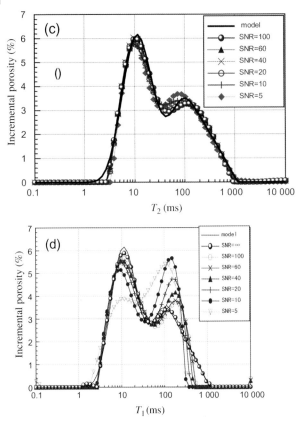

**Figure 31.2** *Continued*

The SIRT method provides an alternative analysis to the traditional SVD and BRD methods because it not only possesses the merits of simplicity and easy programming but also avoids many of the complex preset inversion control parameters in the processing. Accordingly, the artificial errors for the inverted results are reduced. In addition, the process is faster than the SVD system when all of the measured relaxation signals take part in the inversion processing.

Figure 31.3 is an example of NMR well logging in shale-sands formation in the Liaohe China oilfield. Track 2 is the inversion results of MAP (which is a data-processing method of MRIL tool by Halliburton) and track 3 is the inversion results of SIRT method. The $T_2$ spectra inverted by MAP with 10 bins from 4 ms to 2048 ms have a single peak in general, even if the beds have a good reserve capability, for example, in the interval from 2910 m to 2915 m. The $T_2$ spectra do not tally with the aperture distribution feature of sand rocks that have two or more peaks. In contrast, the $T_2$ spectra inverted by the new method with 20 logarithmically spaced bins from 0.3 ms to 3000 ms have two or three tracks in sand rocks, and are identical with the core analysis results.

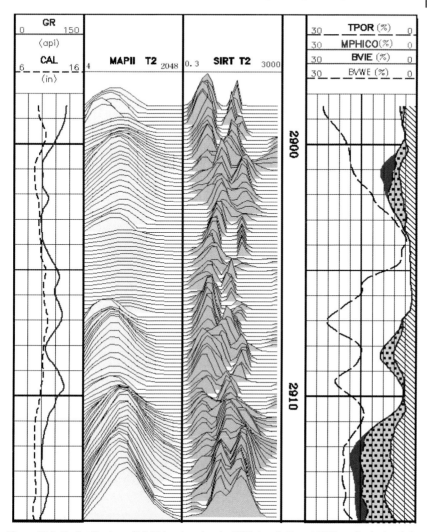

**Figure 31.3** $T_2$ distribution of NMR logging data by MAP (track 2) and SIRT methods (track 3).

## 31.4
## Influence of Multiexponential Inversion

### 31.4.1
### The Number of Bins

The number of preselected $T_2$ relaxation times influences the results of the multiexponential inversion. A set of $T_2$ distributions with different numbers of preselected $T_2$ relaxation times is shown in Figure 31.4a. The SVD algorithm was

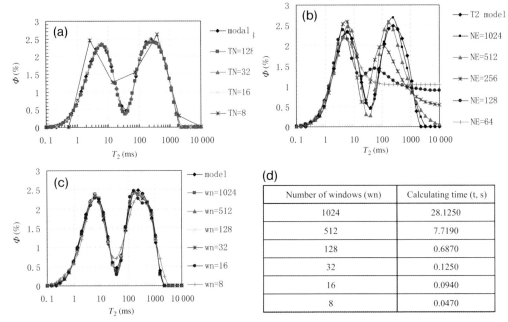

**Figure 31.4** (a) $T_2$ distributions with different number of bins (TN), reducing from 128 to 8; (b) Effects of different echo numbers (NE) to inversion, reducing from 1024 to 64; (c) Comparison of the inversion results with the spin-echo trains compressed, where wn is the number of compressed echoes, reducing from 1024 to 8; (d) Calculation of time at different numbers of compressed echoes

used in the inversion process (the same response is found for the BRD or SIRT method) with the SNR of the data equal to 100. The resolution of the $T_2$ distribution becomes poorer as the number of preselected $T_2$ points becomes smaller. The two peaks cannot be identified distinctively when the number of relaxation times is 8 for a range of $T_2$ from 0.1 ms to 10 000 ms in Figure 31.4a. However, as the number of relaxation times increases, the inverted $T_2$ distribution becomes much closer to the model. although more preselected $T_2$ relaxation times mean a larger matrix **A** in Equation 31.6, and more computation time.

## 31.4.2
### The Number of Echoes

As shown in Figure 31.4b, the second peak of the $T_2$ distribution becomes divergent as the number of echoes decreases, while the $T_2$ range and echo spacing are fixed. Because the signals of long relaxation composition are not acquired completely, there are insufficient echoes to resolve the second peak. The SNR is equal to 100 and the SVD method is used for the inversion, as shown in Figure 31.4b. Hence, in order to resolve the long relaxation time components fully, a sufficiently large number of echoes must be acquired [13].

### 31.4.3
### Data Compression in Time Domain

The data acquired by NMR apparatus in the laboratory always contains thousands of echoes, especially for those samples with long $T_2$ relaxation times. For instance, there are typically 4096 or 8192 echoes for a free water sample. In addition, $T_2$ always has a broad distribution in fluid-saturated porous media, and therefore Equation 31.6 is a large matrix with a slow speed of calculation. In order to increase the computation speed, the time domain data can be compressed into several windows. The echo trains are partitioned by Equation 31.17 and, after compressing the data in each window, Equation 31.7 can be transformed to Equation 31.18 [4, 7, 13],

$$NW_i = N_T^{i/(S-1)}, N_i = NW_i - NW_{i-1} \tag{31.17}$$

$$\min\left\{ \phi(f) = \sum_{i=1}^{S} \frac{1}{N_i \sigma^2} \left( \sum_{j=1}^{n} f_j K_{ij} - G_i \right)^2 \right\} \tag{31.18}$$

where $S$ is the number of windows, $N_T$ is the number of echoes, $N_i$ is the number of echoes in the $i$-th windows, and $N_T = N_1 + \cdots + N_i + \cdots + N_S$, $r_i = N_1 + \cdots + N_{i-1}$, $r_1 = 0$. Equations 31.17 and 31.18 mean that $n$ echoes have been partitioned into $S$ windows ($i = 1, \ldots, s$) with $N_i$ echoes in the $i$-th window. $K_{ij}$, $G_i$ are given by

$$K_{ij} = \sum_{k=r_i+1}^{r_i+N_i} e^{-t_k/T_j}, \quad G_i = \sum_{k=r_i+1}^{r_i+N_i} g_k \tag{31.19}.$$

We can also average the data within each window, assuming that arithmetic averaging is a valid procedure for this type of data.

As shown in Figure 31.4c, after compressing the data in time domain, the shape of the $T_2$ spectra does not change significantly. Only small minor changes occur, when the smallest number of windows is used. This indicates that compression of the data in the time domain is necessary for real-time processing in NMR logging. Figure 31.4d lists the diverse computation times with different numbers of windows. Increasingly less time is required as the number of windows is reduced to 512, 128, 64, 32, 16, 8 from 1024. The SVD algorithm is used in this test, and the same result is obtained for other algorithms, such as BRD and SIRT.

### 31.4.4
### The Correction of Adjustment Parameters

In order to overcome the instability caused by extremely low or high SNR, we introduced different adjustment methods to these algorithms in order to enhance the quality of the inversion [13]. For the SVD method, we add an adjustment parameter in a singular value cut-off formula, that is $\lambda_1 / \lambda_r = A_{dj} \cdot \|g\|_\infty / \sigma$, where $A_{dj}$ is an adjustment parameter. Increasing $A_{dj}$ means reducing the value of $\lambda_r$, which in turn leads to an increase in the number of singular values of $\Lambda_r$ in the equation $f = A_r^{-1}b = V_r\Lambda_r^{-1}U_r^Tg$. Thus, it could enhance the resolution of $T_2$ spectra calculated by the SVD method with the echo data in low-SNR. As shown in Figure 31.5a, the inversion result with $A_{dj}$ equal to 8 is much better than that where $A_{dj}$ is equal to 1 ($A_{dj} = 1$ means no correction). In the same way, decreasing $A_{dj}$ means increasing the value of $\lambda_r$, which leads in turn to a decrease in the number of singular values of $\Lambda_r$. This could suppress the oscillatory behavior of the $T_2$ spectrum calculated using the SVD method for data with a high SNR. As shown in Figure 31.5b, the inversion result with $A_{dj} = 0.1$ is much better than that where $A_{dj} = 1$.

Figure 31.5c shows the influence of the smoothing parameter $\alpha$ in the BRD method. As $\alpha$ increases from 0.01 to 50, the relative amplitude $f_j$ changes from a wildly oscillatory behavior to a smooth curve. With prior knowledge of a reasonable estimate of measurement error ($\sigma$), the optimal $\alpha$ has the linear relationship of $\alpha_{opt}^2(c \cdot c) = n\sigma^2$ [4, 13]. We can search for the optimal $\alpha$ by using a binary search method; this is typically terminated when $\|(M + \alpha I)c - g\|/\|g\| \leq 10^{-6}$, which is an

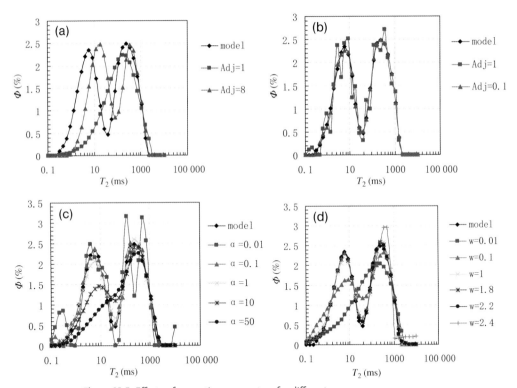

**Figure 31.5** Effects of correcting parameters for different inversion methods. (a) SVD method with different adjustable (Adj) parameters when SNR = 5; (b) SVD method with different adjustable (Adj) parameters when SNR = 1000; (c) Norm smoothing fits with various smoothing parameters ($\alpha$); (d) SIRT method with various weight coefficients (w).

arbitrarily selected tolerance. On occasion, the optimal $\alpha$ can also correct the influence of different SNRs.

In order to lessen the influence of different SNRs, a weighting modification factor ($\omega$) has been added into the SIRT algorithm. The iterative equation can then be written as $\Delta p_j^{q+1} = \omega \sum_{i=1}^{n} \left[ \lambda a_{ij} \Delta y_i^{(q)} \right] / LA(j)$. Figure 31.5d shows that $T_2$ spectra calculated by the SIRT algorithm become smoother as $\omega$ decreases from 2.4 to 0.01. Therefore, reducing the value of $\omega$ would suppress the oscillatory behavior of the $T_2$ spectrum when the SNR of the echo data was extremely low.

## 31.5
## NMR Porosity for Terrestrial Formation in China

One of the most commonly used NMR porosity measurement methods acquires two types of CPMG echo trains [2]: fully and partially polarized. By using a long wait time (TW) of 12 s and a long echo space (TE) of 1.2 ms, the hydrogen nuclei in large-pore fluids are fully polarized and detected, leading to signals from free fluid and a large part of the capillary bound water; in contrast, the signal of the hydrogen nuclei in micropore and clay-bound water is largely lost due to the long echo spacing. While using a short TW of 20 ms and a short TE of 0.6 ms to acquire multiple sets of short CPMG echo trains, clay-bound water is fully polarized and detected; however, other long $T_2$ components are suppressed, leading to a predominantly clay-bound water signal. The $T_2$ distribution of 0.5, 1 and 2 ms is concatenated with the supplemental part of the $T_2$ distribution obtained from the regular mode to form the total porosity $T_2$ distribution. This is defined as the area splicing method (hereafter referred to as method I). The steps are as follows:

(i)   A partially polarized method is used to measure clay-bound water with a short TW of 20 ms and a short TE of 0.6 ms. The data are then fitted with seven bins, as in Equation 31.20:

$$M(t) = \sum_i P_i \exp\left(-\frac{t}{T_{2i}}\right) \quad (i = 1, 2, 3, 4, 5, 6, 10; T_{2i} = 0.5, 1, 2, 4, 8, 16, 256 \, ms)$$

(31.20)

(ii)  A fully polarized method is used to measure capillary-bound water and free fluids with a long TW of 12 s and a long TE of 1.2 ms or 0.9 ms. The data are then fitted with 12 bins, as in Equation 31.21:

$$M(t) = \sum_i P_i \exp\left(-\frac{t}{T_{2i}}\right) \quad (i = 2, 3, \ldots, 12, 13; T_{2i} = 1, 2, \ldots, 1024, 2048 \, ms)$$

(31.21)

(iii) The first four bins of clay-bound water are concatenated with the last nine bins and free fluids, and their sum is described as 'total porosity'– the first three bins as clay-bound water porosity and the last 10 bins as effective porosity, as indicated in Equation 31.22:

$$
\begin{aligned}
&\text{NMR total porosity} && PHIT = \sum P_i \quad (i = 1, 2, 3, \ldots, 12, 13),\\
&\text{clay-bound water porosity} && MCBW = \sum P_i \quad (i = 1, 2, 3),\\
&\text{capillary-bound water porosity} && MBVI = \sum P_i \quad (i = 4, 5, 6, 7),\\
&\text{effective porosity} && MPHE = \sum P_i \quad (i = 4, 5, 6, \ldots, 12, 13),\\
&\text{free fluids porosity} && MBVM = MPHE + MBVI
\end{aligned}
$$

$$(31.22)$$

When the above NMR porosity measurement method is used, the differences between NMR and conventional core porosities in artificial ceramic and Berea sandstone are less than 1 porosity unit (pu), although such differences for terrestrial sediment sandstone and lime-stone were very large (about 2–6 pu). It is believed that the reason for this is the uniform matrix components and pore distributions in artificial ceramic and Berea sandstone, yet quite complex matrix in terrestrial sediments. The latter contain micropores which were not detected when using a short TE of 0.6 ms, and this led to a reduction in the apparent porosity. Therefore, it is suggested that the following, improved, method be used to address the terrestrial formation in China (hereafter referred to as method II) [3]:

(i) A partially polarized method is used to measure clay-bound water with a short TW of 10 ms and a short TE of 0.3 ms. The data are then fitted with six bins, as in Equation 31.23:

$$
M(t) = \sum_i P_i \exp\left(-\frac{t}{T_{2i}}\right) \quad (i = 1, 2, 3, 4, 5, 6, 7; T_{2i} = 0.25, 0.5, 1, 2, 4, 16\,ms)
$$

$$(31.23)$$

(ii) A fully polarized method is used to measure capillary-bound water and free fluids with a long TW of 12 s and a TE of 0.9 ms. The data are then fitted with 12 bins, as in Equation 31.24:

$$
M(t) = \sum_i P_i \exp\left(-\frac{t}{T_{2i}}\right) \quad (i = 2, 3, \ldots, 12, 13; T_{2i} = 1, 2, \ldots, 1024, 2048\,ms)
$$

$$(31.24)$$

(iii) The first four bins of Equation 31.23 are concatenated with the last nine bins of Equation 31.24, and the sum is designated as 'total porosity', with the first three bins as clay-bound porosity and the last 10 bins as effective porosity.

More than 30 samples with a wide range of porosity including artificial ceramic, Berea sandstone, sandstone in Daqing oil field and Xinjiang oil field, and lime-stone in Sichuan oil field have been chosen to carry out the NMR porosity measurements. As indicated in Figure 31.6a, when method I was used, the differences

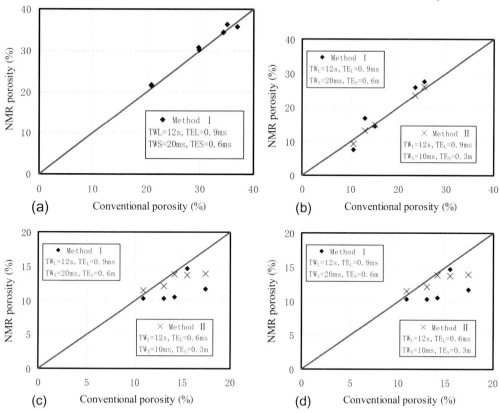

**Figure 31.6** The contrast of NMR porosity and conventional porosity in (a) artificial ceramic and Berea sandstone; (b) sandstones in Daqing oil field; (c) sandstones in Xinjiang oil field; and (d) limestones in Sichuan oil field.

between NMR porosity and conventional porosity in artificial ceramic and Berea sandstone were less than 1 pu. The results of both methods I and II for sandstones in Daqing are shown in Figure 31.6b. The porosity discrepancies were 2–5 pu for method I, but <1 pu for method II. Similarly in Figure 31.6c, the results of methods I and II for sandstones in Xinjiang are displayed. Again, the porosity discrepancies for method I were 2–6 pu, whereas those for method II were reduced to <1 pu in most cases. As shown in Figure 31.6d, for limestone in Sichuan, when simulating the current NMR porosity method (method I), the differences between NMR porosity and conventional porosity were 2–6 pu. This was because the relaxation of fluids in the large pores of limestone is slow, and a long waiting time is needed. If we use a long TW (12 s), a long TE (1.2 ms), the number of echoes (NE) is 500, the number of scanner times (i.e. repeat measuring times) is 32 and $4\,ms \leq T_2 \leq 2048\,ms$, then the difference is less than 1 pu and is acceptable.

From the comparison between conventional porosity and NMR porosity, the porosity differences with method I were found to be between 2 and 6 pu, whereas the accuracy of method II was within 1 pu, except for two core plugs. It is clear that method II, which has been derived from these studies, is suitable for the analysis of terrestrial formations in China, and is capable of obtaining more accurate formation with regards to total porosity.

## 31.6
## Conclusions

The results of these studies may be are summarized as follows:

- A new method of multiexponential inversion of $T_1$ or $T_2$ with core analysis data and simulated echo data has been presented, and tested. The results suggest that the new method is stable, and that the inversion results have a better resolution than the SVD and BRD methods, especially in cases of low SNR.
- More preselected relaxation times would increase the resolution of $T_2$ spectra, but this would involve more computing time. Moreover, the components of long relaxation times would be lost and inversion result would become divergent if insufficient echoes were acquired. Data compression in time domains can increase the computing speed, and does not cause any significant change in the shape of $T_2$ distribution.
- The responses to different SNRs for each inversion method vary, and there are sensitive relationships between inversion parameters and SNR-values, which can be used to increase the accuracy of the inversion.
- The current NMR porosity measurement method is suitable only for large-porosity formations where the matrix components and pore-structure are uniform, but not for terrestrial formations in China. If the same measuring parameters were to be used for different regions, this would lead to major differences between conventional and NMR porosities. Therefore, it is suggested that a core analysis should be carried out before logging, in order to obtain suitable parameters that yield accurate NMR porosity. An empirical approach should not apply simply to terrestrial formations; rather, a set of NMR porosity measurement methods suitable for terrestrial formations in China should be set up in order to make significant improvements in the accuracy of NMR porosity measurements.

## Acknowledgments

These studies were supported by the National Natural Science Foundation of China (40674075, 90510004) and '863' (2006AA06Z215) and '973' (2006CB202306) Projects of Ministry of Science and Technology, China.

# References

1  Xiao, L.Z. (1998) *NMR Image Logging and NMR in Rock Experiments* (in Chinese), Science Press, Beijing.

2  Coates, G.R., Xiao, L.Z. and Prammer, M.G. (1999) *NMR Logging Principles and Applications*, Gulf Publishing Company, Texas.

3  Wang, X.W., Xiao, L.Z., Xie, R.H. *et al.* (2006) Study of NMR porosity for terrestrial formation in China. *Science in China (G)* (in Chinese), **49** (3), 313–20.

4  Bergman, D.J., Dunn, K.J. and LaTorraca, G.A. (2002) *Nuclear Magnetic Resonance: Petrophysical and Logging Applications*, Elsevier Science.

5  Prammer, M.G. (1995) Principles of Signal Processing-NMR Data and $T_2$ distributions, SPWLA 36th Annual Symposium, Paris, France.

6  Dunn, K.J., LaTorraca, G.A. and Warner, J.L. (1994) On the calculation and interpretation of NMR relaxation time distribution. SPE28367, 69th Annual SPE Technical Conference and Exhibition, New Orleans, pp. 45–54.

7  Dunn, K.J. and LaTorraca, G.A. (1999) The inversion of NMR log data sets with different measurement errors. *Journal of Magnetic Resonance*, **140**, 153–61.

8  Press, W.H., Flannery, B.P., Teukolsky, S.A. and Vettering, W.T. (1986) *Numerical Recipes*, Cambridge University Press.

9  Butler, J.P., Reeds, J.A. and Dawson, S.V. (1981) Estimating solutions of first kind integral equations with nonnegative constraints and optimal smoothing. *SIAM Journal on Numerical Analysis*, **18** (3), 381–97.

10  Wang, Z.D., Xiao, L.Z. and Liu, T.Y. (2003) New inversion method with multi-exponent and its application. *Science in China (G)* (in Chinese), **33** (4), 323–32.

11  Gordon, R. (1974) A tutorial on ART (Algebraic Reconstruction Techniques). *IEEE Transactions on Nuclear Science*, **NS-21**, 78–93.

12  Dines, K. and Lyttle, J. (1979) Computerized geophysical tomography. *Proceedings of IEEE*, **67**, 1065–73.

13  Liao, G.Z., Xiao, L.Z., Xie, R.H. *et al.* (2007) Influence factors of multi-exponential inversion of NMR relaxation measurement in porous media. *The Chinese Journal of Physiology* (in Chinese), **50** (3), 932–8.

# Index